Raju S. Davé / Alfred C. Loos

Processing of Composites

Polymer Processing Society

Progress in Polymer Processing

Series Editor: L.A. Utracki

A.I. Isayev
Modelling of Polymer Processing

L.A. Utracki
Two-Phase Polymer Systems

A. Singh/J. Silvermann
Radiation Processing of Polymers

Series Editor: W.E. Baker

I. Manas-Zloczower/Z. Tadmor
Mixing and Compounding of Polymers

T. Kanai/G.A. Campbell
Film Processing

R.S. Davé/A.C. Loos
Processing of Composites

R.S. Davé/A.C. Loos (Editors)

Processing of Composites

With Contributions from

F. Abrams, S.G. Advani, B.T. Åström, V.M.A. Calado,
F.C. Campbell, D. Cohen, R.S. Davé,
B.R. Gebart, B. Joseph, J.L. Kardos, B. Khomami,
S.C. Kim, D.E. Kranbuehl, R.L. Kruse, M-C Li,
A.C. Loos, A.R. Mallow, S.C. Mantell, A.K. Miller,
J.W. Park, L.A. Strömbeck, M.M. Thomas,
K. Udipi, and S.R. White

HANSER

Hanser Publishers, Munich

Hanser/Gardner Publications, Inc., Cincinnati

The Editors:
Raju S. Davé, Morrison & Foerster, 2000 Pennsylvania Avenue NW, Washington, DC 20006-1888, USA
Alfred C. Loos, Department of Engineering Science and Mechanics, Virginia Polytechnic Institute and State University, Blacksburg, VA 24061, USA

Distributed in the USA and in Canada by
Hanser/Gardner Publications, Inc.
6915 Valley Avenue, Cincinnati, Ohio 45244-3029, USA
Fax: (513) 527-8950
Phone: (513) 527-8977 or 1-800-950-8977
Internet: http://www.hansergardner.com

Distributed in all other countries by
Carl Hanser Verlag
Postfach 86 04 20, 81631 München, Germany
Fax: +49 (89) 98 12 64

Library of Congress Cataloging-in-Publication Data
Processing of composites / Raju S. Davé, Alfred C. Loos, editors : with contributions from F. Abrams . . . [et al.].
p. cm. – (Progress in polymer processing)
Includes bibliographical references and index.
ISBN 1-56990-226-7 (hc.)
1. Plastics. 2. Polymeric composites. I. Davé, Raju S. II. Loos, Alfred C. III. Series.
TP1120.P76 1999
668.4—dc21 99-27337

Die Deutsche Bibliothek – CIP-Einheitsaufnahme
Processing of composites / Raju S. Davé/Alfred C. Loos (ed.). With contributions from F. Abrams . . . – Munich : Hanser ; Cincinnati : Hanser/Gardner, 1999
(Progress in polymer processing)
ISBN 3-446-18044-3

Typeset in England by Techset Composition Ltd., Salisbury
Printed and bound in Germany by Kösel, Kempten

PROGRESS IN POLYMER PROCESSING SERIES

Warren E. Baker, *Series Editor*

Advisory Board

Dr. Leszek A. **Utracki**
National Research Council of Canada
CANADA

Dr. George **Vassilatos**
E. I. Du Pont Co.
U.S.A.

Prof. John **Vlachopoulos**
McMaster University
CANADA

Prof. I. M. **Ward**
The University of Leeds
UNITED KINGDOM

Prof. James L. **White**
University of Akron
U.S.A.

Prof. Xi **Xu**
Chengdu University of Science and Technology
CHINA

Foreword

Since World War II, the industry based on polymeric materials has developed rapidly and spread widely. The polymerization of new polymeric species advanced rapidly during the 1960s and 1970s, providing a wide range of properties. A plethora of specialty polymers have followed as well, many with particularly unique characteristics. This evolution has been invigorated by the implementation of metallocene catalyst technology. The end use of these materials has depended on the development of new techniques and methods for forming, depositing, and locating these materials in advantageous ways, which are usually quite different from those used by the metal or glass fabricating industries. The importance of this activity, "polymer processing," is frequently underestimated when reflecting on the growth and success of the industry.

Polymer processes, such as extrusion, injection molding, thermoforming, and casting provide parts and products with specific shapes and sizes. Furthermore, they must control, beneficially, many of the unusual and complex properties of these unique materials. Because polymers have high molecular weights and, in may cases, tend to crystallize, polymer processes are called to control the nature and extent of orientation and crystallization, which, in turn, have a substantial influence on the final performance of the products made. In some cases, these processes involve synthesizing polymers during the polymer processing operation, such as continuous fiber composites processing, which is the topic of this book. Autoclave processing, pultrusion, and filament winding each synthesize the polymer and form a finished part in one step or a sequence of steps, evidence of the increasing complexity of the industry. For these reasons, successful polymer process researchers and engineers must have a broad knowledge of fundamental principles and engineering solutions.

Some polymer processes have flourished in large indutrial units, such as synthetic fiber spinning. However the bulk of the processes are rooted in small- and medium-sized entrepreneurial enterprises in both developed and new developing countries. Their energy and ingenuity have sustained growth to this point, but clearly the future will belong to those who progressively adapt new scientific knowledge and engineering principles to the industry. Mathematical modeling, online process control and product monitoring, and characterization based on the latest scientific techniques will be important tools in keeping these organizations competitive in the future

The Polymer Processing Society was started in Akron, Ohio, in 1985 with the aim of focusing on an international scale on the development, discussion, and dissemination of new and improved polymer processing technology. The society facilitates this by sponsoring several conferences annually and by publishing the journal, *International Polymer Processing*, and this book series, *Progress in Polymer Processing*. This series of texts is dedicated to the goal of bringing together the expertise of accomplished academic and industrial professionals. The volumes have a multiauthored format, which provides a broad picture of the volume topic viewed from the perspective of contributors from around the world. To accomplish these goals, we need the thoughtful insight and effort of our authors and book editors, the critical overview of our Editorial Board, and the efficient production of our publisher.

The book deals with the underlying process fundamentals and manufacturing processes for preparing polymer composites reinforced with continuous fibers. These processes have developed into what is arguably the single largest producer of complex engineered parts, finding significant application in the aerospace industry, for example. The resulting products represent the most significant incursion by polymeric materials into those areas, where high performance traditional materials, such as metals and ceramics, have been used. These achievements are dependent on the complex interplay of chemical kinetics, rheology, and morphology development in a multiphase environment, which leads to the required anisotropic properties. Quite new continuous fiber composite processes have been developed during the last decade, and the complexity and fundamental steps involved signal further imaginative developments in the future. This book includes numerous contributions, industrial and institutional, from America as well as Europe and Asia and, as such, forms a valuable contribution to the field.

Brampton, Ontario, Canada

Warren E. Baker
Series Editor

Contents

Part I Theory

Contributors

Abrams, F., WL/MLBC, Wright Patterson Air Force Base, 011 45433-7750, USA

Advani, Suresh G., Department of Mechanical Engineering, University of Delaware, Newark, DE 19716-3140, USA

Åström, B.T., Department of Lightweight Structures, Royal Institute of Technology, Stockholm, Sweden

Calado, Verônica M.A., Department of Chemical Engineering, University of Rio de Janeiro, Rio de Janeiro 21949-900, Brazil

Campbell, Flake C., Materials Directorate, Wright Laboratories, Charles B. Browning Air Force Base, Dayton, OH 45433, USA

Cohen, D., Hercules Aerospace Company, Magna, UT 84044-0094, USA

Davé, Raju S., c/o Morrison & Foerster, 2000 Pennsylvania Avenue NW, Washington, DC 20006-1888, USA

Gebart, B. Rikard, Swedish Institute of Composites, 8-941 26 Pitea, Sweden

Joseph, Babu, Materials Research Laboratory, School of Engineering and Applied Science, Washington University, St. Louis, MO 63130-4899, USA

Kardos, J.L., Department of Chemical Engineering, Washington University, St. Louis, MO 63130-4899, USA

Khomami, Bamin, Department of Chemical Engineering, Washington University, St. Louis, MO 63130-4899, USA

Kim, S.C., Department of Chemical Engineering, Korea Advanced Institute of Science and Technology, Taejon 305-701, Korea

Kranbuehl, David E., Departments of Chemistry and Applied Science, College of William and Mary, Williamsburg, VA 23187-8795, USA

Kruse, Robert L., 444 Michael Sears Toad, Belchertown, MA 01007, USA

Li, Min-Chung, Impco Technologies, Cerritos, CA 90701, USA

Loos, Alfred C., Department of Engineering Science and Mechanics, Virginia Polytechnic Institute and State University, Blacksburg, VA 24061, USA

Mallow, Andrew R., McDonnell Douglas Aerospace, St. Louis, MO 63146-4021, USA

Mantell, S.C., Department of Mechanical Engineering, University of Minnesota, Minneapolis, MN 55455, USA

Miller, Alan K., Lockheed-Martin Missiles and Space, Sunnyvale, CA 94088, USA

Park, J.W., Department of Chemical Engineering, Korea Advanced Institute of Science and Technology, Taejon 305-701, Korea

Strömbeck, L. Anders, Borealis Industries, 42246 Hisingsbacka, Sweden

Thomas, Matthew M., Department of Chemical Engineering, Washington University, St. Louis, MO 63130-4899, USA

Udipi, Kishore, Monsanto Company, St. Louis, MO 63167, USA

White, Scott R., University of Illinois, Urbana-Champaign, Urbana, IL 61801, USA

Preface

Composite materials have been acclaimed as the "Materials of the Future." A key question is whether composite materials will always remain the materials of the future or if the future is here. Advanced polymer composites, once destined for stealth military aircraft or aerospace uses, are beginning to be used in down-to-earth structures, such as bridges, buildings, and highways. However, there are still considerable impediments to wider use, and composite manufacturers need to make great strides in the development and manufacturing of composite materials.

What makes the fabrication of composite materials so complex is that it involves simultaneous heat, mass, and momentum transfer, along with chemical reactions in a multiphase system with time-dependent material properties and boundary conditions. Composite manufacturing requires knowledge of chemistry, polymer and material science, rheology, kinetics, transport phenomena, mechanics, and control systems. Therefore, at first, composite manufacturing was somewhat of a mystery because very diverse knowledge was required of its practitioners. We now better understand the different fundamental aspects of composite processing so that this book could be written with contributions from many composite practitioners.

This book provides a quick overview of the fundamental principles underlying composite processing and summarizes a few important processes for composite manufacturing. This book is intended for those who want to understand the fundamentals of composite processing. In particular, this book would be especially valuable for students as a graduate level textbook and practitioners who struggle to optimize these processes.

We thank all the chapter authors for their heroic efforts in writting their chapters. Without their contributions this book would be incomplete. In addition, we thank Lloyd Goettler of Monsanto, who is past president of the Polymer Processing Society, for suggesting that we edit this book. Other friends and mentors who had a major influence on our work include Robert L. Kruse, Kishore Udipi, and Allen Padwa, all of Monsanto, and Professor John L. Kardos of Washington University. Professor Warren Baker, Series Editor, has been very helpful in overseeing this project.

Certainly, we may have overlooked others who have helped us on our way to completing this book over a period of four years. Our sincere apologies to them, and we hope they will reflect on their positive contributions when they read this book. Last, but not least, we thank our families who endured through this process. Criticism and comments from readers are most welcome.

Raju S. Davé
Alfred C. Loos

Part I
Theory

1 Chemistry, Kinetics, and Rheology of Thermoplastic Resins Made by Ring Opening Polymerization*

Raj S. Davé[†], Kishore Udipi, and Robert L. Kruse[‡]

The ring opening polymerization of cyclic monomers that yield thermoplastic polymers of interest in composite processing is reviewed. In addition, the chemistry, kinetics, and rheology of the ring opening polymerization of caprolactam to nylon 6 are presented. Finally, the rheo-kinetics models for polycaprolactam are applied to the composite process of reaction injection pultrusion.

1.1 Overview

Ring opening polymerization of cyclic monomers to yield thermoplastic polymers has been studied by a number of investigators [1–19] over the years. A variety of cyclic monomers ranging in structures from the more commonly encountered olefins, ethers, formals, lactones,

* Work done in Monsanto Plastics Division and approved by Monsanto Company for external publication.
† Formerly with Monsanto Plastics and Bayer Polymers and to whom correspondence should be addressed.
‡ Formerly with Monsanto Plastics.

lactams, and carbonates to some of the more esoteric, like the thioformals, thiolactones, iminoethers, siloxanes, cyclic phosphites, cyclic phosphonites, and phosphonitrilic chloride have been polymerized to generate thermoplastics that range in properties from soft elastomeric to hard and crystalline.

All ring opening polymerizations are governed by ring-chain equilibria. Tendency toward polymerization of a cyclic monomer depends upon the existence and extent of ring strain, the initiator used, and the reactivity of the functional group within the ring [20]. Ring strain, which is a thermodynamic property, is generated in a cyclic monomer by the angular distortion of the chemical bonds and the steric effects of the substituents. The lower the ring strain, the more stable is the monomer with lower tendency to polymerize.

The thermodynamics of ring opening polymerization was first proposed by Dainton and Ivin [21] in the form of the following expression:

$$T_{\mathrm{c}} = \frac{\Delta H_p}{\Delta S_{\mathrm{p}} + R\ln[M]} \tag{1.1}$$

where T_{c} is the ceiling temperature (above which polymerization at monomer concentration $[M]$ is not possible), ΔH_{p} and ΔS_{p} are the enthalpy and entropy changes of polymerization, respectively, and R is the universal gas constant. It follows from Equation 1.1 that a lower temperature favors polymerization.

Most cyclic monomers of interest in the field of composites happen to be heterocyclic in nature. Polymerizability of some monomers is summarized in Table 1.1. Ring opening polymerizations invariably follow ionic mechanisms, although a few are known to proceed via the free radical route and some via metathesis involving metallocarbene intermediates.

Among the more common thermoplastics from ring opening polymerization of interest in composite processing are polylactams, polyethers, polyacetals, and polycycloolefins. It has also been shown that polycarbonates can be produced from cyclic carbonates [22]. Anionic ring opening polymerization of caprolactam to nylon 6 is uniquely suited to form a thermoplastic matrix for fiber-reinforced composites, specifically by the reaction injection pultrusion process [23–25]. The fast reaction kinetics with no by-products and the crystalline

Table 1.1 Polymerizability of Some Unsubstituted Cyclic Monomers

	Polymerizability		
	Ring size		
Class of monomer	Five	Six	Seven
Lactam	+	+	+
Lactone	+	+	+
Imide	−	−	+
Anhydride	−	−	+
Ethers	+	−	+

+ = polymerizes
− = does not polymerize

nature of the nylon so produced make anionic polymerization of caprolactam a compelling choice for the reaction injection pultrusion process. In addition to the fast reaction kinetics, low viscosity of the monomer affords superior wetting of the reinforcing fibers, which leads to improved adhesion between the fibers and the matrix polymer, as compared with the conventional thermoplastic composite processes where the melt viscosity of the thermoplastic polymers is too high to afford good wetting of the fibers. Because this chapter will later cover polylactams in greater detail, the chemistry of other thermoplastic polymers by ring opening polymerization will be dealt with here in some detail.

Polyethers are prepared by the ring opening polymerization of three, four, five, seven, and higher member cyclic ethers. Polyalkylene oxides from ethylene or propylene oxide and from epichlorohydrin are the most common commercial materials. They seem to be the most reactive alkylene oxides and can be polymerized by cationic, anionic, and coordinated nucleophilic mechanisms. For example, ethylene oxide is polymerized by an alkaline catalyst to generate a living polymer in Figure 1.1. Upon addition of a second alkylene oxide monomer, it is possible to produce a block copolymer (Fig. 1.2).

Cationic polymerization of alkylene oxides generally produces low molecular weight polymers, although some work [26] seems to indicate that this difficulty can be overcome by the presence of an alcohol (Fig. 1.3). Higher molecular weight polyethylene oxides can be prepared by a coordinated nucleophilic mechanism that employs such catalysts as alkoxides, oxides, carbonates, and carboxylates, or chelates of alkaline earth metals (Fig. 1.4). An aluminum-porphyrin complex is claimed to generate 'immortal' polymers from alkylene oxides that are totally free from termination reaction [27].

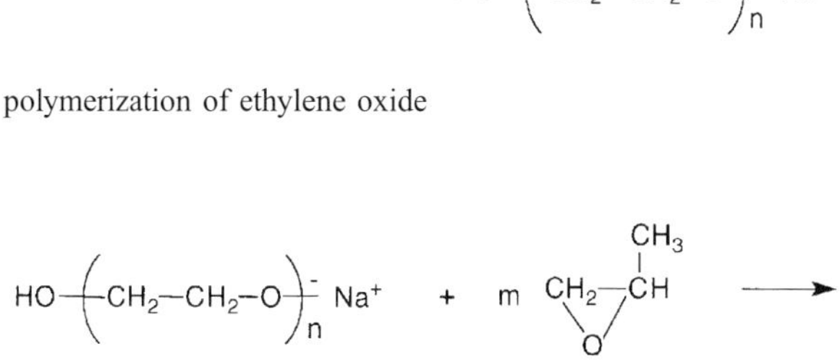

Figure 1.1 Living polymerization of ethylene oxide

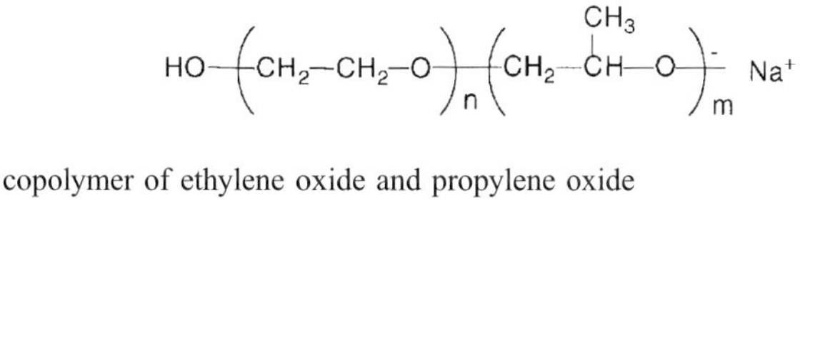

Figure 1.2 Block copolymer of ethylene oxide and propylene oxide

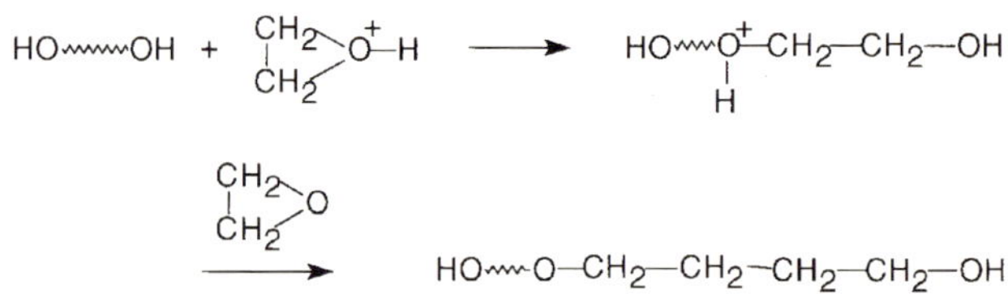

Figure 1.3 Cationic polymerization of ethylene oxide in the presence of an alcohol

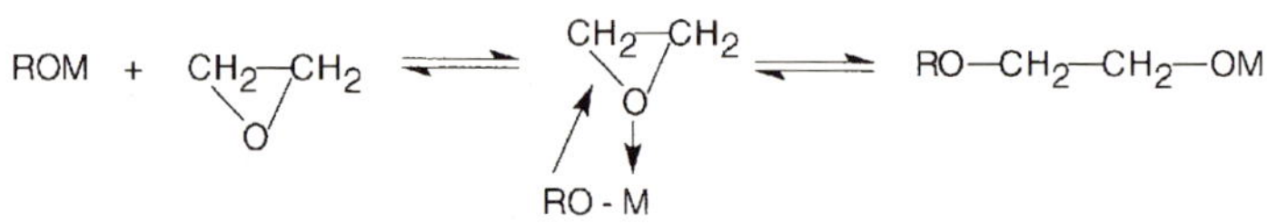

Figure 1.4 Polymerization of ethylene oxide by nucleophilic mechanism

Tetrahydrofuran, a five-member cyclic ether, polymerizes cationically to yield an elastomeric polymer [28]. Oxepane, a seven-member analog, polymerizes to a crystalline polymer.

By organic chemistry formalism, polyacetals are reaction products of aldehydes with polyhydric alcohols. Polymers generated from aldehydes, however, either via cationic or anionic polymerization are generally known as *polyacetals* because of repeating acetal linkages. Formaldehyde polymers, which are commercially known as *acetal resins*, are produced by the cationic ring opening polymerization of the cyclic trimer of formaldehyde, viz., trioxane [29–30] (Fig. 1.5).

Polyacetals are prone to degrade to the monomers at elevated temperatures by an unzipping mechanism. They are either end-capped or copolymerized with low levels of an alkylene oxide to prevent unzipping and impart better processability. Polyacetals from higher aldehydes do crystallize, with the degree of crystallinity depending upon the length of the side chain, *R* (Fig. 1.6). The longer the side chain, the less crystalline is the material and the lower is the melting. Polyformals are prepared by the cationic ring opening polymerization of cyclic formals. These could be regarded as codimers of formaldehyde and cyclic ethers. Thus, polyformals correspond to alternating copolymers of aldehydes and cyclic ethers.

Polycycloolefins are prepared by ring opening metathesis polymerization (ROMP) using transition metal catalysts [31]. By far the most commonly studied monomer is dicyclopentadiene (Fig. 1.7). Cycloolefins with high ring strains like norbornenes and their analogs polymerize very fast and the polymerizations are quite exothermic. Metathesis catalyst systems tend to be sensitive to the presence of polar compounds and the polymerization rates

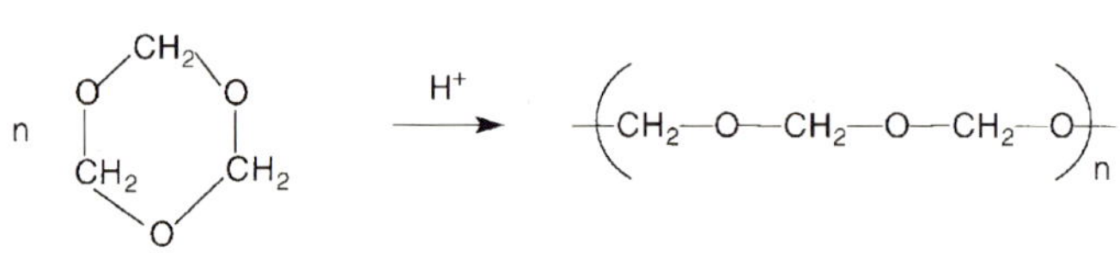

Figure 1.5 Cationic ring opening polymerization of a cyclic trimer of formaldehyde (viz., trioxane)

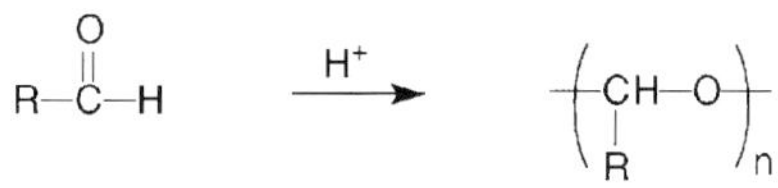

Figure 1.6 Polyacetals from higher aldehydes

are adversely affected. In the case of norbornenes, however, because of the highly strained ring, such catalyst systems appear to be more forgiving.

Polycarbonates, both aliphatic and aromatic, have been prepared by the ring opening polymerization of cyclic monomers or oligomers [22]. Cyclic monomeric precursors are more common in aliphatic polycarbonates, but because of steric reasons aromatic polycarbonates can only be prepared from cyclic oligomers. Both cationic and anionic initiators have been examined and anionic initiators appear to be more efficient.

Although aliphatic polycarbonates have been prepared and studied quite extensively, interest in them has been minimal due to their thermal instability and, in some cases, lack of ductility. Aliphatic polycarbonates with β hydrogens decompose to olefins, alcohol, and CO_2.

Attempts to prepare aromatic polycarbonates from cyclic oligomers had continued through the years without much success. Researchers at General Electric have developed a method that would afford much more control over the composition of the oligomers than ever before [22]. In this process, a bisphenol A-bischloroformate is added slowly to an efficiently stirred mixture of Et_3N, aqueous NaOH, and CH_2Cl_2 to selectively control the hydrolysis/condensation to generate a mixture of essentially cyclic oligomers and high molecular weight polymer (~85/15) with extremely low levels of linear oligomers (Fig. 1.8). This procedure provides a distribution of oligomers of $n = 2–26$ with $>90\%$ species with degrees of polymerization <10.

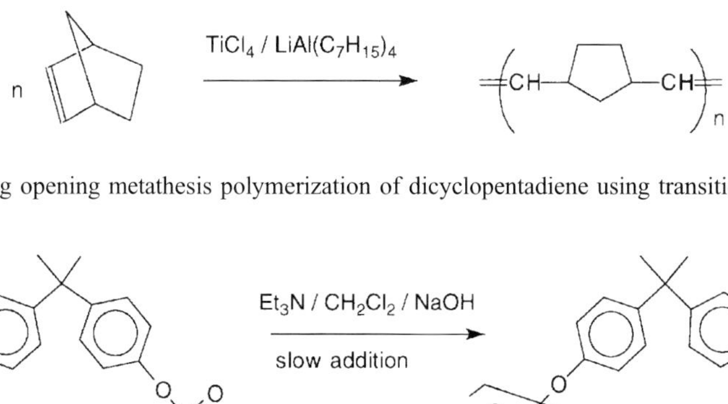

Figure 1.7 Ring opening metathesis polymerization of dicyclopentadiene using transition metal catalysts

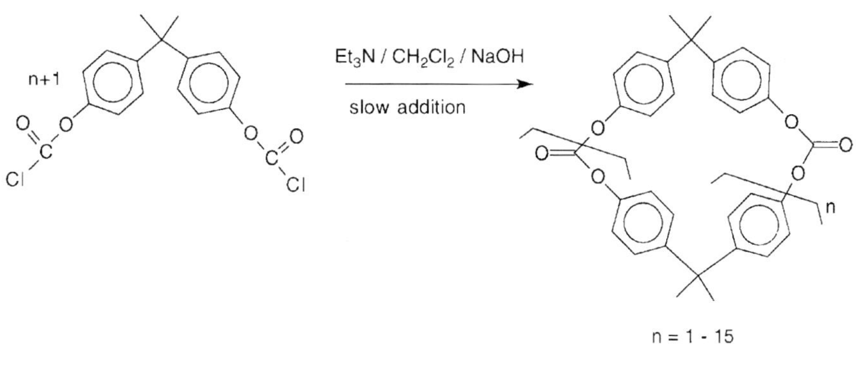

Figure 1.8 Preparation of cyclic oligomers that yield polycarbonates by ring opening polymerization

Heating these cyclic oligomers to 300°C for about 30 minutes results in ring opening polymerization to polycarbonates. The molecular weights of the resulting polycarbonates prepared in the absence of a catalyst are modest ($<20,000$), whereas catalysts like lithium stearate or titanium alkoxides produce polycarbonates of much higher molecular weights (50,000–300,000).

1.2 Chemistry of Anionic Ring Opening Polymerization of Lactams

Anionic ring opening polymerization of lactams to generate polyamides has been studied quite extensively by Sebenda [8–10], Sekiguchi [11], and Wichterle [12–13], among others, in academia, and by Gabbert and Hedrick [14] and by us [23–25] in industry. By far, caprolactam is the most studied lactam and the nylon 6 prepared by this route compares favorably in properties with that prepared by conventional hydrolytic polymerization.

Most of the work reported in the literature employs sodium lactamate salt as catalyst and isocyanate/lactam adducts as initiator. Gabbert and Hedrick [14] preferred to work with acyllactam as the initiator and Grignard salts of caprolactam as the catalyst in view of their ease of handling and fewer side reactions compared with the sodium lactamate and isocyanate system.

Anionic ring opening polymerization of caprolactam (as in other lactams) follows an activated monomer mechanism rather than a conventional activated chain end mechanism. That is, the chain growth reaction proceeds by the interaction of an activated monomer (lactam anion) with the growing chain end (N-acylated chain end in this case). In fact, the anionic attack constitutes the rate-determining step in the propagation. The other characteristic of this mechanism is that activated monomer is regenerated after every unit growth reaction. A typical reaction path for the polymerization of caprolactam is shown in Figure 1.9.

Nucleophilic attack by the amide anion can occur at either the exocyclic or endocyclic carbonyl. The former regenerates the lactamate anion, whereas the latter results in polymerization. Although the locus of nucleophilic attack has no major effects in a homopolymerization, it can exert considerable control over the copolymerizations and on copolymer structure.

From the scheme in Figure 1.9, it is also apparent that the propagation of anionic polymerization requires two active species: lactam anion and N-acyllactam end group. Because the monomer is only consumed via its anion, it is imperative that a certain level of basicity is maintained in the polymerization mixture. The basicity is generally achieved by replacing the proton of the monomer by a less acidic cation such as $MgBr+$. A large number of catalysts have been reported in the literature, including alkali metals [14,32–34], alkali metal hydroxides [32,35–37], alcoholates [38], carbonates [12,39], Grignard reagents [40], alkylaluminums [41], alkalialuminum hydrides [42] and their partial or total alkoxides [43–45] or their lactam salts [46], quaternary ammonium salts of lactams [47] or of other compounds [48–52], and guanidinium salts of lactams [53].

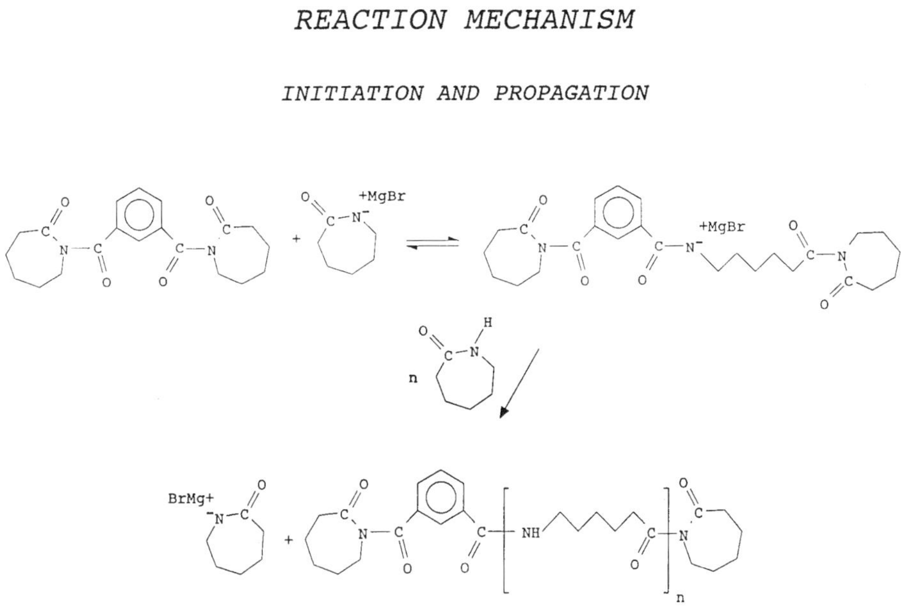

Figure 1.9 Mechanism of anionic polymerization of caprolactam

The anionic catalysts listed earlier react with lactam monomer to first form the salt, which in turn will dissociate to the active species, namely, the lactam anion. A strongly dissociating catalyst in low concentrations, therefore, is always preferable to weakly dissociating catalysts in higher concentrations. The catalytic activity of the various alkali metal and quaternary salts of a lactam generally follows the extent of their ionic dissociation that is controlled by the cation. Activity of a salt decreases with increasing size of the cation due to restricted mobility and decreased ionization potential.

Among the various catalyst systems described earlier, the alkali metal (sodium) lactamate is perhaps the most widely reported in the literature. The Grignard system (caprolactam–magnesium–bromide), despite its many advantages like ease in handling and less acidic MgBr+ cation, has received much less attention. Other catalyst systems may have equal or better reactivity than do the Grignard systems, but they have not been examined at Monsanto to the same extent as the Grignard. Two of the catalyst systems that have received continued investigation outside Monsanto [54] are complexes derived from caprolactam–magnesium–bromide and aluminum alkyls combined with the quaternary salts of lactams, which are often referred to as the *onium salts*. Results to date are quite inconclusive to claim that these systems have a definite advantage over the Grignard.

As initiators, the N-acyllactams have proved to be the most efficient among those investigated by far. Acyllactam end groups in a growing chain, which contain an imide linkage, possess strong acylating power, especially when it involves an already strong nucleophile in a lactam anion. N-acyllactams are generally prepared by the reaction of a lactam with either an anhydride, acid chloride, or an isocyanate. By and large, most of the work reported in the literature is based on monofunctional initiators, although, in principle, multifunctional initiators can be employed. Initiators with functionality higher than 2 can be expected to produce branched polymers.

The effect of the type and level of initiator on polymerization time, monomer conversion, and polymer molecular weight, the effect of polymerization temperature on the crystallization behavior of the polyamide generated, and the role of a higher lactam like laurolactam on the moisture absorption characteristics of the copolymers are discussed in our previous publication [23].

1.3 Kinetics of Anionic Polymerization of Caprolactam

The remainder of this chapter focuses on the kinetics and rheology of the ring opening polymerization of caprolactam to nylon 6. Furthermore, we will discuss the application of rheo-kinetics models to composite processing.

Although a large number of initiators are described in the literature [55,56] to polymerize caprolactam anionically, the primary choice of the catalyst has been sodium (Na), except in the studies at the Monsanto Company by Gabbert and Hedrick [14], Greenley et al. [57], and by us [23–25]. In these studies at Monsanto, caprolactam-magnesium-bromide and isophthaloyl-bis-caprolactam were used as the catalyst and initiator for several reasons, such as the stability and ease of handling of caprolactam-magnesium-bromide compared with sodium, and the proven efficiency of isophthaloyl-bis-caprolactam in earlier studies at the Monsanto Company. The catalyst/initiator combinations used in prior published kinetic studies are: Na/tetraacetyl hexamethylene diisocyanate [58], Na/N-acetylcaprolactam [57,59–62], Na/hexamethylene-1,6-bis-carbamidocaprolactam [62–64], Na/phenyliso-cyanate [57,61,64,65], Na/toluenediisocyanate [61], Na/1,4-diphenylmethanediisocyanate [61], Na/triphenylmethanediisocyanate [61], Na/trimer of toluene diisocyanate [61], Na/phenylcarbamoyl caprolactam [62], Na/2,4-toluene-bis-carbamoyl caprolactam [62], Na/4,4-diphenylmethane-bis-carbamoyl caprolactam [62], Na/hexamethylene-bis-carbamoyl caprolactam [62], and caprolactam-magnesium-bromide/N-acetylcaprolactam [57].

1.3.1 Kinetics Model

The reaction mechanism of ring opening homopolymerization of caprolactam consists primarily of two transacylation reactions: initiation and propagation. The initiation occurs by the addition reaction between initiator and catalyst (described in Sec. 1.2). The propagation then occurs by repeating the addition and hydrogen extraction reactions. According to Sebenda [55], such a "regular" reaction scheme is presented "for the sake of simplicity." In reality, deactivation, branching, and a series of reversible transacylation reactions occurring during the anionic ring opening polymerization of caprolactam produce side reaction products, heterogeneities in the resultant polymer structure, and a broad molecular weight distribution [55].

The low temperature (~140°C) anionic ring opening polymerization is further complicated by the crystallinity in nylon 6. Magill [66] has reported that the temperature for maximum crystallization rate in nylon 6 is about 140–145°C. The nucleation rate is low above 145°C, whereas viscous effects hinder crystal growth below this temperature. As a result, at about 140–145°C, heterogeneous reaction conditions can be encountered (as we have seen in our studies) if there is simultaneous polymerization of caprolactam and crystallization of the nylon 6 formed.

The phenomenological kinetics of the isophthaloyl-bis-caprolactam–initiated anionic polymerization of caprolactam was obtained by the adiabatic reactor method. Adiabatic polymerization was conducted in a heavily insulated, 250 ml glass beaker (Fig. 1.10) that was set up in an air-circulating oven. The front door of the oven had two hand-holes, similar to a glove box, for inserting the experimenter's hands into the oven. This enabled the catalyst and initiator solutions to be combined, vigorously stirred, and poured into the reaction vessel by working through the two hand-holes without opening the oven. The glass polymerization vessel and the catalyst and initiator solutions in caprolactam were preheated to the initial reaction temperature in the oven. After pouring the mixture of the catalyst and initiator solutions into the beaker, a lid with a copper-constantan thermocouple (Fig. 1.10) was placed on the beaker. The total time for mixing and pouring the catalyst and initiator solutions and covering the beaker with the lid was minimized to less than a couple of seconds. The temperature of the oven as well as inside of the glass polymerization vessel were continually monitored and stored by a data acquisition equipment described in Reference [23]. The samples remained in the adiabatic reactor for at least 30 minutes to insure complete reaction and crystallization.

Principles of the adiabatic reactor method have been discussed elsewhere [67,68]. Under adiabatic conditions, assuming constant heat capacity, constant heat of reaction, and homogeneous reaction, temperature rise data yields fractional conversion, X [68]:

$$X = \frac{[M]_0 - [M]}{[M]_0} = \frac{H}{H_{\text{tot}}} = \frac{(T - T_0)}{(T_f - T_0)} \tag{1.2}$$

The terms in Equation 1.2 are described in Nomenclature. The condition of constant heat capacity can be relaxed if accurate data is available for heat capacity as a function of both conversion and temperature.

In the past, two approaches to kinetic modeling have been used: mechanistic models [57–59,63,65,69] and overall models [60–62,64]. The mechanistic models have attempted to account for each possible reaction individually. Although propagation reactions in caprolactam polymerization consist of only a few types of transacylation reactions, their detailed mechanism, as well as kinetics, are not well understood [55]. In seeking a better kinetic model capable of describing the polymerization process and reflecting the chemistry as well, Cimini and Sundberg [69] modified a rate equation originally derived mechanistically by Reimschuessel [70]. Provaznik et al. [71] have subsequently shown the fundamental importance of changes of the reaction medium on the individual polymerization reactions. As a result appropriate corrections concerning the detailed reaction mechanism and kinetics

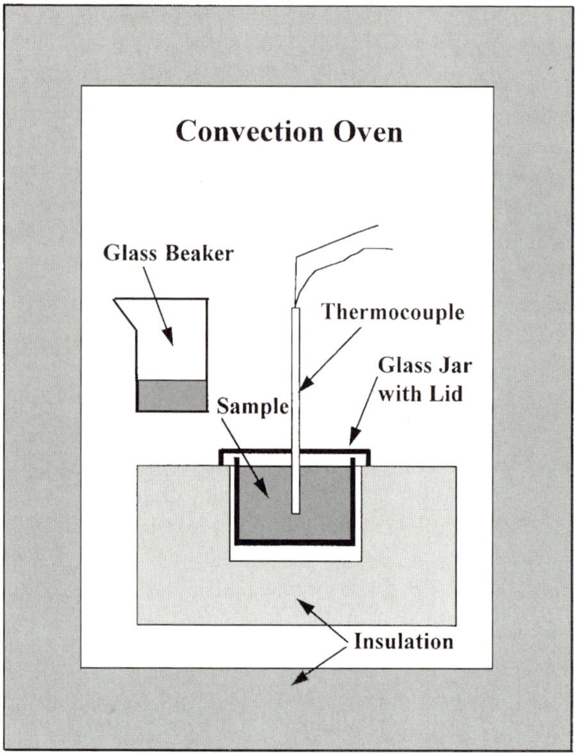

Figure 1.10 Apparatus for measurement of adiabatic temperature rise during anionic polymerization of nylon 6

can be expected in the future; however, the mechanistic models have so far had a limited success in predicting the anionic ring opening polymerization of caprolactam. This approach has been found to be severely hindered by the complex nature of this anionic ring opening polymerization. Although mechanistic models are highly desirable, in the absence of accurate information about the intermediate steps and the possibilities of side reactions like catalyst deactivation and branching, these models are complicated and impractical. On the other hand, the overall models lump all reactions into a single reaction step that accounts for the overall reaction profile like the initial rise in the reaction rate with conversion followed by a decrease in the reaction rate.

We have used an autocatalytic model originally proposed by Malkin et al. [62]. Bolgov et al. [61] found that the originally proposed autocatalytic model [62], which was valid for equal concentration of initiator and catalyst during the anionic polymerization of caprolactam, can be modified for unequal concentration of the initiator and catalyst by an autocatalytic equation of type

$$\frac{dX}{dT} = k \exp\left(\frac{U}{RT}\right)\frac{[A][C]}{[M_0]}(1-X)\left\{1+\frac{b}{([A][C])^{1/2}}X\right\} \tag{1.3}$$

The terms in Equation 1.3 (Malkin's autocatalytic model) are described in Nomenclature. In Malkin's autocatalytic model, the concentration of the activator, $[A]$, is defined as the concentration of the initiator times the functionality of the initiator. For a difunctional initiator [e.g., isophthaloyl-bis-caprolactam, the concentration of the activator (acyllactam) is twice the concentration of the initiator]. The term $[C]$ is defined as the concentration of the metal ion that catalyzes the anionic polymerization of caprolactam. In a magnesium-bromide catalyzed system, the concentration of the metal ion is the same as the concentration of the caprolactam-magnesium-bromide (catalyst) because the latter is monofunctional.

Malkin's autocatalytic model is an extension of the first-order reaction to account for the rapid rise in reaction rate with conversion. Equation 1.3 does not obey any mechanistic model because it was derived by an empirical approach of fitting the calorimetric data to the rate equation such that the deviations between the experimental data and the predicted data are minimized. The model, however, both gives a good fit to the experimental data and yields a single pre-exponential factor (also called the *front* factor [64]), k, activation energy, U, and autocatalytic term, b. The value of the front factor k allows a comparison of the efficiency of various initiators in the initial polymerization of caprolactam [62]. On the other hand, the value of the autocatalytic term, b, describes the intensity of the self-acceleration effect during chain growth [62].

1.3.2 Kinetic Model Verification

In our studies, the catalyst and initiator system was comprised of caprolactam-magnesium-bromide and isophthaloyl-bis-caprolactam, respectively. We determined the optimum values of the kinetic parameters in Malkin's autocatalytic model (Eq. 1.3), which consist of k, U, and b, by regression analysis.

Equation 1.3 was linearized by transposing $(1 - x)$ and the autocatalytic term to the left and then taking the logarithms of both sides of the equation. Fixing the value of b, a linear regression was performed for k and U. This procedure was repeated for several values of b, and an optimum value of b was chosen that gave the best fit straight line to the linearized equation. The corresponding values of k and U obtained from the best fit straight line were chosen as the optimum.

The values of the activation energy U, the front factor k, and the autocatalytic term b for the caprolactam-magnesium-bromide/isophthaloyl-bis-caprolactam system, as well as other catalyst/initiator systems, are shown in Table 1.2. The values of the kinetic constants for the caprolactam-magnesium-bromide/isophthaloyl-bis-caprolactam system are based on the adiabatic temperature rise data in Figure 1.11, with initial polymerization temperatures of 117 and 136°C. It is important to note that the activation energy of the magnesium-catalyzed system is considerably lower (30.2 kJ/mol vs. about 70 kJ/mol) than that for the sodium catalyzed system [60,64]. This is because the magnesium cation is less electropositive than the sodium cation. Compared with the sodium cation, the magnesium cation is therefore less tightly bound to the caprolactam anion.

In the only other reported study on the kinetics of anionic ring opening homopolymerization of caprolactam using caprolactam-magnesium-bromide, Greenley et al. [57]

Table 1.2 Kinetic Constants for Anionic Polymerization of Caprolactam with Different Catalyst and Initiator Systems

System (catalyst/initiator)	Source	Model	Analytical method	U (kJ/mol)	k (L/mol s)	b (L/mol)
MgBr+/IBT[a]	Davé et al. [24]	Malkin's autocatalytic model	Adiabatic temperature analyzed by regression analysis	30.2	1.49×10^4	2.17
Na/HMCCl[b]	Malkin et al. [60]	Same as above	Same as above	63 ± 6	4.17×10^8	0.066
Na/HMCCl[b]	Sibal et al. [64]	Same as above	Same as above	63.8 ± 0.5	2.23×10^8	1.15 ± 0.5
MgBr+/NAC[c]	Greenley et al. [57]	Greenley's mechanistic model	Assuming pseudo first-order, isothermal reaction during low conversion	46	N/A	N/A
MgBr+/IBT[a]	Davé et al. [24]	First-order rate dependence on monomer concentration	Same as above	40.6	7.62×10^5	N/A

[a] = Magnesium-bromide-caprolactam/isophthaloyl-bis-caprolactam
[b] = Sodium/hexamethylene-1,6,-bis-carbamidocaprolactam
[c] = Magnesium-bromide-caprolactam/N Acetylcaprolactam
N/A = Not available or not applicable.

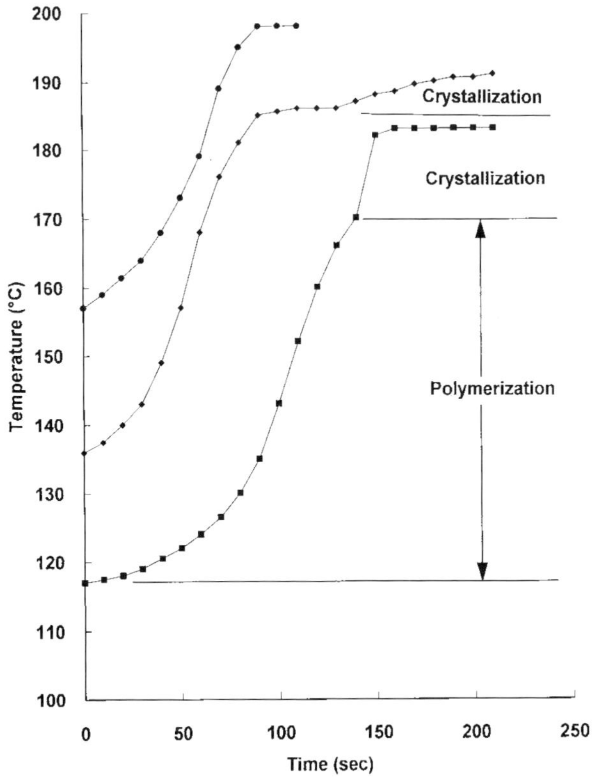

Figure 1.11 Adiabatic conversion of nylon 6: Experimental data for initial polymerization temperatures of 117°C (bottom line), 136°C (middle line), and 157°C (top line) with acyllactam and caprolactam-magnesium-bromide concentrations of 70 and 108 mmol/L, respectively

determined the value of the activation energy by making the following assumptions: (1) the reaction is pseudo–first-order; (2) the reaction is isothermal—consequently, their experimental data were below 20 percent conversion due to the need for pseudoisothermal conditions, and (3) there is a half-order dependence on the initial catalyst concentration. The third assumption was made in the derivation of the rate equation to obtain a better fit. In addition, their rate equation was derived by considering an irreversible degradation reaction with a rate constant k_d in addition to the polymerization reaction with a rate constant, k_p. The degradation rate constant, however, was found to be negligible—k_p was approximately 1500 times larger than k_d [57].

The role of the isothermal and pseudo–first-order reaction assumptions on the observed value of activation energy was assessed to allow comparison of our data to previous work by modifying Malkin's autocatalytic equation so that the autocatalytic term b is equal to zero. The values of the activation energy and front factor were calculated using short-time, low-conversion data. By making the autocatalytic term equal to zero, the modified Malkin autocatalytic model becomes a first-order rate reaction. Table 1.2 shows that by assuming a pseudo–first-order, isothermal reaction during low conversion, the values of the activation energy for the caprolactam-magnesium-bromide catalyzed ring opening homopolymerization of caprolactam are calculated to be nearly the same for Greenley et al. [57] and for us (46 kJ/mol vs. 40.6 kJ/mol). As a matter of fact, even the value of the activation energy calculated by Greenley et al. [57] for the sodium/N-acetylcaprolactam system assuming pseudo–first-order, isothermal reaction during low conversion is much larger than the activation energy reported by other investigators [59–61] for the same catalyst/initiator system (92 kJ/mol vs. about 70 kJ/mol). Based on the preceding calculations, the implication of assuming a pseudo–first-order reaction (i.e., neglecting the autocatalytic term and using only the low conversion data in the determination of the activation energy) is likely to result in gross overprediction in the value of the activation energy.

It appears that most activation energy values in the literature for sodium catalyzed anionic ring opening homopolymerization of caprolactam are in the range of 63–71 kJ/mol despite the variety of initiators used [58–6,64]. This indicates that the value of the activation energy is probably independent of the initiator used and dependent only on the catalyst used in the anionic ring opening polymerization of caprolactam. The results of this study, as well as the study by Greenley et al. [57], add further credence to the last statement that the activation values for a caprolactam-magnesium-bromide catalyzed system is much lower than the activation energy values for a sodium catalyzed system (30 kJ/mol versus about 70 kJ/mol).

We calculated the values of U, k, and b for caprolactam-magnesium-bromide/isophthaloyl-bis-caprolactam system to be 30.2 kJ/mol, 1.49×10^4 L/mol, and 2.17 L/mol, respectively. We have used these optimized values for comparing model predictions with experimental data obtained from adiabatic polymerization to study the effect of initial polymerization temperature and effects of initiator and catalyst concentrations [24].

1.4 Viscosity Growth During Anionic Polymerization of Caprolactam

In order to understand the time-dependent growth of complex viscosity, $|\eta^*|$, during the anionic polymerization of caprolactam, we have developed a concurrent polymerization and rheological test methodology. In this test, two monomer streams—one containing catalyst and the other containing initiator—are mixed essentially instantaneously as they enter the gap between the parallel plates in a Rheometrics Mechanical Spectrometer. The polymerization is thus carried out under isothermal conditions, over the range of 120–160°C, while rheological measurements are made. The complex viscosity $|\eta^*|$ was monitored with an end toward defining the rheokinetics of the anionic ring opening homopolymerization of caprolactam using isophthaloyl-bis-caprolactam and caprolactam-magnesium-bromide as initiator and catalyst. Isothermal rheometry at five temperatures was accomplished for one initiator/catalyst concentration with the objectives of: (1) to define the rheology of nylon 6 anionically polymerizing by ring opening polymerization in caprolactam; and (2) to correlate this rheology with the kinetics of polymerization.

1.4.1 Viscosity Model

The following relation was used for determining the complex viscosity of the polymerizing system [64]:

$$|\eta^*| = |\eta_0| \exp(k^* X) \tag{1.4}$$

where $|\eta^*|$ is the complex viscosity of nylon 6 anionically polymerizing in its monomer, $|\eta_0|$ is the complex viscosity of caprolactam monomer, k^* is a constant, and X is fractional conversion. The complex viscosity of the monomer, $|\eta_0|$ follows an Arrhenius temperature dependence [64]

$$|\eta_0|(T) = 2.7 \times 10^{-7} \exp(3525/T) \quad \text{(Pa s)} \tag{1.5}$$

1.4.2 Viscosity Model Verification

Complex viscosity growth during anionic polymerization of caprolactam was measured under isothermal conditions using a Rheometics Dynamic Mechanical Analyzer (DMA), RMS-800. Figure 1.12 shows a schematic diagram of the delivery method for simultaneously injecting two streams through a static mixer into the rheometer platen gap, which is where the polymerization reaction and oscillatory complex viscosity measurement occur simultaneously. Prior to injection of a sample, the instrument was first equilibrated at the desired temperature and the gap was set. The details of the instrument and run conditions are given in Reference [25].

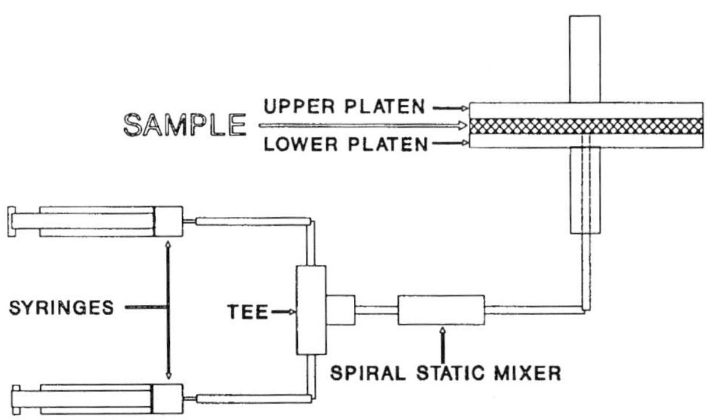

Figure 1.12 Dual stream injection system for in situ rheokinetic study of anionic ring opening polymerization of caprolactam

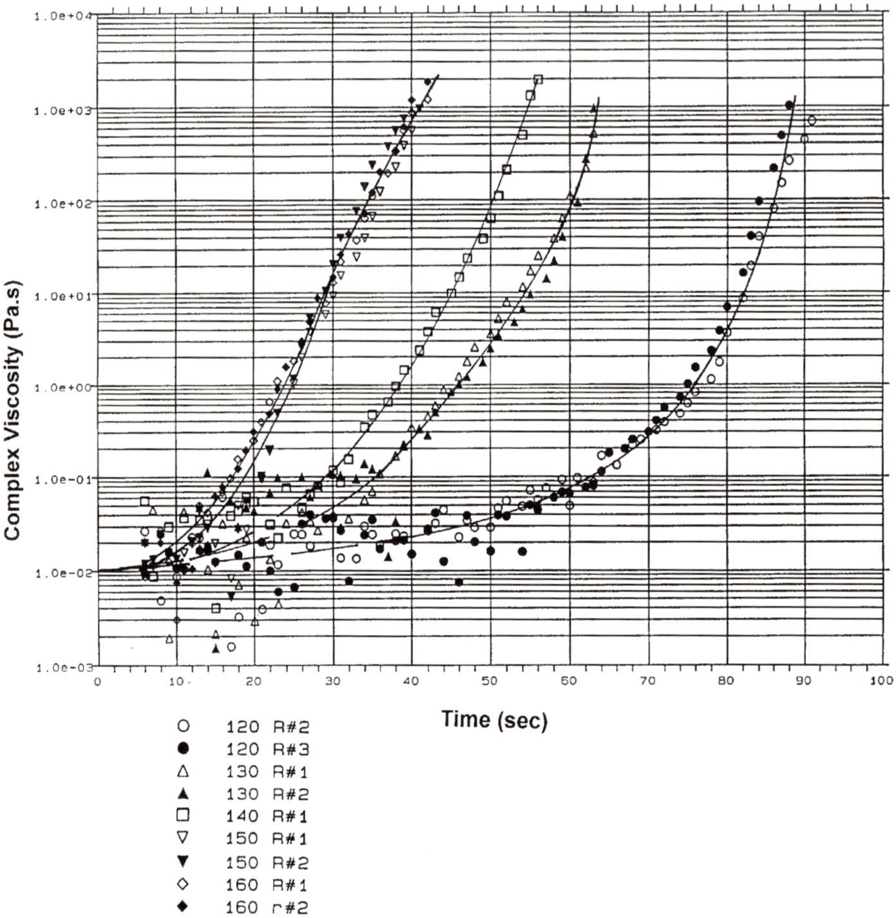

Figure 1.13 Isothermal complex viscosity rise during anionic polymerization of caprolactam using caprolactam-magnesium-bromide/isophthaloyl-bis-caprolactam as the catalyst/initiator system. Run numbers and polymerization temperatures are shown in the legend

The single feed composition investigated consisted of 133 mmols/L of caprolactam-magnesium-bromide and 45 mmols/L of the difunctional isophthaloyl-bis-caprolactam. Note that 45 mmols/L of the difunctional isophthaloyl-bis-caprolactam contain 90 mmols/L of the active acyllactam group, which react with the monofunctional caprolactam-magnesium-bromide to initiate the polymerization reaction.

Upon mixing and injection of the caprolactam monomer streams into the rheological instrument, polymerization was initiated and continued, whereas simultaneously monitoring the complex viscosity and other rheological parameters of the polymerizing system. The maximum measurable complex viscosity levels were achieved in about 100 s or less, depending on temperature.

At each temperature, the expected rapid rise of complex viscosity with conversion, molecular weight build-up, and ultimate solidification was observed and the complex viscosity–time relationship was quantified. The time to achieve maximum permissible torque on the instrument ranged from about 90 s at 120°C to 40 s at 160°C. This corresponds to 10^3 Pa s (10^4 poise) for the geometry used and is on such a steep slope of the complex viscosity–time curve that it is very close to the time for near infinite complex viscosity (i.e., total solidification).

Figure 1.13 shows replicated complex viscosity–time curves for all five temperatures. Some difficulty in reproducibly filling the gap between the parallel plates was encountered, leading to the requirement of measurement replication and averaging of results. In addition, the initial (very early time) viscosities were about 0.01 Pa s (ten times that of water) and are outside the sensitivity range of the instrument, leading to very large scatter at low times. These caveats notwithstanding, the data is relatively good, and provided a quantitative basis for modeling the matrix viscoelastic build-up during the reaction injection pultrusion process.

Figure 1.14 shows time to achieve a given complex viscosity as a function of polymerization temperature. These curves are fitted with a quadratic equation (second-order polynomial).

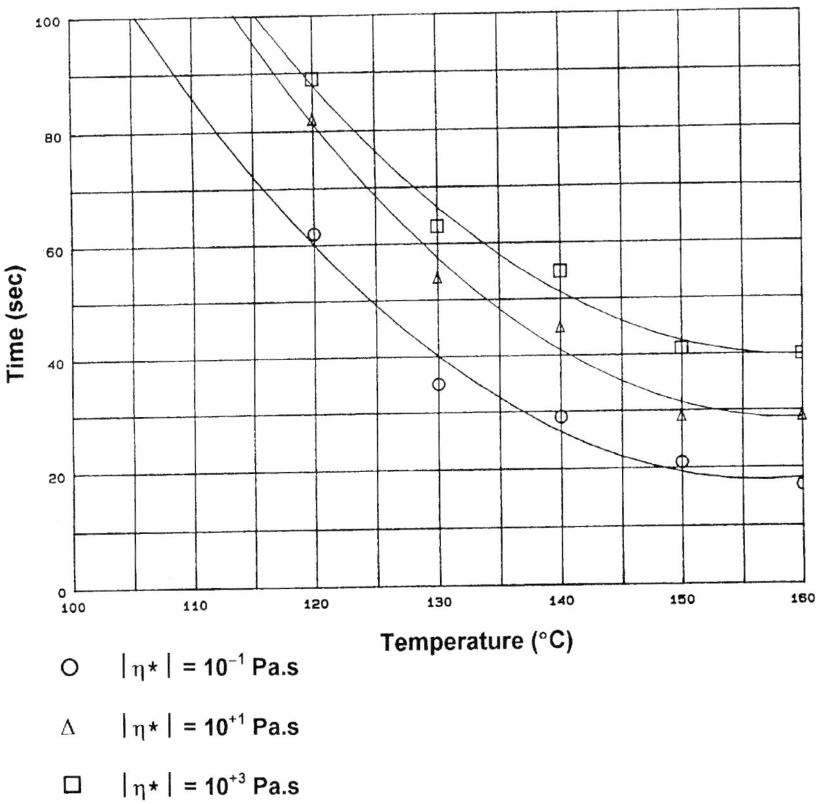

Figure 1.14 Time to reach complex viscosity of 0.1, 10, and 1000 Pa s as a function of temperature during anionic polymerization of caprolactam using caprolactam-magnesium-bromide/isophthaloyl-bis-caprolactam as the catalyst/initiator system. Complex viscosity is shown in the legend

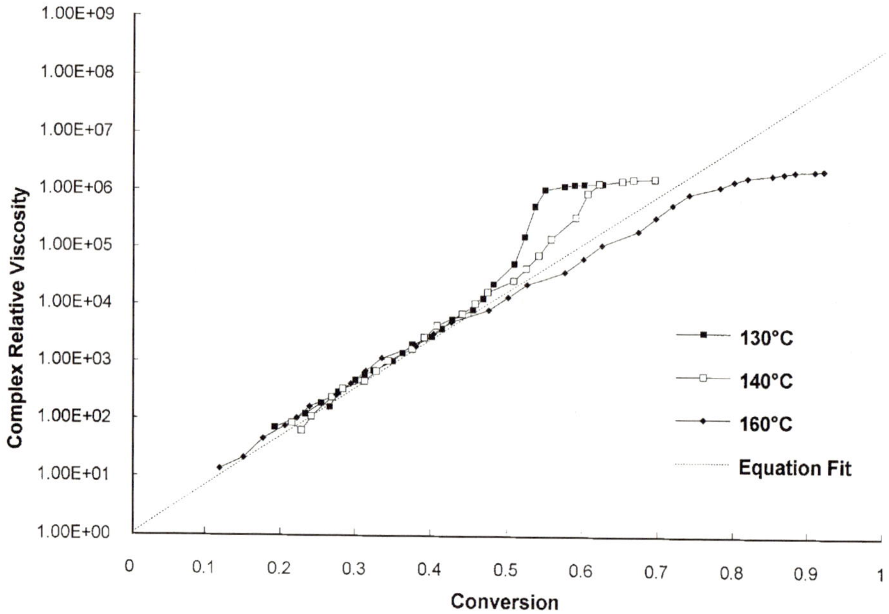

Figure 1.15 Relative complex viscosity ($|\eta^*|/|\eta_0|$) versus calculated conversion for polymerization at 130, 140, and 160°C. Phenomenological equation to fit the data prior to gelation is also shown

Figure 1.15 shows relative complex viscosity ($|\eta^*|/|\eta_0|$) as a function of conversion at 130, 140, and 160°C. Conversion was calculated by the kinetic model described previously [24]. Figure 1.15 shows that between 130 and 160°C (except 150°C) all curves are nearly identical below 50 percent conversion. For polymerization at 130 and 140°C, a sharp increase in relative complex viscosity was observed beyond 50 percent conversion. Sibal et al. [64] suggest that the sharp increase in relative complex viscosity during anionic ring opening polymerization of caprolactam is due to physical "gelation" that probably results from mechanical interlocking of the crystallites being formed during polymerization. This effect is likely not gelation, but simply the result of rigid crystallites on viscosity. The conversion at which the relative complex viscosity curve deviates from the linear, straight-line fit depends on the polymerization temperature because both kinetics of polymerization and kinetics of crystallization are temperature dependent. For polymerization near 145°C, where the crystallization rate is greatest [64], crystallization kinetics strongly competes with chemical kinetics. On the other hand, for polymerization at 160°C, crystallization rate is drastically reduced [66] and, therefore, the 160°C curve does not exhibit the gelation effect.

Even though a mechanistic explanation of the complex viscosity rise and gelation is not available at this time, the information in Figure 1.15 is useful in understanding and modeling reaction injection molding, reaction injection pultrusion, and other fiber-reinforced composite processes that are based on anionic ring opening polymerization of caprolactam. Below

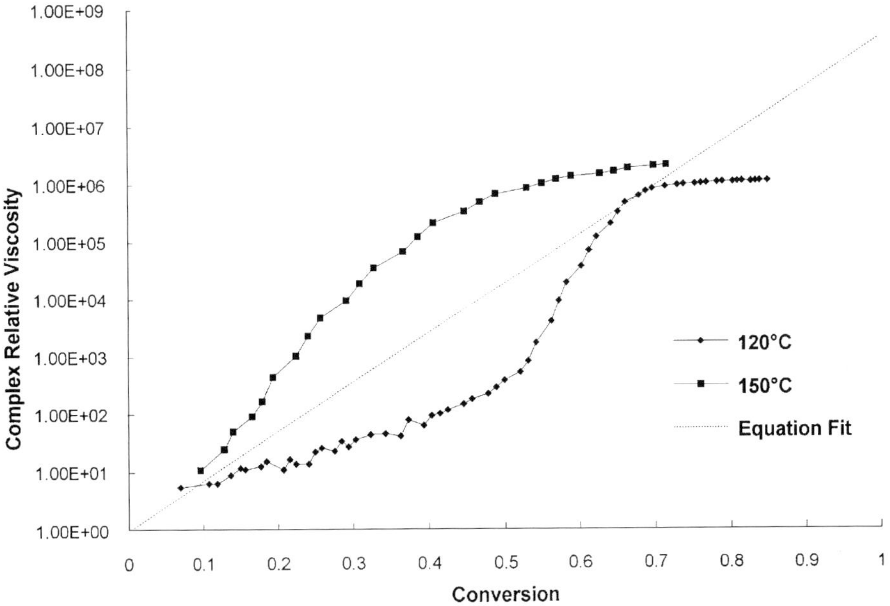

Figure 1.16 Relative complex viscosity ($|\eta^*|/|\eta_0|$) versus calculated conversion for polymerization at 120 and 150°C. Linear, straight-line fit to the phenomenological equation $|\eta^*|/|\eta_0| = \exp(19.6\ X)$ is also shown

50 percent conversion all the curves are almost linear, which suggests a phenomenological equation of the form:

$$|\eta^*|/|\eta_0| = \exp(19.6X) \quad \text{for } X < 0.5 \tag{1.6}$$

where 19.6 is the constant that defines the relative complex viscosity rise during anionic ring opening polymerization of caprolactam using caprolactam-magnesium-bromide/iso-phthaloyl-bis-caprolactam as the catalyst/initiator system. In Equation 1.6, the temperature dependence of $|\eta^*|$ is that for $|\eta_0|$ defined in Equation 1.5.

For sodium/hexamethylene-1,6-bis-carbamidocaprolactam system, Sibal et al. [64] found the value of the constant k^* in Equation 1.4 to be 17.5. Note that the values of the constant k^* in Equation 1.4 that defines the relative complex viscosity rise during anionic ring opening polymerization of caprolactam are comparable for both caprolactam-magnesium-bromide/isophthaloyl-bis-caprolactam and sodium/hexamethylene-1,6-bis-carbamidocapro-lactam as the catalyst/initiator systems even though the kinetic constants for anionic polymerization for these systems are extremely different (see Table 1.2).

Figure 1.16 shows the relative complex viscosity as a function of conversion at 120 and 150°C. The 150°C curve shows a dramatic rise due to the simultaneous crystallization during polymerization. In addition, notice in Figure 1.13 that the complex viscosity–time curves of 150 and 160°C polymerization tend to converge. Both these effects—the nonlinear rise in relative complex viscosity in Figure 1.16 and the convergence of 150 and 160°C curves in

Figure 1.13—occur because of simultaneous crystallization and polymerization at 150°C. This temperature is near the maximum crystallization rate temperature (~145°C) of nylon 6 homopolymer [66]. The presence of solid crystallites increases the complex viscosity of the polymerizing system because of a filler effect.

In Figure 1.16, the 120°C curve deviates from the straight line at a conversion of about 10 percent. This effect is due to "sluggish" polymerization at 120°C like that we have observed in our kinetics study (see Figure 1.5a in Ref. 24). The predicted conversion (x-axis values in Figure 1.16) is based on Malkin's autocatalytic phenomenological model that does not account for either crystallization or diffusion-controlled kinetics. In Figure 1.16, however, the polymerization at 120°C becomes diffusion controlled after about 10 percent conversion. The predicted values of conversion by Malkin's autocatalytic rate equation, therefore, are overestimated values of the true conversion resulting from diffusion controlled kinetics. This error results in the nonlinearity of the 120°C curve.

As a final note, we question why the relative complex viscosity versus conversion curve for the 140°C polymerization did not deviate (up to 50 percent conversion) from the straight-line fit even though Magill's data on crystallization rate of hydrolytic nylon 6 (i.e., nylon 6 made by the conventional, hydrolytic process) versus temperature shows a maximum in the crystallization rate at about 140–145°C. One possibility is that nylon 6 formed by anionic polymerization below the melting point of nylon 6 is a mixture of two crystalline structures: α and γ (see Ref. [23]). The melting points of α and γ structures are about 256 and 228°C, respectively [72]. For anionically formed nylon 6 that is a mixture of α and γ structures, the temperature for maximum crystallization may be higher than the temperature for maximum crystallization of hydrolytic nylon 6 (γ structure). Because Magill's data were obtained on hydrolytic nylon 6, these data may not be directly applicable to the crystallization of nylon 6 polymerizing in molten caprolactam during anionic ring opening polymerization of caprolactam.

1.5 Application of Rheo-Kinetics Modeling to Reaction Injection Pultrusion

Reactive injection pultrusion (RIP) is a means of applying the thermoset pultrusion process, with appropriate changes, to thermoplastic systems. In RIP, glass fibers are pulled through a "feed zone" where the caprolactam monomer, along with catalyst and initiator, are injected. The low viscosity monomers exhibit excellent fiber wetting in the feed zone, and then polymerize around the fibers in the die at 130–200°C. A puller, located beyond the die, pulls the glass fiber reinforced nylon 6 composite. A typical composition of the glass fiber/nylon 6 pultruded composite is 75 percent glass and 25 percent nylon 6, by weight. The typical pultrusion line speed is 1–2 m/min. Figure 1.17 is a schematic of the reaction injection process.

In this section, we will describe the application of the rheokinetic model to adiabatic and isothermal pultrusion by the RIP process. Adiabatic pultrusion is defined as a pultrusion

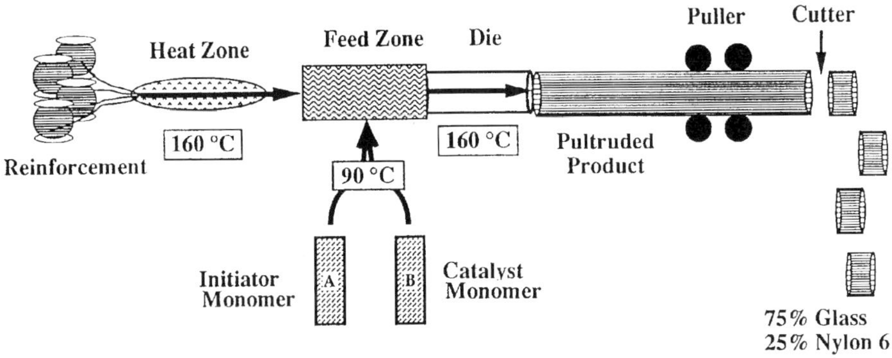

Figure 1.17 Schematic of the reactive pultrusion process

process where no heat transfer occurs across the boundaries (surface) of the pultrudate in the die. On the other hand, an isothermal pultrusion is a pultrusion process where the temperature of the pultrudate is maintained constant from the entrance to the exit of the die. During adiabatic pultrusion, the heat of polymerization, if any, is absorbed by the pultrudate, and the absorbed heat causes the temperature of the pultrudate to increase. In reality, adiabatic and isothermal pultrusion are just hypothetical concepts. In practice, near adiabatic pultrusion conditions can be achieved by maintaining the die wall temperature as close as possible to the predicted temperature of the pultrudate for adiabatic pultrusion. Because RIP of glass fiber nylon 6 is an exothermic process, isothermal operation would require extraction of heat from the die as conversion proceeds. Isothermal pultrusion is impractical because it is very difficult to maintain a constant pultrudate temperature throughout the die. Despite these limitations, modeling adiabatic and isothermal pultrusion processes is extremely valuable in providing insight into the process and to discriminate between weak and strong factors that affect the pultrusion process. There are two alternatives to the preceding approach: (1) Develop an experimental design strategy and conduct experiments to determine the process conditions; (2) Model a real-life pultrusion process. Both of these approaches, however, have major drawbacks. The former approach is expensive and time consuming, whereas the latter approach is tedious because it requires modeling of simultaneous heat, mass, and momentum transfer along with chemical reaction in a multiphase system. In order to simplify the modeling, yet obtain germane information, we have modeled adiabatic and isothermal pultrusion. The adiabatic and isothermal pultrusion models finally reduce to the rheo-kinetics of the matrix resin in the presence of the fiber.

Pultrusion is a steady-state process in which the fiber–resin mass changes its properties as it moves from the entrance to the exit of the die. In order to track the temperature, polymer conversion, and other properties of the fiber–resin mass as it moves along the die, it is useful to define a *representative volume element* (RVE) that rides along the fiber at the line speed of the pultrusion process. An RVE is defined such that it will contain *both* the solid phase (i.e., fibers and resin), irrespective of its location in the composite. In real-life pultrusion, a thermocouple wire that passes through the pultrusion die tracks the temperature of an RVE in the composite.

In the present pultrusion modeling study, the composition of the glass fiber/nylon 6 pultrudate was chosen as 75 percent glass and 25 percent nylon 6 by weight. In addition, the single-feed composition investigated consisted of 108 mmols/L of caprolactam-magnesium-bromide and 35 mmols/L of the difunctional isophthaloyl-bis-caprolactam (i.e., 70 mmols/L of the active acyllactam group), and we performed a parametric study on an adiabatic pultrusion process by changing the following variables:

1. temperature of the incoming glass fiber
2. temperature of the caprolactam monomer feed

In addition, we also compared the effect of adiabatic versus isothermal processing on polymerization time.

Figure 1.18 shows the calculated temperature of an RVE during adiabatic pultrusion as a function of the temperature of the incoming glass fiber when the temperature of the incoming caprolactam monomer is 90°C. At time $t = 0$, the RVE is located at the point of injection of the monomer into the die. In addition, at time $t = 0$, the initial polymerization temperature (IPT) is the average of the incoming glass fiber and monomer temperatures, determined by the rule of mixtures by weight:

$$T_0 = \frac{(1 - w_f)Cp_{\text{caprolactam}} T_{0\text{caprolactam}} + w_f Cp_{\text{fiber}} T_{0\text{fiber}}}{(1 - w_f)Cp_{\text{caprolactam}} + w_f Cp_{\text{fiber}}} \tag{1.7}$$

where T_0 = initial polymerization temperature, $T_{0\text{caprolactam}}$ = temperature of the incoming caprolactam feed, $T_{0\text{fiber}}$ temperature of the incoming fiber, w_f = weight fraction of the fiber in the composite, $Cp_{\text{caprolactam}}$ = specific heat of caprolactam, and Cp_{fiber} = specific heat of the fiber.

The specific heat of E glass fiber is 0.192 cal/gm K (i.e., 0.804 J/gm K) [73]. The specific heat of caprolactam as a function of temperature was fitted to the following equation from experimental data available in Monsanto:

$$Cp_{\text{caprolactam}}\ (\text{J/gm K}) = 0.011114T - 1.7057 \tag{1.8}$$

where T is the temperature of caprolactam in K.

In Figure 1.18, we see that as the temperature of the incoming fiber increases from 100 to 160°C, the time for complete conversion decreases from 240 to 150 s. The upper limit of the temperature of the incoming fiber was set at 160°C because the sizing on most of the available glass fibers burns at temperatures higher than 160°C.

A further leverage to increase the reaction rate was to increase the temperature of the incoming caprolactam monomers. Figure 1.19 is similar to Figure 1.18 except that the incoming monomer feed is at 100°C rather than 90°C. Increasing the incoming monomer temperature from 90 to 100°C reduces the time to complete conversion. For example, for the case where the temperature of the incoming fiber was 120°C, the time to complete conversion was reduced from 240 to 210 s. In addition, one should note from Figures 1.18 and 1.19 that the adiabatic temperature rise in the composite with 75 percent glass by weight is only about 30°C, instead of the 50°C rise seen typically during anionic polymerization of caprolactam into unreinforced nylon 6 (see adiabatic temperature rise data in Figure 1.11 with initial polymerization temperatures of 117 and 136°C). The smaller increase in the temperature of the composite as compared with unreinforced nylon 6 is due to the absorption of some of the

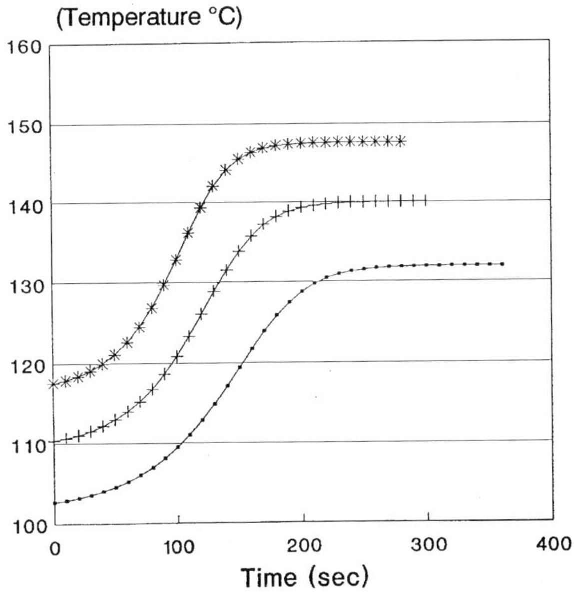

Figure 1.18 The temperature of an RVE during adiabatic pultrusion when the temperature of the incoming caprolactam monomer was 90°C and the temperature of the incoming glass fiber was 120°C (bottom line), 140°C (middle line), and 160°C (top line)

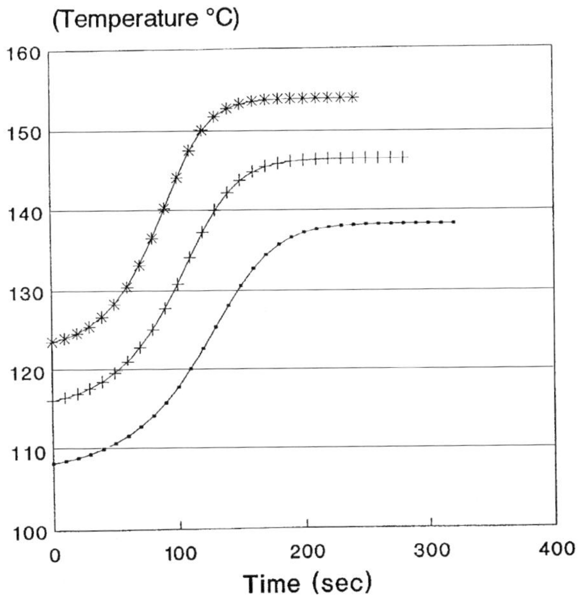

Figure 1.19 The temperature of an RVE during adiabatic pultrusion when the temperature of the incoming caprolactam monomer was 100°C and the temperature of the incoming glass fiber was 120°C (bottom line), 140°C (middle line), and 160°C (top line)

heat of polymerization by the glass to increase the temperature of the glass to that of the reacting monomer surrounding the glass.

Figures 1.20 and 1.21 show calculated conversion as a function of time during adiabatic and isothermal pultrusion of glass fiber/nylon 6 composites with 75 percent glass fiber by weight and monomer feed concentrations stated earlier. The isothermal pultrusion temperature and the initial temperature for adiabatic pultrusion are 136 and 157°C, respectively, in Figures 1.20 and 1.21. Because the adiabatic temperature rise of the composite is about 30°C, the final temperatures are 166 and 187°C, respectively, in Figures 1.20 and 1.21. Figure 1.20 shows that complete conversion (indicated as 1.0, but is about 0.95 in reality) is achieved in 125 and 190 s, respectively, for adiabatic and isothermal processing when initial polymerization temperature was 136°C. Figure 1.21 shows complete conversion times of 90 and 120 s, respectively, for adiabatic and isothermal processing when initial polymerization temperature was 157°C. (Note that in the case of isothermal RIP processing, the polymerization temperature is constant within the full length of the die.) Based on the rheo-kinetic modeling it is clear that the adiabatic pultrusion process is more desirable because the times for complete conversion is shorter than in the case of the isothermal RIP process. The faster conversion by the adiabatic RIP process resulted in increased throughput of the pultruded composite. In addition, it has been our experience that it was extremely difficult to maintain a constant temperature along the full length of the die. On the other hand, we found it relatively easy to set the temperatures along the die to be nearly equal to the adiabatic polymerization temperatures.

From Equation 1.4, the complex viscosity at 57 percent conversion of the caprolactam monomer, which may be defined as solidification or “gel” point, is 100 Pa s (1000 Poise). Combining the information on conversion at the gel point with the data presented in Figures 1.20 and 1.21, the time to gel formation during pultrusion can be estimated.

In a real-life RIP of glass fiber nylon 6, the process is neither isothermal nor adiabatic, as stated earlier. In real-life RIP processes, external heaters are mounted on the die to keep all the exothermic heat within the pultruded product and, sometimes, even add external heat into the pultruded product to enhance the reaction rate further. The adiabatic temperature rise plots shown in this section are valuable tools for determining the minimum temperature along the length of the die and the desired residence time in the die for adiabatic conversion. The pulling speed is, in turn, estimated by dividing the die length by the residence time in the die. The temperature along the die length, however, may be maintained higher than adiabatic conversion temperature for faster processing.

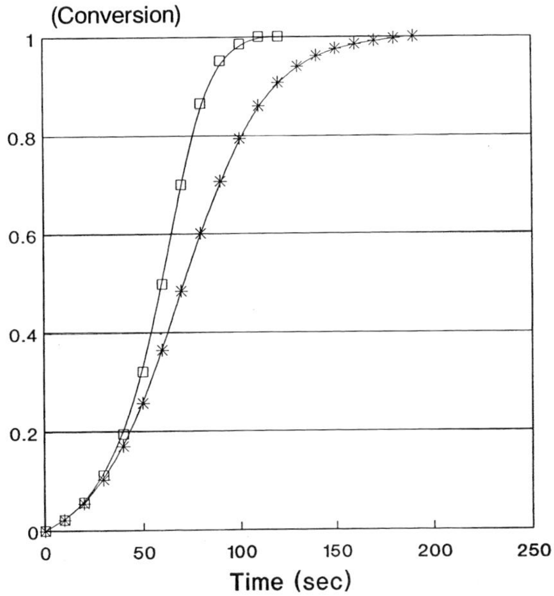

Figure 1.20 Conversion in an RVE during adiabatic (top line) and isothermal (bottom line) pultrusion of glass fiber nylon 6 composite when the initial temperature for adiabatic pultrusion and the isothermal pultrusion temperature were 136°C

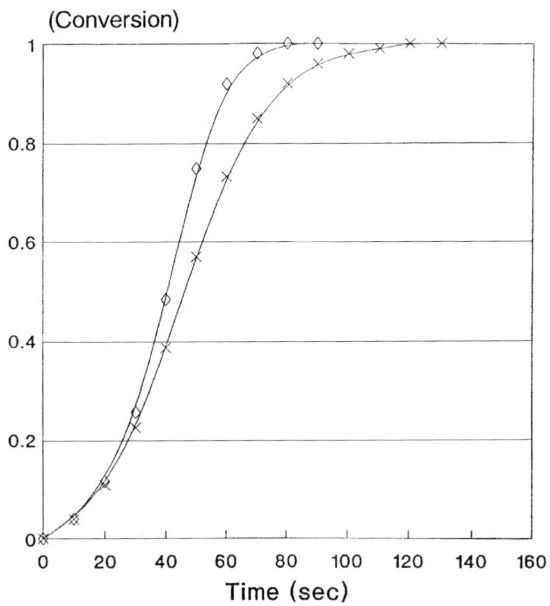

Figure 1.21 Conversion in an RVE during adiabatic (top line) and isothermal (bottom line) pultrusion of glass fiber nylon 6 composite when the initial temperature for adiabatic pultrusion and the isothermal pultrusion temperature were 157°C

1.6 Concluding Remarks

This chapter reviewed the chemistry of ring opening polymerization of cyclic monomers that yield thermoplastic polymers of interest in composite processing. In addition, this chapter focuses on the chemistry, kinetics, and rheology of the ring opening polymerization of caprolactam to nylon 6. Finally, these rheokinetics models are applied to the reactive injection pultrusion (RIP) process.

The kinetics of anionic ring opening polymerization of caprolactam initiated by isophthaloyl-bis-caprolactam and catalyzed by caprolactam-magnesium-bromide satisfactorily fit Malkin's autocatalytic model below 50 percent conversion. The calculated value of the overall apparent activation energy for this system is 30.2 kJ/mol versus about 70 kJ/mol for Na/hexamethylene-1,6,-bis-carbamidocaprolactam as the initiator/catalyst system.

The rheokinetics of polycaprolactam polymerizing in the monomer shows that below 50 percent conversion, the relative complex viscosity versus conversion of the nylon 6 homopolymerization is defined by the phenomenological equation $|\eta^*|/|\eta_0| = \exp(19.6\,X)$, where $|\eta^*|$ is the complex viscosity of nylon 6 anionically polymerizing in its monomer, $|\eta_0|$ is the viscosity of caprolactam monomer, and X is fractional conversion.

Nomenclature

$[A]$	Activator, acyllactam concentration, mol/L
b	Autocatalytic term, L/mol
$[C]$	Catalyst, caprolactam-magnesium-bromide, concentration, mol/L
$Cp_{\text{caprolactam}}$	Specific heat of caprolactam, J/gm K
Cp_{fiber}	Specific heat of the fiber, J/gm K
H	Heat of polymerization, J/mol
H_{tot}	Total heat of polymerization, J/mol
k	Pre-exponential or front factor, L/mol s in Equation 1.3
k^*	Constant in Equation 1.4
M	Monomer concentration, mol/L
M_0	Initial monomer concentration, mol/L
R	Universal Gas Constant, J/mol K
T	Temperature, K
T_0	Initial polymerization temperature, K
$T_{0\text{caprolactam}}$	Temperature of the incoming caprolactam feed, K
$T_{0\text{fiber}}$	Temperature of the incoming fiber, K
T_f	Final adiabatic temperature, K
t	Time, s or sec
U	Activation Energy, J/mol
w_f	Weight fraction of the fiber in the composite

X Fractional conversion
$|\eta^*|$ Complex viscosity during ring opening polymerization, Pa s
$|\eta_0|$ Complex viscosity of the monomer, Pa s

Acknowledgments

Financial support and release for publication were provided by Monsanto Company, Plastics Division (sold to Bayer Corporation in 1995). The authors wish to thank the following fine individuals: Robert Mendelson and Donald Williams for guiding the rheological measurements at Monsanto's Physical and Analytical Science Center, Springfield, MA; Lionel Stebbins for doing the kinetics experiments; Allen Padwa for making Figures 1.1–1.8; Kamran Tavangar for making Figure 1.17; and Donald Nardi for operating the reaction injection pultrusion process.

References

1. Ivin, K.J., Saegusa, T., eds. *Ring Opening Polymerization* (1984) Elsevier Applied Science Publishers, London
2. Frisch, K.C., Reegen, S.L., eds. *Ring Opening Polymerization* (1969) Marcel Dekker, Inc., New York
3. Furukawa, J., Saegusa, T., eds. *Ring Opening Polymerization of Aldehydes and Oxides* (1963) Wiley-Interscience, New York
4. Penczek, S., Kubisa, P., Matyjaszewski, K., *Adv. Polym. Sci.* (1985) 68/69, p. 1
5. Penczek, S., Kubisa, P., Mayjaszewski, K., *Adv. Polym. Sci.* (1980) 37, p. 1
6. Chujo, Y., Saegusa, T., eds. *Encyclopedia of Polymer Science and Engineering* (1988) 14, John Wiley and Sons, New York, p. 622
7. Joyce, R.M., Ritter, D.M., (1941) U.S. Patent 2,251,519
8. Sebenda, J., *J. Macromol. Sci., Chem.* (1972) A6, p. 1145
9. Sebenda, J., *Prog. Polym. Sci.* (1978) 6, p. 123
10. Sebenda, J., *Pure Appl. Chem.* (1976) 48, p. 329
11. Sekiguchi, H., *J. Chem. Soc. Japan* (1967) 88, p. 577
12. Wichterle, O., Sebenda, J., *Collect. Czech. Chem. Communs.* (1956) 21, p. 312
13. Wichterle, O., Sebenda, J., Kralicek, J., *Fortschr. Hochpolym. Forsch.* (1961) 2, p. 578
14. Gabbert, J.D., Hedrick, R.M., *Polym. Proc. Eng.* (1986) 4(2–4), p. 359
15. Udipi, K., *J. Appl. Polym. Sci.* (1988) 36(1), p. 117
16. Furukawa, J., Tsuruta, T., Sakata, R., Saegusa, T., *Makromol. Chem.* (1959) 32, p. 90
17. Hall, H.K., *J. Am. Chem. Soc.* (1958) 80, p. 6412
18. Tomalia, D.A., Sheetz, D.P., *J. Polym. Sci.* (1966) Part A-1, 4, p. 2253
19. Harwood, H.J., Patel, N.K., *Macromolecules* (1968) 1, p. 233
20. Sawada, H., *J. Macromol. Sci. Rev. Macromol. Chem.* (1970) C5, p. 151
21. Dainton, F.S., Ivin, K.J., *Quart. Rev.* (1958) 12, p. 82
22. Brunelle, D.J., In *Ring Opening Polymerization*. Brunelle, D.J., ed. (1995) Hanser, Munich, Chapter 11

23. Udipi, K., Davé, R.S., Kruse, R.L., Stebbins, L.R., *Polymer* (1997) 38(4), p. 927
24. Davé, R.S., Udipi, K., Kruse, R.L., Stebbins, L.R., *Polymer* (1997) 38(4), p. 939
25. Davé, R.S., Udipi, K., Kruse, R.L., Williams, D.E., *Polymer* (1997) 38(4), p. 949
26. Brzezinska, K., Szymanski, R., Kubisa, P., Penczek, S., *Makromol. Chem.-Rapid Commun.* (1986) 7, p. 1
27. Endo, M., Aida, T., Inoue, S., *Macromolecules* (1987) 20, p. 2982
28. Dreyfuss, P., Dreyfuss, M.P., *Ring Opening Polymerization.* Frisch, K.C., Reegen, S.L., eds. (1969) Marcel Dekker, New York, Chapter 2
29. Furukawa, J., Tada, K., *Ring Opening Polymerization.* Frisch, K.C., Reegen, S.L., eds. (1969) Marcel Dekker, New York, Chapter 3
30. Schulz, R.C., Hellermann, W., Nienburg, J., *Ring Opening Polymerization.* Ivin, K.J., Saegusa, T., eds. (1984) Elsevier Applied Science Publishers, New York, Chapter 6
31. Ivin, K.J., *Ring Opening Polymerization.* Ivin, K.J., Saegusa, T., eds. (1984) Elsevier Applied Science Publishers, New York, Chapter 3
32. Saotome, K., *Kogyo Kagaku Zasshi* (1962) 65, p. 402
33. Taniyama, M., Nagaoka, T., Takata, T., Sayama, K., *Kogyo Kagaku Zasshi* (1962) 65, p. 415
34. Taniyama, M., Nagaoka, T., Takata, T., Sayama, K., *Kogyo Kagaku Zasshi* (1962) 65, p. 419
35. Nagaoka, T., Takata, T., Sayama, K., Taniyama, M., *Kogyo Kagaku Zasshi* (1962) 65, p. 422
36. Shpital'nyi, A.S., Shpital'nyi, M.A., Yablochnik N.S., *J. Appl. Chem. USSR* (1959) 32(3), p. 647
37. Yoda, N., Miyake, A., *J. Polym. Sci.* (1960) 43, p. 117
38. Park, J.H., Jung, B., Choi, S.K., *Taehan Hwahakhoe-Chi* (1980) 24, p. 167 [*Chem. Abstr.* (1980) 93, p. 132897]
39. Cefelin, P., Sebenda, J., *Collect. Czech. Chem. Communs.* (1961) 26, p. 3028
40. Yumoto, H., Ogata, N., *Bull. Chem. Soc. Japan* (1958) 31, p. 913
41. Fukumoto, O., Kobayashi, R., Kunimichi, T., Japanese Patent 6213794 (1962)
42. Solomon, O., Oprescu, Cr., *Makromol. Chem.* (1969) 126, p. 197
43. Mahajan, S.S., Roda, J., Kralicek, J., *Angew. Makromol. Chem.* (1979) 75, p. 63
44. Chuchma, F., Bosticka, A., Roda, J., Kralicek, J., *Makromol. Chem.* (1979) 180, p. 1849
45. Ostaszewski, B., Wlodarczyk, M., Wlodarczyk, K., *Zesz. Nauk. Politech. Lodz., Chem.* (1976) 32, p. 136 [*Chem. Abstr.* (1977) 86, p. 90375]
46. Tani, H., Konomi, T., Japanese Patent 6929069 (1969)
47. Sekiguchi, H., Rapacoulia, P., Coutin, B., French Patent 2138228 (1973)
48. Ney, W.O., U.S. Patent 2973343 (1961)
49. Schmidt, W., German Patent 1203465 (1966)
50. Sakata, A., Mizuno, K., Isozaki, F., Japanese Patent 7226195 (1972)
51. Kobayashi, F., Sakata, A., Mizuno, K., Isozaki, F., Japanese Patent 7201222 (1972)
52. Barnes, A.C., Barnes, C.E., U.S. Patent 4217442 (1980)
53. Fiala, F., Kralicek, J., *Angew. Makromol. Chem.* (1977) 63, p. 105
54. Puffr, R., Sebenda, J., *Europ. Polym. J.* (1972) 8, p. 1037
55. Sebenda, J., *Lactam-Based Polyamides.* Puffr, R., Kubanek, V., eds. (1991) 1, Chapter 2, CRC Press, Boca Raton, Florida, p. 29
56. Sekiguchi, H., *Ring-Opening Polymerization.* Ivin, K.J., Saegusa, T., eds. 2, (1984) Elsevier, London, p. 809
57. Greenley, R.Z., Stauffer, J.C., Kurz, J.E., *Macromolecules* (1969) 2, p. 561
58. Sittler, E., Sebenda, J., *Coll. Czech. Chem. Comm.* (1968) 33, p. 270
59. Rigo, A., Fabbri, G., Talamini, G., *J. Polym. Sci. Polym. Lett. Ed.* (1975) 13, p. 469
60. Malkin, A.V., Frolov, V.G., Ivanova, A.N., Andrianova, Z.S., *Polym. Sci. USSR* (1979) 21, p. 691
61. Bolgov, S.A., Begishev, V.P., Malkin, A.Y., Frolov, V. G., *Polym. Sci. USSR* (1981) 23, p. 1485
62. Malkin, A.V., Ivanova, S.L., Frolov, V.G., Ivanova, A.N., Andrianova, Z.S., *Polymer* (1982) 23, p. 1791
63. Wittmer, P., Gerrens, H., *Makromol. Chem.* (1965) 89, p. 27
64. Sibal, P.W., Camargo, R.E., Macosko, C.W., *Polym. Process Eng.* (1984) 1, p. 147

65. Lin, J.D., Ottino, J.M., Thomas, E.L., *Polym. Eng. Sci.* (1985) 25, p. 1155
66. Magill, J.H., *Polymer* (1962) 3, p. 655
67. Lipshitz, S.D., Macosko, C.W., *J. Appl. Polym. Sci.* (1977) 21, p. 2029
68. Camargo, R.E., Gonzalez, V.M., Macosko, C.W., Tirrell, M.V., *Proc. 2nd Int. Conf. Reactive Polym. Processes.* Lindt, J.T., ed. (1982) Univ. Pittsburgh Press, Pittsburgh, PA, p. 126
69. Cimini, R.A., Sundberg, D.C., *Polym. Eng. Sci.* (1986) 26, p. 560
70. Reimschuessel, H.K., *Ring Opening Polymerization.* Frisch, K.C., Reegan, S.L., eds. (1969) Chapter 7, Marcel Dekker, New York
71. Provaznik, M., Puffr, R., Sebenda, J., *Eur. Polym. J.* (1988) 24, p. 511
72. Puffr, R., Raab, M., Dolezel, B., *Lactam-Based Polyamides.* Puffr, R., Kubanek V., eds. (1991) 1, Chapter 6, CRC Press, Boca Raton, Florida, p. 187
73. Schwartz, M.M., *Composite Materials Handbook* (1984) McGraw Hill, New York, p. 2.27

2 Thermoset Resin Cure Kinetics and Rheology

Verônica M.A. Calado and Suresh G. Advani

This chapter presents the curing reactions of thermosetting resins and reviews and discusses the widely used three approaches to characterize and model the reactions: kinetic models (phenomenological and mechanistic), gelation theory, and rheological models. Tables containing different phenomenological kinetic models for neat resins (epoxies and vinyl esters) and for polymeric composites are listed, along with the kinetic parameters. The equations used to describe the coupled phenomena, such as resin flow, heat transfer, and cure reactions, inherent in any polymer and polymer composites manufacturing process, are also presented and discussed. Some optimization and control strategies for obtaining optimum cure cycles are introduced.

2.1 Introduction

The use of polymer composites (i.e., materials composed of high-strength fibers dispersed in a polymer matrix) in the army, aircraft, naval, automobile, appliance, electronic, adhesive, leisure, encapsulant, building, housing, and medical industries has been increasing over the last three decades because of their many advantages, such as weight reduction, design flexibility, corrosion resistance, thermal stability, electrical properties, reduced noise transmission, and reduced manufacturing costs when compared with traditional materials.

2.1.1 Resins

The polymer matrix can be formed by thermoplastic (e.g., acrylic, nylon, polystyrene, polyethylene, polyetheretherketone) or thermosetting (e.g., epoxy, polyester, phenolic, polyimide) resins. The former are transformed into linear polymers, with weak chemical bonds (van der Waals); the latter are converted into a three-dimensional cross-linked network after polymerization [1–3]. Although thermosetting resins are brittle at room temperature and cannot be reshaped upon heating because of the strong chemical bonds, unlike thermoplastics [4] they have higher tensile strength and modulus, excellent chemical and solvent resistance, dimensional and thermal stability, good creep resistance, and excellent fatigue properties [5]. They are also easy to process because of their low viscosity (about 50–200 times that of water) [6]. Thus, they are preferentially used in the manufacturing of composites. The viscosity of thermosetting resins should not be too high or low to avoid dry spots (areas where resin does not impregnate) and/or microvoids.

The most common thermosetting resins used in composites are epoxies (*diglycidyl ethers of bisphenol A—DGEBA*—and *tetraglycidyl*-4,4′-diaminodiphenylmethane—*TGDDM*), unsaturated polyesters (vinyl ester—*DERAKANE 411-C45*), and phenolics (resoles and novolacs). Table 2.1 shows a qualitative comparison among them [6]. The thermosetting resins are obtained from a highly exothermic polymerization reaction, called *cure*, where a mixture of monomers and other minor components, such as curing agents (or hardeners), initiators, inhibitors and catalysts react under heat. The employment of reinforced materials

Table 2.1 Comparison Among Common Thermosetting Resins

Resin type	Performance	Processing	Cost	Aerospace industry	Automotive industry
Polyester	Low	Ease	Low	X	X
Vinyl ester				X	X
Urethane				X	X
Epoxy				X	X
Cyanate ester				X	
Bismaleimide				X	
Phenolic				X	X
Polyimide	High	Difficult	High	X	

Source: Adapted from Palmese [6].

(e.g. additives or *low profiles*, fillers, and fibers) is to improve physical and mechanical properties of a resin as well as to reduce the processing costs.

During cure, large molecules with high molecular weight are formed by cross-linking bonds. As a result, as the reaction progresses, the available volume in the molecular arrangement decreases, resulting in less mobility of the species, thereby, affecting the resin viscosity. This results in the diffusional effects controlling the process rather than chemical factors [7,8].

2.1.2 Reinforcements

Although the reinforcement in a composite may be particulate or fibrous, the latter ones are widely used. The fibers need to have a high elastic modulus and strength along their length, preferably low density, and to be arranged in suitable architecture in order to have desired properties in the final product. Figure 2.1 presents typical fiber arrangements [9].

The properties of the synthetic fibers depend on the manufacturing processes and the fiber age and size. The most common fibers in composite materials are fiberglass (e.g., *E-glass* and *S-glass*) and carbon/graphite. The former is composed basically of silica (SiO_2). The *E-glass* (aluminoborosilicate) type is used more often because of the price, although the *S-glass* (magnesium aluminoborosilicate)-type possesses higher tensile strength, modulus, and processing temperature. The graphite fibers are mostly used in high performance application, such as primary aircraft structures, due to the fact that they have high strength

Random Mat Unidiretional Plain Weave 3D Woven

Figure 2.1 Typical fiber arrangements in a preform

Table 2.2 Characteristics of Fiberglass and Graphite Fibers

	Cost	Electrical properties	Chemical/ moisture resistance	Elevated temperature resistance	Mechanical properties	Thermal conductivity
Fiberglass	Low	Excellent	Good	Good	Very good	Low
Graphite fibers	High	Excellent	Excellent	Excellent	Excellent	High

and modulus and light weight. Table 2.2 displays some characteristics of fiberglass and graphite fibers.

2.1.3 Manufacturing Process

The composites can be manufactured by many different types of processes, such as Autoclave [10,11], Pultrusion [12], Filament Winding [13], and Liquid Injection Molding {Resin Transfer Molding (RTM) [9,14–17], Structural Reaction Injection Molding (SRIM) [18], Seeman Composite Resin Infusion Molding Process (SCRIMP) [9], and Reaction Injection Molding (RIM) [19]}. Every process has its own characteristics, the description of which may be found in the cited papers. In this chapter, RTM is considered because its utilization in manufacturing of composite parts has increased in the last few years as it is a versatile and attractive process for high-volume, high-performance, and low-cost production [15]. The process consists of placing fiber-reinforced preforms composed of many layers into a mold cavity, closing the mold, and then injecting the cold thermosetting resin into the mold cavity. During injection, a resin displaces air in the mold cavity of the preform and wets all individual fibers before the reaction begins. If some air is entrapped, voids are formed, which results in an unacceptable part. These voids may be of macroscopic (in spaces between the fibers) and microscopic (in spaces among the fiber tows) scales. The void formation theory is still not completely understood, although there are a few models in the literature [20–28]. The main objective is to obtain a composite with high quality, namely, a part with no defect, such as voids [29,30], fracture [31], nonuniform mechanical properties (originated from concentration and thermal stress gradients) during the resin curing process. Because of the highly exothermic character of the curing reactions, large amount of heat is generated inside the part that needs to be removed to prevent the resin from degradation. This problem is nontrivial because the thermal conductivity of thermosetting resins is low, trapping heat inside the part, especially in thick composites. This can result in different physical states (i.e., liquid, rubber, and gelled or ungelled glasses) simultaneously in different sections and lead to nonuniform degree of cure and temperatures, causing residual stresses [32].

2.1.4 Cure Cycles

It is fundamental, therefore, to control the rate of heat absorption and temperature variations during the cure cycle [33,34]. The cure cycle depends on the part geometry, thermal

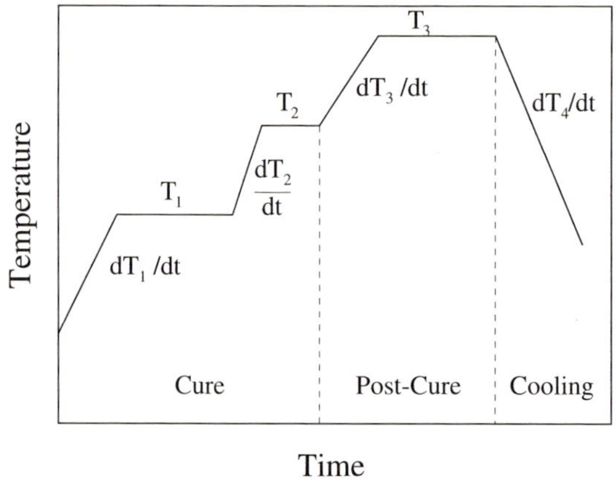

Figure 2.2 An example of cure schedule—two steps with constant temperature and two different heating rates (dT_i/dt) for the curing stage; one step with constant temperature and one heating rate for the postcure stage; one step with one heating rate for the cooling stage

anisotropy, the chemical cure kinetics, and the thermal boundary conditions [35]. Empirical methods may be used to define such a cycle for thin composites, but they are impractical for thick parts [11]. There is a need, therefore, for a rigorous mathematical model that interrelates simultaneous phenomena, such as heat transfer, resin flow, chemical reactions, physical state changes, and void formation [36], in which the optimum cure cycle can be proposed, based on scientific arguments rather than trial and error attempts used in the past. An accurate computer simulation, however, is a formidable task because it is difficult to choose simplifying assumptions to achieve a solvable system while retaining enough physics to describe a realistic process [37]. In addition, one would need to have accurate values of the material parameters used in the constitutive equations that describe the system. An additional issue is the selection of the optimum cure cycles (cure schedules or temperature history) and processing conditions, which relate time, temperature, and pressure. Figure 2.2 illustrates an example of cure cycle, which is composed of three stages: cure, postcure, and cooling. Each stage can combine constant temperature and heating rate, applied during specific periods of time. The difficulty here is to select the optimum cure cycle in order to obtain composites with good mechanical properties that will also permit uniform and complete cure, uniform compaction with all excess resin removed, minimum void contents, no damage of the part, no residual stresses due to large thermal gradients, and cure in the shortest possible time [38–40].

2.1.5 Optimization

An optimization of the cure schedules and processing parameters consists of exploring the most favorable conditions, judged under specified criteria, that will lead to a product satisfying the desired requirements [41]. This is a complicated and interrelated problem [42]. A tradeoff between the curing time and thermal gradients is required because the higher the processing temperature, the faster the cure. Larger thermal gradients, however, can develop inside the part, especially if it is thicker than approximately 2.54 cm, because polymers generally have low thermal conductivity and cannot diffuse the heat very quickly. A target temperature history is usually selected by trial-and-error process, taking into account the degradation temperature of the material. Martinez [43] and Kenny [44] suggested that an optimum cure cycle can be found by minimizing an objective function defined by the absolute value of the difference between the target temperature history and the composite temperature history calculated for a particular cure schedule.

An optimization process requires a control strategy and, consequently, the development of accurate sensors, such as dielectric transducers, acoustic/ultrasonic methods, fluorescence techniques, fiberoptics, and mechanical impedance analyses [45,46]. Research in this area is moving toward the use of control and expert systems [47]. These types of systems are not limited to simple geometries and they do not require detailed knowledge of the material properties, but they do allow for material variabilities and provide real time control of the cure process [47]. They need sensors, rules of thumb or an on-line process model, and a data base and data processing algorithm. Ciriscioli et al. [47] and Pillai [41] developed detailed control strategies for autoclave processes, and similar technology may be transferred to RTM processes.

As a consequence, a high-quality composite can be manufactured with *low cost* if one has a process model and can predict the temperature and time necessary to cure a part completely. To do this, accurate mathematical models are needed to describe the various coupled phenomena of cure kinetics, heat transfer, resin flow, changes in the physical state, and void growth. This chapter aims to discuss some of the many existing studies in the literature, related to these mathematical models, optimization, and control strategies. Thus, the following section addresses some of the cure kinetic models, along with the techniques to characterize the curing processes. Next, the effect of reinforced materials on cure kinetics of neat resins and the resin types for RTM, such as epoxies, vinyl esters, and phenolics are discussed, followed by the coupled phenomena of heat transfer, resin flow, and cure kinetics—with the analysis of composite thermophysical properties, boundary and initial conditions, and numerical methods for solution of the governing equations, optimization, and control strategies.

2.2 Cure Kinetics

There are two types of cure reactions of thermosetting resins: step (e.g., epoxies, phenolics, urethanes) and chain (e.g., unsaturated polyesters, vinyl esters, acrilates) polymerizations. In the first type, the size of the polymer chain increases because of the linking of the oligomers (e.g., monomers, dimers, etc.) to themselves. Short chains can be linked to long ones in a condensation reaction. In the second case, the size of the polymer chain increases because of the additions of monomers to the ends of the chain. In both cases, upon application of heat, thermosetting resins cure chemically to form a cross-linked network, becoming rigid, insoluble, and thermally stable [48].

The curing reactions can be characterized in several ways, such as:

1. kinetic models, relating reaction rate to temperature and degree or extent of cure
2. gelation theory, relating glass transition temperatures to degree of cure
3. rheological models, relating resin viscosity to temperature and weight average molecular weight

In reality, they are all interconnected and the following sections will shed some light on them.

2.2.1 Kinetic Models

The mechanism for cross-linking of thermosetting resins is very complex because of the relative interaction between the chemical kinetics and the changing of the physical properties [49], and it is still not perfectly understood. The literature is ubiquitous with respect to studies of cure kinetic models for these resins. Two distinct approaches are used: phenomenological (macroscopic level) [2,5,50–72] and mechanistic (microscopic level) [3,73–85]. The former is related to an overall reaction (only one reaction representing the whole process), the latter to a kinetic mechanism for each elementary reaction occurring during the process.

The phenomenological models are semiempirical and hence do not provide a clear description of the curing process and its chemistry, which is important for understanding the network formation process. They are based on Equation 2.1 [86–88]:

$$\frac{\partial \alpha}{dt} = \kappa(T) f(\alpha) \tag{2.1}$$

where α is a fractional conversion of the reactive group, called *extent* or *degree of cure*, t is the reaction time, $f(\alpha)$ is a function representing the amount of reacted resin, which is determined for each system, and κ is a rate constant, defined by an Arrhenius type of relationship, Equation 2.2:

$$\kappa(T) = A \exp\left(-\frac{E_a}{RT}\right) \tag{2.2}$$

where A is the frequency factor or preexponential constant, E_a is the activation energy, R is the universal gas constant, and T is the processing temperature expressed in Kelvin. A few functions for $f(\alpha)$ are proposed in the literature, with the most common being Equation 2.3:

$$f(\alpha) = (1 - \alpha)^n \tag{2.3}$$

corresponding to a nth-order reaction [1,50,89–92] and Equation 2.4:

$$f(\alpha) = (1 - \kappa_2' \alpha^m)(1 - \alpha)^n \tag{2.4}$$

related to an autocatalytic reaction [52,56,74,75,93], where $m + n$ represents the reaction order and κ_2' is described by the Arrhenius equation. Variations of $f(\alpha)$ can be found, depending on the type of thermosetting resin and experimental conditions. Because the cure of thermosetting resins is an exothermic process and the only thermal event that exists, the degree or extent of cure, α, can be related to heat (enthalpy) released during the reaction of the components (basically a monomer and a curing agent). Thus, we have Equation 2.5:

$$\alpha = \frac{\Delta H_t}{\Delta H_R} \tag{2.5}$$

where ΔH_t is the enthalpy at a specific time and ΔH_R is the total enthalpy (or the total heat of reaction) at the end of the curing process when $\alpha = 1$. The reaction rate, represented by Equation 2.6, is then:

$$\frac{\partial \alpha}{\partial t} = \frac{1}{\Delta H_R} \frac{\partial H_t}{\partial t} \tag{2.6}$$

because the total heat of reaction is a constant for a particular resin [1]. Figures 2.3–2.5 show general shapes of the curves representing the functions between the degree of cure and the reaction rate with the curing time and the degree of cure for autocatalytic reactions at different temperatures. As the curing temperature increases, the curing time decreases. From Figures 2.3 and 2.4 curves *c* and *d*, respectively, one can see that there are resin systems in which the reaction rate is different from zero at the beginning of the reaction ($t = 0$).

The mechanistic models are more representative of the resin curing kinetics because they are based on stoichiometric balances of reactants involved in the elementary reactions. As a consequence, they are much more complex than the phenomenological models, but they can better represent the kinetics of cure. The physical and mechanical behaviors of the cured resins are determined by the chemical reactions that occur during cure. The understanding of the mechanism and kinetics of cure is one of the most important steps in evaluating the

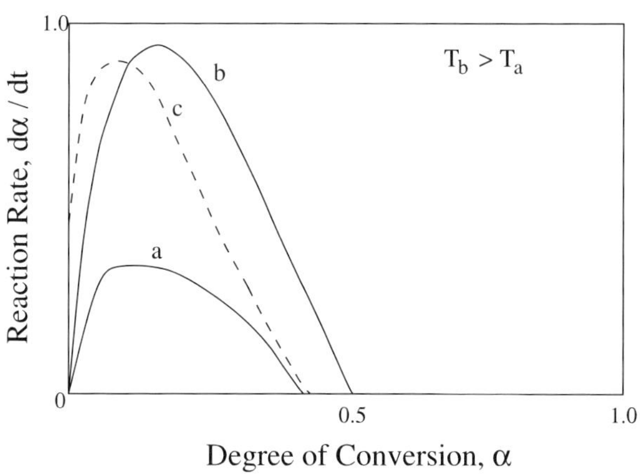

Figure 2.3 Reaction rate versus degree of cure for an autocatalytic model—curves a and b, at temperatures T_a and T_b, respectively, have an initial reaction rate equal to zero, and curve c has an initial reaction rate different from zero

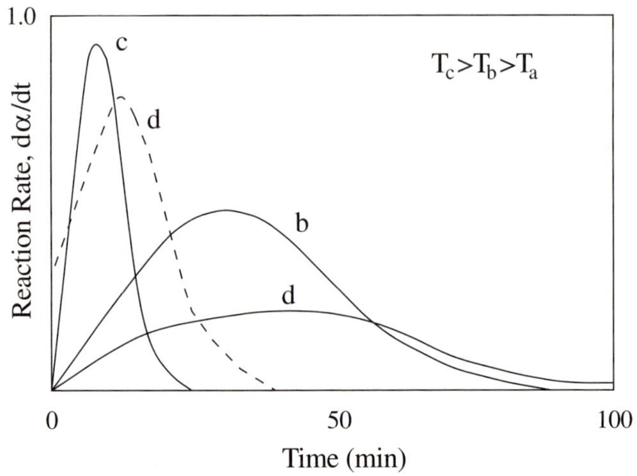

Figure 2.4 Reaction rate versus time for an autocatalytic model—curves a, b, and c, at temperatures T_a, T_b, and T_c, respectively, have an initial reaction rate equal to zero, and curve d has an initial reaction rate different from zero

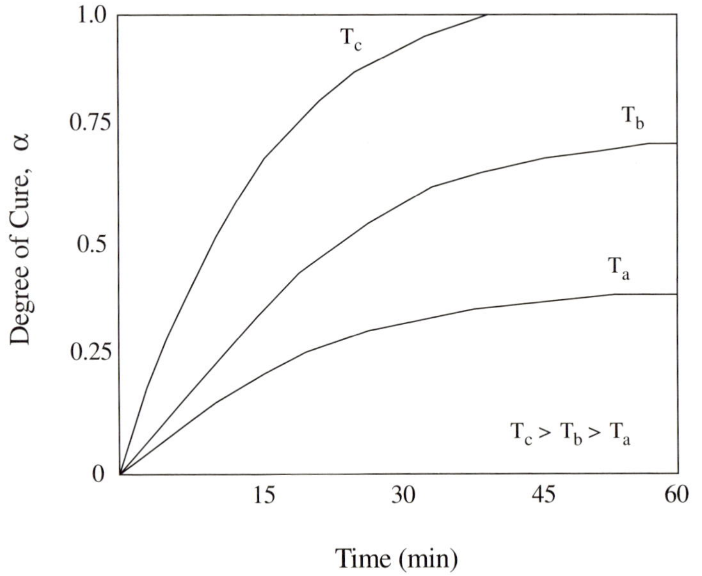

Figure 2.5 Degree of cure time versus time for an autocatalytic model—curves a, b, and c have temperatures T_a, T_b, and T_c, respectively, where $T_c \geq T_b \geq T_a$

structure–property relationship of thermosets [94,95]. Thus, it is helpful to know all possible reactions for a particular system and control the processing conditions in order to favor the specific reaction that gives superior or desired mechanical properties. The chemical reactions that occur during cure of thermosetting resins are so many and so complex, however, that even the mechanistic models do not ensure that the real phenomenon is completely described. Hence, the additional difficulty involved in using mechanistic approach for cure kinetic models is unjustifiable. The kinetic models require a balance between fundamental chemistry and practical empiricism [96] and because of the simplicity of the phenomenological models, they are widely used to describe the curing kinetic of thermosetting resins. It is simpler to compute all chemical effects with a few empirical parameters, even though these parameters are valid only for specific resins over a certain temperature range. The literature is not unanimous regarding the values of parameters, such as A, E_a, m, and n, even for the same systems. Min et al. [65] and Stutz et al. [66] present differences around 20 percent and 10 percent for the activation energies, E_{a1} and E_{a2}, respectively, for the autocatalytic model and the system *Diglycidyl Ether of Bisphenol A (DGEBA)/Diaminodiphenylsulphone (DDS)*, under the same range of curing temperature. The order of reaction, m and n, can be either a function of temperature [3,56,93,97] or not [34,52,61,67,98]. In the strict sense, the order of reaction is constant for each single system; namely, a system where only one reaction takes place. During the cure, however, many reactions can occur simultaneously; as a result, the order of reaction for phenomenological models is in reality an effective value representing all reactions. There is a need for a general model, valid over the entire range of the polymerization of a particular resin system [99].

2.2.2 Gelation Theory

During cure, diverse chemical and physical phenomena occur simultaneously. The cross-linking reactions (chemical phenomenon) cause the physical property changes of the reactants, such as increase of the molecular weight and the resin viscosity. There are two critical points about these physical changes: gelation and vitrification. Gelation is responsible for the transformation of the resin from a liquid to a rubbery state [100], whereas vitrification takes place when the glass transition temperature increases to the temperature of cure, marking the transformation from a rubber to a gelled glass (if gelation has occurred) or from a liquid to an ungelled glass (if gelation has not occurred) [55]. In order to characterize these transformations as a function of temperature and time, Gillham [4] developed a cure phase diagram, called Time-Temperature-Transformation (TTT) and, as shown in Figure 2.6, five states may exist: ungelled glass, liquid, gelled glass, gelled rubber, and char [59,95,101–103]. The gelation and vitrification (*S*-shaped) curves delimit the resin states and three critical temperatures appear: the glass transition temperature of the reactants, T_{go}, the temperature at which gelation and vitrification occur simultaneously, ${}_{gel}T_g$, and the glass transition temperature of the fully cured system, $T_{g\infty}$. Essentially no reaction takes place below T_{go} as the reactive species are immobilized in the glassy state [55]. It has not yet been established if and how the cure kinetics changes above ${}_{gel}T_g$, the gel point; some studies with urethane systems showed an increase [104], no change [105], and a decrease [106] in the reaction rate.

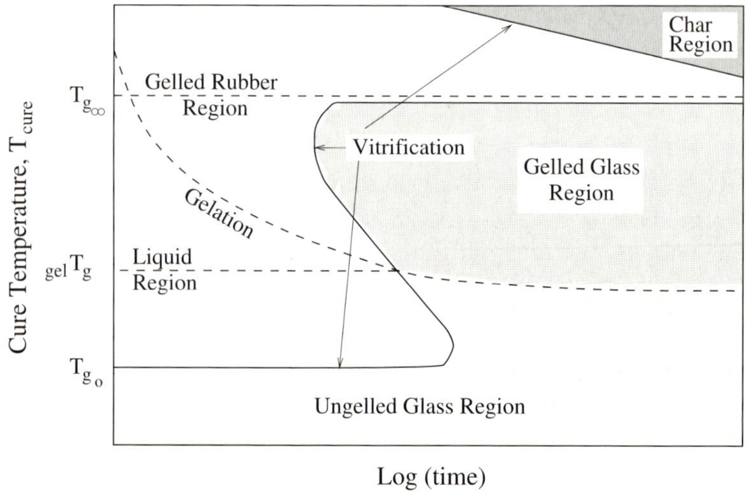

Figure 2.6 Time-temperature-transformation (TTT) cure diagram. Adapted from Enns and Gillham [55]

As depicted in Figure 2.6, a fully cured resin is achieved only if the cure temperature is higher than $T_{g\infty}$, when no vitrification occurs, thus allowing the movement of the molecules and, as a result, the reaction. Nevertheless, the cure temperature can not stay too high for a long period; otherwise, the resin will degradate (char region).

The onset of gelation is of critical importance in processing thermosetting resins, and several models have been developed to predict the extent of reaction up to the gel point [6], α_g. Among them, the well-known Flory–Stockmayer theory [100,107], based on a statistical approach applied to the structural changes (growing of molecular weight) of the resin, is given by Equation 2.7:

$$\alpha_g = \frac{1}{[r + rs(f - 2)]^{1/2}} \tag{2.7}$$

where f is the functionality of the cross-linking groups (e.g., $f = 4$ for tetrafunctional aromatic amine), r is the molar ratio of reacting functional groups (i.e., the molar ratio of epoxy group to amine hydrogen; for stoichiometric mixture of reactants, $r = 1$), and s is the fraction of the amine hydrogen in the multifunctional reactants [59,70]. For the common case of a stoichiometric mixture of diepoxide and tetrafunctional amine, $f = 4$, $r = 1$, and $s = 1$, which results in $\alpha_g = 0.577$. If the reactivities of the primary and secondary amines are unequal, it results in an increase in this value by 4 percent, at most [70,108,109]. As a consequence, the degree of cure at the gel point is in the range of 0.58 to 0.62 [70]. When this value is achieved, the mobility of the molecules is greatly reduced and the diffusional effects should be accounted for describing cure kinetics.

The glass transition temperatures, specific for each thermosetting resin, are used to characterize cure kinetics. They can be measured by many techniques, of which the widely used are Differential Scanning Calorimetry (DSC) and Torsional Braid Analysis (TBA) discussed later. DiBenedetto [110] has derived Equation 2.8 to relate glass transition temperature and extent of reaction before gelation:

$$\frac{T_g - T_{go}}{T_{go}} = \frac{(E_x/E_m - F_x/F_m)\alpha}{1 - (1 - F_x/F_m)\alpha} \tag{2.8}$$

with T_{go} as the glass temperature of the uncross-linked polymer as a reference [60], E_x/E_m as the ratio of lattice energies for cross-linked and uncross-linked polymer and F_x/F_m as the corresponding ratio of segmental mobilities at the glass temperature. These ratios can be determined by fitting Equation 2.8 to experimental data of degree of cure and glass transition temperature and by using Equation 2.9, which is a limiting case of Equation 2.8; when $\alpha = 0,\ T_g = T_{go}$, and when $\alpha = 1,\ T_g = T_{g\infty}$.

$$\frac{E_x/E_m}{F_x/F_m} = \frac{T_{g\infty}}{T_{go}} \tag{2.9}$$

Enns and Gillham [55] report values of E_x/E_m, F_x/F_m, T_{go}, $T_{g\infty}$, and α_g for various systems. After gelation, expressions for the glass transition temperature as a function of extent of cure and molecular weight can be found in Wisanrakkit and Gillham [59].

2.2.3 Rheological Models

The viscosity of a thermosetting resin undergoing a curing reaction is a function of time, temperature, and degree of cure. In case a resin presents a non-Newtonian behavior, a dependence of viscosity on shear rate needs to be accounted for as well [111].

Two concurrent phenomena govern the rheological behavior of a reacting system: one associated with the intensification of the mobility because of the increase in temperature, responsible for decreasing the viscosity, and another one related to the growing size of the molecules during cure, responsible for increasing the viscosity of the resin [44]. The empirical Williams–Landel–Ferry (WLF) equation, Equation 2.10, [112], valid near the glass transition and based on the free-volume theory, is the most used to represent the first phenomenon:

$$\ln\frac{\eta}{\eta_{T_o}} = \frac{-C_1(T - T_o)}{C_2 + T - T_o} \tag{2.10}$$

where $C_1,\ C_2$ are adjustable parameters and T_o is a reference temperature. Mijovic and Lee [113] derived expressions for C_1 and C_2 as a function of temperature, for a mixture of DGEBA and two aromatic amines—methylene dianiline (MDA) and *m*-phenylene diamine (*m*-PDA)—for temperature range of 60–110°C. Although the WLF equation is strictly an empirical expression, it was validated theoretically by Cohen and Turnbull [114].

The underlying concept of free-volume theory is that the movement of the molecules is intrinsically conditioned to the amount of free volume in a molecular ensemble; the less the unoccupied space, the more the collisions among the molecules, which results in a slow response to a perturbation in an equilibrium state [115].

With the objective of accounting for the second phenomena mentioned earlier, Enns and Gillham [55] proposed Equation 2.11:

$$\ln \eta = \ln \eta_\infty + \ln \bar{M}_w + \frac{E_\eta}{RT_o} = -\frac{C_1(T - T_0)}{C_2 + T - T_o} \tag{2.11}$$

with η_∞ as the extrapoled viscosity as $T \to \infty$, E_η as the activation energy, and $\bar{M}_w$ as the weight average molecular weight. The reference temperature, T_o, is arbitrarily chosen to be the temperature above which Arrhenius equation is valid to describe the viscosity–temperature dependence. There are many relations in the literature that describe the weight average molecular weight as a function of degree of cure and the molecular weight of monomers [55,59,116,117].

Many authors [11,13,37,42,118–120] have followed an empirical equation proposed by Stolin et al. [121], Equation 2.12, in order to calculate the resin viscosity:

$$\eta(T, \alpha) = \eta_0 \exp\left(\frac{U}{RT} + K\alpha\right) \tag{2.12}$$

where U is the activation energy of the viscous fluid, R is the gas constant, and K is a constant that accounts for the effect of the chemical reaction on the change in the reacting mass viscosity and, consequently, on the dissipation intensity. It is assumed that U is independent of the degree of cure. The values for these parameters for two systems are listed in Table 2.3.

Other commonly used empirical equation is the one proposed by Castro and Macosko [19,30,122], Equation 2.13:

$$\eta = A_\eta \exp\left(\frac{E_\eta}{RT}\right)\left(\frac{\alpha_g}{\alpha_g - \alpha}\right)^{a+b\alpha} \tag{2.13}$$

The parameters for a few common systems are listed in Table 2.4. Another one, proposed by Lee and Han [9,15,123], Equation 2.14, is:

$$\eta = A_\eta \exp\left(\frac{E_\eta}{RT}\right) \tag{2.14}$$

where the fluid activation energy, E_η, and the frequency factor, A_η, are, Equation 2.15:

$$E_\eta = a + b\alpha \qquad A_\eta = a_o \exp(-b_o\alpha) \tag{2.15}$$

with a, b, a_o, and b_o as specific constants for each resin. For a system composed of a partially cured unsaturated polyester resin, OC-E701 [123], 43 wt% of styrene and *t*-butyl perbenzoate as an initiator, these constants are: $a = 7.8$ kcal/mole; $b = 19.7$ kcal/mole; $a_o = 6.41 \times 10^{-5}$ N.s/m^2; $b_o = 23.1$. Lee and Han [123] pointed out, however, that for these

Table 2.3 Parameters for Equation 2.12

Author	System	η_0 (Pa.s)	U (J/mol)	K (Pa.s)	Range of validity
Lee et al. [118]	Hercules 3501-6	7.93×10^{-14}	9.08×10^4	14.1 ± 1.2	$\alpha < 0.5$
Dusi et al. [120]	Fiberite 976	1.06×10^{-6}	3.76×10^4	18.8 ± 1.2	$\alpha < 0.2$

Table 2.4 Parameters of Equation 2.13

Authors	System	A_η (Pa.s)	E_η (J/mol)	α_g	a	b	Range of validity
J.M. Castro and C.W. Macosko [19] (1980)	RIM2200	10.3×10^{-8}	41.3	0.65	1.5	1.0	—
J.M. Kenny et al. [30] (1990)	Commercial grade thermoset DSM resin for RTM	3.416×10^{-11}	64900	0.088	3.706	−34.62	30–60°C
D.S. Kim and S.C. Kim [122] (1994)	DGEBA and TETA (10 phr)	3.60×10^{-10}	58600	0.765	1.0	5.2	70–90°C

values, at temperatures higher than 156.2°C, the viscosity calculated by Equation 2.14 decreases with an increase in α, which is not physically correct. Hence, they proposed two other equations, valid for the system used: one based on Tajima and Crozier [124] model, Equation 2.16:

$$\log \eta(T, \alpha) = (23.64 + 3.82\alpha) - \frac{26.8(41.63 + T - 90.19\alpha)}{55.03 + T - 90.19\alpha} \tag{2.16}$$

and another based on Hou model [125], Equation 2.17:

$$\log \eta(T, \alpha) = -(6.06 + 4.78\alpha) + \frac{51.6(22.11 + 29.92\alpha)}{T - 143.3 - 33.06\alpha - 281.4\alpha^2} \tag{2.17}$$

The agreement between experimental data and those predicted by Equations 2.16 and 2.17 is very good, but it has been verified only up to 20 percent of conversion and only for a specific type of resin.

A model that is valid for 140–175°C has been proposed by Srinivasan et al. [161] for RTM resins, such as BMI 5250-4-RTM, Equation 2.18:

$$\eta(c_p)\left[1.03 \times 10^{-6} \exp\left(\frac{7253}{T(K)}\right)\right] \exp[1.56 \times 10^{-14}(\exp(0.0696T(K)))t(\text{min})] \tag{2.18}$$

and PR500, Equation 2.19:

$$\eta(cp) = \eta_o a \exp(bt(\text{min})) \tag{2.19}$$

where η_o, a and b are given by Equations 2.20, 2.21, and 2.22, respectively.

$$\eta_o = 1.23 \times 10^{-9} \exp\left(\frac{10441.52}{T(K)}\right) \tag{2.20}$$

$$a = 1.68 \times 10^4 \exp\left(-\frac{4788.48}{T(K)}\right) \tag{2.21}$$

$$b = 4.82 \times 10^{-9} \exp(0.04T(K)) \tag{2.22}$$

The agreement was reasonably well for the initial rise in viscosity.

2.2.4 Diffusion Effects

As previously mentioned, a diffusion or mobility effect should be accounted for in the last stages of a curing process. The viscosity increases because of chain growth and cross-linking and several authors pointed out that, as a consequence, the movements of the reactants slow down or cease near the gel point [55]; thus, the curing process is controlled by these motions rather than by chemical factors [7,8].

In order to quantify diffusional effects on curing reactions, kinetic models are proposed in the literature [7,54,88,95,99,127–133]. Special techniques, such as dielectric permittivity, dielectric loss factor, ionic conductivity, and dipole relaxation time, are employed because spectroscopic techniques (e.g., FT i.r. or n.m.r.) are ineffective because of the insolubility of the reaction mixture at high conversions. A simple model, Equation 2.23, is presented by Chern and Poehlein [3], where a diffusional factor, f_d, is introduced in the phenomenological equation, Equation 2.1.

$$f_d = \frac{\kappa_{ef}}{\kappa_c} = \frac{1}{1 + \exp[C(\alpha - \alpha_c)]} \tag{2.23}$$

where κ_{ef} is the overall effective rate constant [134], k_c is the rate constant for chemical kinetics, C is a constant and α_c is the critical fractional conversion in which a three-dimensional cross-linking network is formed. For chemically controlled reactions, $f_d = 1$; as the conversion approaches α_c, diffusion becomes important. Thus, Equation 2.1 has to be modified to Equation 2.24:

$$\frac{\partial \alpha}{\partial t} = \kappa f(\alpha) f_d \tag{2.24}$$

Khanna and Chanda [67], and Barral et al. [72] successfully applied Equation 2.24 to describe the cure of two different systems: bis-(2,3-epoxypropyl)-1,3-benzenedicarboxylate (IPDGE) with hexahydrophtalicanhydride (HHPA), as a curing agent, and DGEBA with benzyldimethylamine (BDMA), as a catalyst, and 1,3-bisaminomethylcyclohexane (1,3-BAC), as a curing agent, respectively.

2.2.5 Techniques to Monitor Cure

The only way to validate kinetic models is to measure experimentally the degree of cure as a function of time and temperature. It can be done on both macroscopic and microscopic levels by monitoring chemical, physical (refractive index [135], density [136], and viscosity [137]), electrical (electrical resistivity [138,139]), mechanical, and thermal property changes with time [140,141]. The most-used techniques to monitor cure are presented in the next two subsections.

2.2.5.1 Macroscopic Characterization

The techniques to be described here cannot differentiate the multiple reactions that take place during the curing reaction; instead, only an overall kinetics is measured [32,142]. The most commonly used techniques will be described briefly.

Differential Scanning Calorimetry (DSC) This is by far the widest utilized technique to obtain the degree and reaction rate of cure as well as the specific heat of thermosetting resins. It is based on the measurement of the differential voltage (converted into heat flow) necessary to obtain the thermal equilibrium between a sample (resin) and an inert reference, both placed into a calorimeter [143,144]. As a result, a thermogram, as shown in Figure 2.7, is obtained [145]. In this curve, the area under the whole curve represents the total heat of reaction, ΔH_R, and the shadowed area represents the enthalpy at a specific time. From Equations 2.5 and 2.6, the degree and rate of cure can be calculated. The DSC can operate under isothermal or non-isothermal conditions [146]. In the former mode, two different methods can be used [1]:

Method 1: A sample is placed into a calorimeter previously heated (faster equilibrium [139]) or into an unheated calorimeter, whose temperature is raised as quickly as possible up to the curing temperature. It is possible to monitor, simultaneously, α and $\partial\alpha/\partial t$ (Equations 2.5 and 2.6);

Method 2: A sample is cured for various times until no additional curing can be detected. The samples are then scanned (heating rate ranging from 2 to 20°C/min) in order to measure the residual enthalpy, ΔH_{res}. The degree of cure is calculated directly by Equation 2.25,

$$\alpha = \frac{\Delta H_R - \Delta H_{res}}{\Delta H_R} \tag{2.25}$$

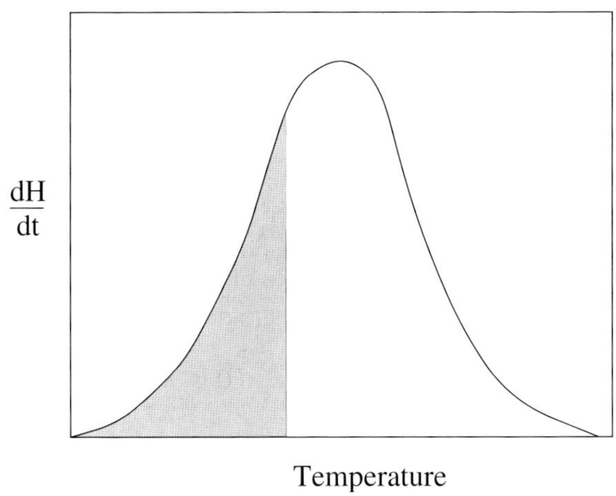

Figure 2.7 Example of a thermogram—rate of enthalpy change with temperature

but not the reaction rate, which is obtained by tangents to the curve of α versus time. For reactions with very small exotherm heat, this method should be utilized instead of Method 1.

In the nonisothermal or dynamic mode, a sample is set into a calorimeter and the temperature is raised at a certain constant heating rate up to the operational temperature. The total heat of reaction is independent of the heating rate (recommended range is 2–20°C/min [1,140]). Three methods are possible:

1. All kinetic information (A and E_a in Equation 2.2) can be obtained from only one experiment. Although this method has been successfully applied for some first order reactions, it is not accurate for others; the heat of reaction is higher (around 10–30%) when compared with isothermal mode [1].
2. A and E_a in Equation 2.2 can be accurately measured from the values of the peak exotherm temperature for various heating rate for all reactions [1]. This is the recommended method.
3. More general than the second method, it measures A and E_a from the values of the temperature necessary to reach a constant conversion for various heating rates. This technique needs considerable effort to be moderately successful.

Although many researchers use DSC for measuring enthalpies over the entire curing process, it is not accurate after the gel point [147–149].

Several authors have pointed out that the parameters of the reaction rate expression for isothermal and dynamic conditions may be different [75,147,150–154]. In general, heat of reaction is higher for the latter condition [1,155]. St John and George [58] mentioned that other exothermic reactions can occur in dynamic DSC, as proved by Bakker et al. [156]. As a matter of fact, there is a controversy regarding the confidence of relating degree of conversion with energy of the system. St John and George [58] affirmed that two serious assumptions are built in these measurements. The first assumption is that all resins cure completely, which may not be the case under isothermal conditions, which require dynamic DSC scan at high temperatures after isothermal run. The second assumption is that all reactions involved during cure have the same enthalpy, which is not always the case [58].

One possible explanation for different results between isothermal and dynamic analysis was first given by Prime [89]. It is related to the fact that the degree of conversion is a function of the curing time and temperature; therefore, we have Equation 2.26:

$$\frac{d\alpha}{dt} = \left(\frac{\partial\alpha}{\partial t}\right)_T + \left(\frac{\partial\alpha}{\partial T}\right)_t \frac{dT}{dt} \tag{2.26}$$

where $d\alpha/dt$ represents the dynamic reaction rate, $(\partial\alpha/\partial t)_T$, the isothermal reaction rate, and dT/dt, the heating rate.

According to some authors [89,92,140,150,152,157,158], the isothermal rate is in reality expressed by Equation 2.1. The physical meaning of the term $(\partial\alpha/\partial T)_t$ was questioned by Simmons and Wendlant [159], Hill [160], as well as Gorbatchev and Lovinenko [161]. They affirmed that $(\partial\alpha/\partial T)_t$ has to be zero because if one fixes the time, the positions of all the particles in the system will be fixed, making α to be constant. By using thermodynamics, Kratochvil and Sestak [157] showed that $(\partial\alpha/\partial T)_t$ exists and is nonzero.

Equation 2.26 can also be expressed in a different form, Equation 2.27:

$$\frac{d\alpha}{dt} = \left(\frac{\partial\alpha}{\partial t}\right)_T \left[1 + a\frac{(\partial\alpha/\partial T)_t}{(\partial\alpha/\partial t)_T}\right] \tag{2.27}$$

where $a = dT/dt$. For isothermal experiments, $a = 0$ and there is no difference between Equations 2.1 and 2.27, however, as the dynamic measurements are simpler and faster than isothermal ones (for the first case, only one experiment is needed to obtain all kinetic parameters), it is important to calculate the term $(\partial\alpha/\partial T)_t$ in order to acquire a relationship between dynamic measurements and isothermal parameters. Dutta and Ryan [152] proposed expressions for $(\partial\alpha/\partial T)_t$ by considering the most common kinetic models, Equations 2.3 and 2.4. For an *n*th order reaction and one specific thermosetting system, Prime [89] found differences as large as 85 percent and 103 percent for the activation energy and the natural logarithm of the frequency factor, respectively, when comparing the results from the isothermal experiments and dynamic experiments without accounting for the second term on the right-hand side of Equation 2.27. The moral is that one should be careful when choosing kinetic parameters from the literature. Another common error is to apply chain rule for degree of cure [71,130,131,162], as shown by Equation 2.28;

$$\frac{d\alpha}{dt} = \frac{d\alpha}{dT}\frac{dT}{dt} \tag{2.28}$$

If this were correct, for isothermal conditions, from Equation 2.28, the extent of cure would be constant with time, which is not the case.

Other thermal techniques are Thermogravimetric Analysis (TGA) [55,68], High Pressure Calorimeter (HPC) [1], Thermomechanical Analysis (TMA) [1,141], and Differential (or Dynamic) Thermal Analysis (DTA) [74]. These are rarely used and will not be discussed here.

Torsional Braid Analysis (TBA) This technique is a variation of Torsion Pendulum (TP) and was developed by John Gillham in 1958 [163]. It consists of measuring the frequency and decay constants that characterize each wave resultant of free torsional oscillations (at approximately 1 Hz) subjected to a sample (a glass braid impregnated with a resin). These measurements are converted into elastic (stored energy) and loss (loss energy) moduli, which are related to transition (gelation and vitrification) times and temperatures. Torsional Braid Analysis is a sensitive technique for determination of physical changes in the resin [4,164–168], occurring at and after the gel point; it has a lower limit of detectability [1]. TBA is not capable of measuring the degree of cure. Other mechanical techniques are Dynamic Mechanical Analysis (DMA) [169,170] and Rheological Dynamic Spectrometer (RDS) [55,171–173].

Due to the fact that there is no accurate technique valid for the entire curing reaction, two techniques should be used to follow cure: one applied to before gelation (DSC) and another one applied after gelation (TBA, DMA, TMA, d.c. conductivity [50,128,174,175]).

2.2.5.2 Microscopic Characterization

Sophisticated techniques are required to measure the concentration of all components in order to trace individual reactions occurring during cure. Some examples are Fourier Transform

Table 2.5 Kinetic Parameter Percentage Errors Between Composite and Neat Resin

	A_1 (min^{-1})	A_2 (min^{-1})	E_{a_1} (J/mol)	E_{a_2} (J/mol)	m —	n —
$\left\lvert \dfrac{\text{Composite} - \text{neat resin}}{\text{Composite}} \right\rvert$	83.2%	50.3%	5.70%	0.344%	11.1%	36.0%

infrared spectroscopy (FT i.r.) [55,57,81,176], solid state ^{13}C nuclear magnetic resonance (n.m.r.) [177–182], near infrared spectroscopy (n.i.r.) [84,142,156,183], and chromatography [179,184].

The spectroscopy methods consist of dispersing a radiation from a source and passing it over a slit system that isolates a narrow frequency range falling on the detector [180]. By using a scanning mechanism, the energy transmitted through a sample as a function of frequency, known as the spectrum, is obtained and compared with the spectrum characteristic for each functional group of thermosetting resins. The mid-infrared (m.i.r.) and n.i.r. regions of the spectrum can be utilized. Although all functional groups involved in cure reactions of epoxy resins have strong characteristic absorptions in the m.i.r. region, the m.i.r. spectra of epoxy resins and hardeners are very complex [183]. The absorption bands of the main functional groups in cure reactions of epoxies are well-isolated in the n.i.r. region of the spectrum (10,000 and 4000 cm^{-1}). A graph of absorbance versus wavenumber, as shown in

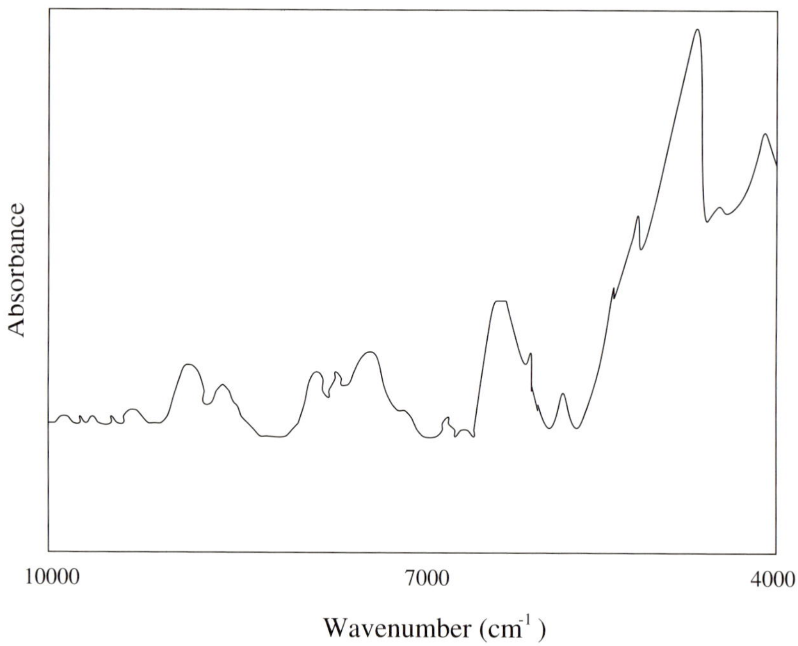

Figure 2.8 Absorbance as a function of wavenumber—every peak is characteristic of a particular group

Figure 2.8, is generated, presenting peaks correspondent to functional groups, such as epoxide, primary and secondary amines, and hydroxyl. The values of the absorption bands for these groups can be found in the papers referenced earlier. The basic principle of all these methods is the comparison between the spectrum of reference substances and spectra of the reactants and products of a curing reaction subjected to radiation. A qualitative and quantitative identification of the components is then possible.

2.3 Effect of Reinforcements

The resin usually flows through the fiber reinforcements. These fibers may be sized, and they may change the kinetics. There is no consensus, however, with respect to the influence of fillers or reinforced materials on the cure kinetics. Inorganic fillers are used in thermosetting resins in order to reduce the cost and the shrinkage and as a heat sink to achieve a better temperature control across a molded part during cure [185]. Structural fibers are treated with chemicals compounds, such as silane, in order to enhance coupling between the fiber and matrix phases [39,186]. It is essential to know to what extent this coupling alters the cure kinetics and thus, if it is possible, to use the kinetic parameters from neat resins to describe the kinetics for resin/reinforced materials. Dutta and Ryan [187] analyzed the influence of carbon black and silica on the cure kinetic of DGEBA/*m*-phenylenediamine, concluding that the heat of reaction is independent of the filler content but dependent on the type of filler, the overall reaction order of the system is not significantly affected and the reaction rate changes because of the variations in the rate constants of the autocatalytic model. This effect is more sensitive at higher temperatures and for carbon black, probably because of chemical complexes on the carbon black surface [187].

Whereas McGee [188] did not verify any effect of particulate fillers (i.e., glass, calcium carbonate, and aluminum) on the reaction kinetics, Lem and Han [185], working with calcium carbonate and clay in an unsaturated polyester resin, concluded that the reaction rate increases by increasing the filler content.

For kaolinite and E-glass in a commercial unsaturated brominated polyester resin with 50 percent by weight styrene, Ng and Zloczower [189] affirmed that although the heat of reaction and the peak temperature are not affected by the presence of the fillers, there is a little retardation effect on the conversion rate, at least in the temperature range analyzed (60–160°C). For environment temperature, Plueddenmann [190] said that glass or other fillers may have a severe inhibiting effect on the cure kinetics of polyesters. Ng and Zloczower [189] applied their systems to pultrusion process and observed that kaolinite and fiberglass act as heat sinks, reducing the peak exotherm [16] and delaying the cure. For carbon fibers, the heat conduction is improved because of its higher thermal conductivity compared with the resin [189]. Rudd and Kendall [16] affirmed that the majority of the additives do not interfere in the polymerization reaction, acting only as heat sinks, which is different from fillers. The fillers affect the flow characterization and the thermal properties of the resin system, which results in increased thermal diffusivity [16].

By working with vinyl ester (specifically DERAKANE 411-C50), Michaud [9] observed a drastic influence of the fibers (E-glass) on the induction time (i.e., the time required for the reaction to become observable at a specific temperature). An induction time of 90 minutes and 5.5 hours was necessary for RTM runs at 53°C and DSC runs at 50°C, respectively. Lee and Lee [33] affirmed that the age of the resin can also alter the induction time: A reduction time of 50 percent was found after 10 months of storage at 70°C. All these effects are more evident at low temperatures.

Han et al. [191] found that the rate of cure of a resin is greatly influenced by the presence of fibers and the type of fibers employed. The rate of reaction for resin–fiber system can be 60 percent different from that of neat resin, after a 10-min cure. A similar conclusion was presented by Mijovic and Wang [192] for graphite–epoxy composites based on TGDDM/DDS (33 phr). They verified large differences (see Table 2.5) in the kinetic parameters when considering an autocatalytic model.

It is clear that the surface of reinforcement can affect the kinetics. The disagreement is regards to the extent. The type of resin and the temperature range are important considerations. In general at low temperatures, one must be cautious about applying the neat resin phenomenological parameters to a system in which the resin will impregnate a network of fibers.

2.4 Epoxy, Vinyl Ester, and Phenolic Resins

Because of the wide application of these resins to diverse industries and their very different kinetic models and mechanisms of cross-linking and reactions, phenomenological kinetic models for epoxy, vinyl ester, and phenolic resins are presented in the next three subsections.

2.4.1 Epoxies

Epoxy resins are thermosetting matrix resins, characterized by the epoxide group [48,193], as shown in Figure 2.9.

Because of the low shrinkage and the excellent adhesion, mechanical properties, and chemical resistance, epoxies have a vast variety of applications, such as adhesives (epoxy-amine), structural materials in high-performance composites (epoxy-amine) (e.g., aircraft primary structure), filament wound pressure vessels, and so forth. The main disadvantage of

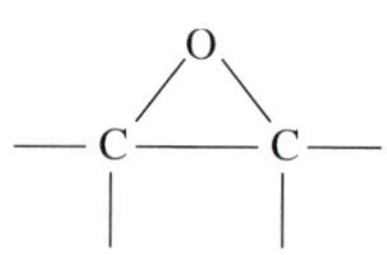

Figure 2.9 Structural formula of epoxide group

epoxy resins is their brittleness on account of their highly cross-linked structure [194]. There are some ways to improve their toughness. One of them is to add a reactive liquid rubber to the epoxy network [195] and the other one is to alter the structure of the monomers [196]. In the first case, if part of the rubber is dissolved in the matrix (when the phase segregation is not complete), then the glass transition temperature and modulus decreases, which is not desirable. In the second case, it is common to use curing agents [194], which are also responsible for promoting a chemical cross-linking bond between them and the epoxy groups, under heat and pressure. They affect properties of a cured system, because they determine the type of curing reaction and influence cure kinetics. The curing agents can be amines, anhydrides, and Lewis acids [6]. The most-used curing agents are all amine-type, such as DDS (4,4′-diaminodiphenylsulphone), *m*-PDA (*m*-phenylenediamine), and MDA (methylene dianiline).

There are many different types of epoxy resins. The most extensively used are based on DGEBA and on tetraglycidyl-4,4′-diaminodiphenylmethane (TGDDM), whose structural formula are found elsewhere [6,58]. The DGEBA resins are known by many different commercial names, such as Shell Epon 825, 826, 828, 1001, 1004, Dow DER 332, 330, 331, 661, 664, and Ciba-Geigy Araldite 6004, 6005, 6010, 6065, 6084, 250. The TGDDM resins have the commercial names of MY 720 and MY 721, which are both from Ciba-Geigy. The former has 63 percent of TGDDM and higher viscosity, whereas the latter has 79 percent of TGDDM. They are the basic component of the common prepregs Hercules 3501-6, Fiberite 934, 976, and Narmco 5208. Hercules 3501-6 is a mixture of epoxides, #1 (100 phr), #2 (16 phr), and #3 (15 phr), curing agent, DDS (44 phr), and BF_3 (2 phr) [118]. Fiberite 976 is composed of MY 720 and DDS [120]. Narmco 5208 is a mixture of TGDDM, DDS, and an epoxy resin based on a bisphenol A Novolac, such as Epi-Rez SU-8 from Celanese [78]. The composite known as Hercules AS4/3501-6 is composed of the resin Hercules 3501-6 and the fibers Hercules AS4 (graphite), with a fiber volume fraction of 58 percent [172]. The Narmco Rigidite 5208/WC 3000-42 (from Narmco Materials) composite consists of woven carbon–fiber reinforcement, impregnated with 42 wt% of Narmco 5208 resin [78].

The curing process of epoxies is mainly dependent on the reactivity characteristics, the structure, and the functional group of resin [61,65]. The basic chemical reactions, catalyzed by hydroxyl group (OH), between the epoxide group and the curing agent are [194]:

$$\text{Epoxide} + \text{Primary Amine} \longrightarrow \text{Secondary Amine}$$
$$\text{Epoxide} + \text{Secondary Amine} \longrightarrow \text{Tertiary Amine}$$

For TGDDM/DDS (<35 wt% of DDS), Morgan and Mones [57] affirmed that all primary groups were consumed at 177°C after 2.5 hours of reaction. The epoxy–amine reactions dominate the early stages of cure and, hence, the composite processing conditions.

Other possible reactions, such as homopolymerization (epoxide+epoxide) and epoxide+hydroxyl group (in the latter stages of cure), can be neglected when the ratio of epoxide to amine is stoichiometric and in the absence of catalyst or accelerator [194]. For TGDDM/DDS resins, the homopolymerization reaction may be neglected at cure temperature below 180°C [84]. At temperatures between 177°C and 300°C, dehydration and/or network oxidation occur, which results in formation of ether cross-linkings with loss of water. Decomposition of the epoxy–OH cure reaction can also take place, which results in propenal formation [57]. Degradation reactions dominate above 250°C. Several mechanistic models exist in the literature [3,5,57,73–78,84,197]. In order to represent the overall cure quantitatively, reaction rate equations should be derived for each basic and parallel reaction, which results in a series of differential equations. When coupling the kinetic model with the resin flow and heat transfer phenomena this makes it difficult to describe the complete curing process. Hence, practice has converged on the use of the phenomenological models due to their simplicity. In general, cure reactions of the amine–epoxy resins [48], at temperatures less than 200°C, show complex kinetics and have been considered as autocatalytic reactions, described by Equation 2.29, between the epoxy and amine groups [65]. Table 2.6 lists the kinetic parameters inherent to this model and the curing temperatures relative to neat epoxy resins and epoxy resin/reinforced material systems from several key investigations in the literature.

$$\frac{\partial \alpha}{\partial t} = (\kappa_1 + \kappa_2 \alpha^m)(1 - \alpha)^n \tag{2.29}$$

with $\kappa_i = A_i \exp(-E_{ai}/RT)$.

One can conclude by analyzing the data of Table 2.6 that for most epoxy systems typical values of heat of reaction, frequency factor, activation energy, and order of reaction are between 34 and 170 cal/g, 6×10^4–10^{13} min^{-1}, 10 and 32 kcal/mol, and 1 and 3, respectively. The glass transition temperature for the fully cured resin varies from 97 to 130°C for DGEBA/*m*-phenylenediamine system [50,60], 85 to 218°C for DGEBA/DDS system [60,194], and 135°C for TGDDM/DDS. The difference mentioned earlier, between the isothermal and dynamic kinetic parameters can be noticed in the papers of Lee et al. [118] and Kenny et al. [32] in Table 2.6, where the activation energy of the latter paper is about 74 percent higher than it is the former for Hercules resin. As the kinetic parameters of the phenomenological models depend on the thermosetting resin systems and the experimental conditions, a data base that includes this information would be very useful in order to avoid repetitive experimental work.

2.4.2 Vinyl Esters

Typical unsaturated polyesters are thermosetting resins, which are prepared by the reaction of a glycol (e.g., ethylene glycol or propylene glycol) and unsaturated or saturated acids, such as fumaric acid (the best one), isophthalic acid (improved strength, good chemical resistance), or an anhydride, such as maleic anhydride and phthalic anhydride (low cost), to give an unsaturated structure [6,48,193]*. They have been used successfully for more than 40 years to build composite parts for corrosion resistant applications [198] and now they can be used in automotive, electrical, and appliance components [34]. Vinyl esters are unsaturated polyester resins, which are a reaction product of an epoxy resin (generally, diglycidyl ether of bisphenol A) with an unsaturated carboxylic acid, such as methacrylic acid [6,88]. The structural formula is: The higher the unsaturated sites, the higher the cross-linking density, which results in better heat resistance, but less tensile strength and elongation [9]. They are

*[6,48,193], see Figure 2.10.

Table 2.6 Cure Kinetic Models—Epoxies

Authors/systems	Kinetic model and experimental conditions	Kinetic parameters
Author: R.B. Prime [89] (1970) *System*: Resin: DGEBA (DOW DER 332) Curing Agent: *m*-phenylenediamine (Aldrich, purissim grade)	1. Kinetic Model: *n*th order 2. Dynamic Experiments: The temperature varied from 5 to 229°C at heating rates of 5, 10, and 20°C/min.	1. Dynamic Experiments; $A = 8.10 \times 10^3\ s^{-1}$ $E_a = 56.09$ kJ/mole $n = 0.9$ $\Delta H_R = 489.76$ J/g 2. Temperatures at which the reaction was fast enough to be detected by *DSC*: $T_{cure} = 64°C$ at $dT/dt = 5°C$ $T_{cure} = 65°C$ at $dT/dt = 10°C$ $T_{cure} = 51°C$ at $dT/dt = 20°C$
Authors: M.A. Acitelli et al. [50] (1971) *System*: Resin: DGEBA (DOW DER 332) Curing Agent: *m*-phenylenediamine (Aldrich, purissim grade), without purification	1. Kinetic Model: *n*th order 2. Isothermal Experiments: The temperature varied from 23 to 157°C.	1. $A = 2.12 \times 10^4\ s^{-1}$ $E_a = 51.5$ kJ/mol 2. The order of reaction, *n*, varied from 2.3 at $T = 23°C$ to 1.0 at $T = 157°C$. 3. The glass transition temperature for the totally cured resin is 130°C.
Authors: R.B. Prime and E. Sacher [51] (1972) *System*: Resin: DGEBA (DOW DER 332) Curing Agent: polyamide (General Mills Versamid 140) without purification	1. Kinetic Model: *n*th order 2. Isothermal Experiments: The temperature varied from 23 to 110°C 3. $\frac{Epoxide}{Polyamide} = EP1 = 2\,g/1.5\,g$, corresponding to 10% excess of polyamide. 4. $\frac{Epoxide}{Polyamide} = EP2 = 2\,g/5\,g$,	1. Epoxide-Primary Amine: at 40°C a. EPI system: $A = 6.34 \times 10^5\ s^{-1}$ $E_a = 60.7$ kJ/mol b. EP2 system: $A = 2.96 \times^6\ s^{-1}$ $E_a = 63.21$ kJ/mol $0.6 \leq n \leq 1.9$ $T_g = -10°C$

(*continued*)

Table 2.6 (*continued*)

Authors/systems	Kinetic model and experimental conditions	Kinetic parameters
	corresponding to 300% excess of polyamide.	2. Epoxide-Alcohol: at 80°C a. EPI system: $A = 1.6 \times 10^6\ s^{-1}$ $E_a = 75.35\,kJ/mol$ $T_g = 85°C$ $0.2 \leq n \leq 1.7$ $T_g = 40°C$ b. EP2 system: $A = 3.3 \times 10^6\ s^{-1}$ $E_a = 73.67\,kJ/mol$ $T_g = 20°C$ 3. Amide-Alcohol: at 110°C a. EP1 system: $A = 4.4 \times 10^9\ s^{-1}$ $E_a = 110.93\,kJ/mol$ $T_g = 30°C,\ 85°C$ b. EP2 system: no data
Author: R.B. Prime [92] (1973) *System 1*: Resin; DGEBA (Dow DER 332) Curing Agent: *m*-phenylene diamine (Aldrich purissim grade) System 2: Resin: DGEBA (Dow DER 332) Curing Agent: polyamide (General Mills Versamid 140)	1. Kinetic Model; *n*th order 2. Dynamic Experiments: The heating rates used were 5, 10, and 20°C/min.	1. Dynamic Experiments: a. System 1: $\Delta H_R = 501.5 \pm 15.9\ J/g$ $n = 0.83$ $A = 1.04 \times 10^4\ s^{-1}$ $E_1 = 54.84\,kJ/mol$ b. System 2: $\Delta H_R = 362.51 \pm 12.56\,J/g$ $n = 1.45$ $A = 5.72 \times 10^4\ s^{-1}$ $E_a = 57.77\,kJ/mol$

		2. Isothermal Experiments: a. System 1: $A = 2.03 \times 10^4 \pm 42.5\ s^{-1}$ $E_a = 51.49 \pm 6.28\ kJ/mol$ $n = 0.98 \pm 0.5$ b. System 2: $A = 2.23 \times 10^6\ s^{-1}$ $E_a = 63.63\ kJ/mol$ $n = 1.13$
Authors: A. Dutta and M.E. Ryan [187] (1979) *System*: Resin: DGEBA (Allied Resin Corp., DER 332) Curing Agent: *m*-phenylenediamine (Aldrich Chemical Co.) Fillers: a. Cabot Furnace-process carbon black (Sterling V, Cabot N-660, Cabot Corporation of Canada Ltd) b. Surface-treated Novacite (silica) (Malverns Mineral Co., NOVAKUP)	1. Kinetic Model: Autocatalytic 2. Isothermal Experiments: The temperature varied from 70 to 170°C. 3. Dynamic Experiments: The samples were scanned from 60 to 327°C at a heating rate of 10°C/min. 4. The reactions were generally complete at a temperature of 297°C. 5. The cure reaction ($\alpha = 1$) took 45 min at $T = 110°C$ for unfilled and filled systems (carbon black) 6. The cure reaction ($\alpha = 1$) took around 24 min and 8 min at $T = 130°C$ and $T = 170°C$, respectively, for Novakup system.	1. Isothermal Experiments: a. Unfilled System: $E_{a1} = 64.46\ kJ/gmol$ $E_{a2} = 45.63\ kJ/gmol$ $0.6 \leq m \leq 1.2$ $m + n = 2$ $m = 7.032 \exp(-B)$ $B = 0.0054T(K)$ b. Filled System: i) Carbon Black $E_{a1} = 64.46\ kJ/gmol$ $E_{a2} = 45.63\ kJ/gmol$ ii. Novakup $E_{a1} = 60.28\ kJ/gmol$ $E_{a2} = 45.63\ kJ/gmol$ 2. Dynamic Experiments: a. Unfilled System and Carbon Black: $\Delta H_R = 619.53 \pm 25.12\ J/g$ resin ($10.8 \pm 4.19\ kJ/gmol$ resin) b. Novakup: $\Delta H_R = 703.25 \pm 12.56\ J/g$ resin ($122.23 \pm 2.51\ kJ/gmol$ resin)

(*continued*)

Table 2.6 (*continued*)

Authors/systems	Kinetic model and experimental conditions	Kinetic parameters
Author: J.M. Barton [54] (1980) *System*: Resin: SHELL Epikote 828 Curing Agent: di(4-aminophenyl)sulphone (Koch-Light purissimum grade)	1. Kinetic Model: $\frac{\partial\alpha}{\partial t} = (A_o + A_1\alpha + A_2\alpha^2)(1-\alpha)^n$ $A_i = \exp(b_i - C_i/RT(k))$ 2. Isothermal Experiments: The temperatures were 142°C, 161°C, 180°C, 190°C, and 200°C. The times needed for $\alpha = 1$ were 125 min at $T = 190°C$ and 67 min at 200°C. 3. Dynamic Experiments: The heating rates were 1, 5 and 10°C/min. The glass transition temperature of the cured resins was determined at 20°C/min.	1. Isothermal Experiments: $n = 1$ $B_1 = 15.222$ $C_1 = 10247.24$ $B_1 = 2.568$ $C_1 = 4080.77$ $B_1 = 1.745$ $C_1 = 3534.08$ 2. Dynamic Experiments: $\Delta H_R = 324.96 \pm 25.45$ J/mol $T_g = 138°C$ at $T = 161°C$ $T_g = 168°C$ at $T = 180°C$ $T_g = 183°C$ at $T = 190°C$ $T_g = 189°C$ at $T = 200°C$
Authors: Woo Il Lee et al. [118] (1982) *System*: Resin: Hercules 3501-6 prepreg Composition in phr: Epoxide #1: 100 Curing Agent (DDS): 44 Epoxide #2: 16 Epoxide #3: 15 BF_3: 2	1. Kinetic Model: a. $\alpha \leq 0.3 \rightarrow$ Autocatalytic (with $m = n = 1$) corrected by the factor: $(B - \alpha)$, b. $\alpha > 0.3 \rightarrow$ 1st order 2. Isothermal Experiments: The temperature varied from 127 to 202°C, with $\Delta T = 25°C$ 3. Dynamic Experiments: The temperature varied from 47 to 317°C at a heating rate of 20°C/min.	1. Dynamic Experiments: $\Delta H_R = 474.27 \pm 5.44$ J/g 2. Isothermal Experiments: a. Autocatalytic Model: $A_1 = 2.101 \times 10^9$ min^{-1} $E_{a1} = 80.71$ kJ/mol $A_2 = -2.014 \times 10^9$ min^{-1} $E_{a2} = 77.90$ kJ/mol $B = 0.47 \pm 0.07$ b. 1st Order Model: $A_3 = 1.960 \times 10^5$ min^{-1} $E_{a3} = 56.68$ kJ/mol

Authors: J. Mijovic et al. [56] (1984) *System 1*: Resin: TGDDM Curing Agent: DDS Composition: Formulation 1: 23 phr Formulation 2: 37 phr	1. Kinetic Model: Autocatalytic 2. Isothermal Experiments: The cure was not complete at temperatures below 205°C for Formulation 2. 3. Dynamic Experiments: The temperature varied from 150 to 300°C at a heating rate of 10°C/min.	1. Isothermal Experiments: a. Formulation 1 $A_1 = 2.22 \times 10^8\ \text{min}^{-1}$ $E_{a1} = 91.21\ \text{kJ/gmol}$ $A_2 = 1.21 \times 10^6\ \text{min}^{-1}$ $E_{a2} = 66.52\ \text{kJ/gmol}$ $0.5 \leq m \leq 0.7$ $m + n = 2$ b. Formulation 2 $A_1 = 9.77 \times 10^{12}\ \text{min}^{-1}$ $E_{a1} = 133.20\ \text{kJ/gmol}$ $A_2 = 6.13 \times 10^4\ \text{min}^{-1}$ $E_{a2} = 49.77\ \text{kJ/gmol}$ $0.5 \leq m \leq 0.7$ $m = n = 2$ 2. Dynamic Experiments: Formulation 1: $\Delta H_R = 725.48\ \text{J/g}$ Formulation 2: $\Delta H_R = 581.69\ \text{J/g}$
Authors: Antonio Moroni et al. [2] (1986) *System*: Resin: a mixture of DGEBA (Shell Epon 826), 80%, and diglycidyl ether of 1,4-butanediol (DGBE) (Celanese's Epirez 5022), 20% Curing Agent: mixture of aromatic di- and polyamines (metaphenylenediamine, 42%, methylenediaminiline, 42%, 2,4-bis(aminobenzyl)aniline and oligomers, 16%)	1. Kinetic Model: a. Isothermal: Autocatalytic b. Dynamic: *n*th order 2. Isothermal Experiments: The temperature varied from 70 to 150°C with the following curing times: $t_{cure} \cong 75$ min at $T = 100$°C with $\alpha_u = 0.95$ $t_{cure} \cong 70$ min at $T = 110$°C with $\alpha_u = 1$ $t_{cure} \cong 60$ min at $T = 120$°C	1. Isothermal Experiments: a. Formulation 1: $A_1 = 1.03 \times 10^6\ \text{min}^{-1}$ $E_{a1} = 56.51\ \text{kJ/mol}$ $A_2 = 7.26 \times 10^5\ \text{min}^{-1}$ $E_{a2} = 49.81\ \text{kJ/mol}$ $1.1 \leq n \leq 1.3$ $m + n = 2$ $T_g = 105$°C b. Formulation 2: $A_1 = 1.16 \times 10^6\ \text{min}^{-1}$

(continued)

Table 2.6 ***(continued)***

Authors/systems	Kinetic model and experimental conditions	Kinetic parameters
	$t_{cure} \cong 45\,\text{min}$ at $T = 130°C$ $t_{cure} \cong 30\,\text{min}$ at $T = 140°C$ 3. Dynamic Experiments: The temperature varied from −60 to 300°C at a heating rate of 10°C/min. 4. Two amine/epoxy ratios were used: Formulation 1: A/E = 1.0 Formulation 2: A/E = 1.1	$E_{a1} = 56.93\,\text{kJ/mol}$ $A_2 = 9.63 \times 10^5\,\text{min}^{-1}$ $E_{a2} = 50.23\,\text{kJ/mol}$ $1.1 \leq n \leq 1.3$ $m + n = 2$ $T_g = 115°C$ 2. Dynamic Experiments: a. Formulation 1: A is a function of the heating rate: $A = 3.69 \times 10^{11}\,\text{min}^{-1}$ at $dT/dt = 2°C/\text{min}$ $A = 1.40 \times 10^{11}\,\text{min}^{-1}$ at $dT/dt = 8°C/\text{min}$ $A = 1.66 \times 10^{11}\,\text{min}^{-1}$ at $dT/dt = 4°C/\text{min}$ $A = 1.02 \times 10^{11}\,\text{min}^{-1}$ at $dT/dt = 12°C/\text{min}$ $A = 5.48 \times 10^{10}\,\text{min}^{-1}$ at $dT/dt = 16°C/\text{min}$ $A = 4.98 \times 10^{10}\,\text{min}^{-1}$ at $dT/dt = 20°C/\text{min}$ $E_a = 93.35 \pm 0.84\,\text{kJ/mol}$ $n = 1.45$

Authors: C. Han et al. [191] (1986) *System 1*: Resin: Epon 9302 Curing Agent: Epon 9350 Fiber: Carbon *System 2*: Epon 9302/Epon 9350/Fiberglass	1. Kinetic Model: Autocatalytic 2. Isothermal Experiments: The temperatures used were 70, 80, and 90°C. 3. Dynamic Experiments: The heating rate used was 10°C/min.	1. Isothermal Experiments: System 1: $m = 0.91$ $n = 1.09$ $A_1 = 1.336 \times 10^{10}\ \mathrm{min}^{-1}$ $E_{a1} = 83.80\ \mathrm{kJ/mol}$ $A_2 = 1.685 \times 10^{9}\ \mathrm{min}^{-1}$ $E_{a2} 66.22\ \mathrm{kJ/mol}$ System 2: $m = 0.93$ $n = 1.07$ $A_1 = 1.165 \times 10^{8}\ \mathrm{min}^{-1}$ $E_{a1} = 69.95\ \mathrm{kJ/mol}$ $A_2 = 2.704 \times 10^{7}\ \mathrm{min}^{-1}$ $E_{a2} = 54.25\ \mathrm{kJ/mol}$
Authors: M.R. Dusi et al. [120] (1987) *System*: Fiberite 976 Resin: TGDDM (MY 720) Curing Agent; DDS	1. Kinetic Model; Autocatalytic 2. Isothermal Experiments: The temperature varied from 140 to 170°C with $\Delta T = 10$°C/min. 3. Dynamic Experiments: The heating rate used was 2°C/min. The ultimate degrees of cure were 0.71, 0.75, 0.80, and 0.84 at 140, 150, 160, and 170°C, respectively.	1. Isothermal Experiments: $A_1 = 2.64 \times 10^{5}\ \mathrm{min}^{-1}$ $E_{a1} = 62.58\ \mathrm{kJ/mol}$ $A_2 = 4.23 \times 10^{5}\ \mathrm{min}^{-1}$ $E_{a2} = 56.89\ \mathrm{kJ/mol}$ $m = 1.03 \pm 0.04$ $n = 1.22 \pm 0.09$ 2. Dynamic Experiments: $\Delta H_R = 530.74\ \mathrm{J/g}$
Authors: J.M. Kenny et al. [32] (1989) *System*: Resin: Hercules and Hexcel, both commercial grade, essentially based on TGDDM Curing Agent: DDS Fiber: Carbon, $\phi_f = 0.6$	1. Kinetic Model; *n*th order 2. Dynamic Experiments: The heating rates varied from 5 to 10°C/min. 3. The processing temperature is 180°C.	1. Hercules; $\log \kappa = 30.04\ (\log(\mathrm{s}^{-1}))$ $E_a = 137.89\ \mathrm{kJ/mol}$ $\Delta H_R = 473.02\ \mathrm{J/g}$ $n = 1.73$ $T_{go} = -3$°C 2. Hexcel: $\log \kappa = 13.45\ (\log(\mathrm{s}^{-1}))$ $E_a = 78.40\ \mathrm{kJ/mol}$ $\Delta H_R = 550.04\ \mathrm{J/g}$ $n = 0.87$ $T_{go} = -3$°C

(*continued*)

Table 2.6 ***(continued)***

Authors/systems	Kinetic model and experimental conditions	Kinetic parameters
Authors: E.M. Woo and J.C. Seferis [303] (1990) *System 1:* Resin: DGEBA (Shell Epon 828) Curing Agent: Trimellitic Anhydride or TMA (Shell Epon 9150) (35–41 phr) *System 2*: Resin: DGEBA Curing Agent: Shell Epon 9150 Accelerator: Onium Salt	1. Kinetic model: nth order 2. Isothermal Experiments: The temperature varied from 100 to 180°C with $\Delta T = 20°C$. 3. Dynamic Experiments: The heating rate used was 10°C/min. For System 1, $\alpha_u \cong 0.95$ and 0.97 at $T = 130°C$ (7 min) and 155°C (10 min), respectively.	1. Isothermal Experiments: a. System 1: $n = 2$ $A = 1.3325 \times 10^5\ min^{-1}$ $E_a = 45.63$ kJ/mol b. System 2: i) For $\alpha < 0.3$: $n = 2$ $A = 1.9879 \times 10^5\ min^{-1}$ $E_a = 48.14$ kJ/mol ii) For $\alpha > 0.3$: $n = 1$ $A = 4.4241 \times 10^5\ min^{-1}$ $E_a = 51.49$ kJ/mol
Authors: J.S. Shim et al. [62] (1991) *System*: Resin: TGAP (EPON HPT 1071) Curing Agent: α, α'-bis(3,5-dimethyl-4-aminophenyl)-p-diisopropylbenzene or DAP (EPON HPT 1062) Composition: DAP/TGAP= 1	1. Kinetic Model: Autocatalytic 2. Isothermal Experiments: The temperature varied from 195 to 225°C with $\Delta T = 10°C$.	1. Isothermal Experiments: $A_1 = 2.43 \times 10^4\ min^{-1}$ $E_{a1} = -93.01$ kJ/mol $A_2 = 1.38 \times 10^5\ min^{-1}$ $E_{a2} = 54.29$ kJ/mol $0.85 \leq m \leq 0.95$ $m + n = 2$
Authors: J.S. Shim et al. [61] (1991) *System*: Resin: N,N,N′N′-tetraglycidyl-α, α'-bis(4-aminophenyl)-*p*-diisopropylbenzene or TGAP (EPON HPT 1071, Shell Chemical Company) Curing Agent: DDS (HT 976, Ciba-Geigy Ltd.) Compositions: a. System 1: DDS/TGAP = 1 b. System 2: DDS/TGAP = 0.71 c. System 3: DDS/TGAP = 0.54	1. Kinetic Model: Autocatalytic 2. Isothermal Experiments: The temperature varied from 195 to 225°C with $\Delta T = 10°C$. 3. Dynamic Experiments: The heating rates used were in the range of 2.5–20°C/min.	1. Isothermal Experiments: a. System 1 $A_1 = 8.14 \times 10^5\ min^{-1}$ $E_{a1} = 68.61$ kJ/mol $A_2 = 7.28 \times 10^5\ min^{-1}$ $E_{a2} = 61.41$ kJ/mol $m = 0.75$ $n \cong 0.25$ b. System 2 $A_1 = 1.70 \times 10^6\ min^{-1}$ $E_{a1} = 73.80$ kJ/mol $A_2 = 5.03 \times 10^7\ min^{-1}$ $E_{a2} = 81.24$ kJ/mol

		$m \cong 0.63$ $n \cong 1.37$ c. System 3 $A_1 = 5.44 \times 10^6\ \text{min}^{-1}$ $E_{a1} = 80.62\ \text{kJ/mol}$ $A_2 = 3.09 \times 10^8\ \text{min}^{-1}$ $E_{a2} = 91.05\ \text{kJ/mol}$ $m \cong 0.56$ $n \cong 1.44$ 2. Dynamic Experiments: $\Delta H_R = 490.01\ \text{J/g}$ for System 1 $\Delta H_R = 557.28\ \text{J/g}$ for System 2 $\Delta H_R = 598.97\ \text{J/g}$ for System 3
Authors: J. Nam and J.C. Seferis [68] (1993) System: TORAYCA T800H/3900-2 prepreg with 35 wt% resin content Resin: TGDDM Curing Agent: DDS Fibers: 145 g/m^2 fiber areal weight	1. Kinetic Model: $\frac{\partial \alpha}{\partial t} = \kappa(a+b)$ $a = y(1-\alpha)$ $b = (1-y)\alpha^m(1-\alpha)^n$ $\kappa = A\exp(-E_a/RT)$ 2. Isothermal Experiments: The temperature varied from 160 to 185°C with $\Delta T = 5$°C. 3. Dynamic Experiments: The heating rates used were 1.02, 2.04, 5.06, 10.09, and 202.31°C/min.	1. Isothermal Experiments: $A = 2.785 \times 10^5\ \text{min}^{-1}$ $E_a = 56.47\ \text{kJ/mol}$ $m = 0.55$ $n = 1.19$ $y = 0.27$ 2. Dynamic Experiments: $\Delta H_R = 142.20 \pm 7\ \text{J/g}$
Authors: U. Khanna and M. Chanda [67] (1993) *System*: Resin: Bis-(2,3-epoxyropyl)-1,3-benzenedicarboxylate (IPDGE)	1. Kinetic Model: Autocatalytic, with $\kappa_1 = 0$, multiplied by a diffusional factor [Eqs. (2.23) and (2.24)]. 2. Isothermal Experiments:	1. Isothermal Experiments: $E_{a2} = 76.19\ \text{kJ/mol}$ $m \cong 0.5$ $n \cong 1.5$

(*continued*)

Table 2.6 (*continued*)

Authors/systems	Kinetic model and experimental conditions	Kinetic parameters
Curing Agent: Hexahydrophthalic anhydride (HHPA) Catalyst: Benzenedimethylamine (BDMA) (0.10; 0.25 and 0.50 phr)	The temperatures varied from 100 to 120°C, with $\Delta T = 5$°C. The ultimate degree of cure was approximately 0.80 for 0.50 phr of BDMA at 100°C and $t_{cure} = 40$ min. 3. Dynamic Experiments: The temperature varied from room temperature to 270°C at a heating rate of 10°C/min. Decomposition of the resin system was observed at temperatures higher than 270°C.	$\alpha_c = 0.830$; 0.747 and 0.837 for 0.10; 0.25, and 0.50 phr of BDMA, respectively. The coefficient C, defined by Equation 18, was an undetermined function of temperature. 2. Dynamic Experiments: $\Delta H_R = 291.35$ J/g; 323.16 J/g, and 313.53 J/g for 0.10, 0.25, and 0.50 phr of BDMA, respectively.
Authors: B.-G. Min et al. [65] (1993) *System*: Resin: DGEBA (Shell Epilote 8283) Curing Agent: DDS (Anchor Chemicals)	1. Kinetic Model; nth-order (valid for $\alpha < 0.7$ and autocatalytic (valid for $\alpha < 0.7$) 2. Isothermal Experiments: The curing temperatures were 130 and 205°C.	1. Isothermal Experiments: a. nth-order Model: $A = 7.313 \times 10^4$ min^{-1} $E_a = 54.08$ kJ/mol $n = 0.914$ at $T = 130$°C $n = 0.833$ at $T = 205$°C b. Autocatalytic Model: $A_1 = 6.617 \times 10^5$ min^{-1} $E_{a1} = 54.08$ kJ/mol $A_2 = 3.278 \times 10^5$ min^{-1} $E_{a2} = 65.09$ kJ/mol $m = 1$ $n = 1$
Authors: H. Stutz et al. [304] (1993) *System*: Resin: DGEBA (Shell Epikote 828) Curing Agent: DDS (Ciba-Geigy)	1. Kinetic Model: Autocatalytic multiplied by the following factor: $\exp\left[-\frac{E_s}{R}\right]\left(\frac{1}{T - T_o} - \frac{1}{T - T_o^i}\right)$ with $T_o = T_g - 45(°C)$ $T^i = T \quad - 45(°C)$	1. Isothermal Experiments: $A_1 = 6.7 \times 10^5$ min^{-1} $E_{a1} = 64.09$ kJ/mol $A_1 = 4.7 \times 10^5$ min^{-1} $E_{a2} = 59.06$ kJ/mol $m = 1$

	$T_g = Q[T_{g\infty} - K_1(1-\alpha)]$ $Q = \left(1 + K_2 \frac{\alpha_c}{1-\alpha_c}\right)$ $\alpha_c = (2\alpha^3(1-\alpha) + \alpha^4)/3$ 2. Isothermal Experiments: The curing temperature used was 120°C. 3. Dynamic Experiments: The heating rate used was 20°C/min.	$n = 2$ $E_s = 300.43$ J/mol 2. Dynamic Experiments: $T_{g\infty} = 110°C$ $K_1 = -156°C$ $K_2 = 0.777$
Authors: C.C. Su and E.M. Woo [5] (1995) *System*: Resin: TGDDM (MY 720, Ciba-Geigy) and poly(ether imide) (PEI, GE Ultem-1010) Curing Agent: DDS (HT976, Ciba-Geigy) Composition: 10, 20, 30, and 50 phr of PEI and 35 phr of DDS	1. Kinetic Model: Autocatalytic 2. Isothermal Experiments: The temperatures used were 177, 187, and 197°C. 3. Dynamic Experiments: The temperature varied from 40°C to 300°C with a heating rate of 10°C/min.	1. Isothermal Experiments: $m \cong 0.5$ $n \cong 1.4$ a. Neat Resin $A_1 = 1.0854 \times 10^7\ \text{min}^{-1}$ $E_{a1} = 80.33$ kJ/mol $A_2 = 5.4036 \times 10^5\ \text{min}^{-1}$ $E_{a2} = 60.99$ kJ/mol b. TGDDM-DDS/PEI (10 phr) $A_1 = 1.2057 \times 10^5\ \text{min}^{-1}$ $E_{a1} = 53.87$ kJ/mol $A_2 = 4.4241 \times 10^5\ \text{min}^{-1}$ $E_{a2} = 58.69$kJ/mol c. TGDDM-DDS/PEI (20 phr) $A_1 = 8.0822 \times 10^4\ \text{min}^{-1}$ $E_{a1} = 51.78$ kJ/mol $A_2 = 5.9720 \times 10^5\ \text{min}^{-1}$ $E_{a2} = 61.20$ kJ/mol

(*continued*)

Table 2.6 (*continued*)

Authors/systems	Kinetic model and experimental conditions	Kinetic parameters
		d. TGDDM-DDS/PEI (30 hr) $A_1 = 5.9874 \times 10^4\ \text{min}^{-1}$ $E_{a1} = 50.69\ \text{kJ/mol}$ $A_2 = 5.9720 \times 10^5\ \text{min}^{-1}$ $E_{a2} = 59.69\ \text{kJ/mol}$ e. TGDDM-DDS/PEI (50 phr) $A_1 = 8.9322 \times 10^4\ \text{min}^{-1}$ $E_{a1} = 52.49\ \text{kJ/mol}$ $A_2 = 6.6000 \times 10^5\ \text{min}^{-1}$ $E_{a2} = 61.58\ \text{kJ/mol}$ 2. Dynamic Experiments: $\Delta H_R = 550.88\ \text{J/g}$ For TGDDM-DDS/PEI 10 phr, $\alpha_u = 0.84$, 0.90 and 0.95 at $T = 177°\text{C}$ (100 min), 187°C (75 min) and 197°C (50 min), respectively. $T_{go} = 3.3$, 5.5, 6.5, 10.9, and 17.0°C for TGDDM-DDS/PEI 0, 10, 20, 30, and 50 phr, respectively. $T_{g\infty} = 189.6$, 190.6, 191.5, 192.4°C for PEI phase with TGDDM-DDS/PEI 10, 20, 30 and 50 phr, respectively. $T_{g\infty} = 223.3$, 219.4, 218.9, 218.1, and 216.9°C for Epoxy phase with TGDDM-DDS/PEI 10, 20, 30, and 50 phr, respectively.
Authors: L. Barral et al. [72] (1995) *System*: Resin: *DGEBA* (Shell Epikote 828) Curing Agent: 1,3-bisaminomethylcyclohexane (1,3-BAC) (Mitsubishi Gas Chemical Co.)	1. Kinetic Model: Autocatalytic, corrected by diffusion factor, Equation 2.23 2. Isothermal Experiments: The temperature varied from 60 to	1. Isothermal Experiments: $A_1 = 5.94 \times 10^3\ \text{s}^{-1}$ $E_{a1} = 43.99\ \text{kJ/mol}$ $A_2 = 1.36 \times 10^6\ \text{s}^{-1}$ $E_{a2} = 57.01\ \text{kJ/mol}$

		$m + n = 2.5–3.0$ $n \cong 2.0$ 2. Dynamic Experiments: $\Delta H_R = 495.50\,J/g$ $\alpha_u = 0.7$ at $T_{cure} = 1800\,s$ for $T = 60°C$ $\alpha_u \cong 0.8$ at $t_{cure} = 1200\,s$ for $T = 80°C$ $\alpha_u = 0.9$ at $t_{cure} = 750s$ for $T = 100°C$ $\alpha_u = 0.9$ at $t_{cure} = 550\,s$ for $T = 110°C$
Author: Rajesh Srinivasan et al. [126] (1995) *System*: Commercial RTM Resins—BMI 5250-4-RTM (from BASF) and PR 500 (from 3M)	1. Kinetic Model: a. BMI 5250-4-RTM: Autocatalytic corrected by the introduction of the ultimate conversion, equation (2.31), given by: $\alpha_u = aT(K) + b$ b. PR500; *n*th order and autocatalytic with $\kappa_1 = 0$ (for the bell shaped part of the curve) for $T \leq 145°C$ and autocatalytic for $T > 155°C$ 2. Isothermal Experiments: a. BMI 5250-4-RTM: the temperature range used was 140–175°C. b. PR 500: the temperature range used was 100–185°C. 3. Dynamic Experiments: a. BMI 5250-4-RTM: the temperature varied from room temperature to 370°C at a heating rate of 1 and 2°C/min. b. PR500: the temperature varied from room temperature to 300°C at a heating rate of 2°C/min.	1. Isothermal Experiments: a. BMI 5250-4-RTM $A_1 = 0.815\,s^{-1}$ $E_{a1} = 29.30\,kJ/mol$ $A_2 = 40.2\,s^{-1}$ $E_{a1} = 37.67\,kJ/mol$ $m = 1 n = 2$ $a = 5.55 \times 10^{-3}$ $b = -1.76$ b. PR 500 for $T \leq 145°C$ i) linear part of the curve: $n = 6.35$ $A = 2.623 \times 10^{-10}\,s^{-1} E_a = 23.23$ kJ/mol ii) bell shaped part of the curve $A_2 = 5.42 \times 10^{-3}\,s^{-1}$ $E_{a2} = 51.36\,kJ/mol$ $m = 2.28$ $n = 1.90$ c. PR500 for $T \geq 155°C$ $A_1 = 3.9017 \times 10^6\,s^{-1}$ $E_{a1} = 87.91\,kJ/mol$ $m = 0.755$ $n = 0.908$

combined with an unsaturated and reactive monomeric diluent (most frequently styrene that has a low cost) to give a low viscosity, liquid resin that possesses enhanced processability [48], and to cross-link with the polyester backbone [9]. Unlike polyesters, they have only terminal unsaturation. Vinyl esters offer better chemical resistance and balance of properties than do typical unsaturated polyesters [6], but they have higher shrinkage than epoxies which can be prevented through the use of additives [88].

The cure of vinyl ester resins is a highly exothermic process (with the reaction heat being proportional to the number of double bonds [199]) and occurs mostly through a free radical cross-linking reaction between the vinyl group of vinyl ester and styrene [88]. This reaction takes place in the presence of organic peroxides, the most common type of catalyst used, which decompose under heat, such as methyl ethyl ketone peroxide (MEKP), or benzoyl peroxide (BPO) [48]. Accelerators (dimethyl aniline, DMA) or promoters (cobalt naphthenate, CoNap) are frequently used along with peroxides. The acceleration of the breakdown of the catalyst [6,48] is the consequence of the generating free radicals. In order to avoid cure at room temperature and increase the shelf-life of the resin, inhibitors are usually added to retard the polymerization [48] and the curing reaction is activated only when the inhibitor is exhausted. The time necessary for it to happen is called *induction time* [30]. The heat from the cure reaction for vinyl ester is much higher than that for epoxy [9]. As a result, the possibility of high thermal gradients and residual stresses when making thick composite materials is inevitable unless one can control the curing process in an intelligent way.

Because of all these minor components (e.g., catalysts and inhibitors, added to major ones) the cure of vinyl ester resins is very complex, involving many competitive reactions. There are some new variables to account for, such as the inhibitor and initiator concentrations and induction time. Several papers [81,96,200,201] use the mechanistic approach, claiming that the phenomenological models do not explicitly include these facts, resulting in a new parameter characterization after each change in resin formulation [96]. Despite these arguments, the phenomenological approach is the most widely used and is based on an autocatalytic model which has been successfully applied to epoxy resins. Many authors [30,34,74,199,202,203] proposed the Equation 2.30 to describe the cure kinetic of unsaturated polyesters:

$$\frac{\partial \alpha}{\partial t} = \kappa_2 \alpha^m (1 - \alpha)^n \tag{2.30}$$

in which the initial cure rate, $\partial\alpha/\partial t = \kappa_1$, is zero [33] and $m + n = 2$ [33,34,199,204,205]. Han and Lem [205] found that κ_1 was different from zero. According to them, it took place because of the fast reaction at the beginning of the cure, provoked by promoters at low temperature (60°C or lower). In the absence of promoters, κ_1 was zero. Since the cure of vinyl esters is diffusion-controlled even at early stages of reaction [33,123,201], partial resin cure

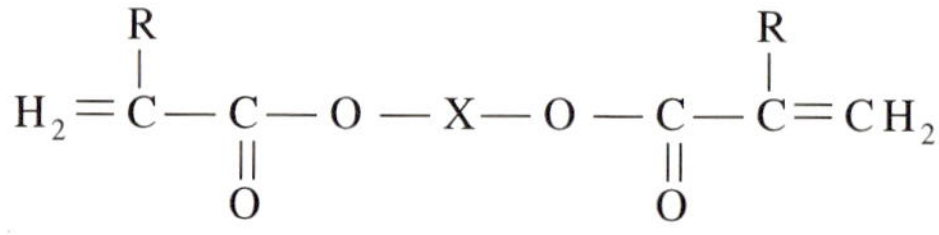

Figure 2.10 General structural formula of vinyl esters

can result. It means that it is very difficult to achieve 100 percent conversion. Because of this, Lam et al. [98] modified Equation 2.30, by introducing a new parameter, α_u, which represents the ultimate conversion; therefore, we have Equation 2.31:

$$\frac{\partial\alpha}{\partial t} = \kappa_2\alpha^m(\alpha_u - \alpha)^n \tag{2.31}$$

Lam et al. [98] verified that the ultimate conversion varies with the cure temperature and the polymer type, but no function was presented. For vinyl esters (DERAKANE 411-45), it is practically constant with $\alpha_u = 0.96$, approximately. Michaud empirically found, however, that α_u changes with temperature as shown by Equation 2.32:

$$\alpha_u = 0.439 + 0.00446T \text{ (in °C)} \tag{2.32}$$

This equation is valid for temperatures between 55 and 105°C. For DSM resin, Kenny et al. [30] presents a linear dependence of α_u on the isothermal cure temperature lower than the glass transition temperature of the fully cured resin, given by Equation 2.33:

$$\alpha_u = -3 + 0.01164T \text{ (in K)} \tag{2.33}$$

for curing temperature between 30°C and 60°C.

Table 2.7 shows some of the kinetic models for vinyl ester and some polyester systems, where one can notice the large differences among the kinetic parameters even for the same resin and range of temperatures. When a reinforced material is added to DERAKANE resin, for example, Palmese [206] found that the heat of reaction, frequency factor, and order of reaction, m, are 63 percent, 295 percent, and 16.5 percent higher, respectively, and that the activation energy as well as the order of reaction, n, are 26.5 percent and 22 percent lower, respectively, than are those from the paper of Lam et al. [98].

2.4.3 Phenolics

Phenolic resins are considered the first fully synthetic polymer materials. It was first patented date in 1899 (A. Smith) [207]. Since then, they have been applied in many commercial applications, such as household and other appliances (e.g., dishwasher, air conditioners, coffee machines, toasters, refrigerators, and iron and pan handles), coatings, electrical engineering (light sockets, switches, transformer components, blower wheels, relays, and connectors), automotive industry (distributor caps, coil towers, commutators, fuse blocks, bulk heads, connectors, and brake components), manufacturing of foam for thermal and acoustic insulation, because of their excellent thermal stability and fire, high temperature, and chemical resistances, the ease and rapidity with which they cure [208], good electrical properties, low cost, high surface hardness [207], low smoke generation, and toxicity [6]. The phenolic-based composites are used extensively as ablative surfaces in rocket nozzles, exit cones and entry vehicles [209].

Phenolics is the general term for representing the product of the reaction between phenol and aldehyde. Although phenol and formaldehyde (high reactivity) [210] are most common,

Table 2.7 Cure Kinetic Models—Vinyl Esters

Authors/systems	Kinetic model and experimental conditions	Kinetic parameters
Authors: M.R. Kamal and S. Sourour [52] (1973) *System*: Resin: Polyester (1 mole phthalic anhydride + 1 mole maleic anhydride + 2.2 moles propylene glycol)/Styrene Catalyst: 60% solution of methyl ethyl ketone peroxide (MEKPO) in dimethyl phthalate (DMP) (Reichold Chemicals Inc.)	1. Kinetic Model: Autocatalytic with $\kappa_1 = 0$ 2. Isothermal Experiments: The temperature varied from 62°C to 127°C. 3. The rate of heat evolution was approximately 0.42 J/gmol s at 60 min for $T = 81$°C and 8.37 J/gmol s at 6.5 min for $T = 118$°C. 4. Dynamic Experiments: The temperature varied from 57°C to 177°C at heating rates of 1.25, 2.5, 5.0, and 10°C/min.	1. Isothermal Experiments: $A_2 = 2.39 \times 10^{10}\ \text{min}^{-1}$ $E_{a2} = 27.28$ kJ/mole $m = 0.25$ $n = 1.75$ 2. Dynamic Experiments: $\Delta H_R = 305.58 \pm 8.37$ J/g
Authors: S.Y. Pusatcioglu et al. [34] (1979) *System*: Resin: a general-purpose polyester resin Curing Agent: Styrene monomer (28% by weight) Initiator: a 60% solution of methyl ethyl ketone (MEK) peroxide in dimethyl phthalate, Lupersol DM, (Pennwalt Co. Lucidol Division)	1. Kinetic Model: Autocatalytic with $\kappa_1 = 0$ 2. Isothermal Experiments: The temperatures were 82, 93, 105, and 122°C. At higher temperatures, the cure time was of order of minutes so that the data could not be obtained with precision. The complete cure was not obtained for the range of temperature used: $\alpha_u \cong 0.80$, 0.85, and 0.90 at $t_{cure} \cong 70$, 35, and 20 min, and $T = 82$, 93, and 105°C, respectively. 3. Dynamic Experiments: The temperature varied from room temperature to 250°C at a heating rate of 10°C/min. 4. The reaction was complete at 180°C.	1. Isothermal Experiments: $A_2 = 2.6 \times 10^{9}\ \text{min}^{-1}$ $E_{a2} = 71.33$ kJ/gmol $m = 0.445$ $n = 1.50$ 2. Dynamic Experiments: $\Delta H_R = 309.35 \pm 2.51$ J/g

Authors: C. Han et al. [191] (1986) *System 1*: Resin: Commercial grade of unsaturated polyester (OC-E701) Initiator 1: Percadox 16N and *t*-butyl perbenzoate Fiber: Fiber Glass *System 2*: Resin: OC-3701 Initiator 2: 16N and benzoyl peroxide (BPO) Fiber: Fiber Glass	1. Kinetic Model: Autocatalytic 2. Isothermal Experiments: The temperatures used were 70, 80, and 90°C. 3. Dynamic Experiments: The heating rate was 10°C/min	1. Isothermal Experiments: a. System 1: $m = 0.569$; $n = 1.44$ $A_1 = 3.170 \times 10^{19}\ \text{min}^{-1}$ $E_{a1} = 143.43\ \text{kJ/mol}$ $A_2 = 1.777 \times 10^{12}\ \text{min}^{-1}$ $E_{a2} = 84.05\ \text{kJ/mol}$ b. System 2: $m = 0.58$ $n = 1.42$ $A_1 = 3.412 \times 10^{14}\ \text{min}^{-1}$ $E_{a1} = 107.04\ \text{kJ/mol}$ $A_2 = 5.167 \times 10^{10}\ \text{min}^{-1}$ $E_{a2} = 75.05\ \text{kJ/mol}$ 2. Dynamic Experiments: $\Delta H_R = 200\ \text{J/g}$ for System 2
Authors: P.W.K. Lam et al. [98] (1990) *System*: DERAKANE 411-45 (45 wt% of styrene monomer)	1. Kinetic Model: Equation (26) 2. Isothermal Experiments: The temperatures used were 100, 110, 115, and 120°C. At 110°C, the ultimate degree of cure was 0.95 after approximately 30 min and the induction time was equal to 10 min. 3. Dynamic Experiments: The temperature varied from room temperature to 250°C at heating rates of 2.5, 5.0, 10, 12.5, and 15°C/min.	1. Isothermal Experiments: $A_2 = 2.435 \times 10^{10}\ \text{min}^{-1}$ $E_{a2} = 104.82\ \text{kJ/mol}$ $m = 0.85$ $n = 1.15$ 2. Dynamic Experiments: $\Delta H_R = 2760.37\ \text{J/g}$

(*continued*)

Table 2.7 ***(continued)***

Authors/systems	Kinetic model and experimental conditions	Kinetic parameters
Authors: S.R. White and C. Kim [172] (1993) *System*: CYCOM 4102 (E-glass/polyester)	1. Kinetic Model: Autocatalytic with $\kappa_1 = 0$ 2. Dynamic Experiments: A heating rate of 20°C/min was used until no further exotherm was observed (350°C).	1. Isothermal Experiments: $A_2 = 3.7 \times 10^{22}\,\text{min}^{-1}$ $E_{a2} = 167.44\,\text{kJ/mole}$ $m = 0.524$ $n = 1.476$ 2. Dynamic Experiments: $\Delta H_R = 77.61\,\text{J/g}$
Authors: J.H. Lee and J.W. Lee [33] (1994) *System*: Resin: DERAKANE 411-45 (Dow Chemical Co.) Initiator: Benzoyl peroxide (BPO) (Fluka Chemie AG) with a concentration of 1 phr	1. Kinetic Model: - $\log\left(\frac{\alpha - u}{\alpha} - 1\right) = \left(\frac{1}{m-1}\right)\log(1-m)\kappa A_t + \left(\frac{1}{m-1}\right)\log\left(\frac{t-t_i}{A_t}\right)$ with t_i as an induction time and A_t as the time shift factor, given by: $\log(A_t) = 0.41772(\Delta 1/T(k))$ The term $(t - t_i)/A_t$ was considered as a reduced time. 2. Isothermal Experiments: The temperatures varied from 86.1 to 127.1°C. The ratio $\alpha/\alpha_{\max}$ is equal to 1 at a reduced time of 25 min.	

	3. Dynamic Experiments: The temperature varied from room temperature to 180°C (no more reaction was observed from this temperature) at a heating rate of 10°C/min.	1. Isothermal Experiments: a. for $\alpha/\alpha_u \leq 0.2$ $\kappa = 7.597 \times 10^8\ \mathrm{s}^{-1}$ $m = 0.654$ $n = 0.346$ b. for $\alpha/\alpha_u \geq 0.2$ $\kappa = 1.616 \times 10^9\ \mathrm{s}^{-1}$ $m = 0.806$ $n = 0.194$ 2. Dynamic Experiments: $\Delta H_R = 307.38\ \mathrm{J/g}$
Author: D. Michaud [9] (1996) *System*: Resin; Dow DERAKANE 411-C50 Initiator: Organic peroxide (Witco USP-245), 1.75 wt% Fiber: E-glass, with $\phi_f \cong 0.43$	1. Kinetic Model: Equation (26) 2. Isothermal Experiments: The temperatures used varied from 55 to 105°C. Palmese [206] worked with the same system using temperatures from 90 to 120°C. 3. Dynamic Experiments: The system temperature was ramped to 200 °C at 5 and 10 °C/min.	1. Isothermal Experiments: a. For 55–105 °C $A_2 = 5.50 \times 10^8\ \mathrm{min}^{-1}$ $E_{a2} = 60.61\ \mathrm{kJ/mol}$ $m = 0.85$ $n = 1.15$ b. For 90–120 °C $A_2 = 9.62 \times 10^{10}\ \mathrm{min}^{-1}$ $E_{a2} = 77.40\ \mathrm{kJ/mol}$ $m = 0.66$ $n = 1.34$ 2. Dynamic Experiments: $\Delta H_R = 425\ \mathrm{J/g}$

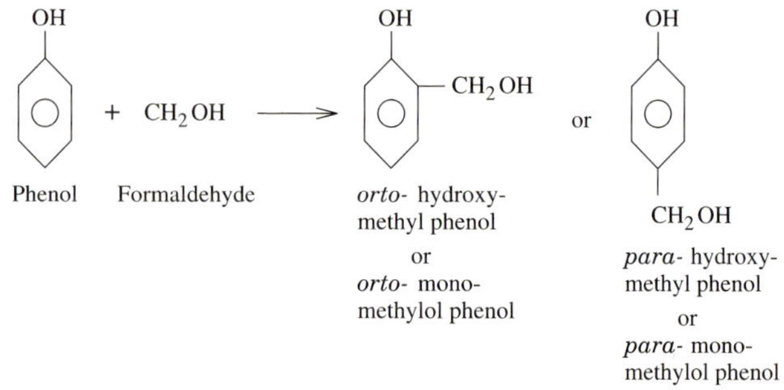

Figure 2.11 Addition reaction of formaldehyde to phenol

other phenols (*m*-cresol, resorcinol) and aldehydes (furfural) can be used [208]. The exothermic reaction, involving phenol and formaldehyde, takes place in these steps:

1. Addition of formaldehyde to phenol (in the positions *ortho* or *para*) to form mono-, di-, and possibly trimethylol phenol [211,212], liberating nearly 20.3 kJ/mol (4.85 kcal/mol) [213] (Figure 2.11). This reaction occurs at temperatures between 100 and 160°C and takes around 2–4 hours. The *para* position is approximately twice more reactive than the *ortho* position [1]. The hydroxymethyl compounds are unstable in acidic media and high temperature, and they are rapidly condensed as outlined in Figure 2.12. They are stable in the basic media, but at high temperature they cross-link by methylene bridges.

2a. Condensation of methylol phenols with a phenolic ring hydrogen to form diaryl methanes (methylene bridges), as shown in Figure 2.12.

2b. Condensation of methylol phenols with themselves (no catalyst is needed) to form dibenzyl ether links (methylene ether bridges), which can be converted into methylene bridges with elimination of formaldehyde [213], according to Figure 2.13.

These condensation reactions (2a and 2b) occur at temperatures above 100°C and characterize the cross-linking (cure) or hardening of the resins. Due to the fact that some volatiles compounds are present, these reactions should take place in equipment under pressure.

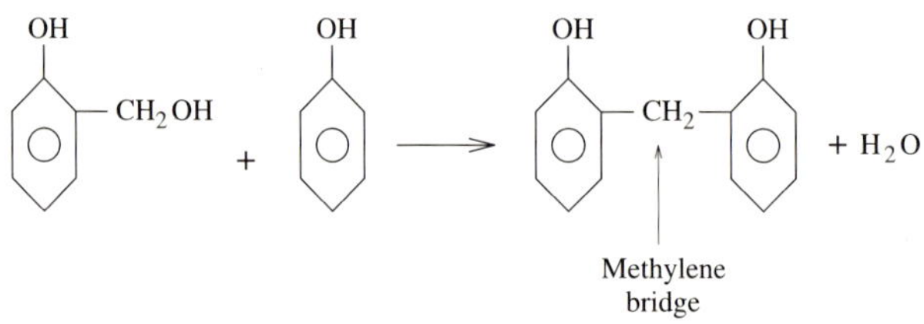

Figure 2.12 Condensation reaction of methylol phenol with a phenol ring hydrogen

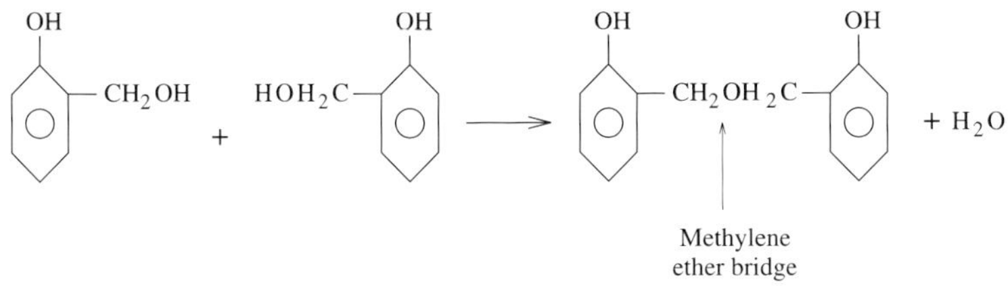

Figure 2.13 Condensation reaction of methylol phenol with themselves

The addition and condensation reactions are very slow and catalysts, acidics, or alkalines, are added to accelerate them [210]. The cure of the phenolic resins, which results in three-dimensional molecules, is extremely complex, involving competing reactions influenced by reaction conditions (e.g., temperature, concentration of reactants, type and concentration of catalysts, condensation time, ratio between the molar concentration of phenol and formaldehyde, pH of the medium, and the structure of the phenol used [reactivity of the different functional groups on the phenol ring]) [207,208,210]. For these reasons, it is understandable that there are no closed equations for describing cure kinetics, similar to those for epoxies and vinyl esters. The literature is scarce and almost all studies present qualitative conclusions [214–218], such as pH is an important parameter for determining the ultimate degree of cure, and the higher the Formaldehyde/Phenol ratio (F/P), the higher the degree of cure of resole resins [219], which is described in the next subsection. The decomposition of phenolics starts at approximately 120–250°C, depending on the molecular structure [213], with water, CO, CO_2, methane, phenol, cresols, and xylenols being eliminated. The phenolic ether resins are stable up to 300°C for a short period. Gel Permeation Chromatography (GPC) [211,214,217,218,220,221], Extraction with Solvent [219], Rockwell Hardness Test [222], and the techniques discussed earlier, such as C 13 n.m.r. [220,223–225], DSC [211], DTA [215,226], and IR [226,227], can monitor the cure of phenolic resins. The most common indexes to characterize the cure of phenolics are viscosity, gel time (or B-time), density, refractive index, solubility, conductivity or dielectric constant, hardness or strength, solvent resistance, and free phenol content [208,228].

With the objective of increasing their properties, the phenolic resins can be modified by addition of fillers (e.g., magnesium oxide), plasticizers, fire retardants, elastomers, and epoxies [228].

Depending on the pH of the reaction environment and the ratio between the molar concentration of formaldehyde (F) and phenol (P), the phenolic resins can be divided into resoles and novolacs.

2.4.3.1 Resoles

Resoles are prepared with excess of formaldehyde, $F/P > 1$, and alkaline catalysts, such as NaOH, KOH, LiOH, $Ba(OH)_2$, $Ca(OH)_2$, and ammonia [208,210] at between room temperature and 100°C [213,228]. They are mostly used for adhesives, castings, binders in the wood products [211,212], interiors of vehicles, decorative laminates, and abrasives [228].

The selection of the F/P ratio is a compromise in balancing mechanical properties (for which F/P = 2.5 is more suitable) and curing states (for which F/P = 1.75 is the optimum). A good ratio, therefore, may be F/P = 2.25 [229].

The cure of resoles can be accomplished by heat or the addition of acids (e.g., hydrochloric, phosphoric, *p*-toluene-sulphonic, and phenolsulphonic acids) for cases where cure should occur at low or room temperatures (e.g., adhesives and casting). They cross-link by either methylene or methylene ether bridges at temperatures between 100 and 180°C [213], preferentially under pressure to avoid bubbles and cavities.

For resoles, unlike novolacs, the reaction between phenol and formaldehyde should be interrupted, by lowering the temperature and adjusting the pH of the solution [208,213], while the resin is still liquid or soluble. The cure can be continued by increasing the temperature and adding acids [208], allowing one to obtain three stages of cured resin, according to Baekeland and Lebach: resoles or A-stage (soluble in alcohols and unstable at room temperature [207]), resitols or B-stage (still fusible, but insoluble in alkalies), and resites or C-stage (insoluble even in strong solvents, melting only above the decomposition temperature [207,208]). The cross-linking of resoles to resites takes a long time [213].

The cure of resoles can be briefly explained by the scheme illustrated in Figure 2.14. Resole resins for RTM contain 7–15% water, which is the by-product from the condensation and is extremely difficult to remove. Curing temperatures above 180°C may cause the thermal and oxidative decomposition of the resin [214].

2.4.3.2 Novolacs

Novolacs are thermoplastic materials, prepared by heating phenol with formaldehyde in the presence of acidic catalysts that are widely accepted in electrical insulators, mechanical parts, shell molds and abrasive wheel binders [222] because of the fast cure, expelled water (superior electrical properties compared to resoles), and good rigidity in the molded parts at elevated temperatures [208]. The ratio F/P must be less than 1 (usually 0.8) [229] in order to avoid cross-linking and gelation during manufacture [213,228].

They are permanently soluble and fusible, necessitating a curing agent to cross-link by the methylene bridges (the only possible kind of linkage), liberating nearly 98.7 kJ/mol (23.6 kcal/mol) [213]. Due to the fact that two stages are needed—preparation of novolac and addition of the curing agent—these resins are known as Two-Stage Resins, unlike resoles, which are known as One-Stage Resins. Hexamethylenetetramine is the most common curing agent, but paraform, quinone, or chloranil, anhydroformaldehyde aniline, and ethylene diamineformaldehyde have been used [208].

The cross-linking reactions of novolacs, which occurs at temperatures around 140–160°C within a few minutes [213], may be schematized as shown in Figure 2.15.

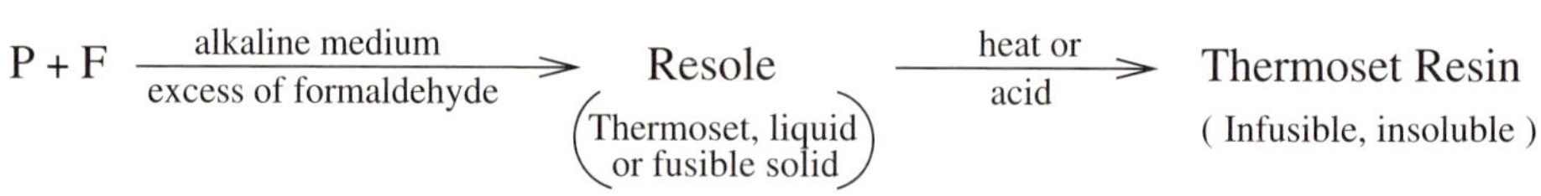

Figure 2.14 Simplified scheme of the cross-linking reactions of resoles

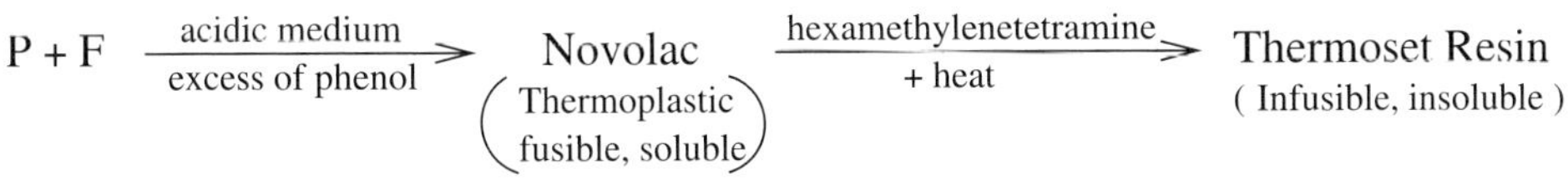

Figure 2.15 Simplified scheme of the cross-linking reactions of novolacs

2.5 The Coupled Phenomena

In all of the manufacturing processes of composite parts mentioned earlier, a resin flows through a preform (or prepreg) placed into a mold. The mold walls can be previously heated, which results in the initiation of resin cure during the filling time, or heated after the filling period. In the first case, a cure model is composed of coupled submodels, resin flow (including viscosity model), heat transfer, and chemical reaction, which have to be solved simultaneously. In the second case, only heat and mass transfer is accounted for to predict the temperature and cure history of the part.

In the following subsections, the balance equations, along with the boundary and initial conditions and their numerical solution, are briefly discussed.

2.5.1 Resin Flow

Several resin flow models are presented in the literature for the RTM [230–235] processes. The resin flow in a preform can be described on both microscopic (unit cell approach) and macroscopic levels [15]. In the former case, flow around an individual fiber or fiber bundle is considered. In the latter case, which has been widely adopted, Darcy's law is used (because Reynolds number is less than 1 for most RTM processes) to model the flow in terms of the permeability of the preform, geometry of the mold as well as the process, and material parameters [15,233].

Darcy's law is expressed by Equation 2.34:

$$\langle \underline{u} \rangle = -\frac{1}{\eta_{eff}} \underline{\underline{K}} \cdot \langle \underline{\nabla} P \rangle \tag{2.34}$$

where $\langle \underline{u} \rangle$ is the volume-averaged velocity vector of the resin, η_{eff} is the effective viscosity of the fluid, $\underline{\underline{K}}$ is the permeability tensor, and $\langle \underline{\nabla} P \rangle$ is the volume-averaged pressure gradient in the fluid, which can be calculated from Equation 2.35, after substituting Equation 2.34 in the continuity equation for incompressible fluid:

$$\underline{\nabla} \cdot \left(\frac{\underline{\underline{K}}}{\eta_{eff}} \cdot \underline{\nabla} P \right) = 0 \tag{2.35}$$

Because of the inherent creeping flow assumption of Darcy's law and a Newtonian behavior of typical resins with initial low molecular weight, the problem is quasisteady state and linear, respectively [231].

The permeability of fiber preforms has proved difficult to measure [236] and many researchers [237–245] have been working on this subject. There is no consensus, however, because of the complex nature of this variable. All microscopic interactions between the fluid and the fiber preform architecture are lumped into this parameter [15]. In addition, a preform is a stack of many layers of diverse materials, and each preform has different geometries or fiber arrangements, offering different resistances to the resin flow [15]. The thickness of most RTM parts has customarily been considered much smaller than the overall dimensions, allowing one to use a two-dimensional version of Equation 2.35 [15,230,232,233]. For thick parts a three-dimensional form of this equation is currently required, which is time consuming because the computational time increases by an order of magnitude with each additional degree of freedom [246]. There are many cases, however, other than a T-joint or change in thickness of a preform [231,247,248], where a two-dimensional simulation will do an adequate job for thin parts.

For a two-dimensional flow, Equation 2.35 can be reduced to Equation 2.36:

$$\frac{\partial}{\partial x}\left(\frac{\bar{K}_{xx}}{\eta_{eff}}\frac{\partial P}{\partial x}\right)+\frac{\partial}{\partial x}\left(\frac{\bar{K}_{xy}}{\eta_{eff}}\frac{\partial P}{\partial y}\right)+\frac{\partial}{\partial y}\left(\frac{\bar{K}_{yx}}{\eta_{eff}}\frac{\partial P}{\partial x}\right)+\frac{\partial}{\partial y}\left(\frac{\bar{K}_{yy}}{\eta_{eff}}\frac{\partial P}{\partial y}\right)=0 \tag{2.36}$$

The effective viscosity, η_{eff}, varies throughout the mold because of its dependence on temperature, degree of cure, and shear rate of the fluid.

By using the finite element/control volume method, Bruschke and Advani [232,233,249] first solved Equation 2.36 for the pressure distribution using the finite element method, and then advanced the flow front using the control volume method. An important feature of this method is that the principal variable for which one solves (i.e., the pressure) is a scalar. This implies that no special techniques are necessary to handle thin-shells, which are nonplanar and curved in three dimensions.

The numerical solution of Equation 2.36, along with the boundary conditions, is given by Advani et al. [15] and Bruschke and Advani [232]. They discretized this equation by using Galerkin finite elements in a flow domain changing with time in order to capture the physics of flow through an arbitrary three-dimensional thin shell geometry [15]. The discretized algebraic equations have to be solved simultaneously with the energy and kinetic equations, which will be presented later, as the viscosity is a function of temperature and degree of cure.

Once the pressure distribution is known, averaged velocities can be calculated from Equation 2.34.

In most polymer reaction molding processes, the filling or flow time is significantly lower than the total cure time [236,250]; therefore, all effects of the curing reaction during mold filling can be neglected. This assumption is valid for RTM processes but not for SRIM, where the reactants are mixed just prior to injection [236]. Another simplification is to ignore the resin flow in the heat and mass balance equations, which reduces greatly the complexity of the mathematical modeling of the process. Details of flow analysis can be found in the book, *Flow and Rheology in Polymer Composites Manufacturing* [15]. To predict cure and temperature of the composite part, however, one must address mass and heat transfer.

2.5.2 Mass Transfer

A mass balance in a differential volumetric element of the resin phase in a porous media, constituted of fibers and resin, results in the following differential equation, Equation 2.37:

$$\frac{\partial\langle c_r\rangle}{\partial t} + \langle \underline{u}\rangle \cdot \underline{\nabla}\langle c_r\rangle = \underline{\nabla} \cdot [(\underline{\underline{D}}_{ef} + \underline{\underline{D}}_d)\underline{\nabla}\langle c_r\rangle] - \langle r_r\rangle \tag{2.37}$$

with $\langle c_r\rangle$ as the volume-averaged volumetric concentration of the resin (resin mass/resin volume), D_{ef} and D_d as the effective and dispersive diffusivity tensors, respectively, and $\langle r_r\rangle$ as the volume-averaged volumetric rate of polymerization. The minus sign for the rate of polymerization is to represent the consumption of the resin during the reaction. The effective diffusivity tensor considers the molecular diffusion and the tortuosity of the porous media [251]. Broyer and Macosko [250] have shown that the time constant for effective diffusion is several orders of magnitude greater than that for reaction in thermosetting polymer systems; therefore, the effective diffusive term in the mass balance may be ignored. Nevertheless, Tucker and Dessenberg [251] claimed that the magnitude of the dispersive term should be evaluated before neglecting it. Studies in this subject are required; the usual approach is to neglect both diffusion terms. As a result, Equation 2.37 simplifies to Equation 2.38:

$$\frac{\partial\langle c_r\rangle}{\partial t} + \langle \underline{u}\rangle \cdot \underline{\nabla}\langle c_r\rangle = -\langle r_r\rangle \tag{2.38}$$

or to Equation 2.39:

$$\langle c_{ro}\rangle\left[\frac{\partial\alpha}{\partial t} + \langle \underline{u}\rangle \cdot \underline{\nabla}\alpha\right] = \langle r_r\rangle \tag{2.39}$$

where α is the degree of cure, defined as given by Equation 2.40:

$$\alpha = \left(1 - \frac{\langle c_r\rangle}{\langle c_{ro}\rangle}\right) \tag{2.40}$$

where $\langle c_{ro}\rangle$ is the initial concentration of the resin.

The term $\partial\alpha/\partial t$ represents the isothermal cure kinetics, models that were reported in earlier sections. A general way to express this rate is to use Equation 2.41:

$$\frac{\partial\alpha}{\partial t} = A\exp\left(-\frac{E_a}{RT}\right)f(\alpha) \tag{2.41}$$

By substituting Equation 2.41 into Equation 2.39, one has Equation 2.42:

$$\langle c_{ro}\rangle\left[A\exp\left(-\frac{E_a}{RT}\right)f(\alpha) + \langle \underline{u}\rangle \cdot \underline{\nabla}\alpha\right] = \langle r_r\rangle \tag{2.42}$$

Equation 2.42 has to be solved along with the resin flow and heat balance equations. The initial and boundary conditions are given by Equations 2.43 and 2.44:

$$\alpha(\underline{x}, 0) = 0 \tag{2.43}$$

$$\alpha(\underline{x}_o, t) = \alpha_{\underline{x}_o} \tag{2.44}$$

The majority of the papers from the literature for RTM process consider the filling time of the mold much lower than the reaction time. Hence, the convective term in Equation 2.42 can be neglected, resulting in a much simplified analysis, Equation 2.45:

$$\langle c_{ro} \rangle A \exp\left(-\frac{E_a}{RT}\right) f(\alpha) = \langle r_r \rangle \tag{2.45}$$

2.5.3 Heat Transfer

Heat transfer in flow through porous media has been analyzed by a variety of authors [251–256]. It should be noted that no universal agreement exists on the analysis of the heat transfer in porous media. Several authors have investigated the role of heat transfer during composite manufacturing processes, using a number of different approaches [235,257,258] that can be grouped into two categories: the two-phase models [235,259–263] and the equilibrium models [12,15,30,32,35,38,40,44,236,264]. In the two-phase model, one takes into account convection between the fiber preform and the fluid. No theoretical means of predicting the heat transfer coefficient between the preform and the resin governing this convection process exists, however, so it has to be determined experimentally for each fluid/fiber preform combination. The equilibrium approach assumes that the fluid and fiber medium have the same temperature at each point. In essence, one assumes the heat transfer coefficient to be infinite [15]. An experimental study [235,265] on flow through fiber mats has shown that the equilibrium model is a valid assumption in slow processes, such as RTM, but not in fast processes such as SRIM [236]. Whitaker [266–268] has analyzed the conditions under which the assumption of local thermal equilibrium will break down, and has established a number of criteria for this in terms of dimensionless variables [236]. Accounting for the thermal equilibrium approach, a heat balance in a differential volumetric element of a system composed of a preform and a resin subject to a curing reaction is given by Equation 2.46:

$$\begin{aligned}(\rho c_p)_c \frac{\partial \langle T \rangle}{\partial t} + (1-\phi_f)(\rho c_p)_r(\langle \underline{u} \rangle \cdot (\underline{\nabla} \langle T \rangle)) &= \underline{\nabla} \cdot [(\underline{\underline{k}}_{cond} + \underline{\underline{k}}_d)\nabla \langle T \rangle] \\ &+ n_{eff} \langle \underline{u} \rangle \cdot \underline{\underline{K}}^{-1} \cdot \langle \underline{u} \rangle + (1-\phi_f)\Delta H_R \langle r_r \rangle (\underline{x}, t)\end{aligned} \tag{2.46}$$

where $\langle T \rangle$ is the volume-averaged equilibrium temperature between the resin and the fibers, $(\rho c_p)_c$ and $(\rho c_p)_r$ are the volumetric heat capacities of the composite and the resin, respectively, $\underline{\underline{k}}_{cond}$ and $\underline{\underline{k}}_d$ are the effective thermal conductivity tensor of the composite and thermal dispersion tensor, respectively, ϕ_f is the fiber volume fraction, ΔH_R is the total heat released per unit mass of resin, and $\langle r_r \rangle(\underline{x}, t)$ is a function representing that part of the temperature change as a consequence of the generation of heat because of the curing mass [269].

Equation 2.46, called herein *coupled equation*, symbolizes the coupled phenomena among heat transfer, flow, and cure. The first term of this equation represents the change in internal energy of the composite, whereas the second term is the contribution of the fluid flow. The first term on the right-hand side represents the heat conduction and dispersion because of the mold walls, and the second is the heat generated because of viscous dissipation. Thermal dispersion is the transport of heat because of local fluctuations in the velocity [236] that is averaged by Darcy's law. For isotropic media, the thermal dispersion tensor has two components: $k_{\parallel}$ and $k_{\perp}$, parallel and perpendicular to the flow directions, respectively [251]. The last term represents the energy generated by the curing reaction, and most researchers [13,32,35–37,40,42,43,119,172,198,270,271] express it as $\rho_c \partial H^*/\partial t$, where $\partial H^*/\partial t$ is the rate of heat generated by the reaction per unit mass of the composite. Equation 2.47 can be obtained from Equation 2.6.

$$\rho_c \frac{\partial H^*}{\partial t} = \rho_r(1-\phi_f)\Delta H_R \frac{\partial \alpha}{\partial t} \tag{2.47}$$

The viscous dissipation is usually neglected and the thermal conductivity and dispersion tensors are interpreted as only one tensor, called the *total effective thermal conductivity tensor*, k_{eff}. On a basis of a scaling analysis, Tucker and Dessenberger [251,264] affirmed that dispersion terms must be included in the heat balance when (see Equations 2.48 and 2.49):

$$Gz = \frac{VH^2}{\alpha_t L} > 1 \tag{2.48}$$

$$\frac{Pe_{eff}}{Gz} = \frac{Ld_p}{H^2} > 1 \tag{2.49}$$

where Gz is the Graetz number, V is a characteristic fluid velocity, H is the half gap height of the mold, L is a characteristic length of the cavity, d_p is the fiber (particle) diameter, Pe_{eff} is the microscopic Peclet number, and α_t is the total thermal diffusivity of the saturated porous medium, given by Equation 2.50:

$$\alpha_t = \frac{k_{cond}}{(\rho c_p)_r} \tag{2.50}$$

The Graetz number is the ratio of the flow direction convection and the transverse conduction [236]. It is the appropriate measure of average convection relative to conduction in a mold filling problem, where the mold thickness is much smaller than the flow length [264]. Pe_{eff} is the ratio between dispersive and conductive transports of heat [264].

By conducting experiments in the RTM process, Dessenberg and Tucker [264] proposed expressions for the perpendicular thermal dispersion coefficient as a function of the local microscopic thermal Peclet number ($Pe = (|\langle \underline{u} \rangle| d_p)/\alpha_r$), for three ranges of Peclet number. The parallel thermal dispersion coefficient was assumed to be zero. They concluded that the experimental data were better described when the thermal dispersion term was included in the heat balance. Despite this fact, the usual approach is not to consider that this term and the upcoming discussion follow this trend, although the present authors recognize the importance of and the need for more studies on the dispersive heat transfer.

2.5.3.1 Numerical Methods

A variety of numerical methods are available to solve the *coupled equation*, Equation 2.46, along with the resin flow, cure kinetic and rheological models discussed earlier. They are finite differences (FD), boundary fitted coordinate system (BFCS) in conjunction with finite differences and finite elements/control volume. The first one is unable to generate a computational grid for complex geometries or for a moving boundary approach [36]. The second one, BFCS/FD, is composed of two steps: the transformation of an irregularly shaped physical domain, x, y, and z, into a regularly (square) shaped domain in the computational space, and the transformation to the computational space of the governing equations, discretized by the finite difference method, followed by their solution [36,231]. In addition to the difficulty in generating the grid, this approach has the disadvantage that the mesh needs to be updated at each time step, involving an additional computational effort [15]. It is also not easy to extend it to multiply connected domains. The finite element method has many features, such as: (1) the ease with which nonuniform and nonrectangular meshes can be employed to model geometrically complex domains accurately with nonstraight boundaries and multiply connected domains; (2) the fact that when problems with different geometries are solved, only the input data are changed, not the computer code; and (3) the possibility to assign parameters (e.g., material properties) on each element, which makes it easy to model laminates and tooling possessing different material properties [270]. This method has also been successfully implemented and used in injection and compression molding [272,273].

Solving Equation 2.46 for a three-dimensional fluid and heat flows will require a computational time of the order of days on a reasonably fast machine. Until the present time, nobody has implemented such a complex system because of the absence of details of material and constitutive parameters and partially on account of the limitations of CPU. Liu, Bruschke, and Advani [274,275] have developed a code for describing the resin flow and cure in RTM processes by considering a two-dimensional version of Darcy's law, along with the boundary conditions and numerical solution presented earlier, and a three-dimensional version of Equation 2.46. For heat transfer, after scaling analysis in a thin-shell geometry, they assumed that the in-plane conduction and out of plane convection are negligible and therefore, ignored. Ignoring convection in the thickness direction is also consistent with the method used to solve for the movement of the free surface flow front. After these assumptions, Equation 2.46 simplifies to Equation 2.51:

$$(\rho c_p)_c \frac{\partial \langle T \rangle}{\partial t} + (1-\phi_f)(\rho c_p)_r \left(\langle u_x \rangle \frac{\partial \langle T \rangle}{\partial x} + \langle u_y \rangle \frac{\partial \langle T \rangle}{\partial y} \right) = k_{eff} \left(\frac{\partial^2 \langle T \rangle}{\partial z^2} \right) + (1-\phi_f) \Delta H_R \langle r_r \rangle (\underline{x}, t) \quad (2.51)$$

The necessary boundary conditions to solve Equation 2.51 are discussed in Advani et al. [15]. The temperature and cure kinetics are coupled with the fluid flow equation by the viscosity. Once the temperature and the degree of cure are known in each control volume, the viscosity may be calculated by the rheological models presented earlier. Bénard and Advani [276,277] have proposed a model for describing the transport phenomena and microstructure development during solidification of semicrystalline polymers and composites. In this case, there is no flow. They applied the finite element and an implicit (Desbiolles et al. [278]) method to solve the spatial and temporal derivatives, respectively, of the heat and mass transfer equations. The future idea is to use the same code, with pertinent modifications, for representing cure in thermosetting resins in RTM processes. Due to the fact that the curing time is much higher than the filling time, no resin flow will be considered, which simplifies the problem. Additional boundary conditions along the plane of the mold walls will be required, however, in order to predict the temperature and cure distributions in three dimensions.

2.5.3.2 Thermophysical Properties

The thermal response of composites to the applied heat during cure is an important aspect for the design of a cure cycle because a lack of understanding of the thermal properties of the system could lead to nonuniformly cured product, which is structurally unsafe [279]. Thus, to solve the governing equations along with the boundary conditions numerically, one needs to specify the material properties for the composites, which are, basically, a mixture of two components: resin and reinforced materials (fiber-glass, graphite). The bulk physical properties, therefore, are obtained by applying the rule of mixtures, Equations 2.52 and 2.53 [32,37,39,43,119]:

$$\rho_c = (1 - \phi_f)\rho_r + \phi_f \rho_f \tag{2.52}$$

$$c_{p_c} = (1 - \phi_f)C_{p_r} + \phi_f c_{p_f} \tag{2.53}$$

in which ϕ_f is the fiber volume fraction and the subscripts r and f represent resin and fibers, respectively. Although some authors [39,191,271,280] use Equation 2.52 multiplied by Equation 2.53 in order to obtain the volumetric heat capacity, Equation 2.54 [32,281] is more correct because it is derived from enthalpy balance.

$$(\rho c_p)_c = (1 - \phi_f)(\rho c_p)_r + \phi_f(\rho c_p)_f \tag{2.54}$$

Fiber-reinforced materials exhibit anisotropy, with the principal directions of the thermal conductivity being parallel and perpendicular to the fiber orientations [13,35]. A tensor transformation between the global coordinate system of Equation 2.46 and the material coordinate system (perpendicular and parallel to the fibers) allows one to determine the anisotropic properties. Details can be found elsewhere [13,35].

Because of the heterogeneous character of the composites, effective properties are desirable, because this allows substantial simplification of the process simulations [281]. The effective density and volumetric heat capacity are calculated by Equations 2.52 and 2.54. Published literature is abundant on expressions of effective thermal conductivity for each direction (i.e., normal and parallel to the fibers, with the latter being much higher than the former) [256,280–288]. Among them, the most used is that developed by Springer and Tsai, Equation 2.55 [282], for the parallel (along the fibers) thermal conductivity

$$\frac{k_{\parallel}}{k_r} = (1 - \phi_f) + \frac{\phi_f k_f}{k_r} \tag{2.55}$$

and for the perpendicular (considering cylindrical filaments in a square array) thermal conductivity, expressed as Equation 2.56:

$$\frac{k_\perp}{k_r} = \left(1 - 2\sqrt{\frac{\phi_f}{\pi}}\right) + \frac{1}{B}\left[\pi - \frac{4}{\sqrt{1 - \left(\frac{B^2\phi_f}{\pi}\right)}}\tan^{-1}\frac{\sqrt{1 - \left(\frac{B^2\phi_f}{\pi}\right)}}{1 + B\sqrt{\frac{\phi_f}{\pi}}}\right] \tag{2.56}$$

where B is given by Equation 2.57:

$$B = 2\left(\frac{k_r}{k_f} - 1\right) \tag{2.57}$$

Other usual expressions for the composite thermal conductivity in the perpendicular direction to the fibers are shown in Equations 2.58–2.60 [15,30,39,188,289];

$$\frac{k_\perp}{k_r} = \frac{1 - B_1 B_2 \phi_f}{1 - B_1 \phi_f} \tag{2.58}$$

$$B_1 = \frac{k_f/k_r - 1}{k_f/k_r + B_2} \tag{2.59}$$

$$B_2 = \frac{1}{4 - 3(1 - \phi_f)} \tag{2.60}$$

derived by Halpin Tsai [290], in Equation 2.61

$$\frac{k_\perp}{k_r} = \left(1 + \frac{3\phi_f(k_f - k_r)}{(1 - \phi_f)(k_f - k_r) + 3k_r}\right) \tag{2.61}$$

derived by Christensen [291] (for cylindrical assemble) and in Equation 2.62

$$\frac{k_\perp}{k_r} = \frac{k_f + k_r + (k_f - k_r)(1 - \phi_f)}{k_f + k_r - (k_f - k_r)(1 - \phi_f)} \tag{2.62}$$

derived by Whitney and McCullough [289]. Torquato [288] has showed that the effective thermal conductivity is rigorously bounded by Equation 2.63:

$$\frac{k_r k_f}{\phi_f k_r + (1 - \phi_f)k_f} \le k_{ef} \le ((1 - \phi_f)k_r + \phi_f k_f) \tag{2.63}$$

The simple rule of mixtures, as used for density and volumetric heat capacity, would result in substantial errors for perpendicular thermal conductivity. For example, for the Hercules graphite/epoxy prepreg, AS4/3501-6 [40], $k_r = 0.167\,\mathrm{W/mK}$, $k_f = 26\,\mathrm{W/mK}$ and $\phi_f = 0.6$; by using Equations 2.56 and 2.60–2.62, $k_\perp/k_r$ is equal to 3.87, 1.94, 5.29, and 2.31, respectively, whereas using the rule of mixtures, $k_\perp/k_r = 93.8$. By using Equation 2.63, the lower and upper bounds of k_{ef}/k_r are 2.48 and 93.8, respectively. The values calculated by Equations 2.60 and 2.62, therefore, are out of this range. It is also worth noting the difference among the values of $k_\perp/k_r$. Hence, there is a need to address this issue of which is the correct effective thermal conductivity as there is no consensus, which results in a widespread variation in the values.

Some authors affirm that the thermophysical properties of the composites vary with degree of cure [203,271] and temperature [52,147,271,292], according to equations presented in Table 2.8. Due to the fact that nearly all thermosetting materials are amorphous, however,

Table 2.8 Values and Expressions for Thermophysical Properties

	Composite $\rho_c,\ c_{pc},\ k_c$	Resin $N\rho_r,\ c_{pr},\ k_r$	Fibers $\rho_f,\ c_{pf},\ k_f$	Authors/ system
Density (kg/m^3)	Rule of mixtures	1260	1790	*Authors*: A.C. Loos and G.S. Springer [40] (1983) *System*: *Hercules AS4/3501-6* Resin: TGDDM Curing Agent: DDS Fiber: Graphite
Specific heat (J/kgK)	Rule of mixtures	1260	712	
Thermal conductivity (W/mK)	Equations 2.56 and 2.57	0.167	26	
Density (kg/m^3)	$\frac{1}{\rho_c}=\frac{w_r^o}{\rho_r}(1-\alpha)+\frac{w_r^o}{\rho P r^o}\alpha\frac{\rho_r^0}{\rho_{cr}}+\frac{w_f^o}{\rho_f}$ $\phi_f=\frac{w_f^o\rho_c}{\rho-f}\quad \phi_{cr}=\frac{w_r^o\alpha\rho_c}{\rho_{cr}}$ $\phi_r=\frac{w_r^o(1-\alpha)\rho_c}{\rho_f}$	a. For polyester: $\rho_r=1100\quad \rho_{cr}=1200$ b. For epoxy: $\rho_r=1130\quad \rho_{cr}=1170$ with ρ_{cr} as the density of a cured resin	a. For fiberglass: 2540 b. For carbon: 1790	*Author*: C. Han et al. [191] (1986) *System 1*: Resin: Commercial grade of unsaturated polyester (OC-E701) Initiator: Percadox 16N and *t*-butyl perbenzoate Fiber: Fiber glass

(*continued*)

Table 2.8 ***(continued)***

	Composite $\rho_c,\ c_{pc},\ k_c$	Resin $N\rho_r,\ c_{pr},\ k_r$	Fibers $\rho_f,\ c_{pf},\ k_f$	Authors/ system
Specific heat (J/kgK)	$w_r c_{pr} + w_{cr} c_{pcr} + w_f c_{pf}$	a. For polyester: $c_{pr} = -c_{pcr} = 1883.70$ b. For epoxy: $c_{pr} = 10758.02 + 2.50\text{T(K)}$ $c_{pcr} = -5651.10 + 0.624\text{T(K)}$	a. For fiberglass: 833.01 b. For carbon: 711.62	*System* 2: Resin: OC-E701 Initiator 2: 16N and benzoyl peroxide (BPO) Fiber: Fiber glass
Thermal conductivity (W/m°C)	$\frac{1}{k_c} = \frac{\phi_r}{k_r} + \frac{\phi_{cr}}{k_{cr}} + \frac{\phi_f}{k_f}$	a. For polyester: $k_r = 0.1695$ $k_{cr} = 6.446 \times 10^{-2} + 3.96 \times 10^{-4}\text{T(K)}$ b. For epoxy: $k_r = 8.288 \times 10^{-2} + 1.80 \times 10^{-4}\text{T(K)}$ $k_{cr} = 0.1921 + 7.2418 \times 10^{-5}\text{T(K)}$	a. For fiberglass: 0.8665 b. For carbon: 8.665	*System* 3: Resin: Epon 9302 Curing Agent: Epon 9350 Fiber: Carbon *System* 4: Epon 9302/Epon 9350/Fiber Glass
Density (kg/m^3)	Rule of mixtures	$1234 - 0.1T(°\text{C}) + 90\alpha \quad \alpha < 0.45$ $1274 - 0.1T(°\text{C})$ for $\alpha \geq 0.45$	1790	*Authors*: Mijovic and Wang [192] (1988)

Specific heat (J/kgK)	$(1 - w_f)c_{pr} + w_f c_{pf}$ where $\phi_f = \frac{w_f \rho_c}{\rho_f}$	$1958 + 2.5T(^\circ C) - 589.94\alpha$	$933.4 + 0.913T(^\circ C) - \frac{4.081 \times 10^7}{T^2(K)}$	*System*: Resin: TGDDM Curing Agent: DDS (33 phr) Fibers: graphite
Thermal conductivity (W/m°C)	Rule of mixtures	$0.161 + [0.00146T(^\circ C) - 0.017]\alpha$	2.510	
Density (kg/m^3)	Rule of mixtures	1100	2540 (fibers) and 2580 (filler)	*Authors*: H. Ng and I.M. Zloczower [189] (1989) *System*: Resin: a commercial unsaturated brominated polyester resin with 50% by weight styrene (Dion 8200) *Initiator*: bis(4-t-butyl cyclohexil) peroxydicarbonate (Percadox 16N) Fibers: E-glass (BB-225) Fillers: kaolin clay (ASP-400)
Specific heat (J/kgK)	$w_r c_{pr} + w_f c_{pf} + w_{fi} c_{pfi}$ with w as the weight fraction and subscript *fi* representing filler	1881	833.01 (fibers) 920.92 (fillers)	
Thermal conductivity (W/mK)	$\frac{1}{k_c} = \frac{1 - \phi_f - \phi_{fi}}{k_r} + \frac{\phi_f}{k_f} + \frac{\phi_{fi}}{k_{fi}}$ with ϕ_f as the filler volume fraction	0.169	0.867 (fibers) 1.967 (fillers)	

(*continued*)

Table 2.8 ***(continued)***

	Composite $\rho_c,\ c_{pc},\ k_c$	Resin $N\rho_r,\ c_{pr},\ k_r$	Fibers $\rho_f,\ c_{pf},\ k_f$	Authors/ system
Density (kg/m^3)	1300	—	—	*Authors*: J.M. Kenny et al. [32] (1989) *System*: Resin: Hercules and Hexcel, both commercial grade, essentially based on *TGDDM* *Curing Agent*: DDS Fiber: Carbon, $\phi_f = 0.6$
Specific heat (J/kgK)	710.6	—	—	
Thermal conductivity (W/mK)	Equations 2.56 and 2.57	0.348	17.4	
Volumetric heat capacity (J/m^3K)	$1.23 \times 10^6 + 4.79 \times 10^3 T(K) + 0.121 \times 10^6 \alpha$	—	—	*Authors*: E.P. Scott and J.V. Beck [279] (1992) *System*: Carbon-fiber/ EPON 828-*m*PDA
Thermal conductivity (W/mK)	$0.662 + 9.53 \times 10^{-4} T(K) + 0.0792\alpha$	—	—	
Density (kg/m^3)	Rule of mixtures	1300	Al_2O_3: 3400 *BN Flake*: 1950 Si_3N_4 Whisker: 3180 Si_2O_3 Particle: 2220	*Authors*: H. Hatta et al. [280] (1992)

Specific heat (J/kgK)	Rule of mixtures	1380	Al_2O_3: 1005 *BN Flake*: 750 Si_3N_4 Whisker: 710 Si_2O_3 Particle: 837	*Systems*: Resin: Kerimid-601 Fibers: Al_2O_3, *BN Flake*, Si_3N_4 Whisker and Si_2O_3 Particle
Thermal conductivity (W/mK)	1. For 2D random short fiber reinforced composite (Al_2O_3 short fiber/Kerimid, Si_3N_4 whisker/Kerimid): $\frac{k_\parallel}{k_r} = 1 + \phi_f \frac{k_f - k_r)B}{H}$ $B = (k_f - k_r)(S_{11} + S_{33}) + 2k_r$ $H = 2(k_f - k_r)^2(1 - \phi_f) \times S_{11}S_{33} + k_r(k_f - k_r) \times (2 - \phi_f)(S_{11} + S_{33}) + 2k_r^2$ $S_{11} = S_{22} = \frac{\beta B_1}{2(\beta^2 - 1)^{3/2}}$ $B_1 = \beta(\beta^2 - 1)^{1/2} - \cosh^{-1}\beta$ $S_{33} = 1 - 2S_{11}$ $k_\perp / k_r = 1 + \phi_f / B_3$ $B_3 = S_{33}(1 - \phi_f) + \frac{k_r}{k_f - k_r}$ with β, aspect ratio, as the ratio of the diameter to thick ness of the flake, which is assumed to be a disk	0.23	Al_2O_3: 30.7 *BN Flake*: 500 Si_3N_4 whisker: 25.1 Si_2O_3 Particle: 13.8	

(*continued*)

Table 2.8 ***(continued)***

Composite ρ_c, c_{pc}, k_c	Resin $N\rho_r$, c_{pr}, k_r	Fibers ρ_f, c_{pf}, k_f	Authors/ system
2. For an in-plane aligned flake reinforced composite (*BN flake*/Kerimid): $k_{ci}/k_r = 1 + \phi_f/B_3$ $B_3 = S_i(1-\phi_f) + \frac{k_r}{k_f - k_r}$ where the subscript i denotes the direction of thermal conduction; namely, $i = z$ for transverse conduction and $i = xy$ for in-plane direction $S_z = 1 - (\pi\beta)/2$ $S_{xy} = (\pi\beta)/4$ 3. For a 3D random short fiber reinforced composite (Si_3N_4 whisker/Kerimid): $k_c/k_r = 1 + (\phi_f B_4)/J$ $B_4 = (k_f - k_r)(2S_{33} + S_{11}) + 3k_r$ $J = 3(1-\phi_f)(k_f - k_r)$ $S_{11}S_{33} + k_r = [3(S_{11} + S_{33})$ $-\phi_f(2S_{11} + S_{33})] = \frac{3k_r^2}{k_f - k_r}$ 4. For a spherical particle rein forced composite (SiO_2 particle/Kerimid): $k_c/k_r = 1 + \phi_f/B_5$ $B_5 = \frac{(1-\phi_f)}{3} + \frac{k_r}{k_f - k_r}$			

Aspect Ratio	—	—	Al_2O_3: 15 *BN Flake*: 0.18 Si_3N_4 whisker: 17.5 Si_2O_3 Particle: 1	
Density (kg/m^3)	1890	—	—	*Authors*: S.R. White and C. Kim [172] (1993) *System*; CYCOM 4102 (E-glass/polyester)
Specific heat (J/kgK)	1260	—	—	
Thermal conductivity (W/mk)	0.2163	—	—	
Density (kg/m^3)	1690	1040 (liquid) 1126 (solid)	2565	*Author*: D. Michaud [9] (1996) *System*: Resin: Dow DERAKANE 411-C50 Initiator: Organic Peroxide (Witco USP-245), 1.75 wt% Fiber: E-glass $\{\phi_f \cong 0.43)$
Specific heat (J/kgK)	1205	1640 (liquid) 1410 (solid at 70°C) 1950 (solid at 140°C)	825	
Thermal conductivity (W/mK)	0.25	0.1005 (solid)	1.040	

these properties change little with the reaction from liquid to solid, within the temperature range of a typical cure process [250,293,294]. As a consequence, density, specific heat, and thermal conductivity of the materials can be considered constant [32,39,198,269]. Pusatcioglu et al. [269] affirmed that the accuracy of the simulation seems to be more dependent upon the values of the kinetic parameters than the thermal properties because the variations in thermal properties with cure had little effect on his results.

2.6 Cure Cycles

A processing cure cycle involves three stages: cure, postcure, and cooling of the parts. An example of a cure diagram is given in Figure 2.16, which plots the temperature of the mold platen and the temperature at the center of a composite-part during the cure. In order to avoid or reduce the peak temperature, called *exotherm*, in the center of the part, which minimizes the stress and uneven cure, it may be necessary to heat the mold surfaces at a certain heating rate (dT/dt), and then hold it at a constant temperature for fixed period of time. This may distribute the heat more efficiently inside the part. With the objective of obtaining good mechanical properties, postcure is recommended for resins that do not totally cure in a normal

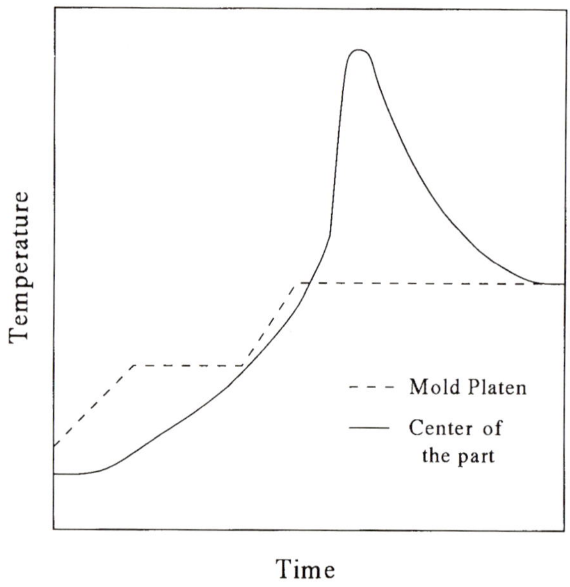

Figure 2.16 Cure cycle—two stages of constant temperature, with two stages of heating rates for the mold platen (curve - - -); temperature in the center of the composite part (curve ——).

cycle [148,295]. The last stage is the cooling of the parts, where large temperature gradients can develop, causing residual stresses and fracture.

The selection of the boundary conditions for thermal fields is intrinsically related to the cure cycles or processing conditions; that is, variation of temperature and pressure with time. The profiles of the temperatures and pressures applied during the curing process may provoke diverse paths of curing reaction, which affects the performance of the finished product significantly [38]. The maximum temperature inside a composite depends on the value of the cure temperature, on the heating rate (i.e., the rate at which the temperature increases), and on the thickness of the part [40,188]. This maximum temperature should be less than the degradation temperature of the system. For Hercules 3501-6/fiberglass system, for instance, the degradation temperature is around 180°C [35,40,42,43].

Before choosing a profile for imposed temperature profile for *optimum* cure schedule, one must take certain factors and rules of thumbs into account, such as:

1. Curing from the outside to the inside of the part may potentially entrap voids and volatile by-products of the chemical reaction and increase residual stresses [35]
2. Resin-rich and void zones can form when the surfaces cure more quickly than the center, trapping resin in the center of the part [37]
3. The lower the curing temperature, the lower the induced residual stresses and the higher the processing time
4. The isothermal condition inside the laminate should be maintained in order to achieve a uniform cure. For autoclave processes, it is possible because the heating rate is low (approximately 1.5°C/min) [41]; however, this heating rate is much higher (5 to 10°C/min) for RTM [9], and the isothermal conditions are difficult to maintain
5. For the RTM processes, curing temperature and time should be less than 175°C and 1 hour, approximately [6]

Broyer and Macosko [250] have shown that if the thickness of a composite is increased above a critical value, then the imbalance between the heat generation and the thermal diffusivity of the system leads to a virtually adiabatic situation (Neuman condition) [32], which should be avoided because of the high temperature in the center of the part, initiating material degradation. The higher the thickness of the part, the lower the thermal conduction in the composite and the higher the relative heat accumulation inside the part [30]. The critical half thickness, H_{cr}, proposed by Broyer and Macosko [250], is given by Equation 2.64:

$$H_{cr} = \sqrt{\frac{k_c}{(\rho c_p)_c A \exp(-E_a/RT)}} \tag{2.64}$$

Nonetheless, note that this is based on constant mold wall conditions. If one wanted to manufacture higher thicknesses, then one would have to select temperature profiles that would potentially cool the mold walls after the cure initiates or introduce heat sink in the center of the part.

There are many proposals for cure cycles for autoclave processes [10,38,40,270,296–299]. The most used is the one developed by Loos and Springer [40], where they proposed *optimum* cure cycles which considers composites with thickness up to 2.1 cm [42] and fiber volume fractions of 39 percent and 42 percent. Some other authors have followed them [42,271], but they have each generally proposed their own cure cycle and maximum temperature; there is no unique solution.

The majority of the papers related to the RTM processes do not address cure cycles. Michaud [9] has analyzed some cure cycles for vinyl ester (DERAKANE 411-C50)/E-glass system in RTM mold heated by recirculating water at a specific temperature. He applied different (asymmetric condition) and identical (symmetric condition) temperatures on the top and bottom surfaces of a stainless steel mold. Although the peak temperature was reduced, an asymmetric condition is not recommended by the author because of the higher processing time and the difficulty in maintaining two distinct platen temperatures.

This suggests that one needs to use simulations of coupled phenomena to develop a methodology for temperature profiles for the mold platen that will meet the objectives of obtaining a desirable composite part. Tools such as inverse algorithms, neural networks, and on-line sensing could help in establishing an optimal cure cycle.

2.7 Optimization and Control Strategies

In all composite part manufacturing processes, optimum cure cycles (or cure schedules or temperature history) need to be selected in order to obtain parts with high quality as well as desired mechanical properties and *low cost* in a *minimum* possible processing time. The selection of these *optimum* cure cycles has been made by trial and error in the past, being expensive, inefficient and time consuming [12,36,41,198]. Over the last 15 years, however, researchers have followed a scientific method through the use of mathematical modeling of the three governing phenomena—resin flow, heat transfer, and cure kinetics—to predict temperatures, pressures, and cure. What initial and boundary conditions will give the uniform cure and low temperature gradients, however, is still more of an art than a science. Optimization seems to be a useful tool that could move us in that direction.

An optimization strategy involves two steps: a selection of an objective function followed by its minimization. Working with cure of thick parts of graphite/epoxy laminates in autoclave, Martinez [43] defined an objective function as the absolute value of the difference between a target temperature history and the composite temperature history, calculated for a particular cure schedule. This target temperature history was accomplished by a trial-and-error process in which the temperature in the center of the fully cured part was calculated for various cure cycles. The target temperature was then chosen based on the shortest possible curing time. The objective function was minimized by using a piecewise linear discretization of the autoclave temperature profile. The optimization algorithm proposed by Kenny [44] is based on an objective function that penalizes prescribed autoclave temperatures out of the specification for one or more parts and prefers prescribed autoclave temperatures that minimizes the time at which all parts achieve the final processing temperature. The drawback of the preceding two methods is that they need to know a priori a target temperature, which requires a trial-and-error method and, consequently, is time consuming. In order to solve this, Pillai [41] proposed an optimization strategy that consists of finding an optimal autoclave temperature profile necessary to complete the cure in minimum time, to limit the exotherm peak and to minimize process-induced residual stresses. He used the Local Criteria Optimization (LCO) method [41] and defined a global objective function as a linear combination of the curing time, the exotherm, the thermal and degree of cure gradients, and the magnitude of the residual stresses. In order to obtain parts with good mechanical properties, some variables, such as temperatures of the tooling, the surfaces, and the center of the composite and the pressure, should be kept under control during the curing reaction [47].

Any control strategy requires the development of a set of control laws, named algorithm, and the sensors or transducers to measure these variables in situ. Data provided by the sensors are interpreted by the rules which, in turn, adjust in real time the process controllers (heater, cooler, pressure) [47].

There are, basically, two control strategies [41,47]: the traditional one, which is a combination of the sensor outputs with the results of the mathematical models described earlier, and the new approach, called *expert systems*, which combines the sensor outputs with the laws. In this last case, little prior information of the material properties and geometry is required [47]. Rules should be applied to each variable v to be controlled; for example:

If $v_1 < v_{lim}$ and $v_2 < v_{lim}$, then perform a specific action, where v_{lim} is the allowable limit value of v, and v_1 and v_2 are the values of v in two different locations.

By using *expert systems*, Ciriscioli et al. [47] developed detailed algorithms for temperatures of the autoclave, surfaces, and center of the composite, compaction, void growth, end of cure, residual stress, and end of manufacturing process.

Pillai [41] pointed out some drawbacks of *expert systems*. He affirmed that there is no quantitative justification for why the rules chosen are necessarily the right ones; the strategy is only as good as the process knowledge of the engineer and there is no predictive capacity. In addition, *expert systems* are expensive and specific to part and resin [45].

The control strategy used by Pillai in autoclave processes does not minimize any sum of squares of differences between an optimal profile and processing profiles at discrete time steps. It is based instead on the guarantee that the trend of the processing variables is the same as that given by the optimization. This strategy is called Process Trend Analysis (PTA) and it is frequently used in chemical industries. It consists of three steps:

1. To define a minimum and maximum limit for the processing variables
2. To select *critical* points in the optimized curve that one wants to reproduce
3. To associate with each of these *critical* points three pieces of information: the first and second derivatives (if negative or positive) and the value of the variable (if below, between, or above the limit range). A series of discrete functions is then defined, permitting the controller to take action (increase or decrease the autoclave temperature or pressure, for example). Pillai claimed that this strategy reduces the amount of temporal data, since it is not necessary to store the continuous function but only some strategical points. This type of control has yet to be extended for RTM processes because RTM is inherently more complex process with many more variables.

2.7.1 Sensors

The efficacy of the mathematical modeling and optimization, as well as control strategies used to describe the curing process, needs to be evaluated, and the only way to do this is to measure in situ temperature, pressure, flow front position, and physical changes. Analytical instruments, such as DSC, FTi.r., and DMA, can be employed to characterize the cure reaction chemistry but they can not be used as a sensor for in situ cure monitoring on account of the sample size constraints and their destructive or intrusive nature [46]. A complete characterization of the cure states in real time is essential for successful manufacturing of composites.

The temperature can be measured easily with thermocouples, which are cheap and reliable, but the other properties and variables, such as degree of cure and resin viscosity, need sophisticated sensors. Some possible sensors being explored for in situ sensing will now be described briefly.

2.7.1.1 Microdielectric Transducers

Microelectric sensors are tiny electrodes placed into the surface of a probe immersed in the resin and connected to a microelectrometer. This technique is based on the movement of the ions and dipoles and is present in most resins, when subjected to a potential difference [45]. Hence, it is used to measure qualitatively the degree of cure and resin viscosity, because chemical bonds (cross-linking) formed during curing reaction reduce or even hinder the movement of the ions. The higher the viscosity and the degree of cure, the lower the electric conductivity. The temporal variations of the electric conductivity and viscosity are close enough that they can be used for processing control [300].

2.7.1.2 Acoustic-Based Transducers

Acoustic emissions (AE) are mechanical stress waves emitted by a material subjected to a load. In order to detect these emissions, small fibers (0.15 cm in diameter [45]) are placed longitudinally through a preform to act as guides for the acoustic waves, becoming then an integral part of the finished part. Energy escapes from these fibers relative to the acoustic impedance between the wave guide and the resin. The sonic output (magnitude of the signal that gets through the wave guide) and the transit time (speed of the signal) are the quantities measured. Studies are being undertaken to overcome the poor reproducibility and to correlate AE data with material physical properties.

2.7.1.3 Fluorescent Transducers

The fluorescence technique is based on the energy emitted by a special compound (used as a probe) mixed with the resin. The fluorescence intensity depends on the local mobility of the molecules [41,45,47]. When the mobility is reduced, the rotation of the molecules is hindered, and the energy can not be released, increasing the fluorescence intensity. Because the mobility is closely related to viscosity, this method can be used to detect chemorheological events, such as gelation and vitrification.

2.7.1.4 Mechanical Impedance Analysis

This technique, basically, involves quantitative measurement of the effect of a vibratory force on a structure. The principle for the mechanical impedance measurement of a moving mass was proposed by Ferry et al. [301] and the use of MIA for monitoring cure reaction in thermosetting resins was suggested by Jang and Zhu [302]. A sample is continuously excited by an electromagnetic shaker with a random input during one experiment. The output signals are fed into the Fast-Fourier-Transform (FFT) analyzer, and the frequency response function (FRF) spectra are displayed by a computer in real-time mode. The cure states can be characterized by the material properties, dynamic modulus, and loss tangent, and calculated from the peaks of the FRF spectra related to the storage Young's modulus, and, hence, to the molecular weight. All these techniques are still under development and accurate expressions relating the measured signals to material properties and cure state are required.

2.8 Summary and Outlook

The wide use of composites in diversified industries requires an optimization of the manufacturing processes, with the objective of obtaining parts with *excellent* mechanical properties with *low cost* and *short* processing times. It can currently be achieved scientifically through a mathematical modeling of the various phenomena, such as resin flow, heat transfer, cure kinetics, void formation, and stress, which occur simultaneously during the manufacturing of the composites. Some theoretical and empirical models were revised herein for cure kinetics and its coupling with resin flow, heat transfer, and viscosity of thermosetting resins in RTM process. The mechanistic ones are more representative of the complex cure phenomena involved; however, even they do not ensure that all chemical reactions taking place among the many reactants can be counted. Simpler models, called *phenomenological models*, are therefore adopted because they lump all reactions into a few parameters that are curve fitted to the experimental data. They are not intended to elucidate the chemistry of the cure; rather, they only describe concisely heat released or the degree of cure by one equation.

The lack of well-established theories and the great complexity of chemical reactions, flow, such as the percolation in a macro- and microscales of a resin through bundles of fiber tows with diverse orientations, and the conductive, the convective, as well as the dispersive heat transfer in all directions of the preform, lead to the use of *effective* parameters. All unknown and complex phenomena are lumped into such material parameters. The serious disadvantage is the fact that the kinetic and *effective* parameters of the phenomenological models are restricted to specific resins, processes, and experimental conditions used. As presented, the literature is abundant on values for these parameters for many resin and preform systems; however, there is divergence among them for the same system and experimental conditions.

It would be very useful if there were a data base that contained values of the degree of cure and the reaction rates at many different curing times as well as the inherent kinetic parameters of the common phenomenological models, for a large range of temperature and reinforced materials used. Although there are some studies in the literature, important issues still need to be addressed in future work. The experimental determination of three-dimensional permeability tensors of preforms has a high degree of complexity and difficulty because of the various preform orientations. As a result, a two-dimensional simulation of the resin flow in RTM parts is widely adopted. Nevertheless, for thick parts, a fluid flow in the thickness direction should not be neglected, requiring a three-dimensional (more complex) or an alternative (easier) solution to capture the influence of the thickness-direction flow on the whole system. This alternative solution can be to average the in-plane permeabilities in the thickness direction, thereby obtaining a reduction in the dimensional order of the problem. This and other alternative solutions need more investigation.

Another important issue to be addressed in future work is related to the selection of the *optimum* boundary conditions and cure cycles, and to the control of important variables, such as injection pressure, resin volumetric flow, processing temperature, and curing time, in order to obtain composites in RTM processes with good mechanical properties. Many authors, based on a mathematical modeling of the main coupled phenomena—resin flow, void formation, heat transfer, and cure reaction—have proposed some optimization and control strategies for autoclave processes. Nevertheless, there is still a need to address these issues systematically and scientifically with respect to RTM processes for thick parts. The large thermal gradients inside the composite are characteristics in this case and should be avoided. One way to accomplish this is to analyze different thermal boundary conditions (constant temperature or convective heat transfer) applied to the mold surfaces. What initial and boundary conditions will give the uniform cure and low temperature gradients, however, is still more of an art than science. A correct modeling of the main phenomena and an optimization strategy is required in order to decide which cure cycle should be adopted for that particular composite system. A control strategy should then be used to send the correct instructions to the controllers (heater and/or cooler of the mold platens as well as pressure vessel).

It is fundamental, however, to have accurate measurements, in situ and in real time, of processing variables such as pressure, temperature, velocity, degree of cure, and viscosity in order to compare the experimental and the simulated data and decide on corrective actions. Among a variety of sensors, the fiberoptics are promising because of their accuracy; however, the drawback is the high price. Research in this field is under development and accurate expressions relating the measured signals to material properties and cure states are required.

In summary, we can say that manufacturing of thick composite parts with good mechanical properties and low cost in RTM processes requires continuing research in: (1) accurate experimental techniques to characterize the architecture and the permeability tensor of the preforms as well as the state of resin cure during the polymerization reaction; (2) mathematical modeling of the resin flow, void formation, heat transfer, and cure kinetics at both macro- and microscopic levels, considering conductive, convective, and dispersive transfer inside the part; (3) strategy for the selection of the ideal initial and boundary conditions in order to obtain the *optimum* cure cycles; (4) strategy for the control of the processing variables during the process; and (5) development of reliable and inexpensive sensors for monitoring the mold filling as well as resin cure stages and the mathematical expressions correlating the output signals from the sensors with the variables of interest. This research, if successful, will make the manufacturing process produces net shape parts at *low cost* and tailored properties for specific application.

Nomenclature

a	Constant of Equations 2.13, 2.15, and 2.19
A	Frequency factor or pre-exponential factor
A_η	Constant of Equations 2.13 and 2.14
b	Constant of Equations 2.13, 2.15, and 2.19
B	Constant of Equations 2.56 and 2.58
c_{p_c}	Specific heat of the composite
$c_{p_{cr}}$	Specific heat of the cured resin
C_{p_f}	Specific heat of the fiber
C_{p_r}	Specific heat of the resin
$\langle c_r \rangle$	Volume-averaged volumetric concentration of the resin
$\langle c_{ro} \rangle$	Volume-averaged volumetric initial concentration of the resin
C	Constant of Equation 2.23
C_i	Adjustable parameters of Equations 2.10 and 2.11
d_p	Fiber diameter
$\underline{\underline{D}}_d$	Dispersive diffusivity tensor
$\underline{\underline{D}}_{ef}$	Effective diffusivity tensor
E_a	Activation energy for kinetic models
E_η	Activation energy for rheological models
E_x/E_m	Ratio of lattice energies for cross-linked and uncross-linked polymer
f	Functionality of the cross-linking groups
f_d	Diffusional factor
F_x/F_m	Ratio of segmental mobilities at the glass temperature
Gz	Graetz number
h_h	Heat transfer coefficient between the heating fluid and the pipe
h_m	Heat transfer coefficient between the resin and the mold wall
H	Half gap height of the mold
H_{cr}	Critical half thickness
ΔH_{res}	Residual Enthalpy
ΔH_R	Heat of reaction
ΔH_t	Enthalpy at a specific time
k_c	Thermal conductivity of the composite
k_{cr}	Thermal conductivity of the cured resin
k_f	Thermal conductivity of the fiber
k_m	Thermal conductivity of the mold material
k_r	Thermal conductivity of the resin
$\underline{\underline{k}}_{cond}$	Effective thermal conductivity tensor
$\underline{\underline{k}}_d$	Thermal dispersion tensor
$k_\parallel$	Parallel thermal conductivity
$k_\perp$	Perpendicular thermal conductivity
K	Constant of Equation 2.12
$\underline{\underline{K}}$	Permeability tensor
K_{ij}	Components of the permeability tensor
L	Characteristic length of the cavity

m	Order of reaction
$\bar{M}_w$	Weight average molecular weight
n	Order of reaction
$\langle \nabla P \rangle$	Volume-averaged pressure gradient in the fluid
Pe_{eff}	Microscopic Peclet number
r	Molar ratio of reacting functional groups
$\langle r_r \rangle$	Volume-averaged volumetric rate of polymerization
R	Gas constant
s	Fraction of the amine hydrogen in the multifunctional reactants
S_{ij}	Constants defined in Table 2.8
t	Time
T	Temperature
T_g	Glass temperature
T_{go}	Glass temperature of the uncross-linked polymer
$T_{g\infty}$	Glass temperature of the cross-linked polymer
T_∞	Temperature of the heated fluid flowing through the tube
$\langle \underline{u} \rangle$	Volume-averaged velocity vector
$\langle u_i \rangle$	Components of the volume-averaged velocity vector
U	Activation energy of the viscous fluid
V	Characteristic fluid velocity
w_{cr}	Cured resin mass fraction
w_f	Fiber mass fraction
w_r	Resin mass fraction
x	Coordinate
y	Coordinate
α	Extent of cure or degree of cure
α_c	Critical fractional conversion
α_g	Extent of cure at the gel point
α_t	Total thermal diffusivity
α_u	Ultimate extent of cure
β	Aspect ratio
η	Viscosity of the resin
η_{eff}	Effective viscosity of the fluid
η_{To}	Viscosity of the resin at temperature T_o
η_∞	Extrapolated viscosity as $T \to \infty$
ϕ_{cr}	Cured resin volume fraction
ϕ_f	Fiber volume fraction
ϕ_r	Resin volume fraction
κ_i	Rate constant
κ_c	Rate constant for chemical kinetics
κ_{ef}	Overall effective rate constant
ρ_c	Density of the composite
ρ_{cr}	Density of the cured resin
ρ_f	Density of the fiber
ρ_r	Density of the resin

Acknowledgments

The support of the University Research Initiation program of the Army Research Office (ARO/URI) at the University of Delaware and CAPES, a Brazilian Agency (process number: 0317/94-2), is gratefully acknowledged.

References

1. Prime, R.B. in *Thermal Characterization of Polymeric Materials* (1981) Turi, E.A. (Ed.) 2, Academic Press, New York, p. 435–569
2. Moroni, A., Mijovic, J., Pearce, E.M., Foun, C. C. *J. Appl. Polym. Sci.* (1986) 32 p. 3761–3773
3. Chern, C.-S., Poehlein, G.W. *Polym. Eng. Sci.* (1987) 27(11) p. 788–795
4. Gillham, J.K. *Polym. Eng. Sci.* (1979) 19(10) p. 676–682
5. Su, C.C., Woo, E.M. *Polymer* (1995) 36(15) p. 2883–2894
6. Palmese, G.R. Cure of Thermosetting Resins in RTM. *Joint Composites Workshop*, September 1995
7. Lunák, S., Vladyka, J., Dusek, K. *Polymer* (1978) 19 p. 931–933
8. Gordon, M., Simpson, W. *Polymer* (1961) 2 p. 383–391
9. Michaud, D.J. Investigation of Curing Behavior in Thick Thermoset Composites Manufactured by Resin Transfer Molding. Master's thesis, Chemical Engineering Department, University of Delaware, Newark, DE, Spring 1996
10. Dave, R., Kardos, J.L., Dudukovic, M.P. *Polym. Comp.* (1987) 8(1) p. 29–38
11. Ciriscioli, P.R., Wang, Q., Springer, G.S. *J. Comp. Mat.* (1992) 26(1) p. 90–102
12. Pitchumani, R., Yao, S-C. *J. Comp. Mat.* (1993) 27(6) p. 613–636
13. Tzeng, J.T., Loos, A.C. *Comp. Manufact.* (1993) 4(3) p. 157–165
14. Kendall, K.N., Rudd, C.D. *Polym. Comp.* (1994) 15(5) p. 334–348
15. Advani, S.G., Bruschke, M.V., Parnas, R.S. In Resin Transfer Molding Flow Phenomena in Polymeric Composites. Advani, S.G. Ed., *Flow and Rheology in Polymer Composite Manufacturing*, volume 10 of *Composite Materials Series*, 10, Chapter 12, Elsevier Science, Amsterdam, p. 465–515, 1994
16. Rudd, C.D., Kendall, K.N. *Proc. Instn. Mech. Engrs.* (1992) 206 p. 77–91
17. Calhoun, D.R., Yalvaç, S., Wetters, D.G., Raeck, C.A. *Polym. Comp.* (1996) 17(1) p. 11–22
18. Coulter, J.P. *Resin Impregnation during the Manufacturing of Composite Materials*. PhD thesis, Mechanical Engineering Department, University of Delaware, Newark, DE, 1988
19. Castro, J.M., Macosko, C.W. *Soc. Plast. Eng. Tech. Papers* (1980) 26 p. 434
20. Damani, S.G., Lee, L.J. *Polym. Comp.* (1990) 11 p. 174–183
21. Patel, N., Lee, L.J. *Polym. Comp.* (1996) 17(1) p. 96–103
22. Patel, N., Lee, L.J. *Polym. Comp.* (1996) 17(1) p. 104–114
23. Patel, N., Lee, L.J. *Polym. Comp.* (1995) 16 p. 386
24. Mahale, A.D., Prud'homme, R.K., Rebenfeld, L. *Polym. Eng. Sci.* (1992) 32 p. 319
25. Parnas, R.S., Phelan, Jr., F.R. *SAMPE Q.* (1991) 22(2) p. 53–60
26. Chan, A.W., Morgan, R.J. *SAMPE Q.* (1993) 24(4) p. 49–53
27. Chan, A.W., Morgan, R.J. *Polym. Comp.* (1993) 14(4) p. 335–340
28. Elmendorp, J.J., During, F. *SPE ANTEC* (1990) 36 p. 1361–1364
29. Kardos, J.L., Dudukovic, M.P., Dave, R. In *Advances in Polymer Science—Epoxy Resins and Composites IV*, Dusek, K. (Ed.) 80 (1986) Springer-Verlag, Berlin, p. 101–123
30. Kenny, J.M., Maffezzoli, A., Nicolais, L. *Comp. Sci. Tech.* (1990) 38 p. 339–358
31. Kinloch, A.J. In *Advances in Polymer Science—Epoxy Resins and Composites I*, Dusek, K. (Ed.) 72 (1985) Springer-Verlag, Berlin, p. 45–67
32. Kenny, J.M., Apicella, A., Nicolais, L. *Polym. Eng. Sci.* (1989) 29(15) p. 973–983

33. Lee, J.H., Lee, J.W. *Polym. Eng. Sci.* (1994) 34(9) p. 742–749
34. Pusatcioglu, S.Y., Fricke, A.L., Hassler, J.C. *J. Appl. Polym. Sci.* (1979) 24 p. 937–946
35. Bogetti, T.A., Gillespie, Jr., J.W. *J. Comp. Mat.* (1991) 25 p. 239–273
36. Saliba, T.E., Servais, R.A., Snide, J.A., Lee, C.W. *Adv. Comp. Conf. Proc. ASM* (1985) p. 163–168
37. Twardowski, T.E., Lin, S.E., Geil, P.H. *J. Comp. Mat.* (1993) 27(3) p. 216–250
38. Springer, G.S. *J. Comp. Mat.* (1982) 16 p. 401–410
39. Sanford, W.M. *Cure Behavior of Thermosetting Resin Composites*. PhD thesis, University of Delaware, Newark, DE, 1987
40. Loos, A.C., Springer, G.S. *J. Comp. Mat.* (1983) 17 p, 135–169
41. Pillai, V.K. *Use of Simulations in Optimization of, and Design of a Knowledge Based Control System for a Composite Manufacturing Process*. PhD thesis, Chemical Engineering Department, University of Delaware, Newark, DE, 1993
42. Hjellming, L.N., Walker, J.S. *J. Comp. Mat.* (1989) 23 p. 1048–1064
43. Martinez, G.M. *Chem. Eng. Sci.* (1991) 46(2) p. 439–450
44. Kenny, J.M. *Comp. Struct.* (1994) 27 p. 129–139
45. Holl, M.W., Rehfield, L.W. *Comp. Mat. Test. Des.* (1992) 10 p. 308–319
46. Jang, B.Z., Shih, W.K., *Mat. Manufact. Proc.* (1990) 5(2) p. 301–331
47. Ciriscioli, P.R., Springer, G.S., Lee, W.I. *J. Comp. Mat.* (1991) 25 p. 1542–1587
48. Browning, C.E. New Applications from New Materials. In Advanced Thermoset Composites, edited by James M. Margolis (1986) Van Nostrand Reinhold Company, New York, pages 1–37
49. Kim, W.G., Lee, J.Y., Park, K.Y. *J. Polymer Sci. Part A: Poly. Chem.* (1993) 31 p. 633–639
50. Acitelli, M.A., Prime, R.B., Sacher, E., *Polymer* (1971) 12 p. 335–343
51. Prime, R.B., Sacher, E., *Polymer* (1972) 13 p. 455–458
52. Kamal, M.R., Sourour, S., *Polym. Eng. Sci.* (1973) 13(1) p. 59–64
53. Kamal, M.R. *Polym. Eng. Sci.* (1974) 14(3) p. 231–239
54. Barton, J.M. *Polymer* (1980) 21 p. 603–606
55. Enns, J.B., Gillham, J.K. *J. Appl. Polym. Sci.* (1983) 28 p. 2567–2591
56. Mijovic, J., Kim, J., Slaby, J., *J. Appl. Polym. Sci.* (1984) 29 p. 1449–1462
57. Morgan, R.J., Mones, E.T. *J. Appl. Polym. Sci.* (1987) 33 p. 999–1020
58. St John, N.A., George, G.A., *Prog. Poly. Sci.* (1994) 19(5) p. 755–795
59. Wisanrakkit, G., Gillham, J.K., Enns, J.B., *J. Appl. Polym. Sci.* (1990) 41 p. 1895–1912
60. Stutz, H., Illers, K.-H., Mertes, J. *J. Polym. Sci. Part B: Polym. Phys.* (1990) 28 p. 1483–1498
61. Shim, J.S., Lee, W., Jang, J. *Poly. J.* (1991) 23(7) p. 903–910
62. Shim, J.S., Lee, W., Jang, J. *Poly. J.* (1991) 23(7) p. 911–916
63. Shim, J.S., Lee, W., Jang, J. *Polym. Bull. (Berlin)* (1991) 25(6) p. 669–675
64. Min, B.-G., Stachurski, Z.H., Hodgkin, J.H. *Polymer* (1993) 34(23) p. 4908–4912
65. Min, B.-G., Stachurski, Z.H., Hodgkin, J.H. *Polymer* (1993) 34(21) p. 4488–4495
66. Stutz, H., Mertes, J. *J. Polym. Sci., Part A: Polym. Chem.* (1993) 31 p. 2031–2037
67. Khanna, U., Chanda, M., *J. Appl. Polym. Sci.* (1993) 49 p. 319–329
68. Nam, J-D., Seferis, J.C. *J. Appl. Polym. Sci.* (1993) 50 p. 1555–1564
69. Kenny, J.M. *J. Appl. Polym. Sci.* (1994) 51 p. 761–764
70. Huang, M.L., Williams, J.G. *Macromolecules* (1994) 27 p. 7423–7428
71. Montserrat, S., Flaqué, C., Pagé, P., Málek, J., *J. Appl. Polym. Sci.* (1995) 56 p. 1413–1421
72. Barral, L., Cano, J., López, A.J., López, J., Nogueira, P., Ramírez, C., *J. Appl. Polym. Sci.* (1995) 56 p. 1029–1037
73. Smith, I.T., *Polymer* (1961) 2 p. 95–108
74. Horie, K., Sawada, M., Mita, I., Kambe, H. *J. Polym. Sci. Part A-1* (1970) 8 p. 1357–1472
75. Riccardi, C.C., Adabbo, H.E., Williams, R.J.J. *J. Appl. Polym. Sci.* (1984) 29 p. 2481–2492
76. Chiao, L., Lyon, R.E. *J. Comp. Mat.* (1990) 24 p. 739–751
77. Cole, K.C. *Macromolecules* (1994) 24(11) p. 3093–3097
78. Cole, K.C., Hechler, J.-J., Noël, D. *Macromolecules* (1994) 24(11) p. 3098–3110
79. Grenier-Loustalot, M.-F., Mouline, G., Grenier, P., *Polymer* (1987) 28 p. 2275
80. Zukas, W.X., Schneider, N.S., MacKnight, W.J. *J. Polym. Mater. Sci. Eng.* (1983) 49 p. 588
81. Stevenson, J.F. *Polym. Eng. Sci.* (1986) 26 p. 746–759

82. Ng, H., Manas-Zloczower, I. *Polym. Eng. Sci.* (1989) 29 p. 1097
83. Muzumdar, S.V., Lee, L.J. *Polym. Eng. Sci.* (1991) 31 p. 1647
84. St John, N.A., George, G.A. *Polymer* (1992) 33(13) p. 2679–2688
85. Kamon, T., Furukawa, H. In *Advances in Polymer Science—Epoxy Resins and Composites IV*, K. Dusek (Ed.) 80 (1986) Springer-Verlag, Berlin, p. 173–202
86. González-Romero, V.M., Casillas, N., *Polym. Eng. Sci.* (1989) 29(5) p. 295–301
87. Piloyan, G.O., Ryabchikov, I.D., Novikova, O.S., *Nature* (1966) 212 p. 1229
88. Hong, J.-L., Wang, C-K., Lin, R-H. *J. Appl. Polym. Sci.* (1994) 53 p. 105–112
89. Prime, R.B. in *Analytical Calorimetry*, Porter, R.S., Johnson, J.F. (Eds.) 2 (1970) Plenum Press, New York, p. 102–210
90. Donnellan, T., Roylance, D. *Polym. Eng. Sci.* (1982) 22(13) p. 821–825
91. Schneider, N.S., Sprouse, J.F., Hagnauer, G.L., Gillham, J.K. *Polym. Sci. Eng.* (1979) 19 p. 304–312
92. Prime, R.B. *Polym. Eng. Sci.* (1973) 13(5) p. 365–371
93. Ryan, M.E., Dutta, A. *Polymer* (1979) 20 p. 203–206
94. Lee, S-N., Yuo, W-B. *Polym. Eng. Sci.* (1987) 27(17) p. 1317–1322
95. Pang, K.P., Gillham, J.K. *J. Appl. Polym. Sci.* (1990) 39 p. 909–933
96. Batch, G.L., Macosko, C.W. *J. Appl. Polym. Sci.* (1992) 44 p. 1711–1729
97. Foun, C.C., Moroni, A., Pearce, E.M., Mijovic, J. *Polym. Mat. Sci. Eng.* (1984) 51 p. 411
98. Lam, P.W.K., Plaumann, H.P., Tran, T. *J. Appl. Polym. Sci.* (1990) 41 p. 3043–3057
99. Huguenin, F.G.A.E., Klein, M.T. *Ind. Eng. Chem. Prod. Res. Dev.* (1985) 24 p. 166–171
100. Flory, P.J. *Principles of Polymer Chemistry* (1953) Cornell University Press, Ithaca, NY
101. Aronhime, M.T., Gillham, J.K. in *Advances in Polymer Science—Epoxy Resins and Composites I*, 78, Dusek, K. (Ed.) (1986) Springer-Verlag, Berlin, p. 83–114
102. Wisanrakkit, G., Gillham, J.K. *J. Appl. Polym. Sci.* (1990) 41 p. 2885–2929
103. Gillham, J.K. *Polym. Eng. Sci.* (1986) 26(20) p. 1429–1443
104. Mussati, F.G. *Rheology of Network Forming Systems*. PhD thesis, University of Minnesota, Minneapolis, 1975
105. Vilesova, M.S., Spasskova, N.P., Lesnevskaya, L.V., Guseva, G.N., Izrailev, L.G., Zolotarev, V.M. *Polym. Sci. U.S.S.R.* (1972) 14 p. 1883
106. French, D.M., Strecker, R.A.H., Tompa, A.S. *J. Appl. Polym. Sci.* (1970) 14 p. 599
107. Stockmayer, W.H. *J. Chem. Phys.* (1943) 11 p. 45
108. Agrawal, A., Uhlmann, D.R. *Polymer* (1991) 32 p. 290–296
109. Rozenberg, B.A., *Adv. Polym. Sci.* (1985) 75 p. 113–165
110. Nielsen, L.E., *J. Macromol. Sci., Rev. Macromol. Chem.* (1969) C3 p. 69
111. Hwang, J.G., Row, C.G., Hwang, I., Lee, S.J. *Ind. Eng. Chem. Res.* (1994) 33(10) p. 2377–2383
112. Williams, M.L., Landel, R.F., Ferry, J.D., *J. Am. Chem. Soc.* (1955) 77 p. 3701
113. Mijovic, J., Lee, C.-H. *J. Appl. Polym. Sci.* (1989) 38 p. 2155–2170
114. Cohen, M.H., Turnbull, D. *J. Chem. Phys.* (1959) 31 p. 1164
115. Sanford, W.M., McCullough, R.L. *J. Polym. Sci.: Part B: Polym. Phys.* (1990) 28 p. 973–1000
116. Miller, D.R., Macosko, C.W., *Macromolecules* (1978) 11 p. 656
117. Macosko, C.W., Miller, D.R. *Macromolecules* (1976) 9 p. 199–213
118. Lee, W.I., Loos, A.C., Springer, G.S. *J. Comp. Mat.* (1982) 16 p. 510–520
119. Young, W-B. *Comp. Sci. Tech.* (1995) 54 p. 299–306
120. Dusi, M.R., Lee, W.I., Ciriscioli, P.R., Springer, G.S. *J. Comp. Mat.* (1987) 21 p. 243–261
121. Stolin, A.M., Merzhanov, A.G., Malkin, A.Y. *Polym. Eng. Sci.* (1979) 19(15) p. 1074–1080
122. Kim, D.-S., Kim, S.C. *Polym. Eng. Sci.* (1994) 34 p. 625–631
123. Lee, D-S., Han, C.D. *Polym. Eng. Sci.* (1987) 27(13) p. 955–963
124. Tajima, Y.A., Crozier, D. *Polym. Eng. Sci.* (1983) 23 186–274
125. Hou, T.H. in *SPE ANTEC Tech. Papers*, (1985) 31, p. 1253
126. Srinivasan, R., Wang, T.J., Lee, L.J. *ANTEC* (1995) p. 3016–3029
127. Rohr, D.F., Klein, M.T. *Ind. Eng. Chem. Res.* (1990) 29 p. 1210–1218
128. Deng, Y., Martin, G.C. *Macromolecules* (1994) 27 p. 5141–5146
129. Yu, W., von Meerwall, E.D. *Macromolecules* (1990) 23 p. 882–889

130. Kim, D.H., Kim, S.C. *Polym. Bull.* (1987) 18 p. 533–539
131. Adabbo, H.E., Williams, R.J.J. *J. Appl. Polym. Sci.* (1982) 27 p. 1327–1334
132. Matsuoka, S., Quan, X., Bair, H.E., Boyle, D.J. *Macromolecules* (1989) 22(10) p. 4093–4098
133. Pascault, J.P., Williams, R.J.J. *J. Polym. Sci., Part B: Polym. Phys.* (1990) 28 p. 85
134. Rabinowitch, E., *Trans. Faraday Soc.* (1937) 33 p. 1225
135. Dannesburg, H. *Soc. Plast. Eng. J.* (1959) 15 p. 875
136. Fisch, W., Hoffman, W., Schmid, R., *J. Appl. Polym. Sci.* (1969) 13 p. 295
137. Willard, P.E. *Polym. Eng. Sci.* (1972) 12(2) p. 120–124
138. Aukward, J.A., Warfield, R.W., Petrie, M.C., *J. Polym. Sci.* (1958) 27 p. 199
139. Miller, D.R., Macosko, C.W. *Macromolecules* (1978) 11(4) p. 656–662
140. Sourour, S., Kamal, M.R. *Thermochimica Acta* (1976) 14(41) p. 41–59
141. Creedon, J.P. *Anal. Calorim.* (1970) 2 p. 185–199
142. Min, B.-G., Stachurski, Z.H., Hodgkin, J.H., Heath, G.R. *Polymer* (1993) 34(17) p. 3620–3627
143. Barrall, E.M. II, Johnson, J.F. In Slade, P.E., Jenkis, L.T. (Eds.) *Thermal Characterization Techniques* (1970) 2, Marcel Dekker, New York, p. 1–39
144. Richardson, M.J. In Dawkins, J.V. (Ed.) *Developments in Polymer Characterization* (1978) 1, Applied Science Publishers, London, p. 205–244
145. Barret, K.E.J. *J. Appl. Polym. Sci.* (1967) 11 p. 1617–1626
146. Vergnaud, J.-M., Bouzon, J. *Cure of Thermosetting Resins—Modelling and Experiments* (1992) Springer-Verlag, Berlin, Germany
147. Fava, R.A. *Polymer* (1968) 9 p. 137–151
148. White, S.R., Hahn, H.T. *J. Comp. Mat.* (1993) 27(14) p. 1352–1378
149. Arridge, R.G.C., Speake, J.H. *Polymer* (1972) 13 p. 443–449
150. MacCallum, J.R., Tanner, J. *Nature* (1970) 225 p. 1127
151. Sestak, J., Kratochvil, J., *J. Thermal Anal.* (1973) 5 p. 193
152. Dutta, A., Ryan, M.E. *Thermochimica Acta* (1979) 33 p. 87–92
153. Ozawa, T. *Thermochimica Acta* (1986) 100 p. 109–118
154. Letton, A., Chiou, P.B. In *Polymer Rheology and Processing*, Collyer, A.A., Utracki, L.A. (Eds.) (1990) chapter 16, Elsevier, Amsterdam, p. 431–458
155. Abolafia, O.R. In *Technical Papers, 27th Annual Technical Conference* (1969) Society of Plastics Engineers (Ed.), Chicago, Illinois, p. 610
156. de Bakker, C.J., St John, N.A., George, G.A. *Polymer* (1993) 34(4) p. 716–725
157. Kratochvil, J., Sestak, J. *Thermochimica Acta* (1973) 7 p. 330–332
158. Draper, A.L. In *Proc. Toronto Symp. Therm. Anal.* (1970) 3, p. 63
159. Simmons, E.L., Wendlandt, W.W. *Thermochimica Acta* (1972) 3 p. 498–500
160. Hill, R.A.W. *Nature* (1970) 227 p. 703
161. Gorbatchev, V.M., Logvinenko, V.A., *J. Therm. Anal.* (1972) 4 p. 475
162. Petrovic, Z.S., Zavargo, Z.Z. *J. Appl. Polym. Sci.* (1986) 32 p. 4353–4367
163. Lewis, A. *Polym. Eng. Sci.* (1979) 19(10) p. 662–663
164. Heijboer, J. *Polym. Eng. Sci.* (1979) 19(10) p. 664–675
165. Gillham, J.K. in *Developments in Polymer Characterization*, Dawkins, J.V. (Ed.) (1978) 3, Applied Science Publishers, London, p. 159–227
166. Babayevsky, P.G., Gillham, J.K. *J. Appl. Polym. Sci.* (1973) 17 p. 2067–2088
167. Gillham, J.K., Roller, M.B. *Polym. Eng. Sci.* (1971) 11(4) p. 295–304
168. Gillham, J.K. In *Thermal Characterization Techniques* (1970) Slade, P.E., Jr., Jenkis, L.T. (Eds.) 2, Marcel Dekker, New York, p. 225–292
169. Wingard, C.D., Beatty, C.L. *J. Appl. Polym. Sci.* (1990) 40, p. 1981–2005
170. Wingard, C.D., Beatty, C.L. *J. Appl. Polym. Sci.* (1990) 41 p. 2539–2554
171. Mussati, F.G., Macosko, C.W., *Polym. Eng. Sci.* (1973) 13(3) p. 236–240
172. White, S.R., Kim, C., *J. Reinforced Plast. Comp.* (1993) 12 p. 520–535
173. Shimazaki, A. *J. Appl. Polym. Sci.* (1968) 12 p. 2013–2021
174. Learmonth, G.S., Pritchard, G. *J. Appl. Polym. Sci.* (1969) 13 p. 2119–2127
175. Maistros, G.M., Partridge, I.K. *Comp. Sci. Tech.* (1995) 53 p. 355–359

176. Bellenger, V., Verdu, J., Francillette, J., Hoarau, P., Morel, E. *Polymer* (1987) 28 p. 1079–1086
177. Sabra, A., Lam, T.M., Pascault, J.P., Grenier-Loustalot, M.F., Grenier, P. *Polymer* (1987) 28 p. 1030–1036
178. Attias, A.J., Bloch, B.J. *J. Appl. Polym. Sci. Part A: Polym. Chem.* (1990) 28 p. 3445–3466
179. Rogers, M.G. *J. Appl. Polym. Sci.* (1972) 16 p. 1953–1958
180. Mertzel, E., Koenig, J.L. In *Advances in Polymer Science—Epoxy Resins and Composites II*, Dusek, K. (Ed.) (1986) 75, Springer-Verlag, Berlin, p. 73–112
181. Saunders, T.F., Levy, M.F., Serino, J.F., *J. Polym. Sci. A1* (1967) 5 p. 1609–1617
182. Batzer, H., Zahir, S.A. *J. Appl. Polym. Sci.* (1975) 19 p. 601–607
183. Xu, L., Fu, J.H., Schlup, J.R. *J. Am. Chem. Soc.* (1994) 116 p. 2821–2826
184. Banks, P., Peters, R.H. *J. Polym. Sci.: Part A-1* (1970) 8 p. 2595–2610
185. Lem, K.-W., Han, C.D. *J. Appl. Polym. Sci.* (1983) 28 p. 3185–3206
186. Palmese, G.R., Karbhari, V.M. *Polym. Comp.* (1995) 16(4) p. 313–318
187. Dutta, A., Ryan, M.E., *J. Appl. Polym. Sci.* (1979) 24 p. 635–649
188. McGee, S.H. *Polym. Eng. Sci.* (1982) 22(8) p. 484–491
189. Hendra, N.G., Manas-Zloczower, I. *Polym. Eng. Sci.* (1989) 29(5) p. 302–307
190. Plueddenmann, E.P. *Silane Coupling Agents* (1982) Plenum Press, New York
191. Han, C.D., Lee, D.S., Chin, H.B. *Polym. Eng. Sci.* (1986) 26(6) p. 393–404
192. Mijovic, J., Wang, H.T. *Sampe J.* (1988) 24 p. 42
193. Matthews, F.L., Rawlings, R.D. *Composite Materials: Engineering and Science* (1994) Chapman and Hall, London
194. Grillet, A.C., Galy, J., Pascault, J.P., Bardin, I. *Polymer* (1989) 30 p. 2094–2103
195. Sultan, J.N., McGarry, F.J. *Polym. Eng. Sci.* (1973) 13(3) p. 29–34
196. Delvigs, P. *Polym. Comp.* (1986) 7(2) p. 101–105
197. Tanaka, Y., Kakiuchi, H. *J. Appl. Polym. Sci.* (1963) 7 p. 1063–1081
198. Mohan, R., Grentzer, T.H. *J. Reinforced Plast. Comp.* (1995) 14 p. 73–84
199. Lam, P.W.K. *Polym. Comp.* (1987) 8(6) p. 427–436
200. Lee, L.J. *Polym. Eng. Sci.* (1981) 21 p. 483–492
201. Han, C.D., Lee, D-S. *J. Appl. Polym. Sci.* (1987) 33 p. 2859–2876
202. Storey, R.F., Sudhakar, D., Hogue, M.L. *Polym. Prep.* (1986) 27(2) p. 167
203. Pusatcioglu, S.Y., Fricke, A.L., Hassler, J.C. *J. Appl. Polym. Sci.* (1979) 24 p. 947–952
204. Kamal, M.R., Sourour, S., Ryan, M. *Soc. Plast. Eng. Tech. Papers* (1973) 19 p. 187
205. Han, C.D., Lem, K.W. *J. Appl. Polym. Sci.* (1983) 28 p. 3155
206. Michaud, D. *Investigation of Curing Behavior in Thick Thermoset Composites Manufactured by Resin Transfer Molding* (1996) M.Sc. Thesis, Chemical Engineering Department, Newark, Del
207. Knop, A., Scheib, W. *Chemistry and Application of Phenolic Resins* (1979) Springer-Verlag, Berlin
208. Martin, R.W. *The Chemistry of Phenolic Resins* (1956) Wiley, New York
209. Stokes, E.H. *AIAA Journal* (1992) 30(6) p. 1597–1601
210. Carswell, T.S. *Phenoplasts* (1947) Interscience Publishers, Inc., New York
211. King, P.W., Mitchell, R.H., Westwood, A.R. *J. Appl. Polym. Sci.* (1974) 18 p. 1117–1130
212. Christiansen, A.W. *J. Appl. Polym. Sci.* (1985) 30 p. 2279–2289
213. Wolfgang Hesse. Phenolic Resins. Personal communication
214. So, S., Rudin, A. *J. Appl. Polym. Sci.* (1990) 41 p. 205–232
215. Katović, Z. *J. Appl. Polym. Sci.* (1967) 11 p. 95–102
216. Waage, K., Gardner, D.J., Elder, T.J. *J. Appl. Polym. Sci.* (1991) 42 p. 273–278
217. Wagner, E.R., Greff, R.J. *J. Polym. Sci.: Part A-1* (1971) 9 p. 2193–2207
218. Rudin, A., Fyfe, C.A., Vines, S.M. *J. Appl. Polym. Sci.* (1983) 28 p. 2611–2622
219. So, S., Rudin, A. *J. Appl. Polym. Sci.* (1990) 40 p. 2135–2149
220. Kim, M.G., Amos, L.W., Barnes, E.E. *Ind. Eng. Chem. Res.* (1990) 29 p. 2032–2037
221. Batzer, H., Zahir, S.A. *J. Appl. Polym. Sci.* (1975) 19 p. 585–600
222. Tonogai, S., Hasegawa, K., Inada, M. *Polym. Eng. Sci.* (1980) 20(9) p. 608
223. So, S., Rudin, A. *J. Polym. Sci. Polym. Lett. Ed.* (1985) 23 p. 403–407

224. Hirst, R.C., Grant, D.M., Hoff, R.E., Burke, W.J. *J. Polym. Sci.: Part A* (1965) 3 p. 2091–2105
225. Woodbrey, J.C., Higginbottom, H.P., Culbertson, H.M. *J. Polym. Sci.: Part A* (1965) 3 p. 1079–1106
226. Katović, Z. *J. Appl. Polym. Sci.* (1967) 11 p. 85–93
227. Antony, R., Pillai, C.K.S. *J. Appl. Polym. Sci.* (1994) 54 p. 429–438
228. BP Chemicals. *Phenolic Resins—An Introduction*, personal communication
229. Odian, G., *Principles of Step Polymerization* (1981) John Wiley and Sons, New York
230. Davé, R. *J. Comp. Mat.* (1990) 24 p. 22–41
231. Friedrichs, B., Güçeri, S.I. *Polym. Eng. Sci.* (1995) 35(23) p. 1834–1851
232. Bruschke, M.V., Advani, S.G. In *Proceedings of Transport Phenomena in Material Processing* (1990) New York 132 p. 149–158
233. Bruschke, M.V. *A Predictive Model for Permeability and Non-Isothermal Flow of Viscous and Shear-Thinning Fluids in Anisotropic Fibrous Media*. PhD thesis, Mechanical Engineering Department, University of Delaware, 1992
234. Phelan, F.R. In *Proc. Am. Soc. Compos. 7th Tech. Conf. on Composite Materials* (1992) ASC, Washington
235. Lin, R., Lee, L.J., Liou, M.J. In *SPE ANTEC'91* (1991) 37 p. 815
236. Tucker, C.L., III. *Polym. Comp.* (1996) 17(1) p. 60–72
237. Woerdeman, D.L., Phelan, F.R., Parnas, R.S. *Polym. Comp.* (1995) 16(6) p. 470–480
238. Gebart, B.R., Lidström, P. *Polym. Comp.* (1996) 17(1) p. 43–51
239. Rudd, C.D., Long, A.C., McGeehin, P., Smith, P. *Polym. Comp.* (1996) 17(1) p. 52–59
240. Ferland, P., Guittard, D., Trochu, F. *Polym. Comp.* (1996) 17(1) p. 149–158
241. Gauvin, R., Trochu, F., Lemenn, Y., Diallo, L. *Polym. Comp.* (1996) 17(1) p. 34–42
242. Bruschke, M.V., Luce, T.L., Advani, S.G. In *Proceedings of the American Society for Composites 7th Technical Conference on Composite Materials* (1992) Washington, p. 103–112
243. Luce, T.L., Advani, S.G., Howard, J.G., Parnas, R.S. *Polym. Comp.* (1996) 16(6) p. 446–458
244. Parnas, R.S., Howard, J.G., Luce, T.L., Advani, S.G. *Polym. Comp.* (1996) 16(6) p. 429–445
245. Sadiq, T.A.K., Advani, S.G., Parnas, R.S. *Int. J. Multiphase Flow* (1995) 21 p. 755–774
246. Calado, V., Advani, S.G. *Comp. Sci. Techn.* (1996) 56 p. 519–531
247. Trevino, L., Rupel, K., Young, W.B., Liou, M.J., Lee, L.J. *Polym. Comp.* (1991) 12 p. 20–29
248. Young, W.B., Rupel, K., Han, K., Lee, L.J., Liou, M.J. *Polym. Comp.* (1991) 12 p. 30–38
249. Bruschke, M.V., Advani, S.G. *Polym. Comp.* (1990) 11 p. 291–304
250. Broyer, E., Macosko, C.W. *AIChE J.* (1976) 22(2) p. 268–276
251. Tucker, C.L., III, Dessenberger, R.B. In *Flow and Rheology in Polymer Composites Manufacturing*, Advani, S.G. (Ed.) (1994) 10 Elsevier, Amsterdam, p. 257–323
252. Tien, C.L. *Adv. Appl. Mech.* (1990) 27 p. 225–281
253. Vafai, K., Sozen, M. *Annu. Rev. Heat Transfer* (1990) 3 p. 145–162
254. Whitaker, S. *Adv. Heat Transfer* (1977) 13 p. 119–203
255. Nield, D.A., Bejan, A. *Convection in Porous Media* (1992) Springer, New York
256. Kaviany, M. *Principles of Heat Transfer in Porous Media* (1995) Springer-Verlag, New York
257. Uenoyama, M., Güçeri, S.I. Technical Report 91-09, University of Delaware, Newark, DE, 1991
258. Scott, F.N. *Processing Characteristics of Polyester Resin for the Resin Transfer Moulding Process*. PhD thesis, University of Nottingham, 1988
259. Chan, A.W., Hwang, S.T., *Polym. Eng. Sci.* (1992) 32 p. 310
260. Aoyagi, H., Uenoyama, M., Güçeri, S.I. *Int. Polym. Proc.* (1992) 7 p. 71
261. Lin, R., Lee, L.J., Liou, M.J. In *Advanced Composites Materials; New Developments and Applications Conference Proceedings* (1991) Detroit, p. 165
262. Lin, R., Lee, L.J., Liou, M.J. *Int. Polym. Pro.* (1991) 6 p. 356
263. Khomami, B. In *Processing of Continuous Fiber Reinforced Composites*, Davé, R.S., Loos, A. (Eds.) (1996) Hanser, Munich.
264. Dessenberger, R.B., Tucker, C.L., III. *Polym. Comp.* (1995) 16(6) p. 495–506
265. Owen, M.J., Middleton, V., Rudd, C.D., Scott, F.N., Hutcheon, K.F., Revill, I.D. *Plast. Rubber Process. Appl.* (1989) 112(4) p. 221–225
266. Whitaker, S. *Chem. Eng. Sci.* (1986) 41 p. 2029–2039

267. Whitaker, S. *Ind. Eng. Chem Res.* (1991) 30 p. 983–997
268. Whitaker, S. *Chem. Eng. Commun.* (1987) 58 p. 171–183
269. Pusatcioglu, S.Y., Hassler, J.C., Fricke, A.L., McGee, H.A. *J. Appl. Polym. Sci.* (1980) 25 p. 381–393
270. Tredoux, L., van der Westhuizen, J. *Comp. Man.* (1995) 6(2) p. 85–92
271. Williams, R.J.J., Benavente, M.A., Ruseckaite, R.A. *Polym. Eng. Sci.* (1990) 30(18) p. 1140–1145
272. Hieber, C.A., Shen, S.F. *J. Non-Newtonian Fluid Mech.* (1980) 7(1) p. 1–31
273. Osswald, T.A., Tucker, C.L. *Polym. Eng. Sci.* (1988) 28(7) p. 413–420
274. Liu, B., Bruschke, M.V., Advani, S.G. *Liquid Injection Molding Simulation—LIMS* (1994) Department of Mechanical Engineering, University of Delaware, Newark, DE
275. Liu, B., Advani, S.G. *Comp. Mech.* (1995) 38 p. 74–82
276. Bénard, A. *Transport Phenomena and Microstructure Development during Solidification of Semicrystalline Polymers and Composites* (1995) PhD thesis, Mechanical Engineering Department, University of Delaware, Newark, Del
277. Bénard, A., Advani, S.G. *Polym. Eng. Sci.* (1996) 36 p. 520
278. Desbiolles, J.L., Droux, J.J., Rappaz, J., Rappaz, M. *Comp. Phys. Report* (1987) 6 p. 371–383
279. Scott, E.P., Beck, J.V. *J. Comp. Mat.* (1992) 26(1) p. 20–36
280. Hatta, H., Taya, M., Kulacki, F.A., Harder, J.F. *J. Comp. Mat.* (1992) 26(5) p. 612–625
281. Pitchumani, R., Yao, S.C. *Int. J. Heat Mass Transfer* (1992) 35(9) p. 2185–2194
282. Springer, G.S., Tsai, S.W. *J. Comp. Mat.* (1967) 1 p. 166–173
283. Kim,I.C., Torquato, S. *J. Appl. Phys.* (1990) 68(8) p. 3892–3903
284. Han, L.S., Cosner, A.A., *J. Heat Transfer* (1981) 103 p. 387–392
285. Pitchumani, R., Yao, S.C. *Trans. ASME, J. Heat Transfer* (1991) 113 p. 788–796
286. McCullough, R.L. *Comp. Sci. Techn.* (1985) 22 p. 3–21
287. Progelhof, R.C., Throne, J.L., Ruetsch, R.R. *Polym. Eng. Sci.* (1976) 16(9) p. 615–625
288. Torquato, S. *Appl. Mech. Rev.* (1991) Lancaster, 4 p. 37
289. Whitney, J.L., McCullough, R.L. In *Delaware Composites Design Encyclopedia* (1990) 2. Technomic Publishing, Lancaster p. 231
290. Halpin, J. *Primer on Composite Materials: Analysis* (1984) Technomic Publishing, Lancaster
291. Christensen, R.M. *Mechanics of Composite Materials* (1979) Wiley Interscience, New York
292. Khreahling, R.P., Kline, D.E. *J. Appl. Polym. Sci.* (1969) 13 p. 2411–2425
293. van Krevelan, D.W. *Properties of Polymers: Correlations with Chemical Structures* (1972) Elsevier, Amsterdam
294. Brandrup, J., Immergut, E.H. *Polymer Handbook* (1975) John Wiley and Sons, New York, V7–12
295. Gebart, B.R., Stömbeck, A. In *Processing of Continuous Fiber Reinforced Composites* Raju S., Davé, Loos, A. (Eds.) (1996) Hanser, Munich.
296. Springer, G.S. *Progr. Sci. Eng. Compos.* (1982) 1 p. 23–35
297. Young, W.-B. *Polym. Comp.* (1996) 17(1) p. 142–148
298. Gutowski, T.G., Cai, Z., Bauer, S., Boucher, D., Kingery, J., Wineman, S. *J. Comp. Mat.* (1987) 21 p. 650–669
299. Gutowski, T.G., Morigaki, T., Cai, Z. *J. Comp. Mat.* (1987) 21 p. 172–188
300. Day, D.R. In *Proceedings of the 33rd International SAMPE Symposium* (1989) p. 594–602, Covina, Cal
301. Ferry, J.D. *Viscoelastic Properties of Polymers* (1970) John Wiley and Sons, New York
302. Jang, B.Z., Zhu, G.H. *J. Appl. Polym. Sci.* (1986) 31 p. 2627–2646
303. Woo, E.M., Seferis, J.C. *J. Appl. Polym. Sci.* (1990) 40 p. 1237–1256
304. Stutz, H., Mertes, J., Neubecker, K., *J. Polym. Sci.: Part A: Polym. Chem.* (1993) 31 p. 1879–1886

3 Phase Separation and Morphology Development during Curing of Toughened Thermosets

J.W. Park and S.C. Kim

The phase separation behavior during curing of polyetherimide (PEI) modified diglycidyl ether of bisphenol A (DGEBA) epoxy and PEI modified bisphenol A dicyanate (BPACY) were studied using SEM, light scattering, and dynamic mechanical analyzer.

At low PEI content (< 10 wt%), the blends exhibited a sea-island morphology formed via nucleation and growth mechanism. Above 25 wt% PEI content, the phase separation proceeded via spinodal decomposition mechanism and nodular morphology was formed. At intermediate compositions dual phase morphology with both sea-island and nodular morphology was observed. This morphology was formed via primary spinodal decomposition and secondary phase separation within the dispersed and the matrix phases.

3.1 Introduction

Thermosets are generally used in advanced composites due to their excellent thermal and dimensional stability, high modulus, and good mechanical properties. Because thermoset resins are inherently brittle, however, some applications require improved fracture resistance. Toughening of thermosets has been achieved through various methods, such as incorporation of reactive liquid rubber [1–9], elastomer [10], or rigid thermoplastics [11–25], and IPN formation with ductile component [26].

Among these methods, incorporation of thermoplastics has been highlighted as a new approach to enhance the toughness of thermosets without significantly lowering the desirable properties. Rigid thermoplastics with high T_g, such as polyethersulfone [11–16], polyetherimide [17–19], and polysulfone [20–23], were frequently used in this approach.

Properties are so closely related to the morphology that the analysis on the phase separation during cure is needed. Many works have shown the improved fracture resistance in thermoplastic-modified thermosets with phase-separated morphology [12–18, 25–29], whereas the homogeneous system did not exhibit the significant increase in the fracture resistance [30]. Thus, phase separation during cure is regarded to be essential to produce the toughened thermosets, and the degree of toughening is very dependent on the degree of phase separation [31].

Relatively few papers have reported the phase separation mechanism during curing of toughened thermosets [1–3, 7–9, 32–40], and the phase separation mechanisms is not well understood.

3.2 Phase Separation in Terms of Thermodynamics and Kinetics

Polymer–polymer systems exhibit phase behavior similar to other mixtures, such that an initially uniform system separates into two or more phases as a result of small change in thermodynamic variable. Two mechanisms can be envisioned to explain this phenomenon: nucleation and growth (NG), and spinodal decomposition (SD).

Nucleation is initiated by local fluctuations of concentration within a metastable region. The activation energy of nucleation depends on the value of the interface energy required to create a nucleus. The droplet grows by diffusion of macromolecules into the nucleate domains. The natural form of the phase separation through NG mechanism is the sea-island type.

In the unstable region, the concentration fluctuations are delocalized and there is no thermodynamic barrier to phase growth. Thus, separations that take place spontaneously lead to long range phase separation. This process is called spinodal decomposition (SD). In this mechanism, decomposition starts with a co-continuous structure and gradually shifts to a droplet morphology because of the breakdown of the continuous structure [41].

In a dynamic and cross-linkable system, such as the curing of a thermoset that contains a thermoplastic, the phase separation is more complicated than nonreaction system. The phase separation is controlled by the competing effects of thermodynamics and kinetics of phase separation and cure rate of thermoset resin (i.e. time dependent viscosity of the system).

Figure 3.1 shows the change of UCST-type phase diagram during curing at isothermal cure condition. The increase in thermoset molecular weight shifts the spinodal and binodal curves upward while the T_g and the viscosity of the mixture also increase. At the beginning of the curing reaction, P_1 and P_2 are located above the binodal curve in a state where the system is miscible with the conversion x_1. As the reaction proceeds and thermoset conversion is progressively increased to x_2, phase separation starts by the NG mechanism when the binodal curve moves upward and passes point P_1. In this metastable region, any further phase separation will be halted if the mixture viscosity is greater than a critical value. In such a case, phase separation by the NG mechanism will start and finish only in this metastable region. It is also possible that P_1 will fall into the unstable region by passing the spinodal curve at the time of structural freeze-in by vitrification or by gelation. In such a situation, the phase separation actually takes place by two mechanisms in sequence: first NG, then SD. If phase separation starts in the metastable region, as in point P_2, then the phase separation will proceed only by SD mechanism.

Teng et al. [42] showed that the final cured product in thermoplastic/thermoset system is in one of four classes: (1) in miscible region and single phase; (2) in immiscible region and multiple phase; (3) in miscible region and multiple phase; and (4) in immiscible region and single phase with the curing rate.

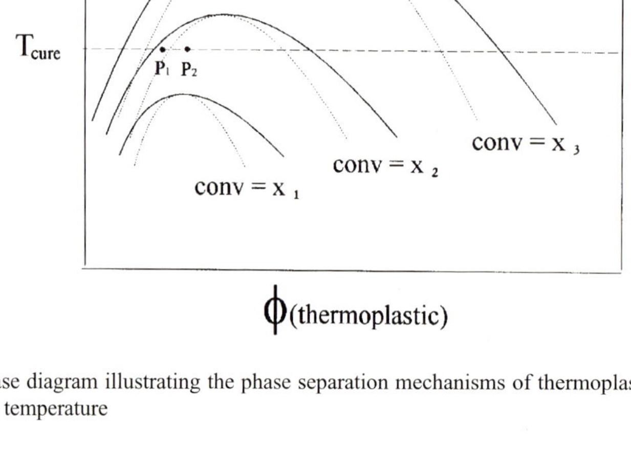

Figure 3.1 Phase diagram illustrating the phase separation mechanisms of thermoplastic/thermoset blends at constant cure temperature

The extent of phase separation can be measured directly by the scanning electron microscope (SEM), transmission electron microscope (TEM), optical microscope, and light or X-ray scattering technique. It is also investigated indirectly by measuring certain physical properties, such as glass transition temperature.

3.3 Literature Review

The phase separation behavior during curing of toughened epoxy has been studied by Inoue and coworkers [32–35]. They stated that the NG mechanism is not expected to take place in any case mainly because of the fact that nucleation is recognized to be a very slow process and the sea-island morphology as well as nodular morphology is developed by only SD decomposition during curing. They supported it by the characteristic change of light scattering profile and light micrographs with curing time.

Inoue et al. also explained the development of the sea-island and connected globular morphology as follows. After the temperature jumps to the curing temperature, the homogeneous mixture starts to phase separate by SD, which results in the development of a co-continuous structure, such as schematically shown in Figure 3.2a. When phase separation proceeds, the periodic distance increases. At the same time phase connectivity will be interrupted by the increase in interfacial tension, which results in a dispersed droplet-type morphology (b). If a network is already established in the epoxy rich region, however, complete interruption cannot be realized (d) and it eventually results in the connected-globule structure (e). Once interruption has taken place, if the dispersed droplets grow in size without changing their loci (c) because they are dispersed in a matrix of PES-rich phase with low

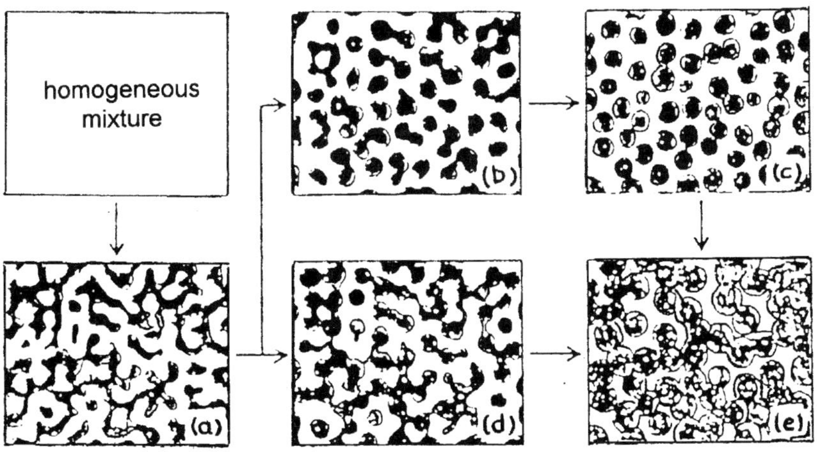

Figure 3.2 Schematic representation of phase separation scheme resulting in the connected globule structure (Reproduced from [32])

mobility (high T_g), then, by further growth, the droplets finally contact with each other to yield the connected globular structure (e).

Even with the aid of a highly reactive curing agent, however, they failed to obtain a co-continuous structure in the composition region showing the sea-island morphology at any curing temperature. They could obtain the co-continuos two-phase structure by lowing the curing temperature (i.e. decreasing the rate of phase separation). It might be an error to assume that the phase separations at different reaction temperature develop by the same phase separation mechanism.

In their paper, Inoue et al. [36] proposed a new approach about phase separation mechanism under successive increases in quench depth. Figure 3.3 shows schematic

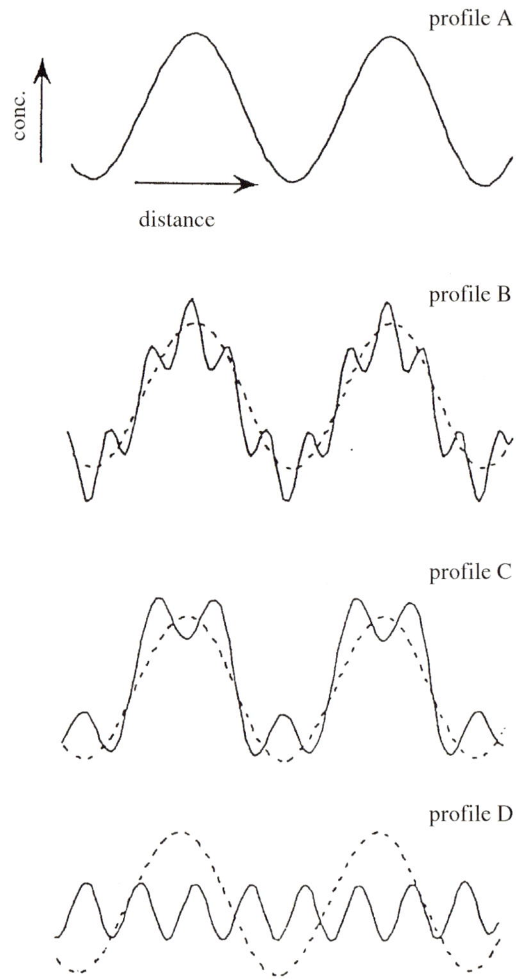

Figure 3.3 Schematic concentration profiles developed by the isothermal SD (profile) and after a second jump to a deeper quench (profile B–D). (Reproduced from [36])

concentration profiles developed by the two-step SD. First, at a shallow quench the SD proceeds to give concentration profile (A). When the system is thrust into a deeper quench, three conceivable situations are shown: short waves expected at a deeper quench overlap with the long waves (profile B), short waves appear between long waves (profile C), and long waves degenerate while short waves develop (profile D).

To evaluate the demixing process under the nonisoquench depth condition, they carried out computer simulations of the time dependent concentration fluctuation using the Cahn–Hilliard nonlinear diffusion equation.

$$\frac{\partial c}{\partial t} = \frac{\partial}{\partial x}\left(D\frac{\partial c}{\partial x}\right) - 2Mk\left(\frac{\partial^4 c}{\partial x^4}\right)$$

where c is the concentration, x is the distance, t is the time, M is the mobility of the molecules, k is the energy gradient coefficient, and D is the diffusion coefficient defined by

$$D = M\left(\frac{\partial^2 f}{\partial c^2}\right)$$

where f is the local free energy of the mixture.

The simulations revealed that the SD under successive increases in quench depth yields the regular two-phase structure as does the case of the SD under isoquench. The structure coarsening is suppressed by the increases in quench depth, and the final morphology is highly dependent on the quench rate. The results adequately describe the characteristic features of structure development in the reaction-induced SD. The following serves as an example of the simulated result. In the case where quench rate $= 0.6$ K/s (Fig. 3.4), a new fluctuation appears between the original ones at a large quench depth (c), both sets of peaks grow to establish a trapezoid/triangle alternating profile (e). This simulated result can explain the results for the low reactivity epoxy/CTBN/piperidine system [33], which offers a light-scattering profile with two peaks. This system actually shows a bimodal distribution of CTBN particles, as expected from the concentration profile in Figure 3.4b.

On the contrary, many authors describe the sea-island morphology was caused by the nucleation and growth mechanism [1–3,7–9,37–40]. Williams et al. [1–3,37–40] proposed a theoretical model to predict the fraction, composition, and particle size distribution of dispersed phase segregated via NG during curing of a rubber-modified epoxy system. The model is based on a thermodynamic description through a Flory–Huggins equation and constitutive equations for the polymerization and phase separation rate. Equations for nucleation, coalescence, and growing rates was derived and analyzed.

They suggested that three possible trajectories of the matrix and domain composition during curing of the modified epoxy (Fig. 3.5) [39]. The trajectory is determined by the relative rates of phase separation and polymerization.

$$k = \text{phase separation rate/polymerization rate}$$

If $k \to \infty$ then equilibrium is instantaneously reached and the system evolves along the binodal curve (trajectory a and a'). On the other hand, if $k \to 0$, then no phase separation will

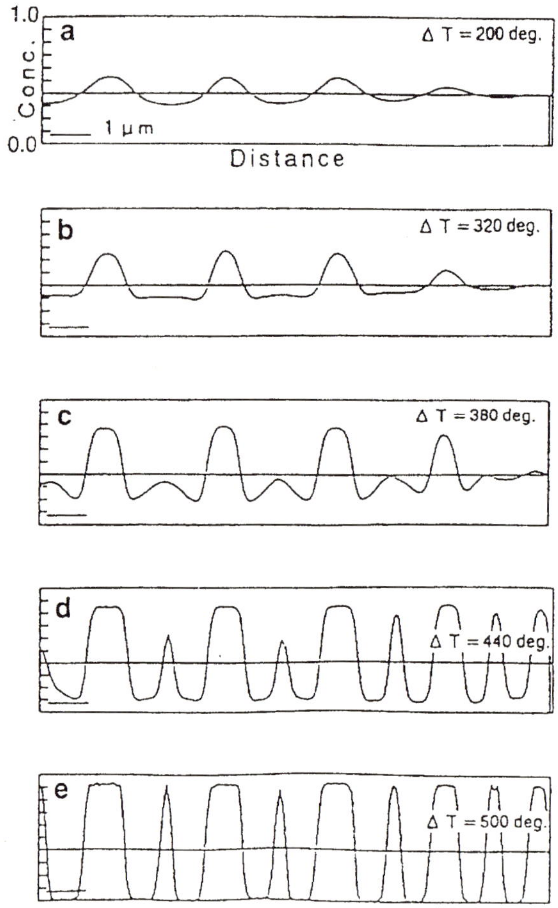

Figure 3.4 Time dependence of the concentration fluctuation during demixing with successive increases in the quench depth (quench rate = 0.6 K/s). (Reproduced from [36])

be produced until the spinodal curve is reached (trajectory *c*), at which point separation takes place by SD, leading to co-continuous morphology. Trajectory (*b*) represents the general case in which phase separation takes place by NG at rate insufficient to achieve equilibrium condition. Composition of the dispersed phase is indicated by trajectory b', which lies outside the metastable region [40].

The necessary condition to have a NG mechanism is that the second phase can nucleate from the homogeneous solution. The nucleation rate depends on interaction parameter, interfacial tension between both phases, and diffusion coefficient.

The nucleation rate (NR) is proportional to

$$NR = N_o D_{AB} \exp(-\Delta G_c / kT)$$

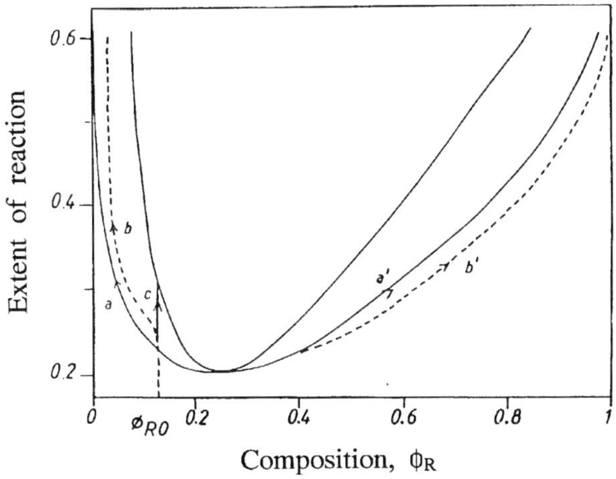

Figure 3.5 Possible trajectories in the metastable region of the conversion versus composition phase diagram: a' and b' represent the compositions of the dispersed phase corresponding to trajectories a and b, respectively. (Reproduced from [39])

where N_o is an adjustable preexponential factor, D_{AB} is the diffusion coefficient, k is the Boltzman constant, and

$$\Delta G_c = 16\pi\sigma_o^3/3|\Delta G_N|^2$$

where σ_o is the interfacial tension between both phases, and ΔG_N is the free energy change per unit volume.

Figure 3.6 shows trajectories for several values of the interfacial tension σ_o in the metastable region that are predicted by using a NG model [2]. For $\sigma_o \leq 0.01$ mN/m, all the trajectories are equivalent (i.e., the value of σ_o has no effect on the phase separation process); however, if σ_o is higher than 0.1 mN/m, then spinodal decomposition is rapidly attained. For $\sigma_o = 0.4$ mN/m practically no phase separation occurs by the NG mechanism, and demixing proceeds by SD.

In the same manner, with decreasing of diffusion coefficient and interaction parameter, the spinodal is reached during the evolution of the system in the pregel stage. The very low values of interfacial tension in rubber modified epoxies (interfacial tension of polymer–polymer–solvent system were reported in range of 10^{-4}–10^{-1} mN/m) therefore lead to an NG mechanism for phase separation.

Phase separation through NG mechanism cannot be observed for polymer–polymer blend systems that show interfacial tension lying in the range 0.5–11 mN/m. In addition, they predicted that a secondary phase separation could take place inside dispersed rubber particles in the case when the average composition of dispersed domains lies in the unstable region at the end of the phase separation [2]. They were not able to observe a phase separation inside dispersed domains with TEM micrographs; however, they concluded that there are two phases inside the dispersed domains by the fact that the glass transition temperature of the rubber-

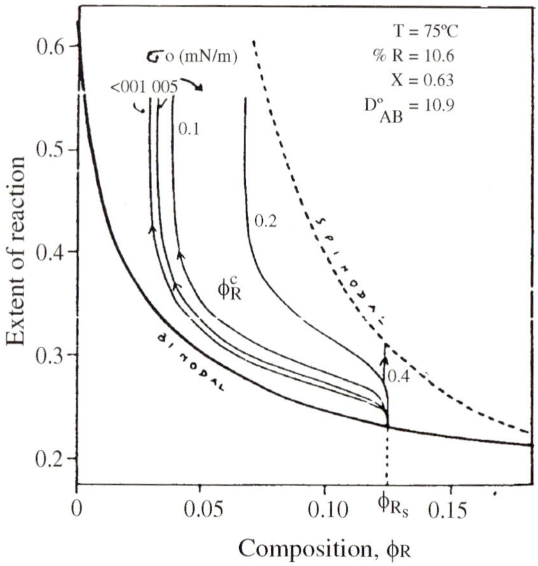

Figure 3.6 Trajectories in the metastable region predicted by using a nucleation growth model for several values of the interfacial tension σ_o. (Reproduced from [2])

rich domain is lower than pure rubber. It is due to the presence of unrelaxed thermal stresses that result from differences in the thermal expansion coefficient of the micro-rubber domain and epoxy matrix in the rubber-rich domain during cooling from the cure temperature to room temperature.

Other papers reported the phase separation behavior for the composition showing dual phase morphology [7,20,21,43–45]. Delides et al. [43] proposed that the viscosity at the point of phase separation is sufficiently large enough to inhibit diffusion of the epoxy through the rubber (CTBN) and result in the generation of the occluded phase, which is the inclusion of epoxy domains within the rubbery phase.

Similar arguments explaining the phase separation were employed by Chou et al. [44]. The dynamics of phase separation was observed using an optical microscope during the course of polyurethane-unsaturated polyester IPN formation at different temperature. Chou et al. suggested that an interconnected phase formed through the spinodal decomposition mechanism developed quickly and was followed by the coalescence of the periodic phase to form a droplet/matrix type of morphology. The secondary phase separation occurred within both the droplet and the matrix phases. Chou et al. did not explain, however, why secondary phase separation occurred.

An increase in curing temperature resulted in UPE gelation, which restricted the secondary phase separation and the coalescence of spinodal decomposition. To be specific the longer the demixing period, the larger or the more interconnected the dispersed phase became. Min et al. [20,21] observed the dual phase morphology at 15 wt% polysulfone modified epoxy system. They employed the phase separation model by Inoue and reported that the formation of the heterogeneous epoxy network and slowdown of phase separation prevented a uniform precipitation of the modifier, finally resulting in a heterogeneous partially phase-inverted structure.

3.4 Experimental

3.4.1 Materials

Diglycidyl ether of bisphenol-A (DGEBA), epoxy resin (YD128, Kuk Do Chem., $\bar{M}_n = 378$), and bisphenol-A dicyanate (BPACY, Arocy B-10, Ciba-Geigy) were used as the thermoset resin. 4,4′-diaminodiphenyl sulfone (DDS, Aldrich Chem. Co.) was used as a curing agent for epoxy. Polyetherimide (PEI, Ultem 1000, General Electric Co., $\bar{M}_n = 18{,}000$) and 2-methyl imidazole (2MZ, Aldrich Chem. Co.) were used as the thermoplastic modifier and catalyst.

3.4.2 Blending and Curing Procedure

Mixtures of PEI/epoxy monomer and PEI/BPACY monomer were prepared by dissolving PEI and thermoset monomers in methylene chloride at room temperature and evaporating the solvent. The epoxy monomer/PEI solution was heated on a hot plate to 140°C, and stoichiometric amounts of DDS were added while stirring for about 7 min. For the catalyzed system, 2-methyl imidazol catalyst was also added.

The mixtures of PEI/(epoxy monomer+DDS) and PEI/BPACY monomer cured isothermally in the aluminum mold according to the schedule shown in Table 3.1.

Table 3.1 Cure Schedule for the Curing of PEI/Thermoset Blend

PEI cont. (wt%)	PEI/Epoxy		PEI/BPACY		
	Precure	Postcure	Precure	Postcure	Curing
0–35	150/6[a] 170/2 190/1.5	240/1.5	180/24 210/9 240/3	285/1	Casting in air-circulated oven
40–80			210/9	260/4	Compression molding
100[b]					Compression molding

[a] Curing temp. in °C/cure time in hours.
[b] The specimen of pure PEI was prepared at 320°C/40 atm for molding time of 30 min.

3.4.3 Phase Separation Behavior

The miscibility between PEI and thermoset monomer or partially cured thermoset resin was analyzed by the cloud point measurement. The cloud point were determined by the onset of the transmitted light intensity change with an optical microscope. In order to maintain the thermal equilibrium during cooling, a slow cooling rate of 0.2°C/min was used during the cloud point measurement. The SEM was used to study the phase separation mechanism at given PEI composition. The specimens were taken out at different time intervals during curing in the oven at 150°C, quenched immediately, and fractured in liquid nitrogen. The electromicrographs showed development of domains during the curing process.

T_g of blend was measured at the scan rate of 5°C/min by Dynamic Mechanical Analyzer (Du Pont DMA 981), which was a resonant frequency measuring type. T_gs were determined by the peaks of the loss modulus peaks.

The light source used in the light scattering experiment was He–Ne laser, attenuated to produce a power of 10 mW, and the wavelength was 0.6328 μm. The thickness of the specimen for the light scattering experiment was 20 μm.

3.4.4 Morphology

The final morphology of specimens cured at different curing temperatures and composition was observed by SEM. Fractured surfaces of postcured specimens prepared in liquid nitrogen, were etched with methylene chloride before examining by SEM.

3.5 Results and Discussion

3.5.1 Phase Diagram

The cloud point curves of the epoxy monomer/PEI blend and BPACY monomer/PEI blend exhibited an upper critical solution temperature (UCST) behavior, whereas partially cured epoxy/PEI blend and BPACY/PEI blend showed bimodal UCST curves with two critical compositions. It is attributed to the fact that, at lower conversion, thermoset resin has a bimodal distribution of molecular weight in which unreacted thermoset monomer and partially reacted thermoset dimer or trimer exist simultaneously. The rubber/epoxy systems that shows bimodal UCST behavior have been reported in previous papers [40,46]. Figure 3.7 shows the cloud point curve of epoxy/PEI system. With the increase in conversion (molecular weight) of epoxy resin, the bimodal UCST curve shifts to higher temperature region.

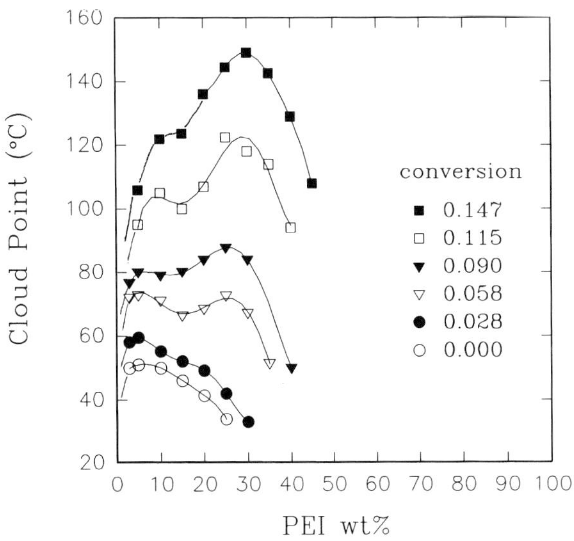

Figure 3.7 Cloud point of the mixture of PEI and epoxy

3.5.2 Morphology

The solvent-etched fracture surface of fully cured PEI modified epoxy with different composition is shown in Figure 3.8. For PEI content smaller than 10 wt%, the PEI-rich phase is dispersed in a continuous epoxy-rich matrix [i.e., sea-island morphology is observed (a and b)]. Above 25 wt% PEI content, nodular structure was observed (e and f) where the epoxy-rich phase forms spherical nodules and the PEI rich phase forms the matrix. With PEI content between 15 wt% and 20 wt%, dual phase morphology, where sea-island morphology and epoxy nodular structure coexist, is present (c and d). Similar morphology was observed in PEI/BPACY blend [47].

3.5.3 Phase Separation Mechanism

The phase separation behavior during curing of PEI/epoxy system was observed by SEM and light scattering under isothermal conditions of 150°C. SEM micrographs of the fracture surface of the 10 wt% PEI blend show typical phase separation pattern of nucleation and growth mechanism (Fig. 3.9). The nucleation (a) and growth (b), as well as the coalescence and ripening in which large particles grow in size at the expense of the smaller ones (c) are well shown in the micrographs. When the conversion of epoxy reaches 0.4 (curing time = 51 min) (d), smaller particles nearly disappear, while the larger particles still grow continuously by coalescence with lower growth rate (e).

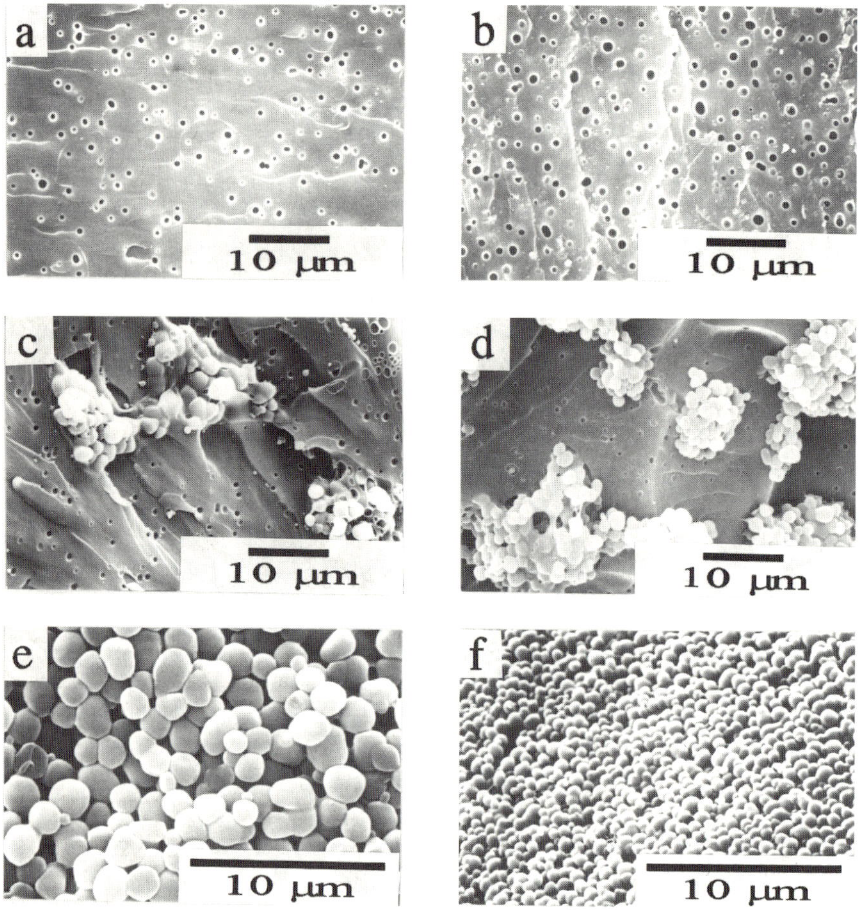

Figure 3.8 Morphology of PEI modified epoxy with various compositions cured at 170°C, PEI = (a) 5 wt%; (b) 10 wt%; (c) 15 wt%; (d) 20 wt%; (e) 25 wt%; (f) 50 wt%

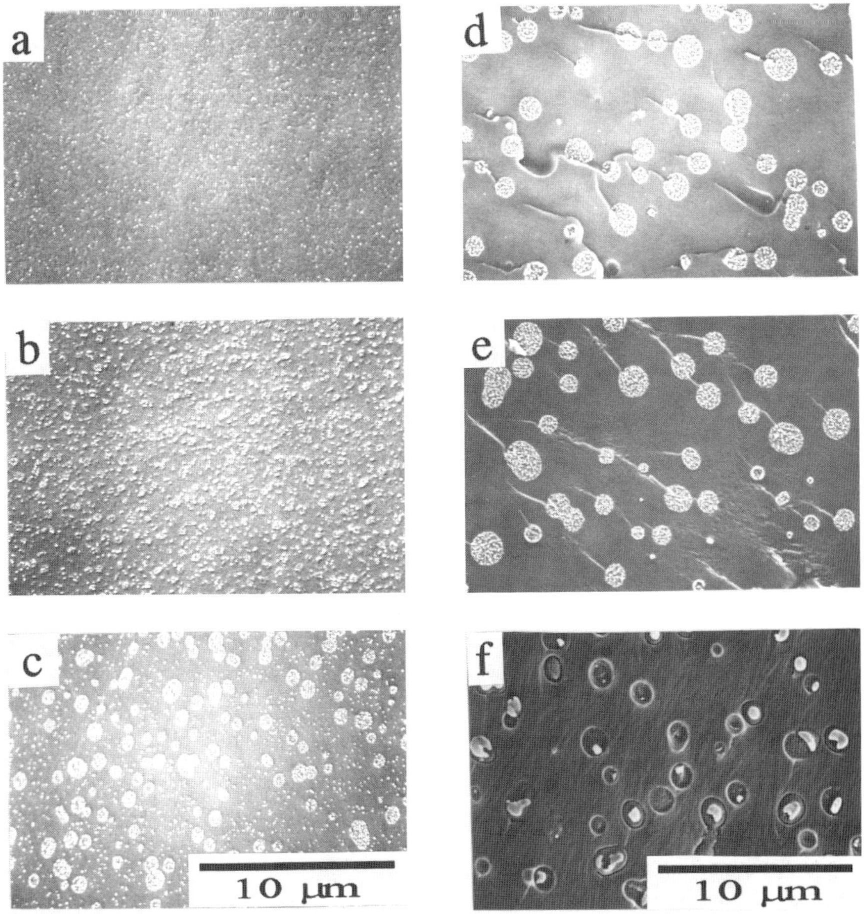

Figure 3.9 Evolution of phase separation of 10 wt% PEI blend via nucleation and growth during cure at 150°C. (a) 37 min ($x = 0.30$); (b) 41 min ($x = 0.33$); (c) 47 min ($x = 0.38$); (d) 51 min ($x = 0.40$); (e) 60 min ($x = 0.46$); (f) 160 min ($x = 0.77$)

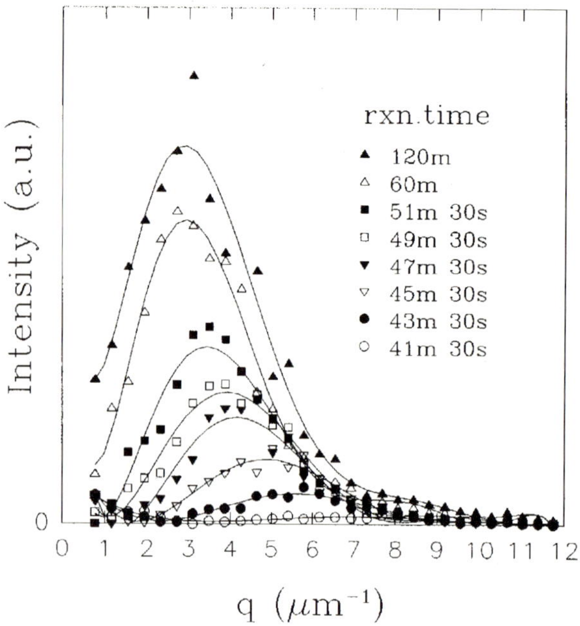

Figure 3.10 Variation of light scattering intensity during cure at 150°C in 10 wt% PEI blend (a.u.: arbitrary unit)

The scattered light intensity change of 10 wt% PEI blend during curing at 150°C is shown in Figure 3.10. When the curing time is 41 min, the maximum intensity is observed at the scattering vector (q_m) of 6.5 μm^{-1}. As the curing reaction proceeds, the maximum peak moves gradually to smaller scattering vector. Figure 3.11 shows SEM micrographs of the fracture surface of 25 wt% PEI blend cured at 150°C with the different curing time.

When epoxy conversion is higher than 0.3, this blend system becomes opaque; however, at low conversion ($x < 0.34$) the contrast between phases is too small to observe the morphology for the initial stage of phase separation by SEM. We can, however, freeze the morphology of the system at initial stage of the phase separation by using the 2MZ-catalyst that shortens gel time. Figure 3.11a shows the fracture surface of 200 ppm 2MZ-catalyzed epoxy/PEI blend. This co-continuous morphology indicates that the phase separation of this blend system goes through spinodal decomposition mechanism. All the systems tend to minimize its interfacial free energy by minimizing the amount of interfacial area. The viscosity of this system at low conversion of epoxy is sufficiently low; thus, the interwoven structures coarsen, and the spherical particle of the epoxy-rich phase is formed. With increasing conversion of epoxy, the particle size increases by coalescence between particles. The volume fraction of an epoxy-rich particle rapidly increases from 0.183 (at $x = 0.34$) to 0.851 (at $x = 0.61$) because the critical composition shifts to higher PEI composition with increasing conversion of epoxy. Finally, the particles come to contact with each other to yield the interconnected nodular structure.

The scattered light intensity change of 25 wt% PEI blend during curing at 150°C is shown in Figure 3.12. For the initial stage of phase separation, the maximum scattered intensity

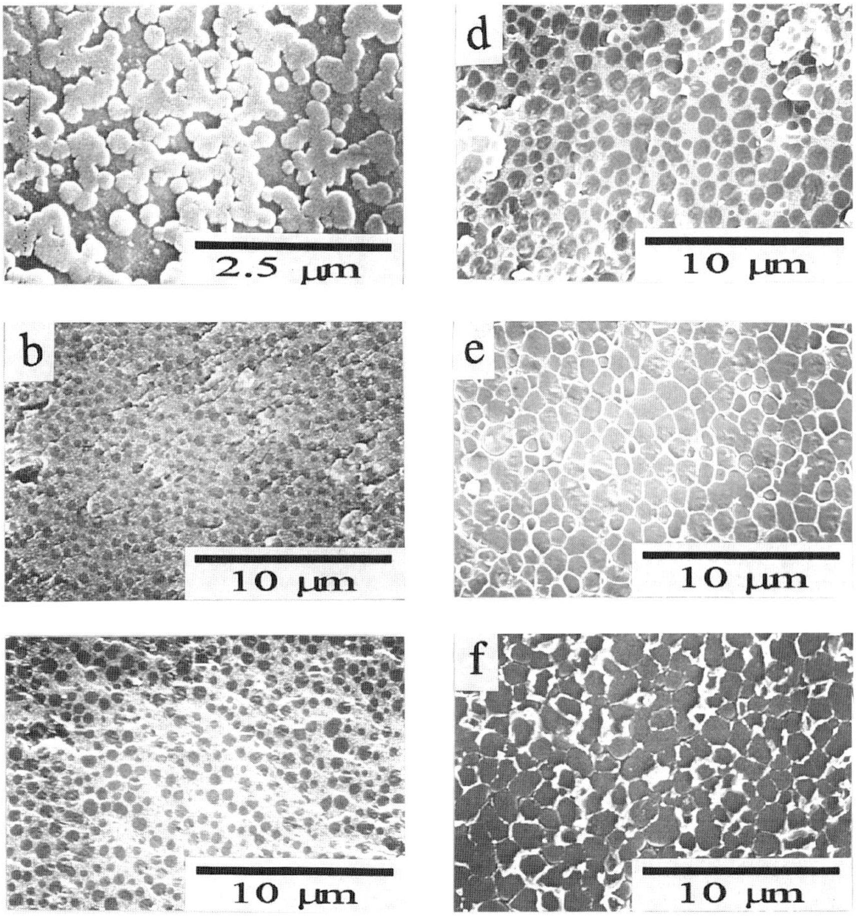

Figure 3.11 Evolution of phase separation of 25 wt% PEI blend via spinodal decomposition during cure at 150°C. (a) 200 ppm 2MZ catalyzed; (b) 43 min ($x = 0.34$); (c) 51 min ($x = 0.40$); (d) 65 min ($x = 0.50$); (e) 90 min ($x = 0.61$); (f) 120 min ($x = 0.71$)

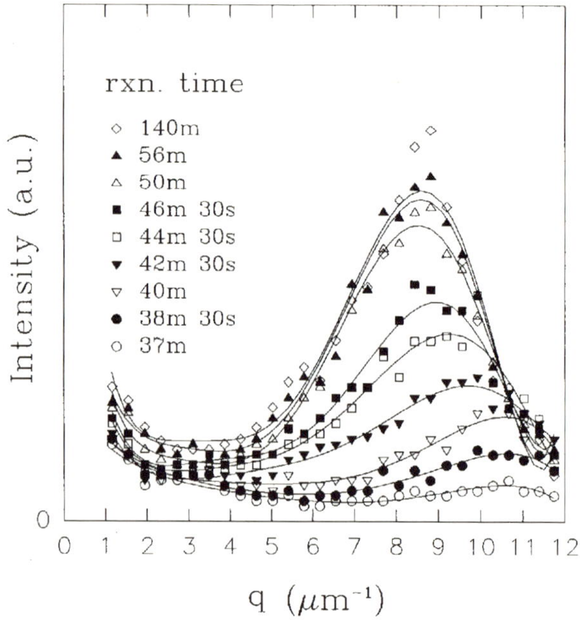

Figure 3.12 Variation of light scattering intensity during cure at 150°C in 25 wt% PEI blend (a.u.: arbitrary unit)

increases at the same scattering vector of 10.7 μm^{-1}. The domain correlation length (d_m = 0.6 μm) estimated by using the q_m is almost equal to that in Figure 3.11a. As the curing reaction proceeds, the q_m decreases with curing time. These characteristic changes in scattering profile are good evidences to show the spinodal decomposition.

Figure 3.13 shows the fracture surface of the 15 wt% PEI blend cured at 150°C for various curing times. The initial co-continuous morphology (a), which means the phase separation through spinodal decomposition, coarsen into dispersed domains (b) and rapidly grow until they form macro-scale phase separation morphology (c, d). Then, as secondary phase separation proceeds, the epoxy rich particles and PEI rich particles are formed in the PEI rich dispersed phase and epoxy rich continuous phase, respectively, and continue to grow (e). Finally, the dual phase morphology showing epoxy nodular structure and sea-island morphology simultaneously are formed (f).

The scattered light intensity and the domain correlation length change of 15 wt% PEI blend during curing show the same result (Figs. 3.14 and 3.15). As the curing reaction proceeds, in the initial stage of phase separation the maximum scattered intensity appears in the scattering vector range from 7.5 to 8.5 μm^{-1} (dm = 0.84–0.74 μm). The maximum scattering vector then rapidly decreases and the maximum intensity increases. When the curing time is about 60 min, a new peak of scattered light intensity appears at q_m = 5.76 μm^{-1} and moves to 3.5 μm^{-1} with elapsing reaction time. It shows that macroscale phase separation proceeds initially in short reaction time; the secondary phase separation then occurs.

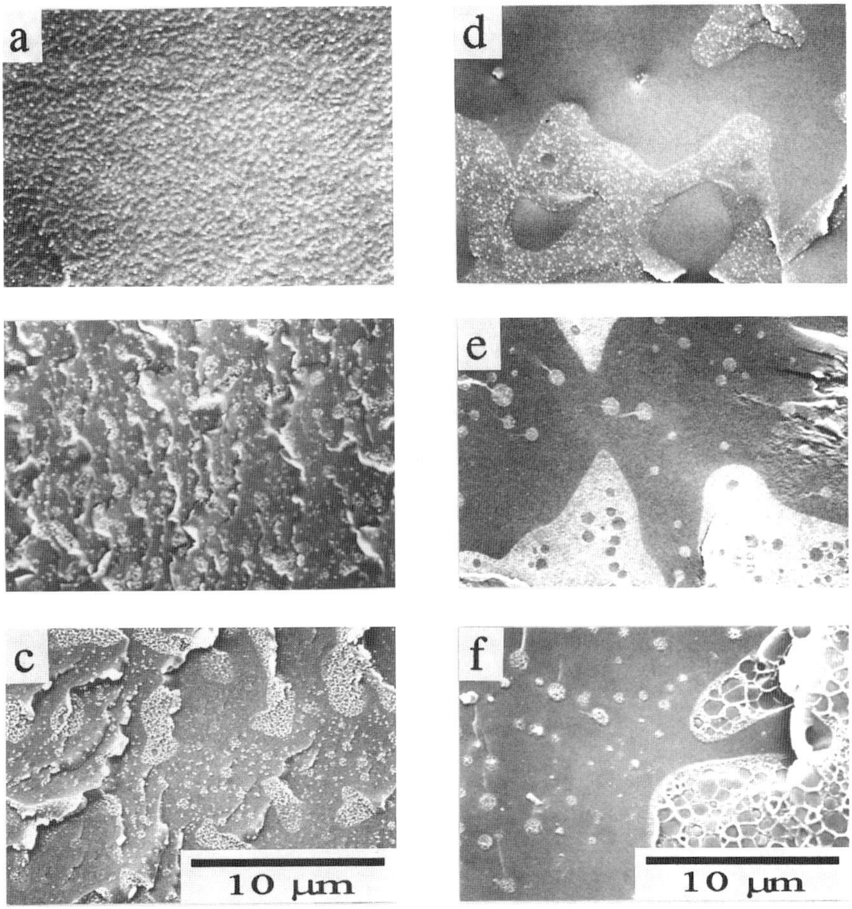

Figure 3.13 Evolution of phase separation of 15wt% PEI blend via spinodal decomposition and secondary phase separation during cure at 150°C. (a) 39 min ($x = 0.31$); (b) 41 min ($x = 0.33$); (c) 43 min ($x = 0.34$); (d) 49 min ($x = 0.39$); (e) 55 min ($x = 0.43$); (f) 120 min ($x = 0.71$)

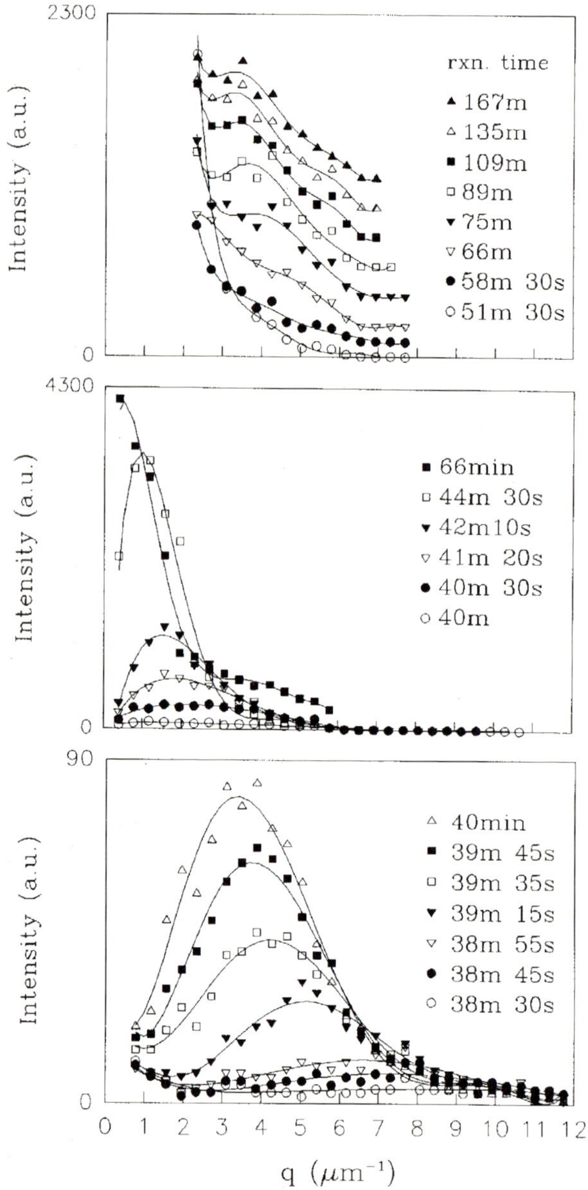

Figure 3.14 Variation of light-scattering intensity during cure at 150°C in 15 wt% PEI blend (a.u. = arbitrary unit)

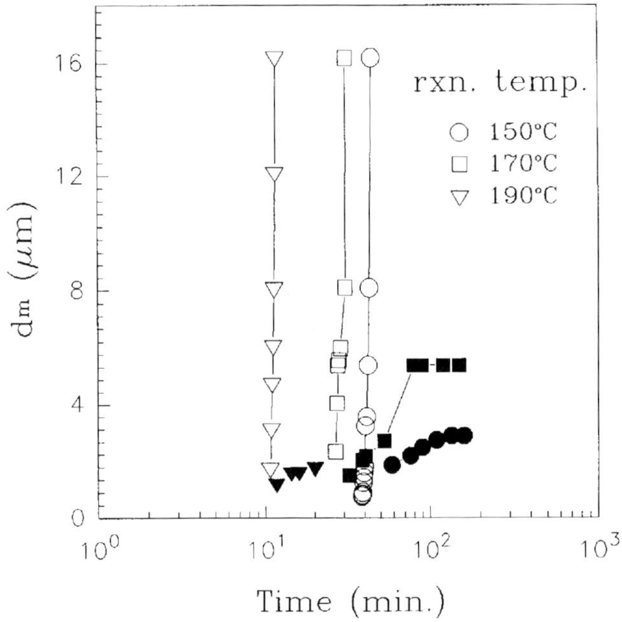

Figure 3.15 Variation of domain correlation length during cure at 150°C in 15 wt% PEI blend (unfilled symbols = primary phase separation; filled symbols = secondary phase separation)

The dual-phase morphology is presumed to be formed in two cases in thc reaction-induced phase separation mechanism. One obvious case is with mixtures that show bimodal UCST behavior that has two critical compositions as this system. Figure 3.16 shows the schematic illustration of mechanism in dual-phase morphology formation for the mixture showing bimodal UCST behavior. At composition ϕ_1 (volume fraction of component 1), phase separation starts at conversion x_1 when the curing temperature is T_1. Following spinodal decomposition mechanism, and due to the low viscosity of the medium (with high concentration of the monomeric thermoset resin at low conversion), the system will have macroscale phase separation of domains having ϕ_1'' and ϕ_1''. As the reaction proceeds and the conversion is now x_2, we see an abrupt change in the equilibrium composition of the thermoplastic-rich phase from ϕ_1'' to ϕ_1''' in a very short time. This abrupt change is similar to the effect of the two-step temperature quenching in polymer blends that have UCST behavior, which results the dual phase morphology (Fig. 3.16b) [48]. In other words, the first phase separated thermoplastic-rich composition of ϕ_1'' now has higher viscosity and experiences the jump of the UCST curve, which induces a secondary phase separation within both the domains and the matrix.

The second case is when the viscosity changes abruptly during the curing reaction. If the curing temperature is high and the initial composition is near the spinodal composition within the unstable region (i.e., a high concentration of low-viscosity thermoset monomer within the unstable region), then due to the fast diffusion rate of the monomer, the initial phase separation occurs at small ΔT (temperature difference of the spinodal temperature at given

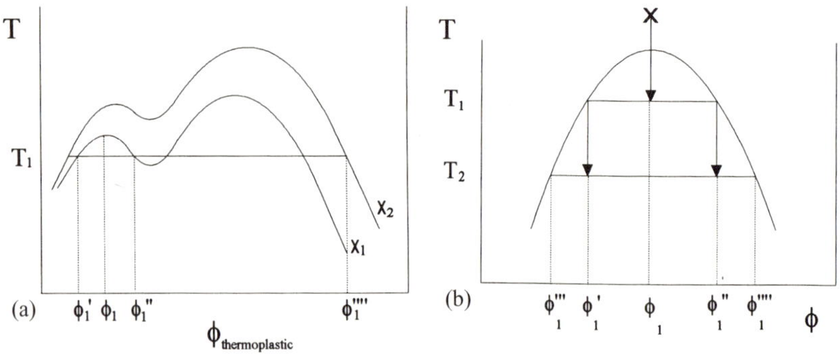

Figure 3.16 Schematic illustration of mechanism of dual phase morphology formation. (a) bimodal UCST; (b) two step temperature quenching in polymer blend

conversion and the reaction temperature), which results in spinodal structure with large domains. The viscosity in the phase-separated PEI-rich domain abruptly increases due to the loss of the thermoset monomer; thus, the secondary phase separation occurs at high ΔT due to the low diffusion rate and relatively fast reaction rate. As a result, secondary phase separation with smaller domain size occurs within the grossly phase separated domains.

We observed the same result by analyzing the phase separation behavior during curing with glass transition temperature (T_g) behavior of fully cured PEI-modified BPACY. The T_g behavior of the PEI/BPACY system cured at 210°C was shown in Figure 3.17. The T_g change of the phase separated domains in each composition of semi-IPNs cured at 210°C shows discontinuity at 15 wt% PEI composition. As the PEI concentration was increased, T_g of the BPACY-rich phase decreased slightly; however, the slope changed abruptly at 15 wt% PEI.

In contrast to the T_g behavior of the BPACY-rich phase, that of the PEI-rich phase increased, showing a maximum at 15 wt% PEI. In addition, the difference between T_gs of both phases was relatively large in the composition range of 1 to 14 wt% PEI as compared with the other compositions.

These features imply that the dominant phase separation mechanism was different in each region below 15 wt% and 15–50 wt% PEI. When there are two different phase separation mechanisms in the composition range where the separated T_gs are observed, it is reasonable to infer that NG occurs in lower PEI composition range and that SD occurs in the midcomposition range. That is, NG is the dominant phase-separation mechanism in the composition range of 1–14 wt% PEI and SD is dominant in the composition range of 15–50 wt% PEI. In the composition range of 50–75 wt% PEI, the phase separation mechanism could not be determined by T_g analysis because T_g of each separated phase could not be determined, although the phase separation was observed. Some extent of phase separation through NG may occur before SD. In this case, the dominant phase separation mechanism is dependent upon the residence time within the metastable region.

Phase separation through NG generates the second phase with the equilibrium composition on the phase diagram that corresponds to the moment of phase separation through concentration jump as described in Figure 3.18a. The T_g difference in each phase, therefore,

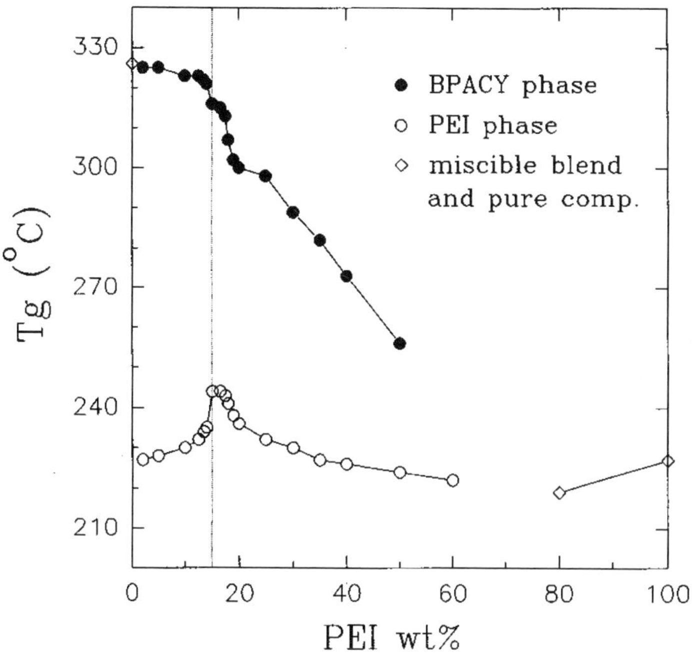

Figure 3.17 T_g behavior of semi-IPNs cured at 210°C: T_gs of (O) BPACY-rich phase, (O) PEI-rich phase and (◇) homogeneous state

is larger compared with the T_g difference of phases formed through spinodal decomposition. Increase in T_g of the PEI-rich phase and decrease in T_g of the BPACY-rich phase in the composition range of 1–14 wt% PEI as the PEI composition increased was attributed to the fact that the equilibrium compositions moved closer to each other as the PEI concentration was increased (Fig. 3.18a). That is, the phase separation of the mixture with lower PEI concentration (ϕ_2) began at higher conversion (x_2) and stopped at the pinning point (x_3). Thus, the semi-IPN cured from this mixture had the second phase with the average composition between ϕ_2'' and ϕ_3''. In contrast to this mixture, the mixture with higher PEI concentration (ϕ_1) began the phase separation at lower conversion (x_1) and produced the second phase with the average composition between ϕ_1'' and ϕ_3''. As a consequence, the semi-IPN cured from the mixture with higher PEI concentration showed higher T_g in PEI-rich phase.

In contrast to the phase separation through NG, phase separation through SD is a process in which the concentration changes continuously and the compositions of separated phases are constrained by a tie line. If a pinning occurs chemically and/or physically during phase separation through this mechanism, then phase separation stops, so that each phase has T_g corresponding to the composition at the moment of pinning. T_g behavior in the composition range of 15–50 wt% PEI in Figure 3.17 resulted from this spinodal decomposition process and could be analyzed as depicted in Figure 3.18b.

The mixture with higher PEI concentration (ϕ_1) underwent phase separation along the tie-line at conversion x_1 in Figure 3.18b. The mixture with lower PEI concentration (ϕ_2) was also

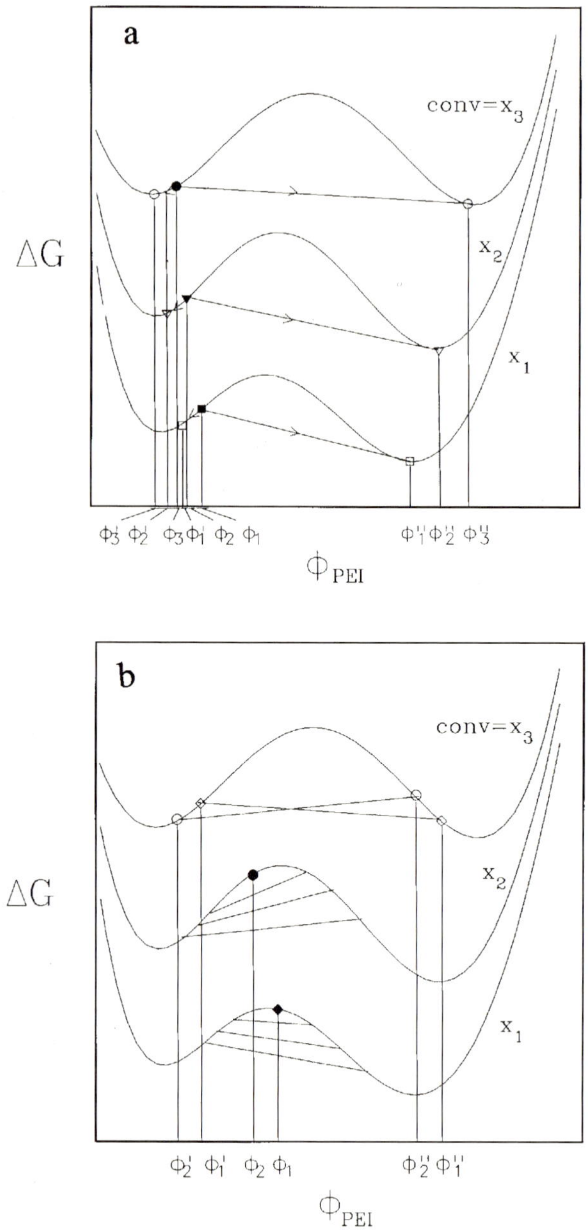

Figure 3.18 Schematic representation of free energy of mixing for the phase separation during cure (a) via nucleation and growth and (b) via spinodal decomposition

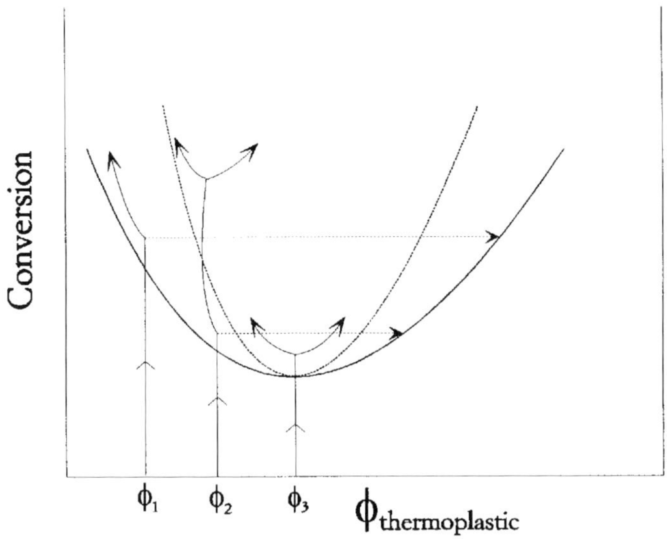

Figure 3.19 Reaction paths undergoing phase separation during isothermal cure

separated along the tie line at conversion x_2. (The mixture with the composition closer to the critical point begins the phase separation at lower conversion.) If the pinning occurred at conversion x_3, then the semi-IPNs had T_gs of BPACY-rich phase corresponding to ϕ_1' and ϕ_2' and T_gs of PEI-rich phase corresponding to ϕ_1'' and ϕ_2'', respectively. This feature resulted in T_g behavior like Figure 3.17 [i.e., the mixture with lower PEI concentration (ϕ_2) was separated to the PEI-rich phase and the BPACY-rich phase showing higher T_g than the mixture with higher PEI concentration (ϕ_1).

Summarizing the phase separation pathway, Figure 3.19 shows the schematic illustration of the reaction paths undergoing phase separation during isothermal cure. When the reaction proceeds at isothermal cure temperature, the phase separation path can be divided into following three cases: staying only the metastable region (ϕ_1) or unstable region (ϕ_3) and starting at the metastable region falling into the unstable region (ϕ_2). Where the dual phase morphology can be observed at ϕ_2 or ϕ_3. The phase separation behavior for each case is shown in Figure 3.20.

3.5.4 Effect of Composition

Table 3.2 shows the influence of the initial PEI concentration on the resulting particle size distribution. Both the size and the number of PEI domains increased in the sea-island morphology, whereas the size of thermoset nodule decreased and the number of nodules increased in the nodular morphology as the initial PEI concentration was increased.

In the composition range where the sea-island morphology was shown, the increase of the PEI concentration caused an increase in the PEI phase volume and a lowering of the

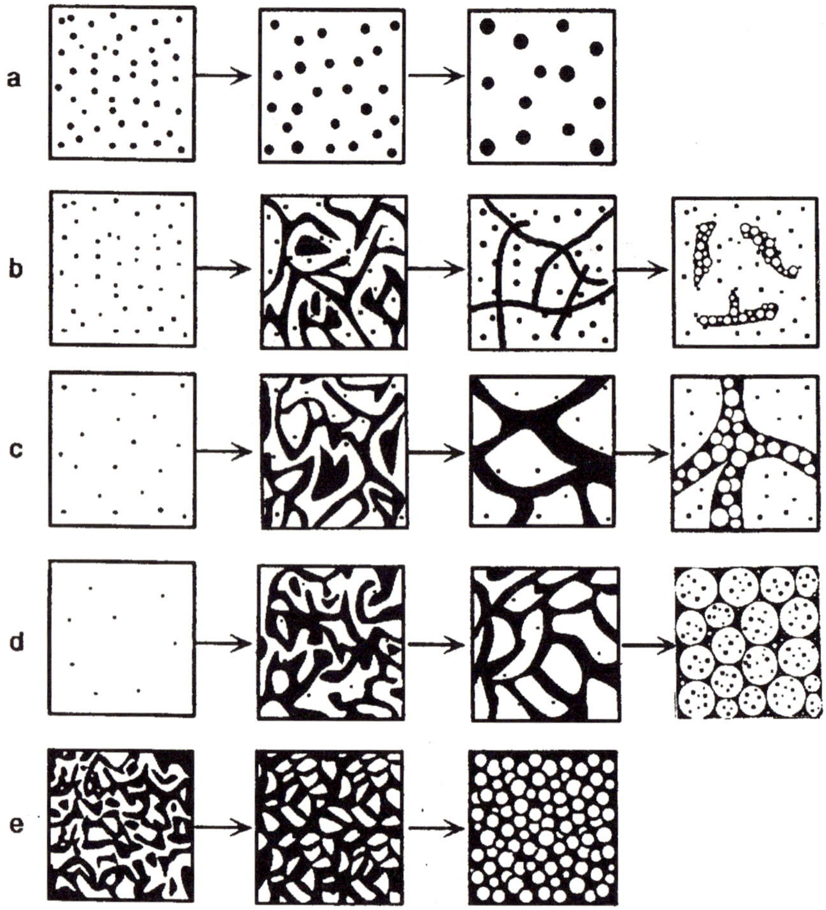

Figure 3.20 Model for the morphology formation during curing of thermoplastic/thermoset blend via; (a) NG only; (b) NG followed by SD: dispersed thermoplastic rich phase; (c) NG followed by SD: bicontinuous, and (d) NG followed by SD: dispersed thermoset rich phase; (e) SD only

conversion of the onset of phase separation (Fig. 3.7), which resulted in an increase in both the size and the number of the PEI domains.

In the composition range showing the nodular morphology, the increase in the PEI concentration increased the viscosity of the system and the PEI phase volume, thus reducing the rate of coalescence of the epoxy nodules in the late stage of spinodal phase separation. Smaller epoxy nodules, therefore, were formed at higher PEI concentration.

In the composition range where the dual phase morphology was shown, the volume fraction of the PEI-rich phase increased with the increasing initial PEI content because the volume fraction of the PEI-rich phase at the onset point of secondary phase separation increases.

Table 3.2 Domain Size and Volume Fraction of Dispersed Phase for PEI Modified Epoxy Cured at Different Precure Temperature

System	5 wt% PEI	10 wt% PEI	15 wt% PEI	20 wt% PEI	25 wt% PEI (μm)	30 wt% PEI (μm)
150°C cure	0.816 ± 0.323[a]	0.980 ± 0.198	1.278 ± 0.322[b] 33.4%[c] 0.566 ± 0.165[d]	2.916 ± 0.778	1.750 ± 0.418	0.866 ± 0.422
170°C cure	0.714 ± 0.184	0.974 ± 0.340	1.962 ± 0.480 20.9% 0.596 ± 0.119	1.503 ± 0.526 32.4% 0.598 ± 0.732	1.784 ± 0.24	1.106 ± 0.210
190°C cure	0.626 ± 0.152	0.844 ± 0.24	1.826 ± 0.392 16.2% 0.656 ± 0.191	1.600 ± 0.390 27.9% 0.600 ± 0.136	2.304 ± 0.632	1.646 ± 0.350

[a] The $\pm x$ values show standard deviation.
[b] Domain size of epoxy nodule within dispersed phase in dual phase system.
[c] Volume fraction of dispersed phase.
[d] Domain size of PEI particle within matrix in dual phase system.

3.5.5 Effect of Cure Temperature

In the composition range showing sea-island morphology, previously reported results showed three cases: (1) increasing domain size [1,43], (2) decreasing domain size [31,34,46], and (3) the presence of a maximum [2,48,49] with an increase in cure temperature. This result was attributed to the combined effect of change in the phase separation rate (viscosity of the matrix) and the rate of curing affecting the pinning process.

If the phase separation proceeds much faster than the cure rate, then the morphology is controlled by the phase separation rate rather than by the cure rate. The diffusivity of the thermoplastic in the thermoset becomes the most important factor. The increase of the diffusivity with increasing cure temperature causes the increase in the particle size.

On the other hand, if the cure rate is much faster than the phase separation, then the morphology is controlled by the cure rate through a chemical pinning process. In this system, phase separation is mainly controlled by the cure rate of the epoxy matrix. Faster curing rates and shorter gel times lead to smaller PEI-rich particles with an increasing cure temperature. The temperature effect on the viscosity of reaction mixture is relatively small (i.e., the complex viscosities measured by Physica are 7 and 4 Pa.s at curing temperatures of 150 and 190°C, respectively).

In the composition range where the nodular morphology was shown, because the matrix (PEI phase) contained relatively small amount of epoxy, the temperature effect on the viscosity was larger than the reaction effect. Thus, the viscosity of the PEI matrix in the nodular morphology became reduced as the cure temperature was increased, which made epoxy nodules coalesce more easily with each other. As the cure temperature is increased, the viscosity of the PEI-rich matrix decreases from 210 Pa.s at 150°C to 50 Pa.s at 190°C.

In the composition range showing dual-phase morphology, with increasing cure temperature, the volume fraction of the PEI-rich macrophase decreases and the PEI-rich macrophase becomes more dispersed. It is caused by the shifting of the critical point at the onset conversion of secondary phase separation to a higher PEI composition region and by shorter gel time with increasing curing temperature. The composition region showing dual phase morphology changes with changing the curing temperature. The dual phase morphology is observed in 13–18 wt%, 14–20 wt%, and 15–20 wt% PEI blend when the curing temperature is 150, 170, and 190°C, respectively. In the case of PEI/BPACY blend, it is observed in 15–19 wt% PEI content at a curing temperature of 210°C.

3.6 Conclusions

In this study, we investigated the phase separation mechanism with the PEI content by observing the change in morphology and light scattering during curing of PEI-modified epoxy. The phase separation mechanism was also analyzed in terms of a glass transition temperature behavior of fully cured PEI-modified BPACY. The PEI/thermoset blends showed sea-island, dual-phase, and nodular morphologies with PEI content.

The dual-phase morphology of the PEI/epoxy blend was observed in the composition range of 13–18 wt%, 14–20 wt%, and 15–20 wt% PEI when curing temperature was 150,

170, and 190°C, respectively. The composition region showing a dual-phase morphology of the PEI/BPACY blend was 15–19 wt% PEI content at a curing temperature of 210°C. The phase-separation mechanism and the morphology of PEI-modified epoxy were dependent on the PEI composition and curing temperature. The sea island morphology was formed via nucleation and growth, but the other morphologies were formed predominantly via spinodal decomposition. Dual-phase morphology was formed via spinodal decomposition, subsequent rapid coalescence, and secondary phase separation within both the dispersed and matrix phases formed by the primary phase separation. The secondary phase of separation presumably occurred through the abrupt changes of equilibrium composition or viscosity during the curing reaction.

Nomenclature

2MZ	2-methyl imidazole
BPACY	bisphenol A dicyanate
CTBN	carboxyl terminated butadiene acrylonitrile
DDS	4,4′-diaminodiphenyl sulfone
DGEBA	diglycidyl ether of bisphenol A
d_m	domain correlation length
G_N	free energy per unit volume
IPN	interpenetrating polymer network
K	phase separation rate/polymerization rate
M_n	number average molecular weight
NG	nucleation and growth
NR	nucleation rate
PEI	polyetherimide
PES	polyethersulfone
q_m	scattering vector
SD	spinodal decomposition
SEM	scanning electron microscope
T_{cure}	curing temperature
TEM	transmission electron microscope
T_g	glass transition temperature
UCST	upper critical solution temperature
x_i	conversion for *i*-component
σ	interfacial tension
ϕ_i	concentration for *i*-component

References

1. Moschiar, S.M., Pascault, J.P., Moschiar, S.M., Riccardi, C.C., Williams, R.J.J. *J. Appl. Polym. Sci.* (1991) 42, 701
2. Moschiar, S.M., Riccardi, C.C., Williams, R.J.J., Verchere, D., Sautereau H. Pascault, J.P. *J. Appl. Polym. Sci.* (1991) 42, 717

3. Verchere, D., Pascault, J.P., Sautereau, H., Moschiar, S.M., Riccardi, C.C., Williams, R.J.J. *J. Appl. Polym. Sci.* (1991) 43, 293
4. Kim, D.H., Kim, S.C. *Polym. Eng. Sci.* (1991) 31, 289
5. Sultan, J.N., McGarry, F.J. *Polym. Eng. Sci.* (1973) 13, 29
6. Viscouti, S., Marchessault, R.H. *Macromolecules* (1974) 7, 913
7. Manzione, L.T., Gillham, J.K. *J. Appl. Polym. Sci.* (1981) 26, 889
8. Manzione, L.T., Gillham, J.K. *J. Appl. Polym. Sci.* (1981) 26, 907
9. Bucknall, C.B., Partridge, I.K., *Polym. Eng. Sci.* (1986) 26, 54
10. Iijima, T., Yoshioka, N., Tomi, M. *Eur. Polym. J.* (1992) 28, 573
11. Rose, J.B. *Polymer* (1974) 15, 456
12. Bucknall, C.B., Partridge, I.K. *Br. Polym. J.* (1983) 15, 71
13. Bucknall, C.B., Partridge, I.K. *Polymer* (1983) 24, 639
14. Hedrick, J.L., Yilgor, I., Hedrick, J.C., Wilkes, G.L., McGrath, J.E. 30th *National SAMPE Symposium* (1985) March 19–21, 947
15. Raghava, R.S., *J. Polym. Sci. Polym. Physics* (1987) 25, 1017
16. Hedrick, J.L, Yilgor, I., Jurek, M., Hedrick, J.C., McGrath, J.E., *Polymer* (1991) 32, 2020
17. Bucknall, C.B., Gilbert, A.H., *Polymer* (1989) 30, 213
18. Gilbert, A.H., Bucknall, C.B. *Makromol. Chem., Macromol. Symp.* (1991) 45, 289
19. Hourston, D.J., Lane, J.M., MacBeath, N.A. *Polym. Int.* (1991) 26, 17
20. Min, B.G., Hodgkin, J.H., Stachurski, Z.H. *J. Appl. Polym. Sci.* (1993) 50, 1065
21. Min, B.G., Stachurski, Z.H., Hodgkin, J.H. *J. Appl. Polym. Sci.* (1993) 50, 1511
22. Mai, K., Zeng, H. *J. Appl. Polym. Sci.* (1994) 53, 1653
23. Wilkinson, S.P., Ward, T.C., McGrath, J.E. *Polymer* (1993) 34, 870
24. Pearson, R.A., Yee, A.F. *J. Apply. Polym. Sci.* (1993) 48, 1051
25. Pearson, R.A., Yee, A.F. *Polymer* (1993) 34, 3658
26. Kim, J.H., Kim, S.C. *Polym. Eng. Sci.* (1987) 27, 1252
27. Butta, E., Levita, G., Marchetti, A., Lazzeri, A. *Polym. Eng. Sci.* (1986) 26, 63
28. Markinnon, A.J. *Polymer* (1993) 34, 325
29. Jang, J., Shin, S. *Polymer* (1995) 36, 1199
30. Chen, M.C., Hourston, D.J., Sun, W.B. *Eur. Polym. J.* (1992) 28, 1471
31. Shimp, D.A., Hudock, F.A., Bobo, W.S. *18th Int. SAMPE Tech. Conf.* (1986) October 7–9, p. 851
32. Yamanaka, K., Inoue, T. *Polymer* (1989) 30, 662
33. Yamanaka, K., Takagi, Y., Inoue, T. *Polymer* (1989) 30, 1839
34. Yamanaka, K., Inoue, T. *J. Mat. Sci.* (1990) 25, 241
35. Kim, B.S., Chiba, T., Inoue, T. *Polymer* (1995) 36, 43
36. Ohnaga, T., Chen, W., Inoue, T. *Polymer* (1994) 35, 3774
37. Vazquez, A., Rojas, A.J., Adabbo, H.E., Borrajo, J., Williams, R.J.J. *Polymer* (1987) 28, 1156
38. Williams, R.J.J., Borrajo, J., Adabbo, H.E., Rojas, A.J. In *Rubber-Modified Thermoset Resins; Advances in Chemistry 208* (1984) Am. Chem. Soc., Washington, DC, p. 195–213
39. Verchere, D., Sautereau, H., Pascault, J.P., Moschiar, S.M., Riccardi, C.C., Williams, R.J.J. In *Toughened Plastics I; Advances in Chemistry 233* (1993) Am. Chem. Soc., Washington, DC, p. 335–363
40. Verchere, D., Pascault, J.P., Sautereau, H., Moschiar, S.M., Riccardi, C.C., Williams, R.J.J. *Polymer* (1989) 30, 107
41. Utracki, L.A. In *Polymer Alloys and Blends* (1986) Oxford University Press, New York, p. 45–52
42. Teng, K.C., Chang, F.C. *Polymer* (1993) 34, 4291
43. Delides, C.G., Hayward, D., Pethrick, R.A., Vatalis, A.S. *J. Appl. Polym. Sci.* (1993) 47, 2037
44. Chou, Y.C., Lee, L.J. *Polym. Eng. Sci.* (1994) 34, 1239
45. Woo, E.H., Su C.C., Kuo, J.F. *Macromolecules* (1994) 27, 5291
46. Kim, D.S., Kim, S.C. *Polym. Adv. Technol.* (1990) 1, 211
47. Lee, B.K., Kim, S.C. *Polym. Adv. Technol.* (1995) 14, 402
48. Kwak, K.D., Okada, M., Chiba, T., Nose, T. *Macromolecules* (1993) 26, 4047
49. Kim, S.C., Ko, M.B., Jo, W.H. *Polymer* (1995) 36, 2189

4 In Situ Frequency Dependent Dielectric Sensing of Cure

David E. Kranbuehl

In situ frequency dependent electromagnetic–impedence measurements provide a sensitive, convenient, automated technique to monitor the changes in macroscopic cure processing properties and the advancement of the reaction in situ in the fabrication tool. This chapter discusses the instrumentation, theory, and several applications of the techniques, including isothermal cure, complex time–temperature cure, resin film infusion, thick laminates, and smart, automated control of the cure process.

4.1 Introduction

With the expanding use of polymeric materials as composite matrixes, adhesives coatings, and films, the need to develop low-cost, automated fabrication processes to produce consistently high-quality parts is critical. Essential to the development of reliable, automated, intelligent processing is the ability to monitor the changing state of the polymeric resin continuously in situ in the fabrication tool during processing. This sensing capability is essential because (1) resin processing properties vary with age, batch, and handling, (2) tool heat transfer properties vary, (3) significant differences exist between one autoclave, press, oven, or pultruder, and (4) operators differ. In addition, there is the need to develop and optimize cure procedures quickly for new tool designs, new resins, and the like. This chapter discusses the application of dielectric sensing to monitor polymeric material cure and provides a fundamental understanding of the underlying science for the use of frequency-dependent dielectric sensors to monitor the cure process.

Frequency-dependent dielectric measurements, often called frequency-dependent electromagnetic sensing (FDEMS), made over many decades of frequency, hertz to megahertz, are a sensitive, convenient, and automated means for characterizing the processing properties of thermosets and thermoplastics [1–21]. Using a planar wafer-thin sensor, measurements can be made in situ in almost any environment. Through the frequency dependence of the impedance, this sensing technique is able to monitor chemical and physical changes throughout the entire cure process. Dielectric sensing techniques have the advantage that measurements can be made both in the laboratory and in situ in the fabrication tool during manufacture. Few laboratory measurement techniques have the advantage of being able to make measurements in a processing tool such as a press or an autoclave, at an adhesive bondline, or of a thin film or coating. Furthermore, because the technique is a direct electrical measurement and involves an inert sensor it can be used to monitor the entire cure process continuously as the resin's state changes from a liquid to a gel to a cured glass. It can be used at temperatures exceeding 400°C and at pressures of 60 atm, with an accuracy of 0.1 percent and a range in magnitude of more than 10 decades. It is difficult for most others in situ techniques, such as acoustics and fiberoptic measurements, to attain this level of sensitivity in harsh processing environments. As such, one major application of this work is measurement of the resin's processing properties in situ during fabrication in the manufacturing mold and as input for on-line intelligent closed-loop process control [2–30].

The FDEMS technique [1,2] and related lower-frequency work [3] can be effective for monitoring a variety of resin cure processing properties such as reaction onset, viscosity, point of maximum flow, degree of cure, buildup in T_g, and reaction completion, as well as detecting the variability in processing properties due to resin age and exposure to moisture [31–41]. The dielectric technique has been shown to be able to monitor similar processing properties in thermoplastics such as T_g, T_m, recrystallization, and solvent–moisture outgassing [2,42]. The technique has the particular advantage over other chemical characterization measurements of being able to monitor these processing properties continuously and in situ as the resin changes from a polymer of varying viscosity to a cross-linked insoluble solid. Another advantage is that measurements can be made simultaneously on multiple samples or at multiple positions in a complex part. Of particular importance is the ability of dielectric sensing to monitor the changing properties of polymeric materials in composites, films, and coatings [43,44] as adhesives in the bondline, and to detect phase separation in toughened systems [20,21,45–47].

At the heart of dielectric sensing is the ability to measure the changes in the translational mobility of ions and changes in the rotational mobility of dipoles in the presence of a force created by an electric field. These variations in molecular mobility due to an electric field force are a very sensitive means of monitoring changes in macroscopic mechanical properties such as viscosity, modulus, T_g, and degree of cure. Mechanical properties reflect the response in displacement on a macroscopic level due to a mechanical force acting on the whole sample. Dielectric sensing is quite sensitive because of the fact that changes on the macroscopic level originate from changes in force displacement relationships on a molecular level. Indeed, it is these molecular changes in force–displacement relationships that dielectric sensing measures as the resin cures and which are the origin of the resin's macroscopic changes in flow, degree of cure, and other mechanical properties.

There is a discussion in the theory section of this chapter of how, from the frequency dependence of measurements of the equivalent capacitance and conductance of the resin, the magnitude of the ionic translational mobility and the dipolar rotational mobility is determined. In the later sections of the chapter, examples are given on how the measurement of the ionic and dipolar mobility and the development of empirical calibrations in the laboratory between quantities such as viscosity, T_g, and modulus can be used to monitor these engineering processing and performance properties in situ in the tool during fabrication. It should be pointed out that there are fundamental physical and chemical relations that explain why measurement of molecular mobility can be used both to monitor changes and often to measure these changing processing and performance properties during cure in absolute units through calibrated theoretical relationships. Applications of the FDEMS in situ sensing technique, however, is the focus of this chapter rather than a detailed fundamental understanding of these relationships [1–21,37,41].

One may ask when and why does one need in situ on-line dielectric sensing. There are a number of important reasons and examples. The first major reason is that dielectric sensing allows one to monitor or see the actual state of the material in the tool at all times during the cure process. Temperature and pressure do not provide direct information about the state of the resin. Thus, dielectric sensing is one of the few means by which the operator can actually monitor the state of the material during processing and tell what the material is doing throughout the entire fabrication process.

Second, by actually monitoring the state of the material, it is possible to control the fabrication process by data rather than a procedure, such as a set time temperature sequence. This means one can have a self-correcting, automated, intelligent cure process that can adapt to variations in material age, fabric permeability, tool heat transfer characteristics, and the like.

Third, in situ sensing is needed to verify the veracity and logic of a model's predictions or an operator's reasoning. Making a composite part that passes mechanical tests does not verify that the modeling equations or operator thinking are correct and can be trusted to make predictions.

Fourth, modeling and individual thinking that lead to a procedure-driven cure cycle is beset with operating difficulties. As has been described by George Springer [30], modeling requires extensive material data characterization of resin properties as well as fabric and tooling properties that are time-consuming to measure and, most importantly, which will vary from day to day and batch to batch. Further, results are generally limited to a particular or a simplified geometry. Fabric preform properties will vary from preform to preform, with lay-up, with bagging, and with position within the preform, and so on. Heat transfer characteristics will similarly vary with the tool, the autoclave, the position within the autoclave, etc. Thus, given the time and material cost, it is critical to monitor or see what is actually happening and to have the potential to detect, verify, and even correct for these processing property changes as the cure proceeds.

In summary, dielectric sensing provides valuable insight in observing the state of the resin during the process, verifying and reducing the time in developing a cure process, as well as providing an automated self-correcting intelligent control system. Further, at the same time, dielectric monitoring has the potential to provide on-line quality verification of the fabrication process, thereby increasing product reliability and reducing postfabrication test costs.

In this chapter the frequency dependence of the complex permitivity in the hertz to megahertz range is used to separate and determine parameters governing ionic and dipolar mobility. The quantitative relationship of the ionic and dipolar mobility to monitoring processing parameters, such as viscosity and degree of cure during the reaction, is discussed. Several applications of in situ sensing are presented. Finally, the application FDEMS sensing for closed loop intelligent process control of the cure process is discussed [22–30].

4.2 Instrumentation

Frequency dependent complex dielectric measurements are made using an Impedance Analyzer controlled by a microcomputer [1–5]. In the work discussed here, measurements at frequencies from 5 Hz to 5×10^6 Hz are taken continuously throughout the entire cure process at regular intervals and converted to the complex permittivity, $\varepsilon^* = \varepsilon' - i\varepsilon''$. The measurements are made with a geometry independent DekDyne microsensor, which has been patented and is now commercially available, and a DekDyne automated dielectric measurement system. This system is used with either a Hewlett Packard or a Schlumberger impedance bridge. The system permits multiplexed measurement of nine sensors. The sensor itself is planar, 2.5 cm × 1.25 cm area and 5-mm thick. This single-sensor bridge microcomputer assembly is able to make continuous uninterrupted measurements of both ε' and ε'' over 10 decades in magnitude at all frequencies. The sensor is inert and has been used at temperatures exceeding 400°C and at more than 60 atm pressure.

4.3 Theory

Frequency-dependent measurements of the materials' dielectric impedance as characterized by its equivalent capacitance, C, and conductance, G, are used to calculate the complex permitivity, $\varepsilon^* = \varepsilon' - i\varepsilon''$, where $\omega = 2\pi f$, f is the measurement frequency, and C_o is the equivalent air replacement capacitance of the sensor.

$$\varepsilon'(\omega) = \frac{C(\omega)\ \text{material}}{C_o}$$
$$\varepsilon''(\omega) = \frac{G(\omega)\ \text{material}}{\omega C_o} \tag{4.1}$$

This calculation is possible when using the sensor whose geometry is invariant over all measurement conditions. Both the real and the imaginary parts of ε^* can have a dipolar and ionic-charge polarization components.

$$\begin{aligned} \varepsilon' &= \varepsilon_{d'} + \varepsilon_{i'} \\ \varepsilon'' &= \varepsilon_{d''} + \varepsilon_{i''} \end{aligned} \tag{4.2}$$

Plots of the product of frequency (ω) multiplied by the imaginary component of the complex permittivity $\varepsilon''(\omega)$ make it relatively easy to determine visually when the low-frequency magnitude of ε'' is dominated by the mobility of ions in the resin and when at higher frequencies the rotational mobility of bound charge dominates ε''. In general, the magnitude of the low frequency overlapping values of $\omega\varepsilon''(\omega)$ can be used to measure the change with time of the ionic mobility through the parameter σ where

$$\begin{aligned} \sigma(\text{ohm}^{-1}\ \text{cm}^{-1}) &= \varepsilon_o \omega \varepsilon_i''(\omega) \\ \varepsilon_o &= 8.854 \times 10^{-14}\ \text{C}^2\ \text{J}^{-1}\ \text{cm}^{-1} \end{aligned} \tag{4.3}$$

The changing value of the ionic mobility is a molecular probe that can be used to quantitatively monitor the viscosity using the relation $\sigma = A\eta^{-x}$ of the resin during cure up to gel. The dipolar component of the loss at higher frequencies can then be determined by subtracting the ionic component.

$$\varepsilon_{d''}(\omega)\ \text{dipolar} = \varepsilon''(\omega) - \frac{\sigma}{\omega\varepsilon_o} \tag{4.4}$$

The peaks in ε_d'' dipolar (which are usually close to the peaks in ε'') can be used to determine the time or point in the cure process when the "mean" dipolar relaxation time has attained a specific value, $\tau = 1/\omega$, where $\omega = 2\pi f$ is the frequency of measurement. The dipolar mobility as measured by the mean relaxation time τ can be used as a molecular probe of the buildup in T_g. The time of occurrence of a given dipolar relaxation time as measured by a peak in a particular high frequency value of $\varepsilon''(\omega)$ can be quantitatively related to the attainment of a specific value of the resin's glass transition temperature.

Finally, the tail of the dipolar relaxation peak as monitored by the changing value of $d\varepsilon''/zdt/\varepsilon''$ can be used to monitor in situ during processing the buildup in degree of cure and related end-use properties such as modulus and hardness during the final stages of cure or postcure. Examples of methodology and application of measurement of σ and τ to monitor the changing magnitude of the viscosity during the pre–gel stage to monitor T_g in the pre–glass stage, and to monitor the late and/or postcure approach to completion of cure follow.

4.4 Isothermal Cure

The variation in the magnitude of ε'' with frequency and with time for the diglycidylether bisphenol A (DGEBA) amine-cured epoxy held at 121°C is shown in Figure 4.1. The magnitude of ε'' changes four orders of magnitude during the course of the polymerization

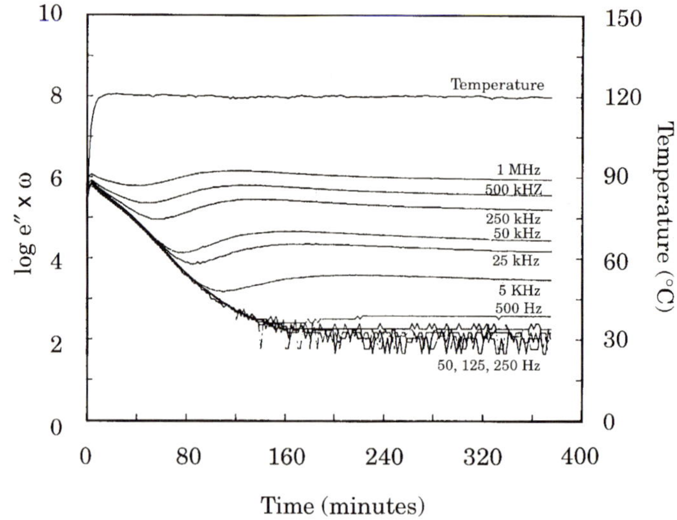

Figure 4.1 Log $\varepsilon''*\omega$ versus time during a 121°C isothermal polymerization

reaction. A plot of $\varepsilon''\times$ frequency is a particular informative representation of the polymerization process because, as discussed from Equations 4.1 to 4.4, overlap of $\varepsilon''(\omega)\times$ frequency for differing frequencies indicates when and at what time translational diffusion of charge is the dominant physical process affecting the loss. The peaks in $\omega\varepsilon''(\omega)$ for individual frequencies similarly indicate when dipolar rotational diffusion processes are contributing to ε''.

The frequency dependence of the loss ε'' is used to find σ by determining from a computer analysis or a log–log plot of ε'' versus frequency, the frequency region where $\varepsilon''\times$ frequency is a constant. In this frequency region the value of σ is determined from Equation 4.3.

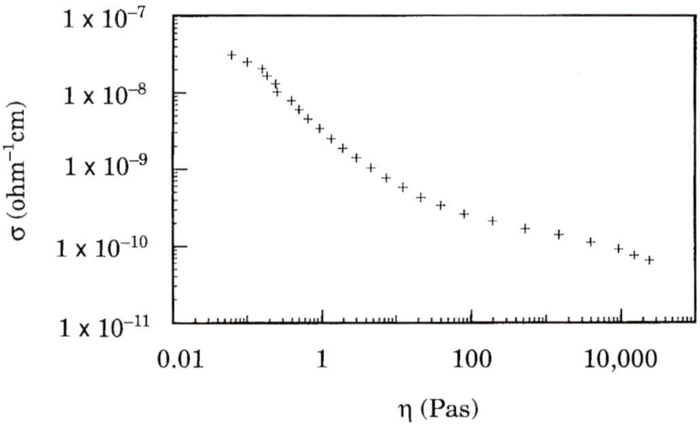

Figure 4.2 Log(σ) versus log(viscosity) during a 121°C polymerization

Figure 4.2 is a plot of log(σ) versus log(viscosity) constructed from dielectric data of Figure 4.1 and measurements on a dynamic rheometer. The figure shows that at a viscosity less than 1 Pa s (10 P), σ is proportioned to $1/\eta$ because the slope of log(σ) versus log(η) is approximately -1. The gel point of the polymerization reaction occurs at 90 min based on the crossover of G' and G'' measured at 40 rad/s. This is very close to the time at which η achieves 100 Pa s, which is also often associated with gel. The region of gel marks the onset of a much more rapid change in viscosity than with σ. This is undoubtedly due to the fact that as gel occurs the viscoelastic properties of the resin involve the cooperative motion of many chains, whereas the translational diffusion of the ions continues to involve motions over much smaller molecular dimensions.

Figure 4.3 is a plot of log(σ) versus degree of cure determined from differential scanning calorimeter measurements. The approximate exponential dependence of σ on α is not surprising because α is approximately exponentially related to viscosity through $\eta = \eta_0 \exp(E/RT + K\alpha)$.

The time of occurrence of a "characteristic" dipolar relaxation time can be determined by noting the time at which $\varepsilon''_{\text{dipolar}}$ achieves a maximum for each of the frequencies measured, where $\tau = 1/(2\pi f)$ at the time at which $\varepsilon''_{\text{dipolar}}$ achieves a maximum for the frequency, f. Values of τ can be measured over a range of frequencies and temperatures.

Figure 4.4 reports values of log(τ) versus $1/T$ at incremental changes in the degree of advancement α for this DGEBA system. Like log(σ), the values of log($1/\tau$) are related to the viscosity and the extent of cure as observed on Figures 4.2 and 4.3. Because the values of τ occur in the postgel, preglass region of cure, 90–300 min, these values of τ are appropriately analyzed as an α-relaxation associated with the T_g to $T_g + 100°$ region.

A Vogel-Fulcher-Tammann-Hesse equation can be used to characterize the temperature dependence of the relaxation times for these six different degrees of cure, 0.70, 0.75, 0.80, 0.825, 0.90, and 0.95:

$$\log \tau = A + \frac{B}{T - T_\infty} \tag{4.5}$$

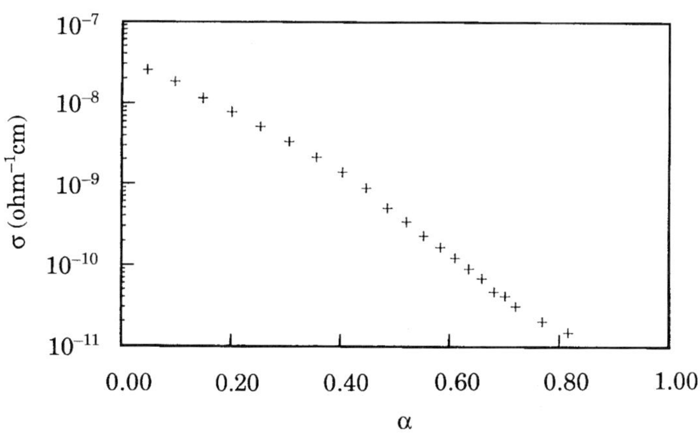

Figure 4.3 Log σ versus degree of cure α during 121°C isothermal cure

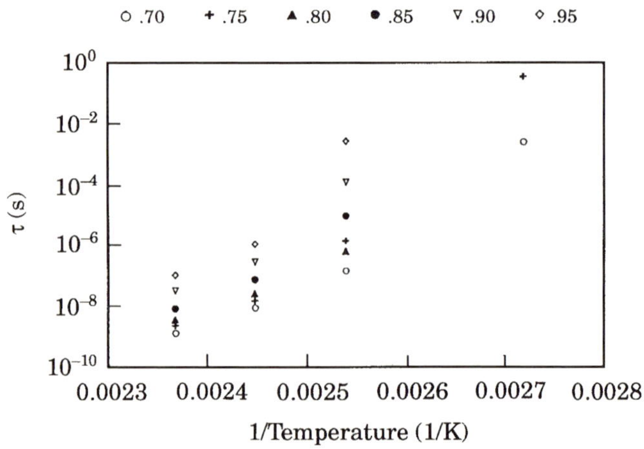

Figure 4.4 Log(τ) versus $1/T$ (°K) at comparable degrees of advancement between 0.70 and 0.95

The best fit occurs at T_∞ values 50°C below the respective values of T_g for each of these values of α. T_g was determined by advancing the reaction to a particular value of α based on the time–temperature DSC kinetic data. The softening temperature, T_g, was determined from the onset of the drop in G'.

Figure 4.5 is a plot of $1/(T - T_\infty)$ versus τ, where $T_\infty(\alpha) = T_g(\alpha) - 50$°C. The observation that the value of τ changes with T_∞ suggests that the parameters A and B show little variation in this preglass region. Further, the results show that the time of occurrence of successive increases in the relaxation time monitors the value ($T_{\text{rx temp}} - T_g$). Thus, through the equation Figure 4.4, the ε'' relaxation peaks can be used to measure the build-up in T_g and degree of cure as the polymerization proceeds.

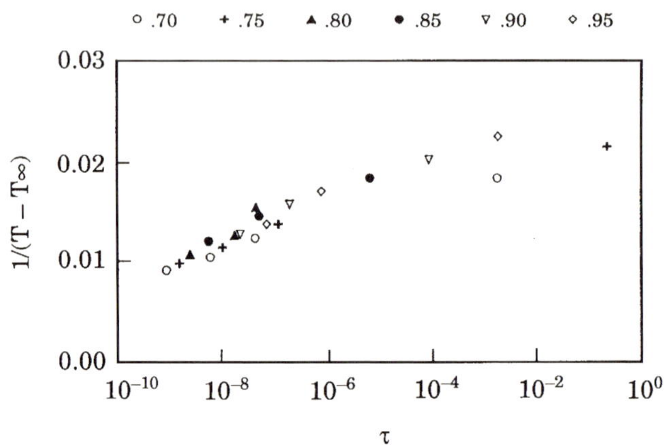

Figure 4.5 Log(τ) versus $1/(T - T_\infty)$, where $T_\infty = T_g(\alpha) - 50$°C

4.5 Monitoring Cure in Multiple Time Temperature Processing Cycles

As a representative example of cure monitoring using a common commercially used aerospace resin and a complex cure cycle, Figure 4.6 displays the output of $\omega\varepsilon''(\omega)$ for a two-stage (i.e., 121°C and 177°C), ramp-hold sequence used to cure a commercial, widely used MY720 aromatic epoxy system. This resin consists of a tetraglycidyl 4,4′-diamino diphenylmethane (TGDDM) and diamino diphenyl sulfone (DDS). This system with catalyst is sold by the Hercules Corporation as 3501-6.

Figure 4.6 shows two peaks of the overlapping $\omega\varepsilon''(\omega)$ lines. These indicate the times and magnitude of maximum flow as monitored by the ionic mobility. The first peak, which is the highest degree of flow, occurs at the beginning of the first hold. The second point of high flow, which shows high ionic mobility, occurs midway up the ramp between the 121°C hold and the 177°C hold. As the temperatures rise, the fluidity increases, until such time as the rate of reaction and thereby the degree of cure, which is also increasing during the temperature ramp, overwhelms the temperature effort on fluidity. At this point the fluidity begins to drop and a peak in $\omega\varepsilon''(\omega)$ occurs, indicating the second occurrence of maximum flow.

The gradual drop in the magnitude of $\varepsilon''(\omega)$ during the final hold and its rate of change $d\varepsilon''/dt$ monitor the buildup in modulus and its rate of buildup. When ε'' attains a constant value, the system has reached its final lowest value of dipolar ionic mobility. Thus, when no further changes in mobility can be detected, $d\varepsilon''/dt = 0$, the system is fully reacted at that hold temperature. Monitored in this way, the changing values of ε'' are a very sensitive means of detecting the final small changes in degree of cure and the buildup in end-use properties

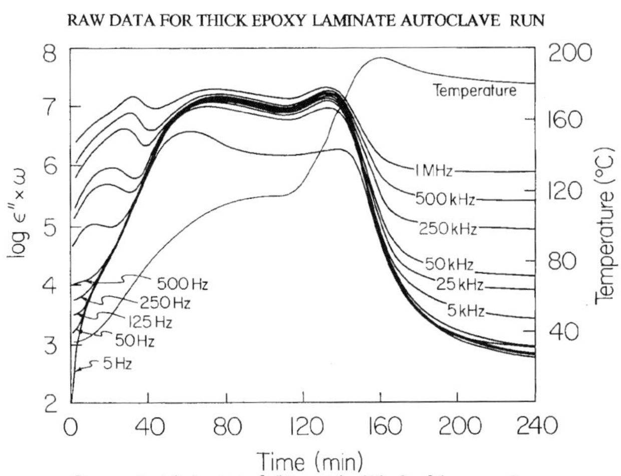

Sensor output during cure of ε″ X ω at the 64th ply of the composite

Figure 4.6 Plot of $\varepsilon'' \times \omega$ versus time of the sensor output at the sixty-fourth ply of the thick epoxy-graphite laminate during cure in the autoclave

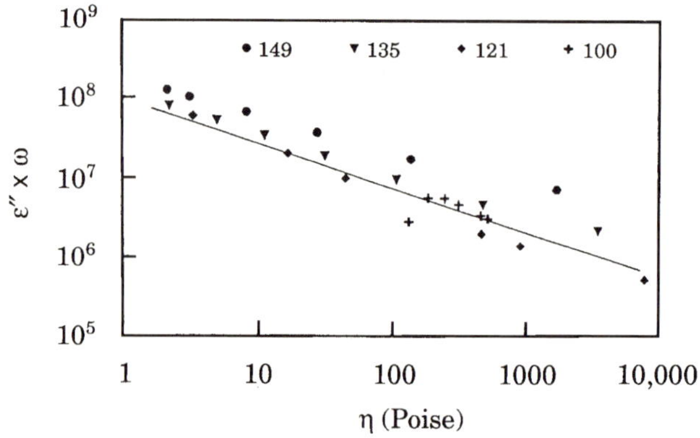

Figure 4.7 Log $\omega\epsilon''(\omega)$ versus log(viscosity) for the TGDDM epoxy-based on four isothermal runs

such as T_g or modulus. Figure 4.6 suggests that even after a 2-hour hold, the mobility is still decreasing for this fresh 3501-6 resin, and the final cure or end-use properties have not been attained.

Figure 4.7 shows the correlation between the viscosity and the ionic mobility based on isothermal runs for this system as monitored by the value of ε'' (5 kHz). A representative calibration curve relating the FDEMS sensor output to degree of cure is shown in Figure 4.8. Unlike viscosity, separate calibration curves for different temperatures must be generated from the isothermal runs because they are temperature dependent.

The buildup in final cure properties, such as degree of cure during the last hold, is monitored with a high degree of sensitivity using the value of $d\varepsilon''/dt/\varepsilon''$. Figure 4.9 is a correlation plot of the normalized rate of change in ε'' (5 kHz) compared with model predicted values (based on numerous DSC runs) of the buildup in the degree of cure.

Figures 4.10 and 4.11, which can be compared to the autoclave cure in Figure 4.6, show two important applications of FDEMS sensing as applied to processing of an epoxy system. Figure 4.10 shows the effect on processing properties as monitored by the value of $\omega\varepsilon''(\omega)$ for

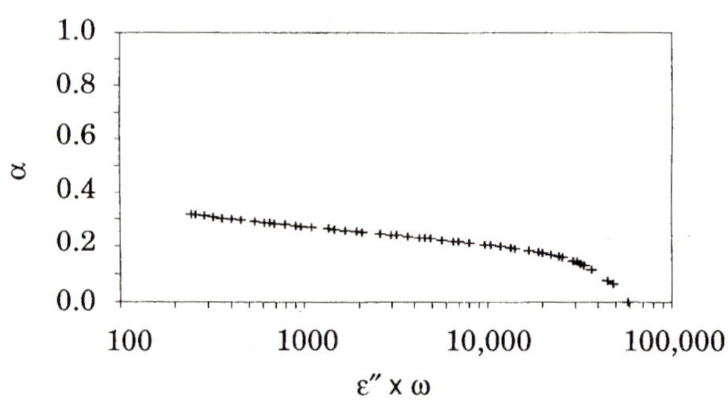

Figure 4.8 Correlation curve relating degree of cure to $\omega\varepsilon''(\omega)$ at 135°C

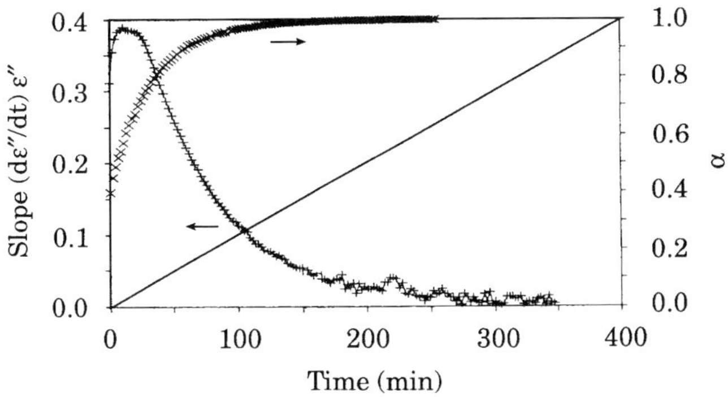

Figure 4.9 Correlation of $d\varepsilon''/dt/\varepsilon''$ with time and degree of cure and with time showing sensitivity of normalized rate of change in ε'' to changes in final degree of cure near end reactions

this system when the first hold temperature drifts 20°C higher (i.e., to 140°C). Figure 4.11 shows the output for a 130°C, 177°C ramp hold sequence, but for a batch of 3501-6 epoxy resin after it has been to left to age at room temperature for 30 days.

Even without the calibration relations the effect on cure processing properties can be clearly seen from the sensor output. In Figure 4.10, the value of ε'' peaks higher, but it drops much more rapidly. The second peak in ε'' during the ramp is much lower. Thus, the effect of the 140°C hold is to cause a higher fluidity initially; however, the high fluidity region lasts for a much shorter time. It is equal important at the second point of high fluidity, which is usually critical for composite–prepreg consolidation, that the fluidity is significantly lower than that

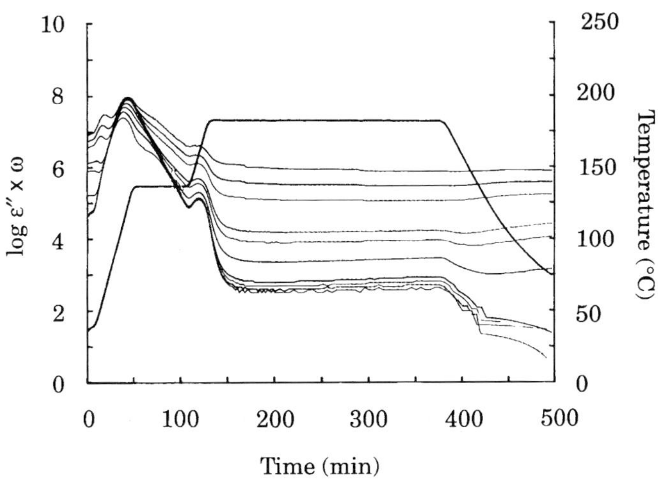

Figure 4.10 TGDDM epoxy cured in a press with a 140°C, 177°C cure cycle. Values of $\omega\varepsilon''(\omega)$ are displayed for frequencies 1 MHz, 500, 250, 125, 50, 5 kHz, and 500, 250, 125, 50 Hz (top to bottom)

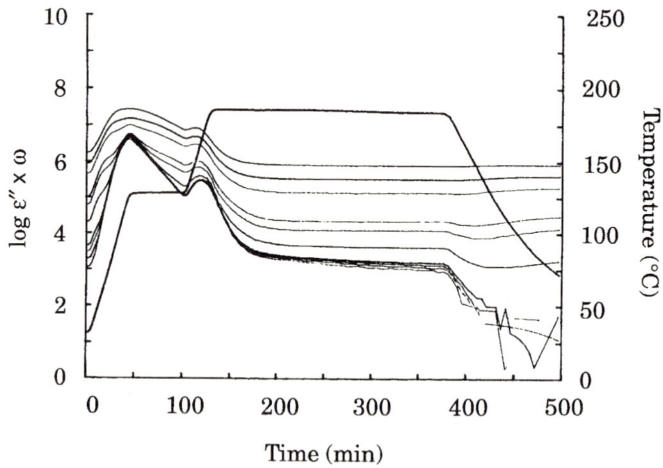

Figure 4.11 TGDDM epoxy that has been aged at room temperature for 30 days after which it is cured in a press using a 130°C intermediate hold

observed for a 130°C hold. The overall effect of aging is seen as decreasing the level of fluidity throughout the cure procedure. Full cure is achieved much sooner in the final hold for the aged resin because the value of $d\varepsilon''/dt$ approaches zero much sooner.

4.6 Monitoring Cure in a Thick Laminate

A major advantage of the FDEMS sensing technique is the capability of monitoring and quantitatively measuring processing properties in situ during cure in complex shaped parts during processing in autoclaves, presses and pultruders.

For example, sensors can be placed in a thick 192 ply Hercules 3501-6 epoxy-graphite laminate at multiple positions (see Fig. 4.12) such as the tool surface, at 32 plies, at 64 plies, and in the center at the ninety-sixth ply. This 192-ply composite layup with the embedded sensors was cured in an autoclave. The output at each sensor was measured automatically at intervals throughout the cure cycle by multiplexing the four sensors through a computer-controlled switch to the impedance bridge. The multiplexed sensor–bridge–computer system can make more than 100 permitivity measurements in less than 2 minutes. Measurements are recorded continuously throughout the cure cycle without interruption over the 10^6–10^{-2} range in magnitudes of ε' and ε''.

Figure 4.13 is a plot of the viscosity determined by the sensor at each of the four positions in the thick laminate during cure in the autoclave. The FDEMS sensor data show that the middle ply achieves its first viscosity minimum 20 minutes after the plies on the surface of the

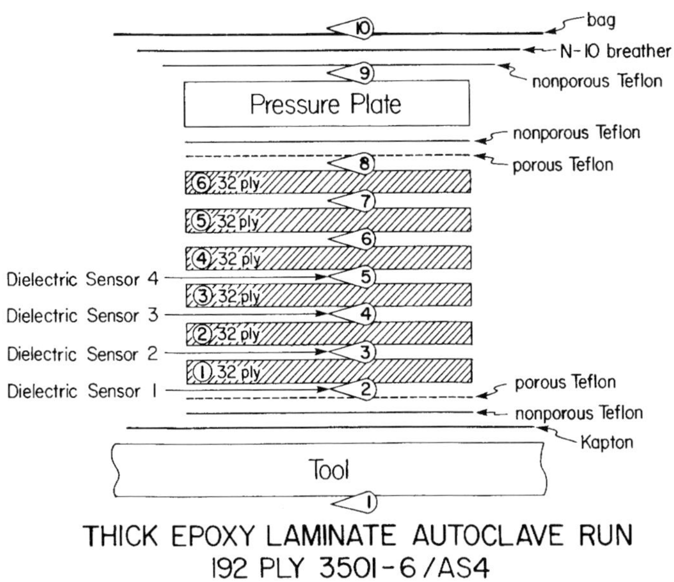

Figure 4.12 The lay up of a thick TGDDM epoxy (Hercules 3501-6) graphite laminate autoclave run showing the positions of the dielectric sensors and thermocouples

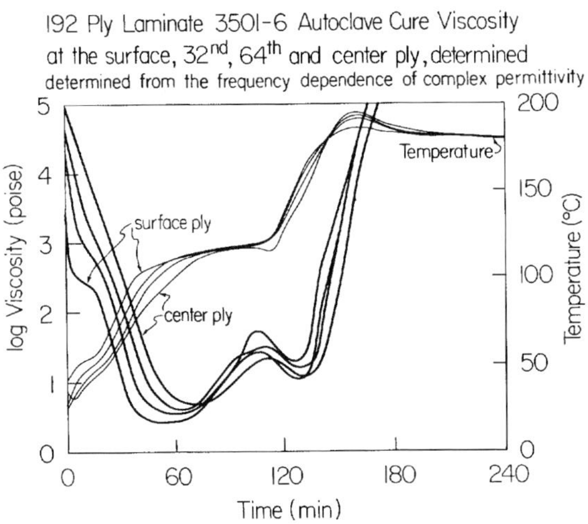

Figure 4.13 The viscosity of the thick laminate as determined from the frequency dependence of the FDEMS sensors at the surface, thirty-second, sixty-fourth, and center plies

tool plate. The middle plies continue to lag the surface ply until the second ramp to 177°C. At this point the exothermic 3501-6 epoxy reaction starts heating the laminate from the inside. As a result, the heat generated at the center ply causes the viscosity to catch up with the viscosity of the surface plies. In this way, FDEMS sensor output measurements of the viscosity at the center and surface can be used to evaluate the cure cycle. In this example the center and surface plies achieve their viscosity minimum in the high temperature ramp at roughly the same time. At all earlier times the center lags the reaction at the surface.

Thus, the sensor both reveals what is occurring during cure throughout the part and also reveals the exact time–temperature–magnitude dependence of processing properties such as viscosity for that tool, geometry and particular run.

The sensor output can be used to test the validity of processing models such as the Loos-Springer model [30]. Sensor measured values of η can be compared with the Loos-Springer model predictions. Figure 4.14 is a comparison of the model's predictions and the measured values at the sixty-fourth ply. Agreement in the viscosity's time dependence and magnitude with the predictions of models is essential if the model is to be verified and used with confidence.

The sensor output can also provide useful input for controlling the cure cycle. For this epoxy the most widely recommended and successful cure cycle forces the second viscosity minima to occur at the same time throughout the laminate's thickness as seen from the results of Figure 4.13. Thus, one can hypothesize that an efficient and effective cure cycle causes the viscosity minima at the surface and center to occur at the same time, and as rapidly as possible. One might therefore propose rapidly raising the air temperature in the FDEMS sensor-controlled run until the exotherm at the center causes the viscosity at the center ply to start to catch up to the surface viscosity. At this point the air temperature could be lowered rapidly to hold the surface viscosity in this high flow condition while the center viscosity catches up. At such point as the center viscosity goes through its viscosity minimum and

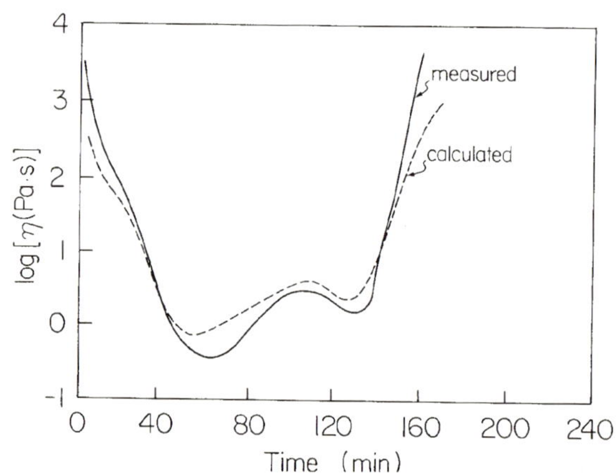

Figure 4.14 Comparison of the viscosity at the sixty-fourth ply as predicted by the FDEMS sensor and the Loos-Springer model

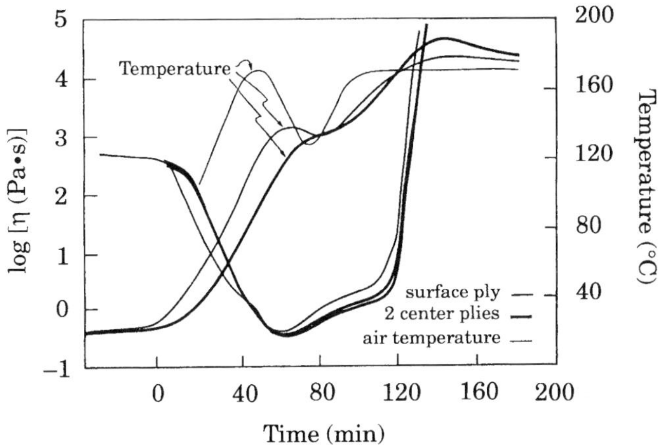

Figure 4.15 The viscosity at each sensor position of a 192-ply graphite-epoxy composite during an FDEMS sensor-controlled autoclave cure

advances beyond the surface plies, the air temperature would be set to the final 177°C hold. The 177°C hold would continue until the sensor output indicated, through $d\varepsilon''/dt$ approaching zero, that the reaction was complete.

Figure 4.15 shows the results from such a sensor controlled run in an autoclave. FDEMS sensor-measured viscosities from two sensors at the center ply and one sensor at the surface ply are shown in Figure 4.15. Air autoclave temperatures and the temperatures at the surface and center ply are also shown. The starting time for the FDEMS sensor-controlled autoclave run and the manufacturer's cure cycle run are defined as the time at which the tool surface temperature starts to increase.

The FDEMS sensor-controlled run significantly reduced the time lag and viscosity difference between the center ply and the tool surface ply. The amount of flow as measured by the magnitude of the viscosity minimum was greater in the FDEMS sensor-controlled run. The approach of $d\varepsilon''/dt$ to zero was used to determine cure completion. The total cure time of 200 min in this FDEMS controlled run is 40 min less than the conventional cure cycle.

4.7 Resin Film Infusion

As an example of cure monitoring in resin film infusion, Figure 4.16 shows the output of $\omega\varepsilon''(\omega)$ from sensor 1 for the MY720 aromatic TGMDA epoxy system commercially sold as 3501-6 by Hercules.

Measurements were taken of ε'' at 3-min time intervals during a resin film infusion (RFI) cure cycle in an autoclave at the Northrup Grumann Corporation Los Angeles plant. The T

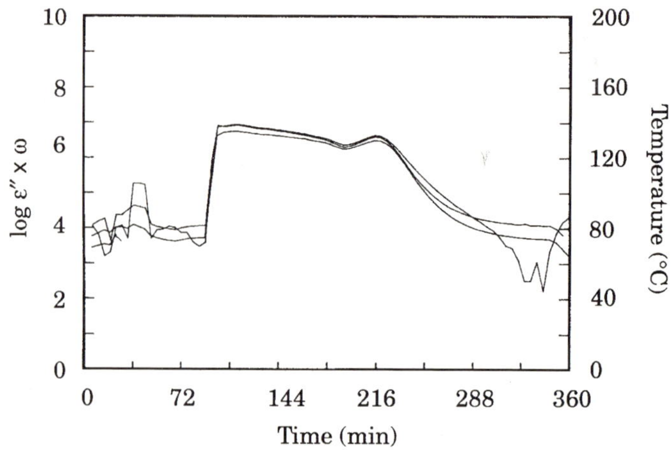

Figure 4.16 Output of $\omega\varepsilon''(\omega)$ from sensor 1 (see Fig. 4.1) for the MY720 aromatic TGMDA epoxy system commercially sold as 3501-6 by Hercules

stiffened graphite preform shown in Figure 4.17 was furnished by the McDonnell Douglas Aerospace. Temperature holds occurred at approximately 120°C and 177°C. Three frequencies (i.e., 250, 500, and 500 Hz) are shown in Figure 4.16. The time the resin front infiltrates to the sensor's location is shown by the large jump in sensor output. The overlapping lines indicate the signal is dominated by ionic mobility until the second hold has been reached. Figure 4.7 shows the correlation between the viscosity and the ionic mobility dominated values of ε'' at 5.0 kHz. Figure 4.7 was generated with data taken at temperatures between 100 and 149°C. It shows $\varepsilon''_{\text{ionic}} \alpha 1/\eta$ for viscosities up to gel, at which point the relation between the "molecular viscosity" governing ionic mobility is no longer directly related to the

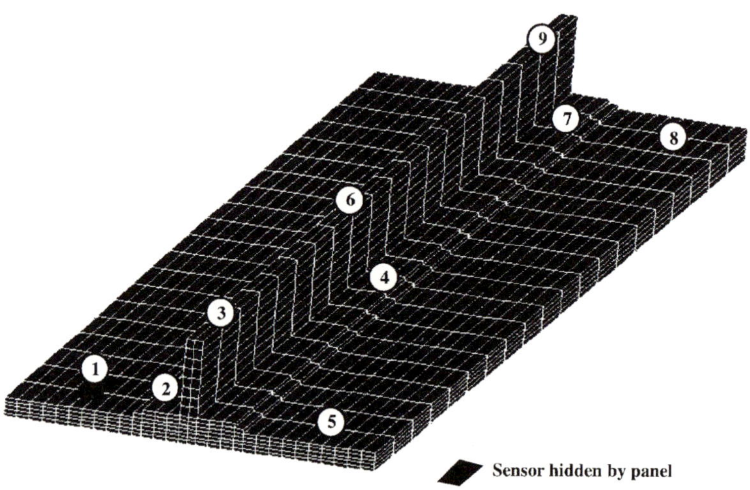

Figure 4.17 Locations of the nine sensors monitoring position of the resin front viscosity and degree of cure in the T stiffened panel

reciprocal of the macroscopic viscosity as measured on a dynamic mechanical viscometer. As discussed, this approximate inverse relation between molecular mobility and viscosity breaks down as the resin begins to cross-link and gel. Thus, FDEMS sensing can be used to monitor viscosity quantitatively in situ during processing. Through correlation with other techniques, FDEMS can be calibrated to detect gel. A calibration curve relating degree of cure and the FDEMS sensor output has also been generated (see Fig. 4.8). Unlike the viscosity calibration curve, it is temperature dependent. The buildup in final cure properties such as degree of cure during the last hold was monitored in situ using the FDEMS $d\varepsilon''/dt/\varepsilon''$ output and the correlation plot shown in Figure 4.9 for 3501-6.

Figures 4.18 and 4.19 show the viscosity and degree of cure of two different positions on the preform, position 1 on the flat panel and position 3 at the top of the *T* stiffener (see Fig.

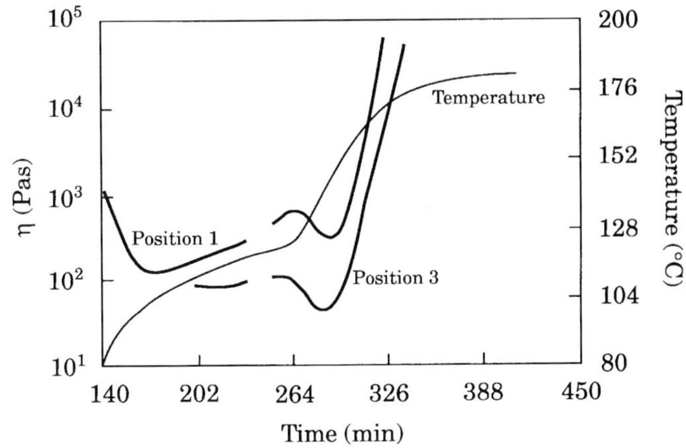

Figure 4.18 Viscosity of two different positions on the preform, position 1 on the flat panel, and position 3 at the top of the *T* stiffeners (see Fig. 4.1)

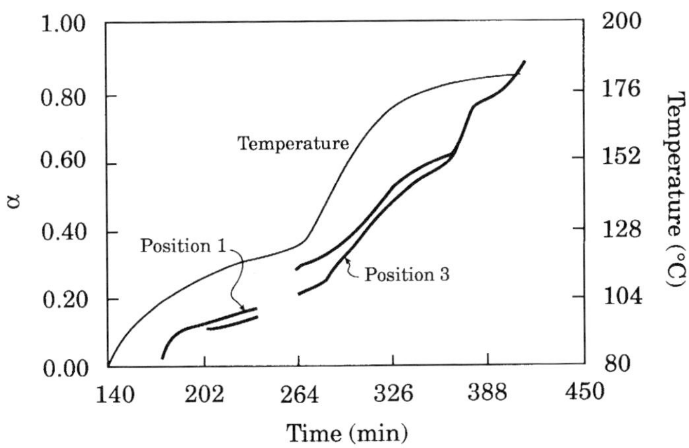

Figure 4.19 Degree of cure of two different positions on the preform, position 1 on the flat panel, and position 3 at the top of the *T* stiffeners (see Fig. 4.1)

4.17). These results are obtained from the FDEMS sensor output as shown in Figure 4.16 and using the calibration plots such as shown in Figure 4.7. Wetout times at the nine sensor locations shown in Figure 4.17 are:

Sensor Location	1	2	3	4	5	6	7	8	9
Wetout Time (min)	63	104	168	114	97	168	110	100	163

4.8 Smart Automated Control

A smart artificial intelligence control system is based on monitoring the actual *state* of the resin, not on a predetermined time–temperature procedure. Thus, the temperature scheme the part experiences is based on the achievement of molecular landmarks in the curing process, and not on time. Because these critical points may not occur at the exact same time for each part being made, time is not wasted holding a part at some stage when the cure process could be advanced. This gives the smart, closed-loop process a flexibility to adjust to variations in resin cure and heat transfer properties. Thus, it is much more efficient and reliable than a rigid time–temperature procedure.

Overall an intelligent closed-loop system produces more consistent parts because the advancement to the next stage of cure is based on the resin achieving a certain molecular state rather than on time. The advancement of the viscosity and degree of cure is monitored. Final cure is defined by a universal degree of cure. This is a more consistent way to produce composite parts than simply subjecting them to the same time–temperature schedule when batch variations and differences in prefabrication handling are present.

A smart closed-loop system is composed of a heating/pressure device, such as a press or autoclave, sensors able to gather data in situ from the composite part, and smart software that collects the sensor information, interprets it, and then makes decisions that control the fabrication device. The smart software combines both the predictions of the processing model and the insight of operator experience. Figure 4.20 is a schematic diagram of the FDEMS

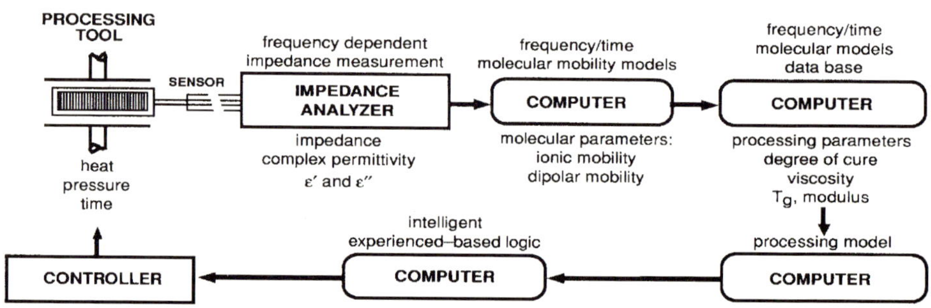

Figure 4.20 Schematic of the FDEMS sensor smart system for monitoring quality and for expert automated process control

sensor smart system for in situ monitoring of cure and for expert automated process control. The expert system was constructed to optimize and control the achievement of six critical stages during the resin film infusion cure process. These were (1) achieve a low resin viscosity, (2) maintain a low viscosity until impregnation is complete, (3) advance the resin reaction at an intermediate temperature to a particular degree of cure that avoids an excessive exothermic effect (this value varies with part thickness), (4) ramp to final cure temperature once reaction has advanced, (5) monitor degree of cure during final hold, (6) determine achievement of proper degree of cure, which is related to attainment of ultimate T_g, and desired use properties, and then (7) turn the process cycle off.

Figure 4.21 shows the sensor output for the smart automated sensor expert system-controlled run. The resin reached the center sensor at 37 min. The viscosity is maintained at a low value by permitting slow increases in the temperature. At 60 min, fabric impregnation was complete. The resin was advanced during a 121°C hold to a predetermined value of degree of cure of 0.35, based on the Loos model's predictions of the extent of the exothermic effect. This value of α is clearly dependent on panel thickness. Then at 130 min, the ramp to 177°C was begun. Achievement of an acceptable complete degree of cure was determined by the sensor at 190 min. Then the cure process was shut down.

As an aside, we note that the FDEMS sensor input information can also be used to detect the onset of phase separation in toughened thermoset systems and to monitor cure in thin film coatings and adhesive bond lines. It is particularly important that the FDEMS sensor is also very sensitive to changes in the mechanical properties of the resin due to degradation. As such, it can be used for accelerated aging studies and as a dosimeter to monitoring the composite part during use to determine the knockdown in the required performance properties with time.

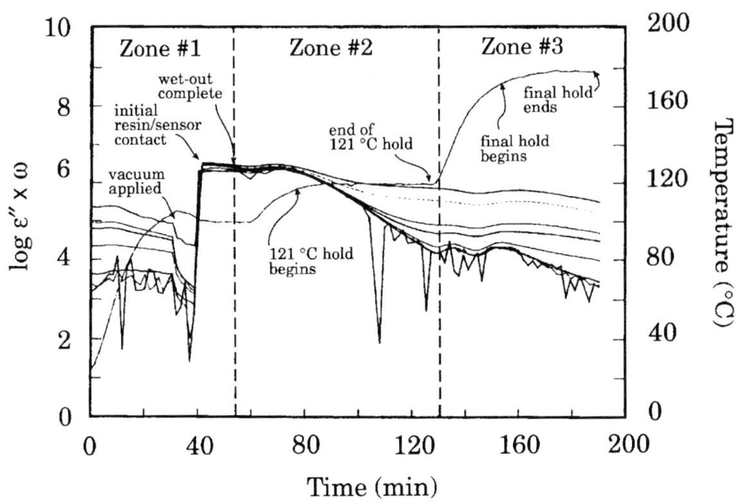

Figure 4.21 Sensor output for the smart automated sensor expert system-controlled run

In summary an FDEMS sensor system can be used to monitor the processing properties in situ during the fabrication process of a composite part. A “smart” sensor control system can be used to monitor resin properties for reproducability-quality assurance, to ensure fabric impregnation, and to control and optimize the composite fabrication process intelligently through in situ sensor feedback.

4.9 Conclusions

In situ FDEMS sensing is a convenient, sensitive, automated technique for monitoring in situ in the fabrication tool the changes in cure processing properties such as viscosity, buildup in T_g, buildup in modulus, and degree of cure continuously throughout the cure process. The technique can be used to monitor the state of the resin at various positions in the mold, at various ply depths in a thick laminate, and during complex time–temperature cure cycles. It can be coupled with cure models and operator expertise to provide continuous in situ online data of the current cure state of the resin for use as input for smart or intelligent automated cure process control.

Acknowledgment

David E. Kranbuehl appreciates partial support from the NSF Science and Technology Center at Virginia Polytechnic Institute and State University under Contract #DMR91-2004, a NASA Langley grant NAGI-23 and support from the Douglas Aircraft Corporation.

References

1. Kranbuehl, D., *Developments in Reinforced Plastics*, Vol. 5 (1986) Elsevier Applied Science Publishers, New York, p. 181–204
2. Kranbuehl, D., *Encyclopedia of Composites* (1989) Lee S.M. (Ed.) VCH Publishers, New York, p. 531–543
3. Senturia, S., Sheppard, S., *Appl. Polym. Sci.* (1986) 80 p. 1–48
4. May, C., Chemorheology of Thermosetting Resins, *Polymer Materials Science and Engineering* (1983) ACS Symposium Series 227, American Chemical Society, Washington, DC
5. Hedvig, P. *Dielectric Spectroscopy of Polymers* (1977) Wiley, New York
6. Mijovic, J., Bellucci, F., Nicolois, L. *Electrochem. Soc.* (1995) 142(4), p. 1176–1182
7. Mijovic, J., Winnie Tee, C.F. *Macromolecules* (1994) 27, p. 7287–7293
8. Bellucci, F., Valentino, M., Monetta, T., Nicodemo, L., Kenny, J., Nicolais, L., Mijovic, J. *J. Poly. Sci. B. Poly, Phys.* (1995) 33, p. 433–443
9. Mangion, M.B.M., Johari, G.P. *Macromolecules* (1990) 23, p. 3687–3695
10. Mangion, M., Johari, G. *J. Poly., Sci. B* (1991) 29 p. 1117–1125
11. Parthun, M.B., Johari, G. *Macromolecules* (1992) 25 p. 3254–3263
12. Boiteux, G., Dublineau, P., Feve, M., Mathieu, C., Seytre, G., Ulanski, J. *Polym. Bull.* (1992) 30 p. 441–447

13. Mathieu, C., Boiteux, G., Seytre, G., Villain, R., Dublineau, P. *J. Non-Crystalline Solids* (1994) 172–174, p. 1012–1016
14. Xu, X., Galiatsatos, V. SPE *Technical Papers (ANTEC '93)* (1993) 39, p. 2875
15. Xu, X., Galiatsatos, V. *Makromol. Symp.* (1993) 76, p. 137
16. Deng, Y., Martin, G. *Macromolecules* (1994) 27 p. 5141–5146
17. Companik, J., Bidstrup, S. *Polymer* (1994) 35 p. 4823–4840
18. Tombari, E., Johari, G.P. *J. Chem. Phys.* (1992) 97 p. 6677–6686
19. Johari, G.P., Pascheto, W.P. *J. Chem. Svc Trans.* (1995) 91 p. 343–351
20. MacKinnon, A., Jenkins, S., McGrail, P., Pethrick, R. *Macromolecules* (1992) 25 p. 3492–3499
21. Maistros, G., Black, H., Bucknall, C., Partridge, I. *Polymer* (1992) 33 p. 4470–4478
22. Hart, S., Kranbuehl, D., Hood, D., Loos, A., Koury, J., Havery, J. *Int. SAMPE Symp. Proc.* (1994) 39(1) p. 1641–1651
23. Kranbuehl, D. Matériaux & Techniques—Advanced Composite Materials (1994) Nov.–Dec. p. 18–22
24. Kranbuehl, D., Kingsley, P., Hart, S., Hasko, G., Dexter, B., Loos, A.C. *Polym. Comp.* (1994) 15(4) p. 297–305
25. Hart, S., Kranbuehl, D., Loos, A., Hinds, B., Koury, J., Harvey, J. *SAMPE* (1993) 38 p. 1009–1019
26. Hart, S., Kranbuehl, D., Loos, A., Hinds, B., Koury, J. *SAMPE Symp.* (1992) 37 p. 225–230
27. Loos, A.C., Kranbuehl, D.E., Freeman, W.T., *Intelligent Processing of Materials and Advanced Sensors* (1987) Wadley, H.N.G. et al. (Ed.) The Metallurgical Society, Inc., Warrendale, PA, p. 197–211
28. Day, D., *35th International SAMPE Symposium* (1990) 35 p. 1507–1516
29. Kranbuehl, D., Eichinger, D., Hamilton, T., Clark, R. *Polym. Eng. Sci.* (1991) 31 p. 56
30. Ciriscioli, P., Springer G. Smart Autoclave Cure *Technomic* (1990) Lancaster, Penn.
31. Kranbuehl, D.E., Delos, S.E., Jue, P.K. *Polymer* (1986) 27 p. 11–20
32. Kranbuehl, D., Delos, S., Hoff, M., Weller, W., Haverty, P., Seeley, J. *Am. Chem. Soc. Symp. Ser.* (1988) 367 p. 101–113
33. Sheppard, N., Gavericle, S., Day, D., Senturia, S. *Proc. 26th Natl. SAMPE Symp.* (1981) p. 65–76
34. Dickie, R., Labana, S., Bauer, R. *Crosslinked Polymers: Chemistry, Properties and Applications* (1988) ACS Symposium Series 367, American Chemical Society, Washington, DC, p. 100
35. Kranbuehl, D. *Plast. Rubber Compos., Process* (1991) 16 p. 213–223
36. Kranbuehl, D., Eichinger, D., Hamilton, T., Clark, R. *Polym. Eng. Sci.* (1991) 31 p. 56–60
37. Kranbuehl, D. *J. Non-Crystalline Solids* (1991) 131–133 p. 930–936
38. Parthun, M.B., Johari, G. *Macromolecules* (1992) 25 p. 3254–3263
39. Mathieu, C., Boiteux, G., Seytre, G., Villian, R., Dublineau, P. *J. Non-Crystalline Solids* (1994) 172–174 p. 1012–1016
40. Martin, G., Tungare, A., Grotto, J. *Am. Chem. Soc. Polym. Characterization* (1990) 227 p. 205–215
41. Deng, Y., Martin, G. *J. Polym. Sci. B Polymer Physics* (1994) 32 p. 2115–2125
42. Kranbuehl, D., Delos, S., Hoff, M., Weller, L., Haverty, P., Seeley, J., Whitham, B. *Proc. 32nd Int. SAMPE Symp.* (1987) p. 338–348
43. Kranbuehl, D. *Polymer Materials Science and Engineering Preprints* (1993) 68 p. 224–225
44. Kranbuehl, D. *Polymer Materials Science and Engineering Preprints* (1990) 68 p. 90–93
45. Kranbuehl, D., Kim, T., Liptak, S.C., McGrath, J.E. *Polym. Preprints* (1993) 34 p. 488–489
46. Maistros, G., Black, H., Bucknall, C., Partridge, I. *Polymer* (1992) 33 p. 4470–4478
47. MacKinnon, A., Jenkins, S., McGrail, P., Pethrick, R. *Macromolecules* (1992) 25 p. 3492–3499

5 A Unified Approach to Modeling Transport of Heat, Mass, and Momentum in the Processing of Polymer Matrix Composite Materials

Bamin Khomami

Based on a local volume averaging concept general balance equations for transport of mass, momentum, and energy in stationary and moving porous media have been developed. In turn, these equations have been used to obtain specific governing equations in a number of important polymer matrix composite manufacturing processes, such as resin transfer molding, injected pultrusion, and autoclave processing. Moreover, appropriate numerical techniques for solution of these coupled partial differential equations have been briefly outlined and a few example simulations have been performed.

5.1 Introduction

Transport of mass, energy, and momentum in porous media is a key aspect of a large number of fiber-reinforced plastic composite fabrication processes. In design and optimization of such processes, computer simulation plays an important role. Recent studies [1–14] have reported on development of a number of computer simulation techniques for modeling various composite fabrication processes. Some of these simulation models are isothermal, whereas others consider heat transfer and chemical reaction. Overall, computer simulation models are intended to provide information regarding filling time, degree of cure, void fraction, and residual stresses to engineers involved in design and optimization of various composite fabrication processes.

Transport of mass, momentum, and energy in porous media is at the heart of any effort aimed at modeling fabrication of fiber-reinforced composite materials. There are, unfortunately, no universally accepted balance equations for transport of mass, momentum, and energy in porous media. Hence, the purpose of this chapter is to provide the basic balance equations for modeling various composite fabrication processes such as injected pultrusion, autoclave processing, and resin transfer molding based on the best available methods from the porous media literature. Kaviany [15] has fortunately compiled the present state of the art on theory of flow in porous media upon which Tucker and Dessenburg [16] have derived the basic balance equations for the resin transfer molding process. Thus, we will not go into as much detail as the previous authors when deriving the basic balance equations; however, sufficient detail about the porous media theory will be provided to give the reader a basic understanding of how the equations were derived. Moreover, we will provide simplifications of the basic balance equations for various composite fabrication processes. This approach should provide an accurate basis for modeling various composite manufacturing processes as well as providing a check to earlier modeling efforts.

The chapter is organized as follows. The concept of local volume averaging for deriving transport equations in porous media is discussed in Section 5.2. Brief derivations of the balance equations for mass, energy and momentum are presented in Section 5.3. Section 5.4 is divided into three subsections on modeling of resin transfer molding, injected pultrusion, and autoclave processing. In each subsection the balance equations for transport of mass, energy, and momentum along with appropriate numerical techniques for solution of the governing equations are provided, and a few example simulations have been performed.

5.2 Local Volume Averaging

It is extremely difficult to model macroscopic transport of mass, energy, and momentum in porous media commonly encountered in various fields of science and engineering based on microscopic transport models that account for variation of velocity and temperature as well as other quantities of interest past individual solid particles. The basic idea of porous media theory, therefore, is to volume average the quantities of interest and develop field equations based on these average quantities.

The first step in applying volume averaging is to consider a representative volume for every point **A** in the porous medium. This volume must be large enough to contain sufficient amount of each phase such that continuum theory for transport of mass, energy, and momentum would apply, but small enough so that long range variations do not affect the averages.[1]

Figure 5.1 depicts a representative volume **V** as a sphere centered on **A** with surface area **S**.

A number of different volume averages can be defined. First we will consider a spatial average, which is the average value of a quantity within V,

$$\langle G \rangle = \frac{1}{V} \int_V G dV \tag{5.1}$$

This average describes the combined property of both phases. A second type of average can be defined which is specific to a given phase. This is called the phase average,

$$\langle G_r \rangle = \frac{1}{V} \int_V G Y_r dV = \frac{1}{V} \int_{V_r} G_r dV \tag{5.2}$$

where Y is a phase function and is defined as:

$$Y_r(A, t) = \begin{cases} 1 & \text{if A lies in the resin} \\ 0 & \text{if A lies in any other phase} \end{cases} \tag{5.3}$$

We have chosen to call the two phases resin and fiber. Each phase will be denoted by subscript "r" and "f," respectively. A similar phase function (i.e., Y_f) can be defined for the fiber phase. It should be noted that if the fiber phase is stationary Y is not a function of time.

As shown by Equation 5.2, phase averages describe the average of quantities in a single phase in volume V. One can also define an average for points lying within a single phase and average them over the volume of that phase. This is called an intrinsic phase average,

$$\langle G_r \rangle^r = \frac{1}{V_r} \int_{V_r} G_r dV = \frac{\int_V G_r Y_r dV}{\int_V Y_r dV} \tag{5.4}$$

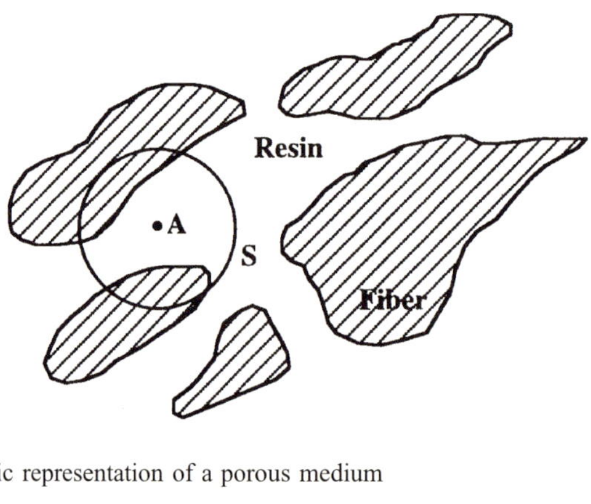

Figure 5.1 A schematic representation of a porous medium

[1] For more detail see [15,16].

The three averages can be related to one another in a simple manner. For example,

$$\langle G \rangle = \langle G_r \rangle + \langle G_f \rangle \tag{5.5}$$

and

$$\langle G_r \rangle = \frac{V_r}{V} \langle G_r \rangle^r = \varepsilon_r \langle G_r \rangle^r \tag{5.6}$$

where ε_r is the resin volume fraction often referred to as porosity.

5.3 Derivation of Balance Equations

5.3.1 Conservation of Mass

The majority of composite fabrication processes are carried out with thermosetting resins. One must therefore keep track of both the cured and uncured resin. In what follows, first the overall mass balance equation for the resin phase will be developed, then a balance equation for the cured portion of the resin will be presented.

By performing a mass balance over a differential stationary volume, one obtains the continuity equation,

$$\frac{\partial \rho_r}{\partial t} + \vec{\nabla} \cdot \left(\rho_r \vec{U}_r \right) = 0 \tag{5.7}$$

phase averaging this equation gives,

$$\left\langle \frac{\partial \rho_r}{\partial t} \right\rangle + \langle \vec{\nabla} \cdot \rho_u \vec{U}_r \rangle = 0 \tag{5.8}$$

The preceding equation can also be written as,

$$\frac{\partial}{\partial t} \langle \rho_r \rangle - \rho_r \frac{\partial \varepsilon_r}{\partial t} + \vec{\nabla} \cdot \langle \rho_r \vec{U}_r \rangle + \frac{1}{V} \int_{S_i} \rho_r \vec{U}_r \cdot \hat{n}_i ds = 0 \tag{5.9}$$

where S_i is the interfacial area between the resin and the fiber phase in volume **V** and $\hat{n}_i$ is the unit normal to that surface pointing toward the fiber phase. It can be shown [15,16] that

$$\frac{1}{V} \int_{S_i} \vec{U}_r \cdot \hat{n}_i ds = \frac{\partial \varepsilon_r}{\partial t} \tag{5.10}$$

by assuming that ρ_r varies much more slowly than $\vec{U}_r$, Equation 5.10 can be used to simplify Equation 5.9,

$$\frac{\partial}{\partial t} (\rho_r \varepsilon_r) + \rho_r \vec{\nabla} \cdot \langle \vec{U}_r \rangle = 0 \tag{5.11}$$

Differentiating the first term and using the assumption that ρ_r is relatively constant, reduces Equation 5.11 to,

$$\frac{\partial \varepsilon_r}{\partial t} + \vec{\nabla} \cdot \langle \vec{U}_r \rangle = 0 \tag{5.12}$$

Equations 5.9–5.12 can be used for conservation of total mass of the resin under various conditions. To keep track of the degree of cure of the resin we need to write a separate mass balance for the cured phase α;

$$\frac{\partial \alpha}{\partial t} + \vec{U}_r \cdot \vec{\nabla} \alpha = \vec{\nabla} \cdot D \vec{\nabla} \alpha + Z_\alpha \tag{5.13}$$

where α is the degree of cure, D is the diffusion coefficient, and Z_α is the rate of production of α per unit volume. It should be noted that in Equation 5.13 we have assumed that the diffusion is Fickian and enhancement of mass transfer due to dispersion has been neglected. For a detailed discussion regarding dispersion the reader is referred to [15,16]. The amount of mass transfer due to diffusion in most composite manufacturing processes is typically small compared with convection. This term, therefore, is usually neglected,

$$\frac{\partial \alpha}{\partial t} + \vec{\nabla} \cdot (\vec{U}_r \alpha) = Z_a \tag{5.14}$$

Taking a phase average of this equation, one obtains

$$\left\langle \frac{\partial \alpha}{\partial t} \right\rangle + \langle \vec{\nabla} \cdot (\vec{U}_r \alpha) \rangle = \langle Z_a \rangle \tag{5.15}$$

One can rewrite this equation as,

$$\frac{\partial \langle \alpha \rangle}{\partial t} + \vec{\nabla} \cdot \langle \vec{U}_r \alpha \rangle + \frac{1}{V} \int_{S_i} \hat{n}_i \cdot (\vec{U}_r \alpha) ds - \frac{1}{V} \int_{S_i} \hat{n}_i \cdot (\vec{U}_r \alpha) ds = \langle Z_a \rangle \tag{5.16}$$

In case of a stationary porous medium, the second surface integral on the left-hand side does not come into play and the first integral vanishes due to the no slip boundary condition. For a moving fibrous media, the two integrals cancel each other. In general the degree of cure and the rate of production of the cured resin measured experimentally are related to the intrinsic phase averages $\langle \alpha \rangle^r$ and $\langle Z_\alpha \rangle^r$. Hence, one should rewrite Equation 5.16 in terms of intrinsic phase averages which is consistent with experimental measurements. When this is done the species balance equation is given by:

$$\varepsilon_r \frac{\partial \langle \alpha \rangle^r}{\partial t} + \langle \vec{U}_r \rangle \cdot \vec{\nabla} \langle \alpha \rangle^r + \langle \alpha \rangle^r (\vec{\nabla} \cdot \langle \vec{U}_r \rangle) = \varepsilon_r \langle Z_\alpha \rangle^r \tag{5.17}$$

In Equation 5.17 the effect of dispersion has been neglected. In general $\langle Z_\alpha \rangle^r$ is measured experimentally and the results are reported as a function of $\langle \alpha \rangle^r$ and $\langle T_r \rangle^r$. One can therefore rewrite Equation 5.17 as,

$$\varepsilon_r \frac{\partial \langle \alpha \rangle^r}{\partial t} + \langle \vec{U}_r \rangle \cdot \vec{\nabla} \langle \alpha \rangle^r + \langle \alpha \rangle^r (\vec{\nabla} \cdot \langle \vec{U}_r \rangle) = \varepsilon_r \langle f(\langle \alpha \rangle^r, \langle T_r \rangle^r) \rangle \tag{5.18}$$

Note that in practice one approximates $\langle f(\langle \alpha \rangle^r, \langle T_r \rangle^r) \rangle$ by $f(\langle \alpha \rangle^r, \langle T_r \rangle^r)$. This is vigorously valid only if f varies linearly in α and T. This is not usually the case, but if variations in α and T_r are small in comparison to f, then this approximation is valid.[2]

[2] See [17] for a more detailed discussion.

5.3.2 Conservation of Momentum

Performing a momentum balance over a differential volume of a homogenous material leads to the Cauchy equation of motion,

$$\frac{\partial(\rho_r\vec{U}_r)}{\partial t}+\vec{\nabla}\cdot(\rho_r\vec{U}_r\vec{U}_r)=\vec{\nabla}\cdot\underline{\underline{\sigma}}_r+\rho_r\vec{g} \tag{5.19}$$

where $\underline{\underline{\sigma}}$ is the total stress tensor and $\vec{g}$ represents the gravitational body force per unit mass. Phase averaging this equation gives,

$$\frac{\partial\langle\rho_r\vec{U}_r\rangle}{\partial t}+\langle\vec{\nabla}\cdot\rho_r\vec{U}_r\vec{U}_r\rangle=\langle\vec{\nabla}\cdot\underline{\underline{\sigma}}_r\rangle+\langle\rho_r\vec{g}\rangle \tag{5.20}$$

Expanding the phase averaged equation one gets,

$$\frac{\partial\langle\rho_r\vec{U}_r\rangle}{\partial t}+\vec{\nabla}\cdot\langle\rho_r\vec{U}_r\vec{U}_r\rangle=\vec{\nabla}\cdot\langle\underline{\underline{\sigma}}_r\rangle+\langle\rho_r\vec{g}\rangle+\frac{1}{V}\int_{S_i}\underline{\underline{\sigma}}_r\cdot\hat{n}_i ds+\frac{\rho_r}{V}\int_{S_i}\vec{U}_r\cdot\hat{n}_i ds \tag{5.21}$$

If we assume that the body force is only due to gravity and use the definition $\underline{\underline{\sigma}}_r=-p_r\underline{\underline{I}}+\underline{\underline{\tau}}$ where p_r is the isotropic resin pressure, $\underline{\underline{I}}$ is the unit tensor, and $\underline{\underline{\tau}}$ is the deviatoric stress as well as assuming a constant density and define a new pressure $P_r=p_r+\rho_r gh$, Equation 5.21 simplifies to,

$$\rho_r\frac{\partial\langle\vec{U}_r\rangle}{\partial t}+\rho_r\vec{\nabla}\cdot\langle\vec{U}_r\vec{U}_r\rangle=-\vec{\nabla}\langle P_r\rangle+\vec{\nabla}\cdot\langle\underline{\underline{\tau}}_r\rangle \\ +\frac{1}{V}\int_{S_i}\underline{\underline{\sigma}}_r\hat{n}_i ds-\frac{1}{V}\int_{S_i}\rho_r gh\hat{n}_i ds+\frac{\rho_r}{V}\int\vec{U}_r\cdot\hat{n}_i ds \tag{5.22}$$

It is more convenient to use an intrinsic phase average for the pressure drop because it is measured that way experimentally. Replacing $\langle P_r\rangle$ by $\varepsilon_r\langle P_r\rangle^r$ and using the fact that [16],

$$\vec{\nabla}\varepsilon_r=-\frac{1}{V}\int_{S_i}\hat{n}_i ds \tag{5.23}$$

one can rewrite Equation 5.22 in the following manner,

$$\rho_r\frac{\partial\langle\vec{U}_r\rangle}{\partial t}+\rho_r\vec{\nabla}\cdot\langle\vec{U}_r\vec{U}_r\rangle=-\varepsilon_r\vec{\nabla}\langle P_r\rangle^r+\vec{\nabla}\cdot\langle\underline{\underline{\tau}}_r\rangle \\ +\frac{1}{V}\int_{S_i}\underline{\underline{\sigma}}_r\cdot\hat{n}_i ds-\frac{1}{V}\int_{S_i}\rho_r gh\hat{n}_i ds+\frac{\langle P_r\rangle^r}{V}\int_{S_i}\hat{n}_i ds+\frac{\rho_r}{V}\int_{S_i}\vec{U}_r\cdot\hat{n}_i ds \tag{5.24}$$

where the four integrals in this equation are due to the interactions between the fiber and the resin. We can define a parameter f_d that represents the resin–fiber drag force associated with motion of the resin through the fibers and rewrite Equation 5.24 as follows:

$$\rho_r\frac{\partial\langle\vec{U}_r\rangle}{\partial t}+\rho_r\vec{\nabla}\cdot\langle\vec{U}_r\vec{U}_r\rangle=-\varepsilon_r\vec{\nabla}\langle P_r\rangle^r+\vec{\nabla}\cdot\langle\underline{\underline{\tau}}_r\rangle-f_d \tag{5.25}$$

At this point, we need to develop an expression for f_d that contains only the averaged field variables. This has been done vigorously by Slattery [18,19], and it has shown that in absence of resin inertia

$$f_d = M\left[\langle\vec{U}_r\rangle^r - \langle\vec{U}_f\rangle^f\right] \tag{5.26}$$

where M is a scalar resistance coefficient that could depend on the viscosity of the resin (μ), ε_r and the characteristic length of a porous medium (ℓ) and $|\langle\vec{U}_r\rangle^r - \langle\vec{U}_f\rangle^f|$. Based on dimensional analysis one can show that [18],

$$f_d = \frac{\varepsilon_r^2 \mu}{\underline{\underline{k}}_p} \cdot \left[\langle\vec{U}_r\rangle^r - \langle\vec{U}_f\rangle^f\right] \tag{5.27}$$

where $\underline{\underline{k}}_p$ is the permeability tensor. There have been numerous experimental studies that have indicated that permeability is only determined by the geometry and dimensions of the fiber phase and it is not a function of resin properties [20,21], which is consistent with what is obtained by the above analysis.

As mentioned earlier, the expression for f_d is obtained under conditions of no inertia. If we further assume the resin is Newtonian (i.e., $\tau = \mu[\vec{\nabla}\vec{U}_r + \vec{\nabla}\vec{U}_r^t])$) and the fiber phase is stationary, then Equation 5.25 can be simplified to the well-known Brinkman equation [22],

$$-\varepsilon_r \vec{\nabla}\langle P_r\rangle^r + \mu\vec{\nabla}\cdot\vec{\nabla}\langle\vec{U}_r\rangle - \frac{\varepsilon_r \mu}{\underline{\underline{k}}_p} \cdot \langle\vec{U}_r\rangle = 0 \tag{5.28}$$

Moreover, if one assumes that the $\langle\vec{U}_r\rangle$ changes very slowly on the length scale of the porous media, (i.e., ℓ), then the viscous stress term in the Brinkman equation can be neglected and this equation reduces to:

$$\langle\vec{U}_r\rangle = \frac{-\underline{\underline{k}}_p}{\mu} \cdot \vec{\nabla}\langle P_r\rangle^r \tag{5.29}$$

which is the anisotropic version of Darcy's law [23].

Up to this point, we have only considered cases where inertia is negligible. If inertia is important, then one can generalize Forchheimer's equation [15,16,24,25] to obtain an expression for f_d as follows:

$$f_d = \frac{\varepsilon_r \mu}{\underline{\underline{k}}_p}\langle\vec{U}_r\rangle + \frac{\varepsilon_r \rho_r b}{\sqrt{k_p}}|\langle\vec{U}_r\rangle|\langle\vec{U}_r\rangle \tag{5.30}$$

but before employing this equation, one needs to assess at what Reynolds number inertial effects become important. First, we define a pore-size Reynolds number as,

$$\mathrm{Re}_p = \frac{\rho_r U_{ave}\sqrt{k_p}}{\mu} \tag{5.31}$$

where U_{ave} is a characteristic resin velocity. At $\mathrm{Re}_p > 1$ inertial effects become noticeable [26,27]. The value of parameter b in Equation 5.30 has been experimentally determined to be in the range of 0.5 to 0.55 [26,27]. Using Equation 5.30 in conjunction with the fact that in absence of inertial dispersion [15],

$$\rho_r \vec{\nabla} \cdot \langle \vec{U}_r \vec{U}_r \rangle = \rho_r \vec{\nabla} \cdot \left[\frac{1}{\varepsilon_r} \langle \vec{U}_r \rangle \langle \vec{U}_r \rangle \right] \tag{5.32}$$

one can rewrite Equation 5.25 as follows:

$$\rho_r \frac{\langle \partial \vec{U}_r \rangle}{\partial t} + \rho_r \vec{\nabla} \cdot \left[\frac{1}{\varepsilon_r} \langle \vec{U}_r \rangle \langle \vec{U}_r \rangle \right] = -\varepsilon_r \vec{\nabla} \langle P_r \rangle^r + \mu \cdot \vec{\nabla} \cdot \vec{\nabla} \langle \vec{U}_r \rangle - \varepsilon_r \left[\frac{\mu}{\underline{\underline{k}}_p} + \frac{\rho_r b}{\sqrt{k_p}} |\langle \vec{U}_r \rangle| \right] \langle \vec{U}_r \rangle \tag{5.33}$$

It should be noted that in the preceding analysis we have assumed that $\underline{\underline{k}}_p$ has an inverse. Although ample support for this assumption exists [19], no vigorous proof is available.

5.3.3 Conservation of Energy

When considering heat transfer in a multiphase system, one can either treat each phase separately or the phases can be assumed to be in local equilibrium. In general, it is more complicated to treat each phase separately; hence, we will use the local equilibrium approach that assumes all the phases have the same local average temperature. By performing an energy balance over a differential resin element one obtains,

$$\rho_r C_{p_r} \frac{\partial T_r}{\partial t} + \rho_r C_{p_r} \vec{\nabla} \cdot (T_r \vec{U}_r) = -\vec{\nabla} \cdot \vec{q}_r + \Phi_r + \Phi_c \tag{5.34}$$

where $\vec{q}_r$ represents the heat flux, C_{p_r} is the resin heat capacity, and Φ_r and Φ_c represent viscous dissipation and dissipation due to the curing reaction, respectively. Assuming that there are no phase changes and that the resin has a constant density and thermal conductivity, and neglecting variations in C_{p_r} in the averaging volume, phase averaging of Equation 5.34 gives,

$$(\rho C_p)_r \frac{\partial}{\partial t} \langle T_r \rangle - \frac{(\rho C_p)_r}{V} \int_{S_i} (T_r \vec{U}_r) \cdot \hat{n}_i ds + (\rho C_p)_r \left\{ \vec{\nabla} \cdot \langle T_r \vec{U}_r \rangle + \frac{1}{V} \int_{S_i} (T_r \vec{U}_r) \cdot \hat{n}_i ds \right\} = -\langle \vec{\nabla} \cdot \vec{q}_r \rangle + \varepsilon_r \langle \Phi_r \rangle^r + \varepsilon_r \langle \Phi_c \rangle^r \tag{5.35}$$

We have rewritten the contribution due to viscous dissipation and chemical reaction in terms of their intrinsic phase averages because they are a more appropriate average for these quantities. For a stationary porous media the first surface integral on the left-hand side does not come into play and the second integral is zero due to the no slip boundary condition. For a moving porous media the two integrals cancel each other.

If we assume that conduction in the liquid phase is governed by Fourier's law, Equation 5.35 simplifies to,

$$\varepsilon_r(\rho C_p)_r \frac{\partial \langle T_r \rangle^r}{\partial t} + (\rho C_p)_r(\langle \vec{U}_r \rangle \cdot \vec{\nabla} \langle T_r \rangle^r + \langle T_r \rangle^r (\vec{\nabla} \cdot \langle \vec{U}_r \rangle))$$
$$= \vec{\nabla} \cdot \{\varepsilon_r K_r \vec{\nabla} \langle T_r \rangle^r\} + \frac{K_r}{V} \int_{S_i} \hat{n}_i \cdot \vec{\nabla} T_r ds + \varepsilon_r \langle \Phi_r \rangle^r + \varepsilon_r \langle \Phi_c \rangle^r \tag{5.36}$$

where, K_r is the thermal conductivity of the resin phase. It should be noted that terms arising due to dispersion have been neglected in the above equation. A similar equation for the fiber phase can be generated by setting $\Phi_r = \Phi_c = 0$,

$$\varepsilon_f(\rho C_p)_f \frac{\partial \langle T_f \rangle^f}{\partial t} + (\rho C_p)_f(\langle \vec{U}_f \rangle \cdot \vec{\nabla} \langle T_f \rangle^f + \langle T_f \rangle^f (\vec{\nabla} \cdot \langle \vec{U}_f \rangle))$$
$$= \vec{\nabla} \cdot \left\{\varepsilon_f K_f \vec{\nabla} \langle T_f \rangle^f\right\} - \frac{K_f}{V} \int_{S_i} \hat{n}_i \cdot \vec{\nabla} T_f ds \tag{5.37}$$

where, K_f is the thermal conductivity of the fiber phase. Adding Equations 5.36 and 5.37, assuming local thermal equilibrium (i.e., $\langle T_r \rangle^r = \langle T_f \rangle^f = \langle T \rangle$), and neglecting thermal dispersion one obtains,

$$\{\varepsilon_r(\rho C_p)_r + \varepsilon_f(\rho C_p)_f\} \frac{\partial \langle T \rangle}{\partial t} + \{(\rho C_p)_f \langle \vec{U}_f \rangle + (\rho C_p)_r \langle \vec{U}_r \rangle\} \cdot \vec{\nabla} \langle T \rangle + \{(\rho C_p)_f (\vec{\nabla} \cdot \langle \vec{U}_f \rangle)$$
$$+ (\rho C_p)_r (\vec{\nabla} \cdot \langle \vec{U}_r \rangle)\} \langle T \rangle$$
$$= \nabla \cdot \{(\varepsilon_f K_f + \varepsilon_r K_r) \vec{\nabla} \langle T \rangle\} + \varepsilon_r \langle \Phi_r \rangle^r = \varepsilon_r \langle \Phi_c \rangle^r \tag{5.38}$$

If one assumes that the flow is governed by Darcy's law [16], then

$$\langle \Phi_r \rangle = \mu \langle \vec{U}_r \rangle \cdot \underline{\underline{k}}_p^{-1} \cdot \langle \vec{U}_r \rangle \tag{5.39}$$

In addition, the rate of heat released per unit volume of resin is given by

$$\langle \Phi_c \rangle = \rho H_c \frac{d\alpha}{dt} \tag{5.40}$$

where H_c is the heat of reaction per unit mass and $d\alpha/dt$ denotes the reaction rate. After substituting these expressions in Equation 5.38, one obtains the final energy balance,

$$\{\varepsilon_r(\rho C_p)_r + \varepsilon_f(\rho C_p)_r\} \frac{\partial \langle T \rangle}{\partial t} + \{(\rho C_p)_f \langle \vec{U}_f \rangle + (\rho C_p)_r \langle \vec{U}_r \rangle\} \cdot \vec{\nabla} \langle T \rangle + \{(\rho C_p)_f (\vec{\nabla} \cdot \langle \vec{U}_f \rangle)$$
$$+ (\rho C_p)_r (\vec{\nabla} \cdot \langle \vec{U}_r \rangle)\} \langle T \rangle = \vec{\nabla} \cdot \{(\varepsilon_f K_f + \varepsilon_r K_r) \vec{\nabla} \langle T \rangle\} + \varepsilon_r \rho_r H_c \langle f(\alpha, T_r) \rangle^r$$
$$+ \mu \langle \vec{U}_r \rangle \cdot \underline{\underline{k}}_p^{-1} \cdot \langle \vec{U}_r \rangle \tag{5.41}$$

In order to use the preceding equations, one requires a constitutive equation that relates the resin viscosity to temperature, degree of cure, and the deformation rate. If the resin can be considered Newtonian, then usually,

$$\langle \mu \rangle^r = f_\mu \left\{ \langle \alpha \rangle^r, \langle T_r \rangle^r \right\} \tag{5.42}$$

however, in the preceding equation one assumes that there exist a one-to-one correspondence between pointwise viscosity and average viscosity. Although this is not true in general, this assumption is necessary in order to obtain a closed set of equations. The validity of this assumption could be tested if viscosity could be expressed as a product of two functions, one in T and one in α [28]. Furthermore, if the resin is assumed to have a shear rate dependent viscosity, then one could use a generalized Newtonian constitutive equation to express the dependence of viscosity on deformation rate,

$$\langle\mu\rangle^r = f_\mu\left\{\langle\alpha\rangle^r, \langle T_r\rangle^r, \langle\dot{\gamma}_r\rangle^r\right\} \tag{5.43}$$

where

$$\dot{\gamma}_r = \left[\tfrac{1}{2}\,\mathrm{trace}(\underline{\underline{\dot{\gamma}}}_r \cdot \underline{\underline{\dot{\gamma}}}_r)\right]^{1/2} \quad \text{and} \quad \underline{\underline{\dot{\gamma}}}_r = (\vec{\nabla}\vec{U}_r) + (\vec{\nabla}\vec{U}_r)^t \tag{5.44}$$

In Equation 5.44, $(\vec{\nabla}\vec{U}_r)^t$ is the transpose of the velocity gradient tensor.

The simplest generalized Newtonian constitutive equation is the power law model that assumes the viscosity has the following dependence on shear rate,

$$\mu(\dot{\gamma}) = \mu_o \dot{\gamma}_r^{n-1} \tag{5.45}$$

where μ_o is approximately the viscosity in the limit of zero shear rate and n is the power law index. Other generalized Newtonian models, such as that by Carreau and Ellis [29], can be used to depict more accurately the shear rate dependence of viscosity of concentrated polymer solutions and melts. These models, however, have more adjustable parameters than the power law model.

Throughout the preceding derivations (i.e., for mass, momentum, and energy) we have neglected transport due to dispersion for the following reason. When dispersion is considered one generally requires knowledge about the specific architecture of the fiber bed, particularly the existence of a unit cell is normally required for accurate calculation of the dispersion tensor. Because most fiber beds encountered in the manufacture of polymer matrix composites do not have a well-defined unit cell, it becomes difficult to estimate the appropriate dispersion tensor. Rather than keeping dispersion terms in the equations and trying to estimate the appropriate dispersion tensors as done by [15] and referred to in [16] we have decided to leave them out. For a complete discussion of the effect of dispersion in porous media transport, the reader is referred to [15,16].

5.4 Specialized Equations for Various Polymer Matrix Composite Manufacturing Processes

In this section a specialized set of equations governing transport of mass, momentum, and energy in resin transfer molding, injected pultrusion, and autoclave processing are obtained from the general balance equations presented in Section 5.3. This involves eliminating unimportant terms in the general balance equations based on the specific nature of the process.

5.4.1 Resin Transfer Molding (RTM)

In RTM a fibrous reinforcement is placed into a mold cavity and impregnated with either a thermosetting or thermoplastic polymeric resin, as schematically depicted in Figure 5.2. Thermosetting resins are mostly used in the RTM process. Hence, during filling of the mold the resin could undergo a curing reaction that leads to an increase in its viscosity. Moreover, at the end of the filling cycle the resin is further cured in the mold until the desired level of cure is achieved.

In the RTM process, one can assume that the resin has a relatively constant density during the filling stage (i.e., filling is usually completed before the resin gels). When taken in combination with the fact that the reinforcement is stationary (i.e., $\langle \vec{U}_f \rangle = \langle \vec{U}_f \rangle^f = 0$), this can be used to simplify Equation 5.12 (i.e., $\partial \varepsilon_r / \partial t = 0$) and obtain a mass conservation equation for this process. An equation for conservation of the cured species during the filling and the postfilling curing process can be obtained by using the conservation of mass equation for RTM processes (i.e., $\vec{\nabla} \cdot \langle \vec{U}_r \rangle = 0$) to simplify Equation 5.18.

Momentum transfer can be described by Equation 5.29 provided $\mathrm{Re}_p < 1$ (which is a reasonable assumption in majority of RTM processes [16]). Finally, combining all of the preceding assumptions plus the assumption of a local equilibrium allows us to simplify Equation 5.41 significantly and obtain an energy equation for this process (i.e., $\langle \vec{U}_f \rangle = \vec{\nabla} \cdot \langle \vec{U}_r \rangle = \vec{\nabla} \cdot \langle \vec{U}_f \rangle = 0$). In summary, the appropriate governing equations for transport of mass, momentum, and energy in the RTM process are:

Overall mass conservation:

$$\vec{\nabla} \cdot \langle \vec{U}_r \rangle = 0 \tag{5.46}$$

Cured species mass conservation:

$$\varepsilon_r \frac{\partial \langle \alpha \rangle^r}{\partial t} + \langle \vec{U}_r \rangle \cdot \vec{\nabla} \langle \alpha \rangle^r = \varepsilon_r \langle f(\langle \alpha \rangle^r, \langle T_r \rangle^r) \rangle \tag{5.47}$$

Momentum transfer:

$$\langle \vec{U}_r \rangle = -\frac{\underline{\underline{k}}_p}{\mu} \cdot \vec{\nabla} \langle P_r \rangle^r \tag{5.48}$$

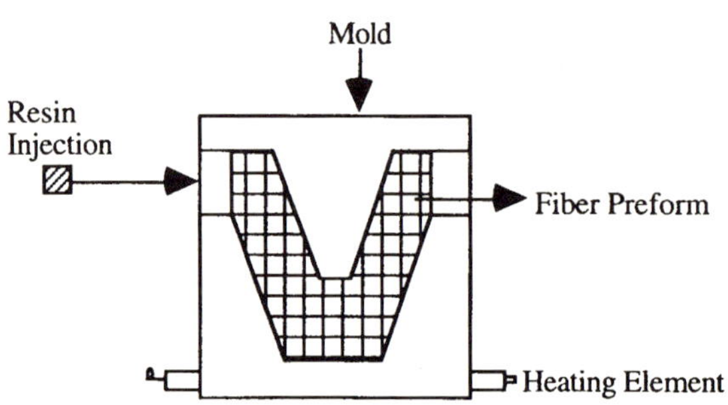

Figure 5.2 A schematic of the RTM process

Energy transfer:

$$\left\{\varepsilon_r(\rho C_p)_r + \varepsilon_f(\rho C_p)_f\right\}\frac{\partial\langle T\rangle}{\partial t} + (\rho C_p)_r\langle\vec{U}_r\rangle\cdot\vec{\nabla}\langle T\rangle$$
$$= \vec{\nabla}\cdot\left\{(\varepsilon_r K_r + \varepsilon_f K_f)\cdot\vec{\nabla}\langle T\rangle\right\} + \varepsilon_r\rho_r H_c\langle f(\alpha, T_r)\rangle^r + \mu\langle\vec{U}_r\rangle\cdot\underline{\underline{k}}_p^{-1}\cdot\langle\vec{U}_r\rangle \quad (5.49)$$

Finally, the viscosity μ can be described by Equations 5.42 or 5.43, depending on whether a Newtonian or a generalized Newtonian viscosity model is required to describe the resin rheology.

To solve the preceding set of governing equations, Equation 5.48 is substituted into Equation 5.46 and a pressure evolution equation is obtained. In arbitrary domain shapes, this equation can be solved using finite differences with boundary fitted coordinates [1], finite elements with a moving grid [10], or finite element control volume techniques [4]. In turn, the velocities are computed followed by solution of the energy transport and cured species balance equations. The latter two equations are difficult to solve with traditional methods for elliptic partial differential equations (i.e., such as central difference finite difference methods or Galerkin finite elements) because the transport is convection dominated. To overcome these difficulties upwind finite differences [30], consistent streamline upwind finite elements [30,31], or Lagrangian techniques [31] are employed to solve these equations. In general, a combination of finite element control volume techniques (solution of pressure and velocity field) with either consistent streamline upwind finite elements or Lagrangian techniques (solution of species and energy balance equations) provide the most accurate and cost-efficient solution to this class of problems [31].

A few sample simulations of the RTM process using the governing equations developed earlier are shown in Figures 5.3 and 5.4. To be specific, we have carried out two-dimensional isothermal simulations in a square cavity using the finite element control volume technique [31]. The finite element mesh and the boundary conditions used in all the simulations are shown at the top of Figure 5.3. At the flow front, we use a zero-gauge pressure condition (i.e., no vacuum pressure is considered and capillary pressure effects have been neglected). At the bottom of Figure 5.3 a sample flow front progression is shown. It can clearly be observed that in this case the flow front becomes flat and the filling process becomes one-dimensional toward the end of the filling process. This is due to the fact that K_{yy} is larger than K_{xx}; hence, the flow front reaches the top and bottom walls more quickly, resulting in a flat front toward the end of the filling process. From a practical point of view such a filling pattern is highly desirable; however, it is generally difficult to cut the preform in the desired shape without fraying along the edges. Then, upon placement of the fabric in the mold and introduction of the permeant, preferential flow along the walls, which is commonly referred to as "race tracking," will occur. This is illustrated in Figure 5.4, where simulations have been performed with the element adjacent to solid surfaces of the cavity have a permeability 1000 times the elements in the core. As also shown by the figure the fluid preferentially wets the walls and then penetrates the core region. The difference between the top and the bottom figure is that in the top figure $K_{xx} = K_{yy}$ in the core, which will result in symmetric radial penetration of the core (i.e., flow fronts are circular), but $K_{xx} = 5K_{yy}$ in the bottom figure in the core so one observes an asymmetric radial penetration of the core (i.e., flow fronts are elliptical with their long axis aligned in the y direction).

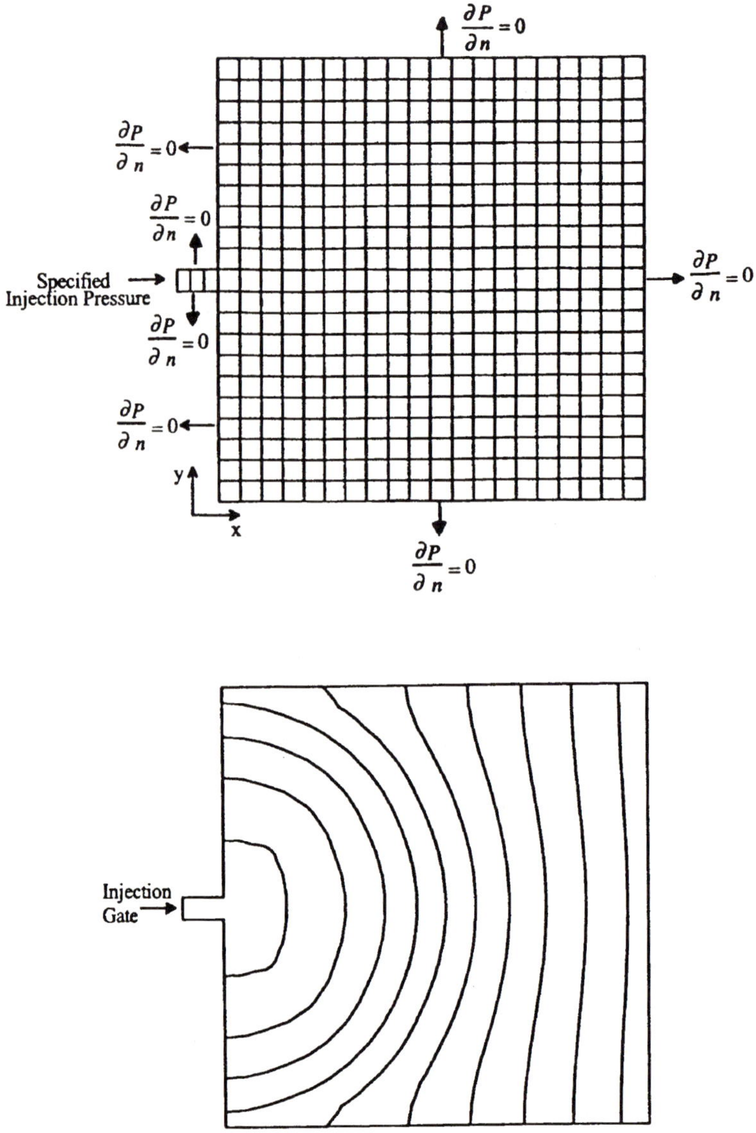

Figure 5.3 Sample two-dimensional simulations of the RTM process. (Top) finite element mesh; (bottom) flow front progression, $K_{yy} = 3K_{xx}$, $K_{xy} = 0.0$

5.4.2 Injected Pultrusion (IP)

Injected pultrusion is a process in which fiber preforms are pulled through a mold where the liquid resin is injected into the fibers and then heated until the desired degree of cure in the

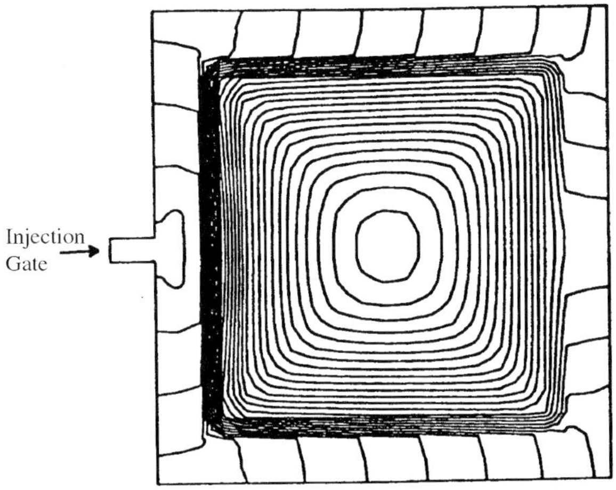

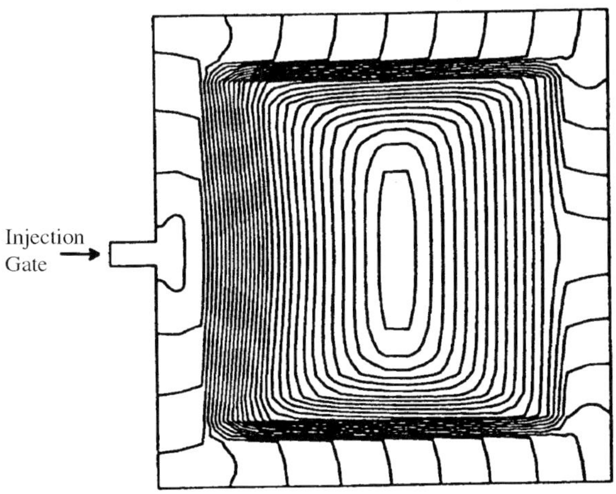

Figure 5.4 Sample two-dimensional simulation of the RTM process. (Top) flow front progression, around the solid boundaries $K_{xx} = K_{yy} = 1000$, $K_{xy} = 0.0$, in the core $K_{xx} = K_{yy} = 1.0$, $K_{xy} = 0.0$; (bottom) flow front progression, around the solid boundaries $K_{xx} = K_{yy} = 1000$, $K_{xy} = 0.0$ in the core $K_{xx} = 5K_{yy} = 5$, $K_{xy} = 0.0$

product is achieved (Fig. 5.5). In IP, the concept of liquid injection molding has been incorporated into conventional pultrusion processes that allow production of composites with improved mechanical and thermal properties, less environmental concerns, and higher production rates.

The reinforcement phase is not stationary in IP; hence, when performing conservation of mass, momentum, and energy this point has to be taken into consideration. In the IP process, the resin has a relatively constant density (i.e., it does not reach gel point until the end of the

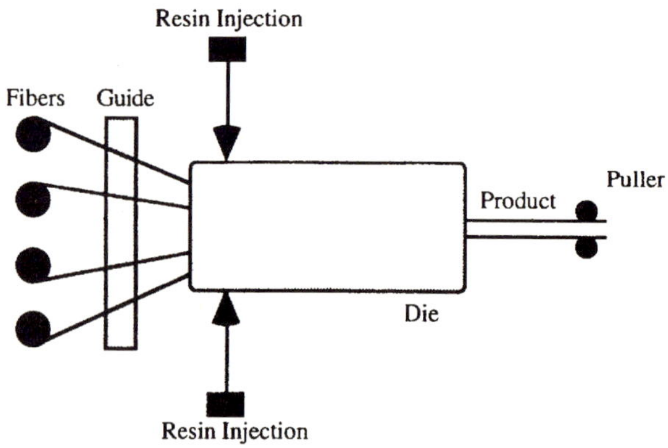

Figure 5.5 A schematic of the injected pultrusion process

die) and $\partial\varepsilon_r/\partial t \cong 0$ because the fiber bed is not being consolidated. Using these assumptions the mass conservation equation can be written by simplifying Equation 5.9:

$$\vec{\nabla}\cdot\langle\vec{U}_r\rangle + \frac{1}{V}\int_{S_i}\vec{U}_r\cdot\hat{n}_i ds = 0 \tag{5.50}$$

Since the fiber phase is not stationary, the surface integral cannot be set to zero without further considerations. As shown earlier, $\partial\varepsilon_r/\partial t = 1/V\int_{S_i}\hat{U}_r\cdot\hat{n}_i ds$ (see Eq. 5.10). Because $\partial\varepsilon_r/\partial t \cong 0$ in the IP process, the contribution of the surface integral to the overall mass balance is negligible. Based on this observation Equation 5.50 can be simplified and the appropriate equation for a conservation of mass in this process can be obtained (i.e., $\vec{\nabla}\cdot\langle\vec{U}_r\rangle = 0$). Using this, Equation 5.18 can be simplified and the appropriate species balance equation for the IP process can be obtained. This equation is similar to the equation obtained for the RTM process.

The movement of the fiber phase has to be specifically taken into consideration in the momentum transfer equation. Hence, in absence of significant inertial forces (i.e., $Re_p < 1$)[3] Equation 5.28 must be modified to account for the movement of the fiber phase,

$$-\varepsilon_r\vec{\nabla}\langle P_r\rangle^r + \mu\vec{\nabla}\cdot\vec{\nabla}\langle\vec{U}_r\rangle - \frac{\varepsilon_r^2\mu}{\underline{\underline{k}}_p}\cdot\left[\langle\vec{U}_r\rangle^r - \langle\vec{U}_f\rangle^f\right] = 0 \tag{5.51}$$

where $\langle\vec{U}_f\rangle^f$ is the fiber pull speed. Again, if one assumes that $\langle\vec{U}_r\rangle$ changes very slowly on the length scale of the porous media, then the viscous stress term can be neglected and the equation reduces to:

$$\langle\vec{U}_r\rangle^r = \frac{-\underline{\underline{k}}_p}{\mu\varepsilon_r}\vec{\nabla}\langle P_r\rangle^r + \langle\vec{U}_f\rangle^f \tag{5.52}$$

[3] In most IP process the fluid inertia is negligible.

If one assumes that the resin has a constant density and thermal conductivity, an energy equation for the IP process can be obtained by simplifying Equation 5.41. First by using the mass balance equation (i.e., $\vec{\nabla} \cdot \langle \vec{U}_r \rangle = 0$) one of the convective heat transfer terms in the resin phase can be neglected. Another term can be neglected by setting $\vec{\nabla} \cdot \langle \vec{U}_f \rangle = 0$. This assumption is justified since in the IP process $\partial \varepsilon_r / \partial t = 0$.

In summary, the appropriate governing equations for transport of mass, momentum, and energy in the IP process are:

Overall mass conservation:

$$\vec{\nabla} \cdot \langle \vec{U}_r \rangle = 0 \tag{5.53}$$

Cured species mass conservation:

$$\varepsilon_r \frac{\partial \langle \alpha \rangle^r}{\partial t} + \langle \vec{U}_r \rangle \cdot \vec{\nabla} \langle \alpha \rangle^r = \varepsilon_r \langle f(\langle \alpha \rangle^r, \langle T_r \rangle^r) \rangle \tag{5.54}$$

Momentum transfer:

$$\langle \vec{U}_r \rangle = \frac{-\underline{\underline{k}}_p}{\mu} \vec{\nabla} \langle P_r \rangle^r + \varepsilon_r \langle \vec{U}_f \rangle^f \tag{5.55}$$

Energy transfer:

$$\left\{ \varepsilon_r (\rho C_p)_r + \varepsilon_f (\rho C_p)_f \right\} \frac{\partial \langle T \rangle}{\partial t} + ((\rho C_p)_r \langle \vec{U}_r \rangle + (\rho C_p)_f \langle \vec{U}_f \rangle) \cdot \vec{\nabla} \langle T \rangle = \vec{\nabla} \cdot \left\{ (\varepsilon_r K_r + \varepsilon_f K_f) \cdot \vec{\nabla} \langle T \rangle \right\}$$
$$+ \varepsilon_r \rho_r H_c \langle f(\alpha, T_r)^r \rangle + \mu \langle \vec{U}_r \rangle \cdot \underline{\underline{k}}_p^{-1} \cdot \langle \vec{U}_r \rangle \tag{5.56}$$

The viscosity μ can be described by Equations 5.42 or 5.43, depending on whether a Newtonian or a generalized Newtonian viscosity model is required to describe the resin rheology. Moreover, the same numerical procedures as outlined in Section 5.4.1 can be used to solve the preceding governing equations. Equations 5.53 to 5.56 can also be used in the conventional pultrusion process. In general, in these processes, if no taper in the die is present, then the flow is a simple drag flow and $\vec{\nabla} \langle P_r \rangle^r$ in Equation 5.55 is identically zero. In most pultrusion dies, however, a tapered section exists to ensure full permeation of the fiber bed [32,33]. Hence, the preceding equations can be used directly to model conventional pultrusion operations as well as injected pultrusion processes [31,34–36].

Using Equations 5.53–5.56 a three-dimensional non-isothermal simulation of a typical IP process has been performed using a finite element control volume technique [31,34–36]. The density specific heat and thermal conductivity of the resin and reinforcement used in the simulations are given in Table 5.1.

The cure kinetics are assumed to be as follows [31]:

$$f(\langle \alpha \rangle^r, \langle T \rangle) = (K_1 + K_2 \langle \alpha^r \rangle)(1 - \langle \alpha \rangle^r)(B - \langle \alpha^r \rangle) \quad \langle \alpha \rangle \leqslant 0.3 \tag{5.57}$$

$$f(\langle \alpha \rangle^r, \langle T \rangle) = K_3 (1 - \langle \alpha \rangle^r) \quad \langle \alpha \rangle \geqslant 0.3 \tag{5.58}$$

where $K_i = A_i \exp(-\Delta E_i / R \langle T \rangle)$, $i = 1, 2, 3$ and B is a constant that is determined through the fitting procedure. Table 5.2 summarizes the values used in the simulation.

The finite element mesh used in the computations is shown in the top of Figure 5.6. Similar to the RTM process $\partial P / \partial n$ is set to zero on all solid boundaries and the pressure is

Table 5.1 Physical Properties of the Resin and Reinforcement

Properties materials	Density (g/ml)	Heat capacity (Cal/g°C)	Thermal conductivity (Cal/cm sec°C)
Resin	1.274	0.468	3.85×10^{-4}
Fiber	1.37	0.962	6×10^{-3}

Table 5.2 Cure Kinetic Parameters

Parameters	Value
A_1	2.101×10^9 min^{-1}
A_2	-2.014×10^9 min^{-1}
A_3	1.960×10^5 min^{-1}
ΔE_1	8.07×10^4 J/mol
ΔE_2	7.78×10^4 J/mol
ΔE_3	5.66×10^4 J/mol
B	0.47

Source: Ref. [31]

specified at the inlet. In addition, $d\alpha/dn$ is set to zero at all solid boundaries and an initial temperature of 80°C has been specified. The wall temperatures in the curing and filling sections have been set to 160°C and 80°C, respectively. At the flow front a zero-gauge pressure condition as well as a no-heat flux condition is used. The viscosity of the resin is given by

$$\langle \mu_r \rangle^r = \mu_\infty \exp(U/R\langle T \rangle + W\langle \alpha \rangle^r) \tag{5.59}$$

where $\mu_\infty = 7.9 \times 10^{-14}$ Pa s, $U = 9 \times 10^4$ J/mol, $R = 8.314\,J/\text{mol}°\text{K}$, and $W = 14.1$. The heat of reaction H_c is assumed to be 473.6 J/g [31]. The values used in the simulations are typical of epoxy/carbon fiber composites. Moreover, the permeability values have been obtained from [20] for a carbon fiber bed of porosity 0.5.

In the bottom of Figure 5.6 a typical flow front progression for the IP process with a low fiber pull speed is shown. Although the fibers are fully wetted prior to the exit and a flat flow front is obtained quickly (i.e., one that allows uniform curing of the resin) significant leakage exists from the back (i.e., as indicated by the large spacing between flow fronts near the back). To avoid the undesirable leakage the fiber pull speed is significantly increased. The result of this simulation is shown at the top of Figure 5.7. As seen in this figure, the selected pull speed is high enough to prevent any back flow but, at the same time, it has resulted in incomplete wetting of the fibers. This is highly undesirable because it will result in a part with a void in its center. Using the simulation code one can obtain the optimum pull speed under given operating conditions such that the fibers are fully wetted as well as obtaining a flat flow front in the curing region without having any leakage from the back (i.e., the closely spaced flow front in the rear of the die are indicative of little or no back flow). The flow front progression for this optimum condition is shown in the bottom of Figure 5.7. Sample temperature profiles

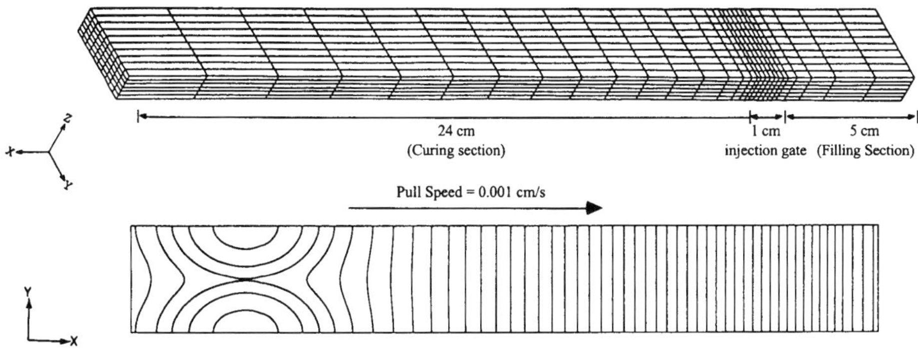

Figure 5.6 Sample three-dimensional simulations of the IP process. (Top) Finite element mesh. Total length 30 cm ($0 \leq X \leq 30$), total height 1 cm ($0 \leq Z \leq 1$), total width 3 cm ($0 \leq Y \leq 3$). Fluid is injected from both sides through the thickness (i.e., in the Z-direction through a 1 cm × 1 cm square). (Bottom) Flow front progression at the midplane (i.e., $Z = 0.5$), $K_{xx} = K_{zz} = 2K_{yy}$; $K_{xy} = K_{xz} = K_{yz} = 0.0$

and degrees of cure in the curing region for this optimum condition are shown in Figure 5.8. The temperature profile is shown at the top figure. As also shown by the figure the wall temperature in the curing section is set to 160°C and a parabolic temperature profile in the thickness direction (i.e., in the z-direction) is observed, which indicates diffusion-dominated heat transfer. Along the flow direction (i.e., in the x-direction), however, the temperature is raised from 80°C, which is the temperature of the fluid at the entrance to the curing section, to approximately 160°C at the exit. Heat transfer is clearly convection dominated along the flow direction. The cure profiles are very similar to the temperature profiles as shown at the bottom of Figure 5.8. At the inlet to the curing section the degree of cure is zero, whereas the desired value of 50 percent cure is achieved at the exit. In this particular simulation the goal was to obtain a void free part with a 50 percent degree of cure at the exit. By varying the injection

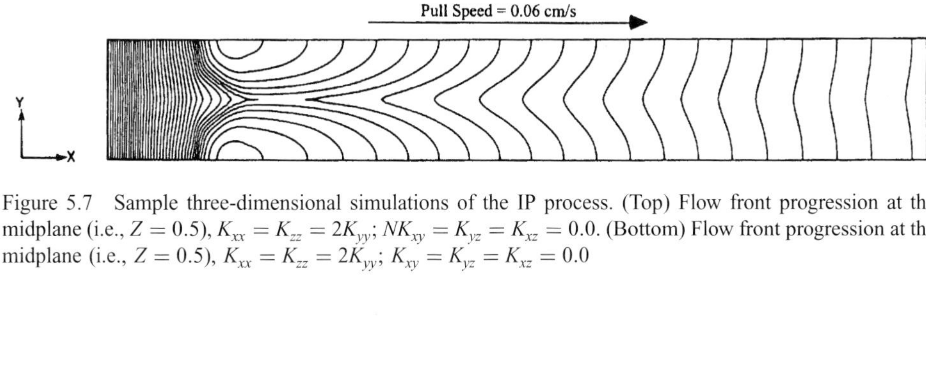

Figure 5.7 Sample three-dimensional simulations of the IP process. (Top) Flow front progression at the midplane (i.e., $Z = 0.5$), $K_{xx} = K_{zz} = 2K_{yy}$; $NK_{xy} = K_{yz} = K_{xz} = 0.0$. (Bottom) Flow front progression at the midplane (i.e., $Z = 0.5$), $K_{xx} = K_{zz} = 2K_{yy}$; $K_{xy} = K_{yz} = K_{xz} = 0.0$

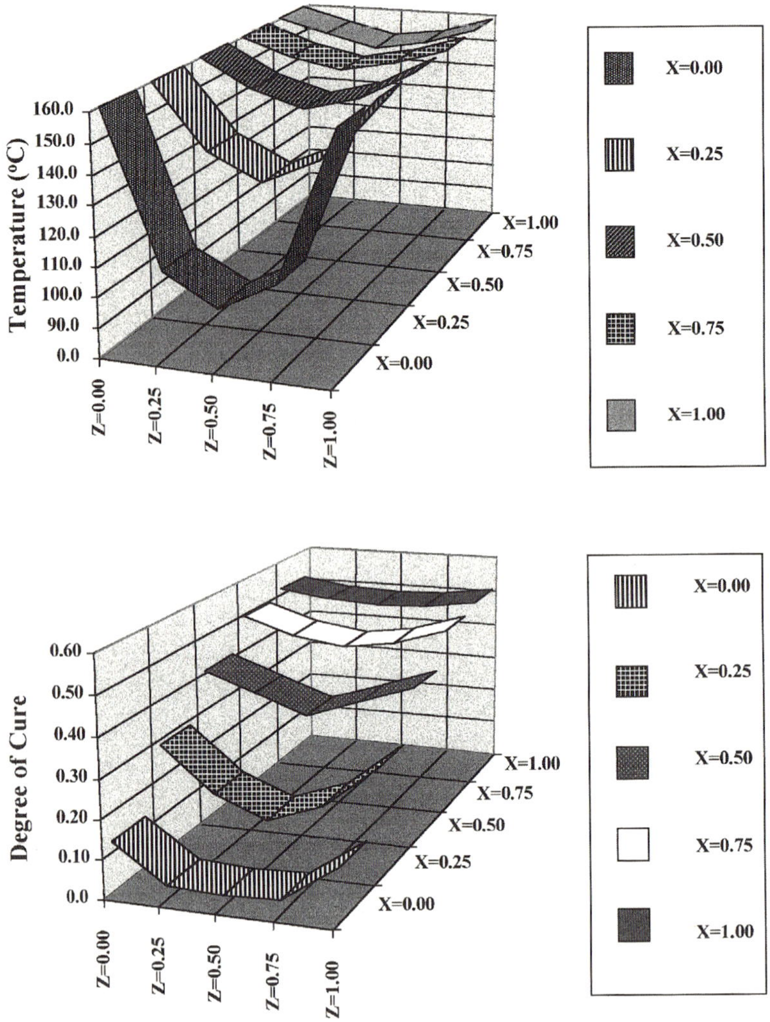

Figure 5.8 Temperature and degree of cure profiles in the curing section at $Y = 1.5$. X is the fraction of the total length of the curing section (i.e., $X = 0$ denotes the entrance to the curing section). (Top) temperature profile; (bottom) degree of cure profile

pressure, pull speed, and the temperature of the wall and length of the curing section we were able to obtain our desired specifications. It should be noted that the simulation code can be used to address other important issues, such as determining the optimum taper angle (i.e., in many IP processes the filling section has tapered walls in order to aid in rapid attainment of a flat flow front) as well as the pulling force required to pull the assembly out of the die. The interested reader is referred to References [34–36] for more details regarding design and optimization of the IP process.

5.4.3 Autoclave Processing (AP)

Autoclave processing is a process in which individual prepreg plies are laid up in a prescribed orientation to form a laminate (Fig. 5.9). The process involves consolidation of the laminate, which generally results in a three-dimensional flow field. Similar to the IP process the fiber bed is not stationary in the AP process; hence, its movement has to be specifically considered when the appropriate conservation equation for this process are developed. If it is assumed that the resin has a relatively constant density (i.e., the excess resin is squeezed out before the gel point is reached) then the appropriate conservation of mass equation for this consolidating system is Equation 5.12.

Using Equation 5.12, the species mass conservation equation can be obtained by simplifying Equation 5.18 (i.e., $\vec{\nabla} \cdot \langle \vec{U}_r \rangle = -\partial\varepsilon_r/\partial t$). The appropriate momentum conservation equation in absence of inertia (i.e., inertial effects are minimal in the AP process [32]) for this process is similar to the IP process (i.e., Eq. 5.56). Unlike the IP process, however, the fiber velocity is not known. Hence, either another equation for the fiber phase velocity as a function of the consolidation force must be developed or the magnitude of this velocity should be compared with the liquid phase velocity to see if it can be neglected. In general, because the consolidation process is relatively slow and the fiber interspacing is small compared with the consolidating speed and the thickness of the laminate, contribution of the fiber movement to the liquid velocity can be safely ignored.

An energy balance equation for this system can be obtained by simplifying Equation 5.41 using the same assumptions used earlier (i.e., $\langle \vec{U}_f \rangle = 0$, $\vec{\nabla} \cdot \langle \vec{U}_r \rangle = -\partial\varepsilon_r/\partial t$ and $\vec{\nabla} \cdot \langle \vec{U}_f \rangle = \partial\varepsilon_r/\partial t$).

In summary, the appropriate governing equations for transport of mass, momentum, and energy in the AP process are:

Overall mass conservation:

$$\frac{\partial\varepsilon_r}{\partial t} + \nabla \cdot \langle \vec{U}_r \rangle = 0 \tag{5.60}$$

Cured species mass conservation:

$$\varepsilon_r \frac{\partial\langle\alpha\rangle^r}{\partial t} + \langle \vec{U}_r \rangle \cdot \vec{\nabla}\langle\alpha\rangle^r - \langle\alpha\rangle^r \frac{\partial\varepsilon_r}{\partial t} = \varepsilon_r \langle f(\langle\alpha\rangle^r, \langle T_r\rangle^r)\rangle \tag{5.61}$$

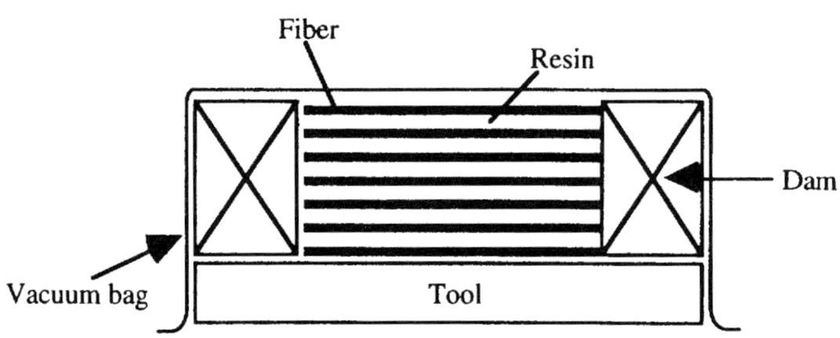

Figure 5.9 A schematic of the autoclave process

Momentum transfer:

$$\langle \vec{U}_r \rangle = \frac{-\underline{\underline{k}}_p}{\mu} \vec{\nabla} \langle P_r \rangle^r \tag{5.62}$$

Energy transfer:

$$\left\{ \varepsilon_r (\rho C_p)_r + \varepsilon_f (\rho C_p)_f \right\} \frac{\partial \langle T \rangle}{\partial t} + (\rho C_p)_r \langle \vec{U}_r \rangle \cdot \vec{\nabla} \langle T \rangle + ((\rho C_p)_f - (\rho C_p)_r) \frac{\partial \varepsilon_r}{\partial t} \langle T \rangle$$
$$= \vec{\nabla} \cdot \left\{ (\varepsilon_f K_f + \varepsilon_r K_r) \vec{\nabla} \langle T \rangle \right\} + \varepsilon_r \rho_r H_c \langle f(\alpha, T_r)^r \rangle + \mu \langle \vec{U}_r \rangle \cdot \underline{\underline{k}}_p^{-1} \cdot \langle \vec{U}_r \rangle \tag{5.63}$$

The viscosity μ can be described by Equations 5.42 or 5.43, depending on whether a Newtonian or a generalized Newtonian viscosity model is required to describe the resin rheology.

To solve the preceding set of equations, Equation 5.62 is plugged into Equation 5.60. By separately determining the compaction properties of the fiber bed [32] an evolution equation for the pressure can be obtained. Because this is a moving boundary problem the derivative in the thickness direction can be rewritten [32] in terms of an instantaneous thickness. The pressure field can then be solved for by finite difference or finite element techniques. Once the pressure is obtained and the velocity computed, the energy and cured species conservation equations can be solved using the methodology outlined in Section 5.4.1.

5.5 Conclusions

In this review, a set of balance equations for transport of heat, mass, and momentum in stationary and moving porous media has been derived based on a local volume averaging approach. The advantage of this method is that it allows precise definition of average temperature, velocity, and pressure. Moreover, equations are derived rigorously from first principles.

Specific balance equations for various polymer matrix composites manufacturing processes (i.e., RTM, IP, and AP) have been obtained by simplifying the balance equations. Particular attention has been paid to state all the assumptions used to arrive at the final equations clearly in order to clearly show the range of applicability of the equations. Moreover, appropriate numerical techniques for solution of these coupled partial differential equations have been briefly outlined and a few example simulations have been performed.

Throughout this summary we have neglected the effect of dispersion on the overall transport of mass and heat. This is due to the fact that if dispersion is included, dispersion tensors must be determined before the equation can be solved. This can be done by solving the appropriate transport equation within a unit cell. Because a unit cell cannot be defined in most reinforcements used in polymer matrix composites, however, dispersion tensors cannot be accurately determined, so we have left dispersion effects out of our equations. In general, we anticipate dispersion to play a minor role in the IP, AP, and RTM processes. This assumption can be checked, however, by evaluating the dispersion terms using an approach similar to [16] where experiments and correlations are used to determine the importance of dispersion.

In this summary, the local thermal equilibrium model has been used to derive the energy equation. This model is much simpler than the two-phase model; however, the local thermal equilibrium model is most likely not adequate to describe the transport of energy when the temperature of the fluid and solid are undergoing extremely rapid changes. Although such extremely rapid temperature changes are not expected, in most RTM, IP, and AP processes the correctness of the local thermal equilibrium assumption can be verified by following the procedure discussed by Whitaker [28].

It should be noted that the effect of fluid viscoelasticity on transport of mass, momentum, and heat in porous media has not been discussed in this summary. Although some preliminary studies have been performed in this area [21], no definitive governing equations exist.

Nomenclature

A_1	Constant in the cure equation
A_2	Constant in the cure equation
A_3	Constant in the cure equation
b	Coefficient of inertial term in Forchheimer model
B	Constant in the cure kinetic equation
C_{p_r}	Specific heat of the resin at constant pressure
C_{p_f}	Specific heat of the fiber at constant pressure
f	Function describing the dependence of chemical reaction rate on degree of cure and temperature
f_d	Fluid–solid drag force
$\vec{g}$	Body force vector per unit mass due to gravity
g	Acceleration due to gravity, scalar function
G_r	Arbitrary scalar quantity associated with the resin
F_f	Arbitrary scalar quantity associated with the fiber
H_R	Heat of reaction per unit mass of the resin
K_1	Constant in the cure kinetic equation
K_2	Constant in the cure kinetic equation
K_3	Constant in the cure kinetic equation
K_r	Thermal conductivity of the resin
K_f	Thermal conductivity of the fiber
$\underline{\underline{k}}$	Permeability tensor
ℓ^p	Length scale of porous medium
$\vec{n}_i$	unit normal vector
P_r	Modified resin pressure
p_r	Pressure in the resin
$\vec{q}_r$	Heat flux in the resin
$\vec{q}_i$	Heat flux in the fiber
R	Gas constant
Re_s	Reynolds number based on the pore diameter
S_i	Resin–solid interfacial surface within the representative volume

T	Temperature
T_r	Temperature in the resin
T_F	Temperature in the fiber
t	time
$\vec{U}_r$	Resin velocity
$\vec{U}_f$	Fiber velocity
V	Representative volume in the porous medium
V_r	The portion of representative volume occupied by the resin
V_f	The portion of representative volume occupied by the fiber
W	Constant in the viscosity equation
Y_r	Phase function of the resin
Y_f	Phase function of the fiber
Z_α	Rate of production of the cured species
α	Degree of cure
$\underline{\underline{\delta}}$	Unit tensor
$\dot{\gamma}_r$	Scalar shear rate in the resin
$\underline{\underline{\dot{\gamma}}}$	Rate of deformation tensor
ΔE_1	Constant in the cure equation
ΔE_2	Constant in the cure equation
ΔE_3	Constant in the cure equation
μ	Resin viscosity
μ_∞	Constant in the viscosity equation
ρ_r	Density of the resin
ρ_f	Density of the fiber
$\underline{\underline{\sigma}}_r$	Total stress of the resin
$\underline{\underline{\tau}}_r$	Extra stress tensor in the resin
$\langle\ldots\rangle$	Volume average or phase average of a quantity
$\langle\ldots\rangle^r$	Intrinsic phase average over the resin
$\langle\ldots\rangle^f$	Intrinsic phase average over the fiber

References

1. Coutler, J.P., Güceri, S.I. *J. Reinf. Plast. Compos.* (1988) 7, p. 200
2. Molnar, J.A., Trevino, L., Lee, L.J. *Polym. Compos.* (1989) 10, p. 414
3. Gonzalez-Romero, V., Macosko, C.W. *Polym. Process Eng.* (1985) 3, p. 173
4. Lin, R., Lee, L.J., Liou, M.J. *Int. Polym. Process* (1991) 6, p. 356
5. Lee, C.C., Tucker, C.L. *J. Non-Newt. Resin Mech.* (1987) 24, p. 245
6. Osswald, T.A., Tucker, C.L. *Int. Polym. Process.* (1989) 5, p. 79
7. Hieber, C.A., Chen, S.F. *J. Non-Newt. Resin Mech.* (1980) 7, p. 1
8. Chiang, H.H., Hieber, C.A., Wang, K.K. *Polym. Eng. Sci.* (1991) 31, p. 116
9. Chiang, H.H., Hieber, G.A., Wang, K.K. *Polym. Eng. Sci.* (1991) 31, p. 125
10. Dupret, F., Vandenschuren, L. *AIChE J.* (1988) 34, p. 1959

11. Dave, R., Kardos, J.L., Duduković, M.P. *Polym. Compos.* (1981) 8, p. 123
12. Couniot, A., Crochet, M.J. *Proc. 2nd Int. Conf. on Numerical Methods in Industrial Forming Processes*, Mattiasson, K., Samuelsson, A., Wood, R.D., Zienkiewics, O.C. (Eds.) (1986) p. 165, Balkema, Rotterdam
13. Fracchia, C.A., Castro, J., Tucker, C.L. (1989) 4th Tech. Conf., p. 157, Technomic, Lancaster, Penn
14. Bruscke, M.V., Advani, S.G. *Polym. Compos.* (1990) 11, p. 398
15. Kaviany, M. *Principles of Heat Transfer in Porous Media* (1991) Springer, New York
16. Tucker, C.L., Dessenberger, R.B. (1994) Advani, S.G. (Ed.) Elsevier, New York
17. Whitaker, S. *Chem. Eng. Sci.* (1986) 41, p. 2029
18. Slattery, J.C. *AIChE J.* (1969) 13, p. 866
19. Slattery, J.C. *Momentum, Energy and Mass Transfer in Continua*, 2nd ed. (1981) Kreiger, Newark
20. Skartsis, L., Khomami, B., Kardos, J.L. *Polym. Eng. Sci.* (1992) 32, p. 221
21. Skartsis, L., Khomami, B., Kardos, J.L. *J. Rheol.* (1992) 36, p. 1377
22. Brinkman, H.C. *Appl. Sci. Res.* (1947) A1, p. 27
23. Darcy, H.P.G. (1856) Victor Dalmont, Paris
24. Hsu, C.T., Cheng, P. *Int. J. Heat Mass Transfer* (1990) 33, p. 1587
25. Vafai, K., Tien, C.L. *Int. J. Heat Mass Transfer* (1981) 24, p. 195
26. Beavers, S.G., Sparrow, E.M., Rodenz, D.E. *J. Appl. Mech.* (1969) 36, p. 771
27. Beavers, S.G., Sparrow, E.M., Rodenz, D.E. *J. Appl. Mech.* (1973) 40, p. 655
28. Whitaker, S. *Chem. Eng. Commun.* (1987) 58, p. 171
29. Bird, R.B., Armstrong, R.C., Hassager, O. *Dynamics of Polymeric Liquids: Volume 1—Resin Mechanics*, 2nd ed. (1987) John Wiley & Sons, New York
30. Tucker, III, C.L. *Polym. Comp.* (1996) 17, p. 59
31. Kommu, S., Khomami, B., Kardos, J.L., *Polym. Compos.* (1998) 19, p. 335
32. Dave, R. *J. Comp. Mat.* (1990) 24, p. 22
33. Batch, G.L. (1988) Ph.D. Thesis, Department of Chemical Engineering, University of Minnesota
34. Kardos, J.L., Yang, B., Khomami, B. (1994) AIChE Annual Meeting, San Francisco, Calif.
35. Kardos, J.L., Khomami, B., Ramachandran, P., Yang, B., Shepard, R. (1996) Proceedings of the Fifth World Congress of Chemical Engineering, San Diego, Calif.
36. Mustafa, I., Khomami, B., Kardos, J.L., *AIChE J.* (1999) 45, p. 151

6 Void Growth and Dissolution

J.L. Kardos

The stability, growth, and transport of voids during composite processing is reviewed. As a framework for this model, the autoclave process was selected, but the concepts and equations may be applied equally effectively in a variety of processes, including resin transfer molding, compression molding, and filament winding. In addition, the problem of resin transport and its intimate connection with void suppression are analyzed.

6.1 Introduction

The process of fabricating a high-performance structural composite part, such as a 64-ply graphite/epoxy laminate, currently lacks a thorough scientific basis. Although based on sound engineering principles, today's technology is not yielding the part-to-part reliability that will be required of larger, more complex structures currently on the drawing boards. One of the major problems in achieving this reliability is the occurrence of voids in the final part. The mechanism governing resin flow during the process, whether it be a closed-mold process like resin transfer molding, or an open-mold process, such as autoclaving, is intimately connected with this problem. The void phenomenon clearly occurs during a very complex fabrication process involving heat, mass, and momentum transfer with simultaneous chemical reaction in a multiphase system with time-dependent material properties and boundary conditions. To model such a process by using first principles of transport phenomena is clearly difficult, and solving complex differential equations in closed form is generally not tenable.

In this review we will first examine an approach to modeling the stability, growth, and transport of voids during any composite manufacturing process. As a framework for this examination, we will concentrate on the autoclave manufacturing process, but the concepts and equations are general in nature and may be applied equally effectively in a variety of processes, including resin transfer molding, compression molding, and filament winding. We will then consider the problem of resin transport and how it is intimately connected to void suppression.

Before presenting the general model formulation and some results, it will be helpful to briefly examine the autoclave process details and some of the evidence for voids.

6.1.1 The Autoclave Process

One of the most often used production procedures for fabricating a high-performance structural laminate is the Autoclave/Vacuum Degassing (AC/VD) laminating process. In this process, individual prepreg plies are laid up in a prescribed orientation to form a laminate. The laminate is laid against a smooth tool surface and covered with successive layers of glass bleeder fabric, Mylar or Teflon sheets, and, finally, a vacuum bag. The entire sandwich is shown schematically in Figure 6.1 for a $[0, \pm 45, 90]_{2s}$ laminate. The lower part of the figure depicts a microscopic view of the tortuous pore structure between the fibers for such a laminate, which might be viewed as a repeating unit in the thickness direction for a much thicker laminate of 48 to 64 plies. In reality there are many more fibers in each ply (only one is shown here for simplicity), so any vertical path is even more tortuous than is depicted here.

After bagging, the entire tool is moved into an autoclave, a vacuum is applied to the bag, and the temperature is typically increased in the manner shown in Figure 6.2 for a T300-5208 system (Thornel 300 fibers/Narmco 5208 epoxy resin system). The 5208 matrix is typical of an aerospace-grade resin and is based on tetraglycidyl methylene dianiline (TGMDA) and diaminodiphenyl sulfone (DDS). A highly complex process occurs in the laminate during temperature increase and pressure application. As the temperature is increased, the resin viscosity decreases rapidly and chemical reaction begins. Somewhere between 93 and 135°C (200 and 275°F) the resin viscosity reaches a minimum and then begins to increase. Up to this point, however, little laminate consolidation has taken place other than that associated with wetting between the plies. During the 135°C (275°F) temperature hold, an autoclave pressure of 5.78 atm (85 psi) is applied and laminate consolidation occurs.

The transfer of autoclave pressure to the resin in the laminate does not occur hydrostatically because the resin is not enclosed in a constant-volume system. Flow can occur initially both vertically (thickness direction) and horizontally. Furthermore, the network of fibers can also eventually act as a network of springs to which the vacuum bag and bleeder assembly transfer the stress from the autoclave pressure. This stress can then be transferred

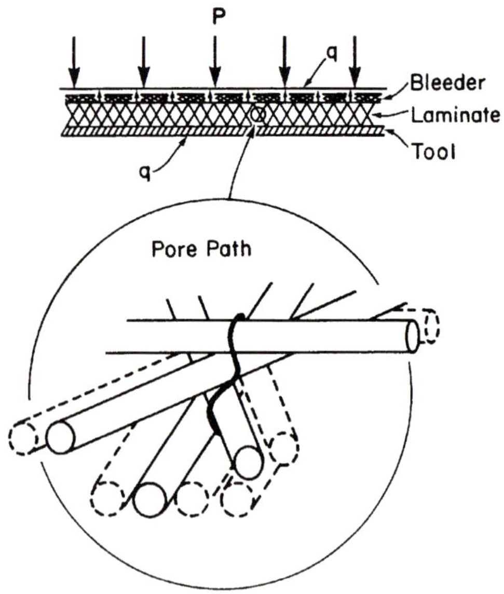

Figure 6.1 Schematic of laminate lay-up. Insert shows serpentine path that matrix resin and voids might take through connected pores formed by the graphite fibers. Each ply is actually many more fibers thick than is shown

through the fiber network to the tool surface with only a slight frictional loss. As the fiber spring network is loaded and compressed, more resin flow occurs; however, the viscosity at this point is increasing, and the permeability is decreasing, making resin flow more difficult throughout the laminate. Nonetheless, results have shown that initial resin flow is high enough so that only a relatively small fraction of the autoclave pressure is ever generated in the resin [1] for a free-bleeding (no dams) situation.

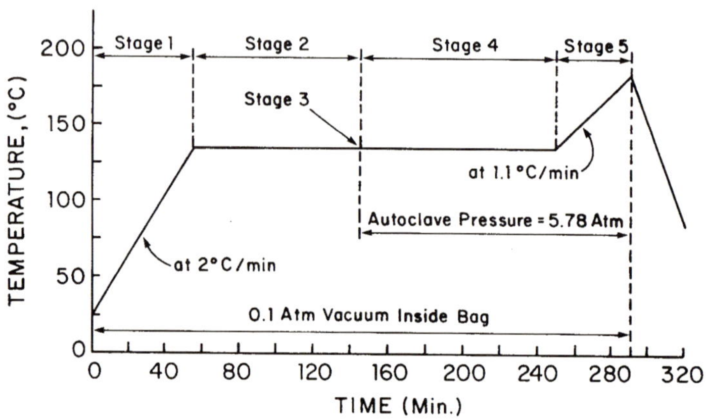

Figure 6.2 Curing cycle temperature–time profile for typical graphite–epoxy composite in a vacuum bag autoclave process. Autoclave pressure is applied during the 135°C (275°F) hold

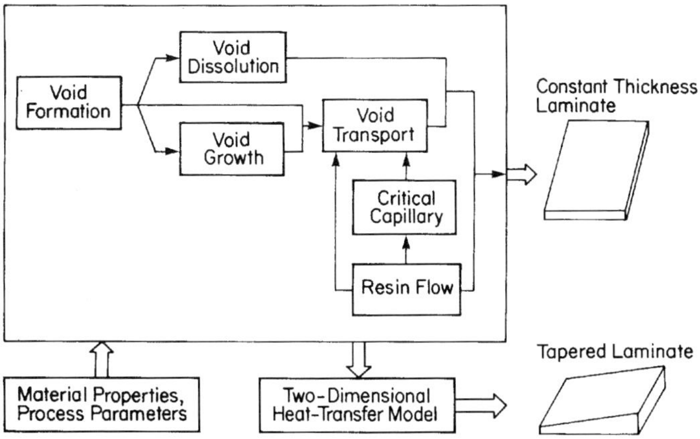

Figure 6.3 Schematic of void model framework

6.1.2 Void Evidence

The occurrence of voids has been thoroughly documented in thick laminates [2]. In almost all cases, they are apparently associated with the prepreg surface. The exact mechanism of void formation depends on the system, but in the most general case it can include mechanical entrapment as well as nucleation of stable voids in the resin phase.

6.1.3 The General Model Framework

The model framework for describing the void problem is schematically shown in Figure 6.3. It is, of course, a part of the complete description of the entire processing sequence and, as such, depends on the same material properties and process parameters. It is therefore intimately tied to both kinetics and viscosity models, of which there are many [3]. It is convenient to consider three phases of the void model: void formation and stability at equilibrium, void growth or dissolution via diffusion, and void transport.

Models and results will be presented for the first two phases in some detail, as well as an initial approach for the void transport problem.

6.2 Void Formation and Equilibrium Stability

Voids can be formed by either entrapment of air mechanically or by one of two nucleation processes. Mechanical entrapment could include (1) entrained gas bubbles from the resin-mixing operation, (2) bridging voids from large particles or particle clusters (quenched DDS curing agent, airborne particles, or paper release agent), (3) voids from wandering tows, fuzz balls, or broken fibers, (4) air pockets and wrinkles created during lay-up, and (5) ply terminations. Nucleation can occur either homogeneously within the resin or heterogeneously at a resin–fiber or a resin–particle interface. In the case of mechanical entrapment, the initial void distribution is dictated by the processing steps prior to the AC/VD step and it must be fed into the model as an experimentally determined parameter.

6.2.1 Nucleation of Voids

Homogeneous nucleation may be described by assuming that critical-size nuclei will be formed from ideal vapor (water or air) at a rate, I, given by classical nucleation theory [4]. The equation is

$$I = \left[\frac{P^*}{(2\pi MkT)^{1/2}}\right] 4\pi r^{*2} n \exp\left[\frac{-\Delta F^*}{kT}\right] \tag{6.1}$$

where

P^* = water vapor pressure, air pressure, or total mixture pressure (water plus air)
M = molecular weight of the vapor phase
k = Boltzmann constant
T = absolute temperature
r^* = radius of the critical nucleus
n = number of molecules/unit volume in the nucleus, and
ΔF^* = maximum free energy barrier for the nucleation process.

The free energy barrier ΔF^* and the critical nucleus size r^* are given by [4]

$$\Delta F^* = \frac{16\pi\gamma_{LV}^3}{3(\Delta F_v)^2} \text{ and } r^* = -\frac{2\gamma_{LV}}{\Delta F_v} \tag{6.2}$$

where ΔF_v is the free energy change per unit volume for the phase transition (dissolved air or water to air or water vapor), and γ_{LV} is the surface energy between the liquid resin and the void nucleus. ΔF_v can conveniently be approximated by

$$\Delta F_v = \Delta H_v \left(\frac{T_0 - T}{T_0}\right) \tag{6.3}$$

where ΔH_v is the heat of transition per unit volume between the two phases, T_0 is the equilibrium transition temperature, and T is the actual temperature of the system.

It is far more likely that heterogeneous nucleation plays the governing role in nucleation of voids. The effect of a particle or substrate is to lower the free energy barrier ΔF^*. Thus, Equations 6.1 and 6.3 remain unchanged, and Equation 6.2 takes on the new form

$$\Delta F^* = \frac{16\pi\gamma_{SV}^3}{3(\Delta F_v)^2} \frac{(2 + \cos\theta)(1 - \cos\theta)^2}{4} \tag{6.4}$$

where γ_{SV} is the surface energy between the void nucleus and the substrate, and θ is the contact angle between the void and the substrate.

Equation 6.1 provides the rate at which stable nuclei will form. It also demonstrates the clear rate dependence on the temperature, surface energy, and transition enthalpy.

6.2.2 Void Stability at Equilibrium

Whether or not stable nuclei and mechanically trapped voids grow or redissolve depends on several factors. Growth may occur via diffusion of air or water vapor or by agglomeration with neighboring voids. Dissolution may occur if the changes in temperature and pressure cause an increase in solubility in the resin, as we shall see.

Let us first consider the synergistic effect that water has on void stabilization. It is likely that a distribution of air voids occurs at ply interfaces because of pockets, wrinkles, ply ends, and particulate bridging. The pressure inside these voids is not sufficient to prevent their collapse upon subsequent pressurization and compaction. As water vapor diffuses into the voids or when water vapor voids are nucleated, however, there will be an equilibrium water vapor pressure (and therefore partial pressure in the air–water void) at any one temperature that, under constant total volume conditions, will cause the total pressure in the void to rise above that of a pure air void. When the void pressure equals or exceeds the surrounding resin hydrostatic pressure plus the surface tension forces, the void becomes stable and can even grow. Equation 6.5 expresses this relationship

$$P_g - P_e = \frac{\gamma_{LV}}{m_{LV}} \tag{6.5}$$

where P_g and P_e are the void and resin pressures, respectively, γ_{LV} is the resin–void surface tension, and m_{LV} is the ratio of void volume to its surface area. The difference in pressure, therefore, is counterbalanced by the surface tension forces. When the temperature rises for a constant volume system, P_g will rise faster than P_e, whereas γ_{LV} will decrease slightly. In addition to P_g increasing in accordance with the perfect gas law, the partial pressure of water in the void can rise exponentially because of the temperature effect on the water vapor pressure.

It is instructive to quantitatively examine how P_g might vary as the temperature is increased under equilibrium conditions. Let us assume first that the air–water vapor mixture is an ideal gas and that a Raoult's Law relationship holds for the partial pressure of the water over the water–resin solution at equilibrium. By Raoult's Law

$$p_1 = x_1 p_{0_1} \tag{6.6}$$

where

x_1 = water concentration in the resin–water solution, mole fraction
p_{0_1} = pure component vapor pressure (in this case, water)
p_1 = partial pressure of water in the gas mixture in the void

The assumption of an ideal solution is obviously not correct, but in the low water concentration range of interest, the resulting error will not be excessively large for a first estimate.

The total pressure in the gas void (P_G) is

$$P_G = p_a + p_1 \tag{6.7}$$

where p_a is the partial pressure of air. At constant temperature,

$$P_G = p_a + x_1 p_{0_1} \tag{6.8}$$

Assuming that the equilibrium water uptake in the prepreg is 2 percent by weight results in a mole percentage of 0.286 for the TGMDA/DDS resin system. Note that a very small weight percentage of the water yields a rather large mole percentage. If p_a is initially 101 kPa (14.7 psia) at lay-up and by taking P_{0_1} from the steam tables at various temperatures, we can calculate the increase in void pressure as the curing cycle proceeds. The results are shown in Figure 6.4 for various equilibrium percentages of water in the resin. For a 2 per cent water uptake in the resin, the void pressure can rise to about 228 kPa (33 psia) before autoclave pressure is applied in a typical commercial cycle. Of course, resin viscosity also plays a role in the decision of when to apply the autoclave pressure. Note also that at 177°C (350°F), more than 414 kPa (60 psia) pressure can be generated in the void.

The calculations shown in Figure 6.4 represent an upper bound, but not an unreasonable one. First, although the void was assumed to remain at constant volume, compaction of the laminate is certainly occurring, and unless resin flow into the bleeder is significant, not much resin volume decrease will occur. Furthermore, if volume were not kept constant, then the voids would grow and likely coalesce, which is a condition of stabilization that is just as bad, if not worse, than that considered in the calculation. Second, it is assumed that Raoult's Law

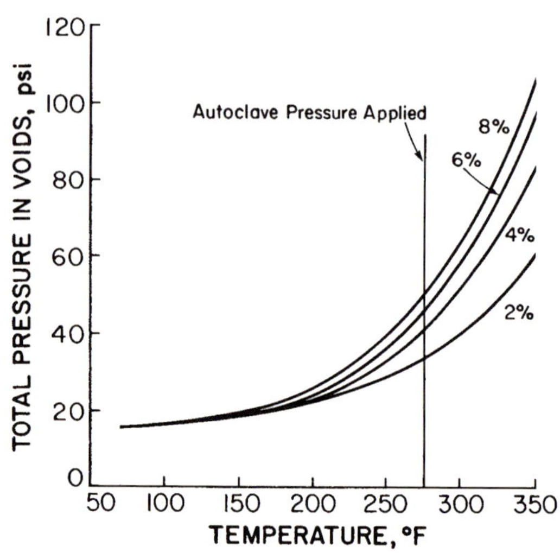

Figure 6.4 Effect of water on the total void pressure as the temperature increases in a typical curing cycle. Percentages are equilibrium water contents by weight in the resin phase

provides the relationship between the water partial pressure in the void and the water dissolved in the resin. It is more likely that Henry's Law, or some other nonideal solution rule, would hold, because the water–resin solution is certainly not an ideal solution; unfortunately, there is no way to estimate a priori what nonideal rule should apply. Thus, using Raoult's Law provides an upper bound on the partial pressure, and the calculation represents a worst-case estimate.

It should be pointed out that the approach outlined above is perfectly general and, while water has been used as the volatile species, any solvent could in principle be used and the analysis can be applied to devolatilization problems. Furthermore, more than one solvent could be considered. Equations 6.7 and 6.8 would then contain partial pressure terms for each solvent.

It is of interest to examine the effect of void size in conjunction with the void pressure results of Figure 6.4. Using Equation 5 and assuming a spherical shape for the voids, we find

$$P_g = P_e + \frac{6\gamma_{LV}}{d_V} \tag{6.9}$$

where d_v is the void diameter. Using a value for γ_{LV} of 50 dynes/cm, one can construct an equilibrium stability map with the resin pressure as a parameter. The results are displayed in Figure 6.5 and present some interesting limits to equilibrium void growth. The lines of constant resin pressure represent boundaries above which void growth can occur at that resin pressure, and below which the voids will collapse and eventually dissolve. The region below

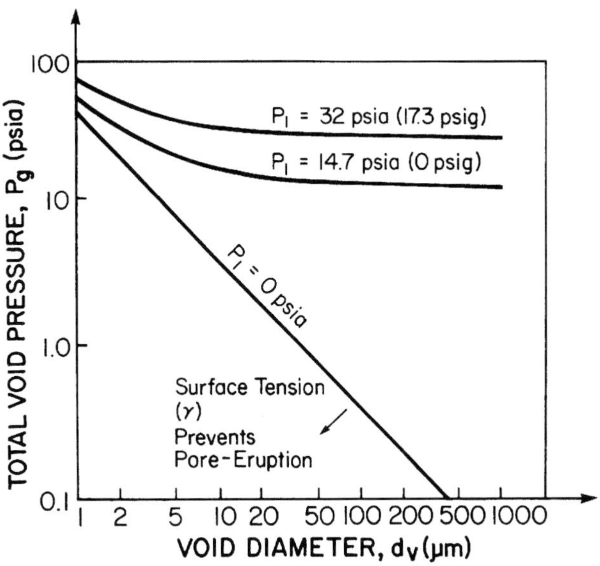

Figure 6.5 Equilibrium void stability map for a typical epoxy resin system. Curves indicate stable void equilibrium states for liquid–resin pressures indicated. Growth takes place above the lines and dissolution occurs below the lines for any given resin pressure

the line $P_1 = 0$ represents the void pressure necessary to overcome only the resin surface tension forces. For the case of atmospheric pressure in the resin, the total pressure in very small voids must exceed about 276 kPa (40 psia), whereas larger ones must exceed about 69 kPa (10 psia) in order to grow. As the resin pressure increases, higher total void pressures must be attained in order to create stable voids, particularly for the larger voids.

Hinrichs [1] has shown that resin pressures can drop to about 103–117 kPa (15–17 psig) even though autoclave pressures of up to 586 kPa (85 psig) are used. This means that void growth will occur for sufficiently high water contents and the problem of transporting the voids out of the laminate is extremely important.

6.3 Diffusion-Controlled Void Growth

6.3.1 Problem Definition

In the preceding discussion we considered equilibrium void stability; however, actual processing conditions involve changing temperature and pressure with time. Whereas equilibrium calculations provide bounds on void growth, it is the time-dependent growth process that is most important from a product quality viewpoint.

One physically realistic possibility involves the growth of a pure water vapor void by diffusion of water from the surrounding liquid resin into the void. Another equally possible mechanism also involves the diffusion process, except that the water enters a void that initially consisted only of entrapped air. In either case, it is assumed that water originally dissolved in the resin during prepreg fabrication or storage is the main component that contributes to void growth. There are two possibilities for void nucleation. In the case of entrapped air, initial nuclei of finite size always exist. Water dissolved in the resin diffuses into these nuclei whenever conditions are such that this diffusion path is favored. In the absence of air nuclei, water vapor nuclei can form at degrees of supersaturation dictated by Equations 6.1 to 6.4. For low molecular weight materials like water, critical-size nuclei contain 50–100 molecules [4] and are so small that their effective diameter is essentially zero.

The exact modeling of void growth would require the solution of many time-dependent coupled partial differential equations. Because many physical properties are not known with sufficient accuracy and because excessive computational time would be required, such an effort does not seem warranted; however, two alternate, simplified approaches seem feasible. One could consider a typical "equivalent" capillary through the plies and examine void growth perpendicular to the plies (see Fig. 6.1). A knowledge of the pore geometry of the fiber–resin medium and the surface tension, contact angle, viscosity, and density of the resin as a function of the temperature are required for this approach. A second approach is to consider void growth in an isotropic pseudohomogeneous medium. This latter approach was selected because experimental evidence indicates that voids do not grow preferentially perpendicular to the plies. This approach then becomes equivalent to bubble growth in a liquid medium.

Numerous papers cover various aspects of bubble growth [5–10]. Scriven's classical analysis [6] accounts for diffusion, convective transport caused by bubble expansion, and the moving boundary condition at the gas–liquid interface. This approach is superior to the quasi-stationary equation that considers only diffusion [5] or to the approaches that neglect convective transport but retain boundary movement [11,12]. Subramanian and Weinberg [10] show that the quasi-stationary approximation [5], which ignores all motion, is more consistent than the approach that retains only the boundary movement. It thus appears that void growth can be most expeditiously predicted by using the quasi-stationary assumption [5] along with Scriven's formulation [6]. Neglecting the time derivative in the quasi-stationary conservation equation leads to the quasi-steady–state approach [7] that was used by Kardos et al. [2] and which leads to the following equation for void growth:

$$\frac{d(d_B^2)}{dt} = 4D\beta \tag{6.10}$$

$$t = 0; \; d_B^2 = 0 \tag{6.11}$$

where d_B(cm) is the void diameter, D (cm^2/hr) is the water diffusion coefficient and β is the growth driving force defined by

$$\beta = \frac{C_\infty - C_{sat}}{\rho_g} \tag{6.12}$$

where ρ_g (gm/cc) is the density of the gas in the void. For constant D and β, Equation (6.10) predicts that the void diameter changes with the square root of time.

$$d_B = 2\sqrt{D\beta t} \tag{6.13}$$

6.3.2 Model Development

6.3.2.1 Assumptions

In developing the void growth model, the following simplifying assumptions were made:

1. The void is stagnant between two plies, and its center is not moving with respect to fixed coordinates in the laminate.
2. The void is approximately spherical and its effective size is calculated based on an equivalent sphere.
3. There is no interaction between voids (no coalescence).
4. The void is considered to be in an infinite isotropic fluid medium.
5. For pure water voids, void nucleation is instantaneous. For air–water voids, air void nucleation or entrapment has already occurred during lay-up or at the beginning of the cure cycle.
6. At any given time the temperature and moisture concentration in the bulk resin are uniform.

7. At each temperature, a pseudo-steady–state is established with respect to the concentration profile (i.e., the profile does not change during diffusion at any particular temperature).
8. The viscous, inertia, and surface tension effects are neglected because they are significant only during the initial expansion of the original void nucleus [6]. As the void grows, its growth rapidly becomes limited by the rate of arrival of the diffusing species.

6.3.2.2 Governing Transport Equations

Under the preceding assumptions, the conservation equation of the diffusing species may be expressed as:

$$\frac{\partial c}{\partial t} + \frac{da}{dt}\frac{a^2}{r^2}\frac{\partial c}{\partial r} = \frac{D}{r^2}\frac{\partial}{\partial r}r^2\frac{\partial c}{\partial r} \tag{6.14}$$

with the initial and boundary conditions

$$C(0, r) = C_\infty \tag{6.15a}$$

$$C(t, \infty) = C_\infty \tag{6.15b}$$

$$C(t, a(t)) = C_{sat} \tag{6.15c}$$

When the density of the gas changes much less rapidly than the radius of the void, then the equation of continuity which the void radius $a(t)$ satisfies is:

$$\frac{da}{dt} = \frac{1}{\rho_g} D \frac{\partial c}{\partial r}(t, a(t)) \tag{6.16}$$

where a(cm) is the void radius and ρ_g (g/cm^3) is the gas density.

The following asymptotic solution for large values of β was developed by Scriven [6]:

$$d_B = 4\beta\sqrt{Dt} \tag{6.17}$$

The difference in the void diameter predictions between Equations 6.13 and 6.17 is a factor of $2\sqrt{\beta}$.

The change in diameter per unit time (growth rate) can be obtained by differentiation of Equation 6.17, which yields

$$\frac{d(d_B^2)}{dt} = 16D\beta^2 \tag{6.18}$$

The initial condition is:

$$\text{at } t \equiv 0;\ d_B^2 = 0 \tag{6.19}$$

Equations 6.18 and 6.19 may now be used to calculate void growth during the cure cycle. The driving force β is adjusted according to the changing temperature and pressure in the laminate, which affect C_{sat} and ρ_g. This is equivalent to assuming that on the time scale of void growth, which is a relatively slow process, all other processes have a chance to reach a new equilibrium. Pressure and temperature are therefore assumed to be uniform throughout the resin at any moment in time, but to vary with time according to the prescribed cure cycle. Equations 6.18 and 6.19 therefore represent the essence of the model. All that remains is to properly evaluate the model input parameters such as the diffusivity, D, the resin water concentration, C_{∞}, the interface water concentration, C_{sat}, and the gas density, ρ_g, as functions of temperature and pressure during the cure cycle. Again, wherever water has been mentioned earlier, any solvent or solvents may be substituted, and the equations may be applied to the devolatilization problem.

6.3.2.3 Evaluation of Input Variables

In the development of the computer code for the void growth model, the following input relationships are provided in the program. These could, of course, be modified to account for different cure cycles or different material systems. The values used in this study are provided as default options in the code.

1. *Vapor Pressure.* The vapor pressure of water within the void is dependent on temperature according to the Clausius–Clapeyron Equation.

$$\frac{dp_{H_2O}}{dT} = \frac{\Delta H_v}{T(v_g - v_L)} \tag{6.20}$$

Assuming that (1) the heat of vaporization, ΔH_v, is constant, and (2) the vapor phase behaves as an ideal gas, the following dependence of vapor pressure on temperature results:

$$p^*_{H_2O} = \left(p^*_{H_2O} \exp\frac{\Delta H_v}{RT_o}\right) \exp\frac{-\Delta H_v}{RT} \tag{6.21}$$

where T_o(°K) is the boiling point of water at 1 atm (373° K); $p^*_{H_2O}$ (atm) is the vapor pressure of water; $p^*_{H_2O}$ is the water vapor pressure at boiling (1 atm); ΔH_v (cal/mol) is the heat of vaporization (9720 cal/mol); R (cal/mol°K) is the ideal gas constant (1.987 cal/mol-K). Substitution yields

$$p^*_{H_2O} = 4.962 \times 10^5 \exp\left(\frac{-4892}{T}\right) \tag{6.22}$$

The pressures inside and outside of the void are effectively equal until the resin viscosity becomes so high that viscous effects become important. As the resin proceeds toward solidification, the pressure in the void can rise significantly above the resin pressure. Surface tension effects are also negligible for voids larger than 100 μm.

2. *Partial Pressure.* For an air–water void, it is assumed that the void initially contains dry air only. In the case of an air–water mixture in the void, the mole fraction of dry air is

$$y_{air} = \frac{p_o T}{p T_o}\left(\frac{d_{B_o}}{d_B}\right)^3 \tag{6.23}$$

where y_{air} is the mole fraction of dry air; p (atm) is the total pressure in the resin; d_{B_o} (cm) is the initial diameter of the void; p_o (atm), T_o(K) are the initial pressure and temperature in the resin, respectively.

The partial pressure of water within an air–water void is

$$p_{H_2O} = (1 - y_{air})p \tag{6.24}$$

where p_{H_2O} (atm) is the partial pressure of water. For a pure water void, the partial pressure of water equals the total pressure (i.e., $p_{H_2O} = p$).

3. *Gas Density*: The density of the gas within the void is given by the following equation.

$$\rho_g = (1 - y_{air})\frac{M_{H_2O}p}{RT} + y_{air}\frac{M_{air}p}{RT} \tag{6.25}$$

where M_{H_2O} (gm/mol) and M_{air} (gm/mol) are the molecular weights of water and air, respectively. For a pure water void, $y_{air} = 0$ in Equation 6.25 results in $\rho_g = M_{H_2O}p/RT$.

4. *Solubility*. A simple parabola fits the water solubility data of the resin in the prepreg equilibrated at different relative humidity exposures as follows [13]:

$$S_o = k_1(RH)^2 = k_1 \times 10^4\left(\frac{p_{H_2O}}{p^*_{H_2O}}\right) \tag{6.26}$$

where S_o is the solubility (wt% of water per unit weight of prepreg), k_1 is a constant, and $(RH)_o$ is the relative humidity. For graphite/epoxy prepregs, it is assumed that water is insoluble in the graphite fibers. For the T300/Narmco 5208 system, $k_1 = 5.58 \times 10^{-5}$ [13]. This constant, k_1, was determined from data obtained at 25°C. At higher temperatures, k_1 was assumed not to change. This is a useful approximation and the error introduced is small.

5. *Water Concentration*. The water concentration in the resin, C (gm/cm^3), is a function of water solubility in the prepreg, S_o (wt%), the weight fraction of the resin in the prepreg, W_R (gm/gm), and the resin density, ρ_R (gm/cm^3) as follows:

$$C = \frac{S_o}{100}\frac{\rho_R}{W_R} \tag{6.27}$$

For the T300/Narmco 5208 system, the water concentrations within the bulk resin and at the void surface were obtained from measured solubility, weight fraction, and density data. The water concentration in the bulk resin is dependent only on the initial relative humidity under which the resin was equilibrated and is given by:

$$C_\infty = 2.13 \times 10^{-6}(RH)_o^2 \tag{6.28}$$

The water concentration at the void surface is a function of both the temperature and partial pressure of water within the void as follows:

$$C_{sat} = 8.651 \times 10^{-14}\exp\left(\frac{9784}{T}\right)p^2_{H_2O} \tag{6.29}$$

The interface concentration is called C_{sat}, even though an air–water mixture may not be saturated.

6. *Diffusivity.* The diffusivity of water in the prepreg (either fresh or cured), follows an Arrhenius-type equation [13]

$$D = D_o \exp\left(-\frac{E_a}{RT}\right) \tag{6.30}$$

where D (cm^2/hr) is the diffusivity; D_o (cm^2/hr) is the preexponential constant, E_a(J/mol) is the activation energy for diffusion per mole, R (J/mol K) is the universal gas constant, and T(K) is the absolute temperature.

   For the system T300/Narmco 5208, $D_o = 0.105$ cm^2/hr and $(E_a/R) = 2817$ K for fresh prepreg [13].
7. *Resin Content.* The fresh prepreg contained 32 wt% resin (the default option in the computer code developed in this study).
8. *Resin Density.* A resin density, ρ_R, of 1.22 g/ml was used in this study.
9. *Initial Relative Humidity* $(RH)_o$. This is the relative humidity at which the prepreg was equilibrated.

   If zero initial void diameter is assumed, then no void growth occurs while $C_{sat} = C_\infty$. The time increment from the start of the cure cycle to the moment when $C_{sat} = C_\infty$ is denoted as t_{BEGIN} and is given by:

$$t_{BEGIN} = \frac{T_{CR} - T_{IN}}{R_T \times 60} \tag{6.31}$$

where T_{CR} (K) is the temperature at which $C_{sat} = C_\infty$, T_{IN} (K) is the initial temperature at the start of the cure cycle, and R_T (K/min) is the heating rate.

6.3.3 Model Predictions for Void Growth

Using the input information described earlier, void behavior was examined during the various stages of the cure cycle shown in Figure 6.2. The specific stages are as follows:

Stage 1: The temperature is increased from 25°C (298 K) to 135°C (408 K) at 2°C/min, yielding a rise time of 55 min. The autoclave pressure is one atm.; the pressure inside the vacuum bag is 0.1 atm and remains so throughout the entire cycle.

Stage 2: The temperature is held at 135°C (408 K) for (1) 15 min, (2) 60 min, and (3) 90 min. The autoclave pressure is still 1 atm.

Stage 3: The autoclave pressure is increased from one to 5.78 atm at a constant temperature of 135°C (408 K).

Stage 4: The temperature continues to be held constant at 135°C (408 K) under a constant autoclave pressure of 5.78 atm for 105 min.

Stage 5: The temperature is increased from 135°C (408 K) to 179°C (452 K) at a rate of 1.1°C/min. under an autoclave pressure of 5.78 atm, and the rise time taken is 40 min.

Because the actual pressure profile in the resin is as yet unknown, it is assumed that the void experiences a resin pressure of 0.1 atm during Stages 1 and 2, which then increases to 5.78 atm during stages 3–5. The resin never actually experiences the total autoclave pressure, so 5.78 atm represents an upper bound.

Figure 6.6 shows the effect of the processing cycle on the void diameter for pure water and air–water voids of 0.1 cm initial diameter under the specified cycle conditions. It was assumed that the air–water void initially consists of pure air, even though there is likely to be a small but finite water partial pressure. This plot can be divided into the various stages of void growth and dissolution and interpreted as follows.

During Stage 1, under constant pressure but increasing temperature, the diffusivity increases exponentially with absolute temperature, but C_{sat} decreases exponentially with absolute temperature. Both these effects, however, favor void growth and a rapid exponential increase in the void diameter results during this stage. During Stage 2, the pressure and temperature are constant, hence the diffusivity, ρ_g, and C_{sat} are constant (for an air–water void, ρ_g and C_{sat} are not quite constant but the variation is slight). Because C_∞ is also fixed, it is dependent only on the initial humidity exposure (Eq. 6.28) and β in Equation 6.10 is constant. Because β and the diffusivity are constant during Stage 2, void growth during this stage is a function of $\sqrt{t}$ only. Thus, in Figure 6.6, we see a square root relation for void growth with time during Stage 2. During Stage 3, the pressure is increased and the volume decrease is calculated using the ideal gas law. Stage 4 is similar to Stage 2 except that now there is a negative driving force for void growth and hence void dissolution occurs. Stage 5 similarly, mimics Stage 1, except that void dissolution occurs rather than void growth.

Although the void behavior is calculated through Stage 5, it is likely that the viscosity rises sufficiently high in Stage 4 so that the model assumptions are no longer valid. The void dissolution calculated in Stages 4 and 5 will therefore probably not occur in reality.

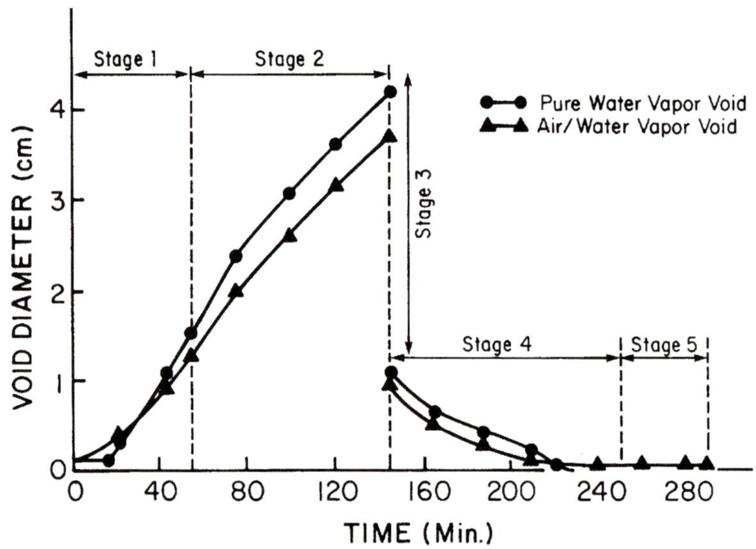

Figure 6.6 Progression of void growth during the cycle of Figure 6.2 for both pure water and air–water voids. Note the change when pressure is applied at the end of Stage 2

An important conclusion from Figure 6.6 is that it is immaterial for a small void whether or not the initial void contains pure water or an air–water mixture. The diameters at any particular time during the cure cycle are nearly identical when the initial void diameters under 0.1 atm are the same. An air–water void initially containing pure air has a very large driving force for diffusion of water vapor from the resin to the void during the first few minutes of the cycle. This results in diffusion into the void of a large amount of water vapor (relative to the original amount of dry air in the void). As a result the mole fraction of water vapor in the air–water void quickly approaches unity; thereafter, the rate of diffusion of water vapor across the interface of the air–water void is nearly the same as that for a pure water void.

It is also noteworthy that after the initial period of growth, the higher density of the air–water vapor mixture produces a void that is slightly smaller than the pure water vapor void produced under identical conditions. The air–water void, however, cannot completely dissolve during the cure cycle, whereas the water void is capable of complete dissolution.

Figure 6.7 shows the final diameter of a pure water void at the end of the cure cycle after it has grown from voids of various initial diameters under the conditions specified. For an initial void diameter of zero, the final diameter is about 1.25 cm under the specified conditions of growth. On the other hand, relatively large initial void diameters (0.5 cm) only triple in size.

The final diameter of a pure water void for different initial relative humidities of the resin is shown in Figure 6.8. The marked increase in the final void size illustrates the pronounced effect that initial relative humidity exposure has on the final void size. This behavior is described by Equation 6.28 in which C_∞ (which is fixed during the cure cycle and determines the driving force) increases with the square of the initial relative humidity exposure. Thus, increasing the initial relative humidity by a factor of 2 would result in a four-fold increase in C_∞. This would in turn increase the driving force 4-fold when the other conditions of growth are kept identical and when $C_{sat} \ll C_\infty$.

Figure 6.9 shows the final diameter of the void for different degrees of vacuum applied during Stage 1. It is apparent that the larger the vacuum applied (lower pressure), the greater is the potential for big voids. Referring to Equation 6.29, C_{sat} is directly proportional to the square of the partial pressure of water in the void. The partial pressure of water in the void is equal to the applied pressure for a pure water void, and nearly equal to the applied pressure after a small initial growth period for an air–water void. Thus, reducing the applied pressure by a factor of 2 would result in a four-fold decrease in C_{sat}. This would in turn would increase the driving force. In addition ρ_g is decreased with decreasing pressure, which further increases β and thereby the driving force.

Figures 6.7 to 6.9 identify the two most important parameters that affect the final void diameter; (1) the initial relative humidity exposure of the prepreg and (2) the applied pressure during the cure cycle. In terms of industrial processing of laminates, the results of Figure 6.8 show that the humidity exposure of the resin before and during processing should be kept below about 52 per cent if an initial vacuum of 0.1 atm is to be applied and void-free product is desired. Maintaining this low relative humidity might well be prohibitively expensive, depending upon the climate of the production facility. The alternative to this solution is not to apply an initial vacuum. Figure 6.9 shows that under the current model assumptions an initial pressure greater than 0.35 atm would prevent voids in the final product.

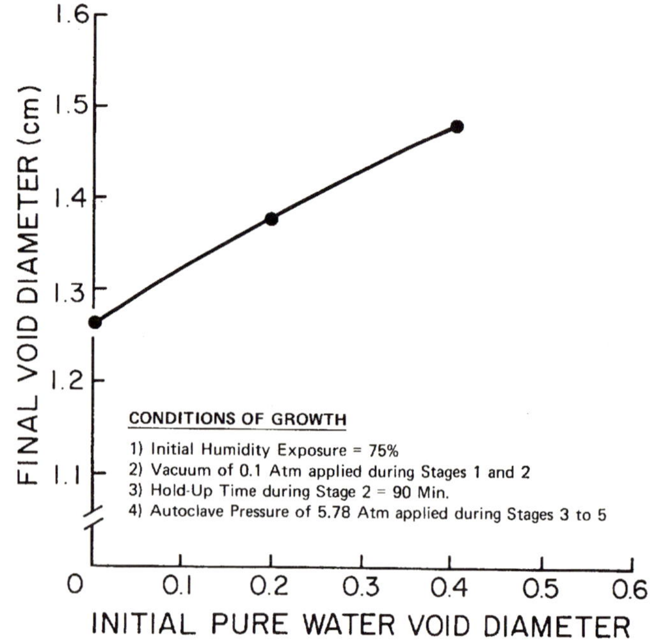

Figure 6.7 Effect of initial pure water void size on the final void size for the process conditions shown

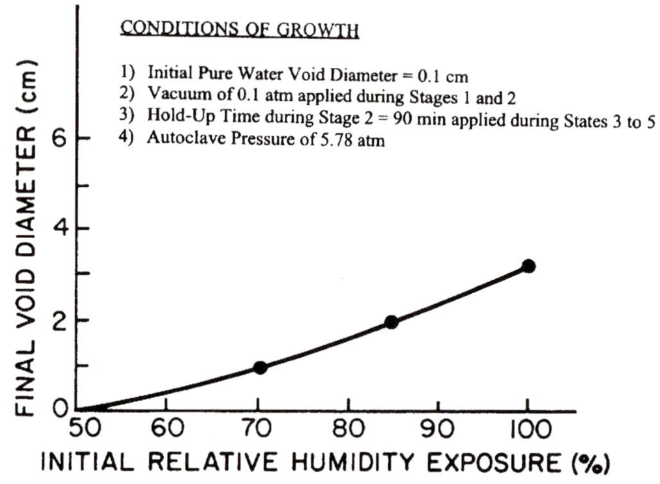

Figure 6.8 Effect of initial relative humidity exposure of the prepreg on the final void size in the laminate for the process conditions shown

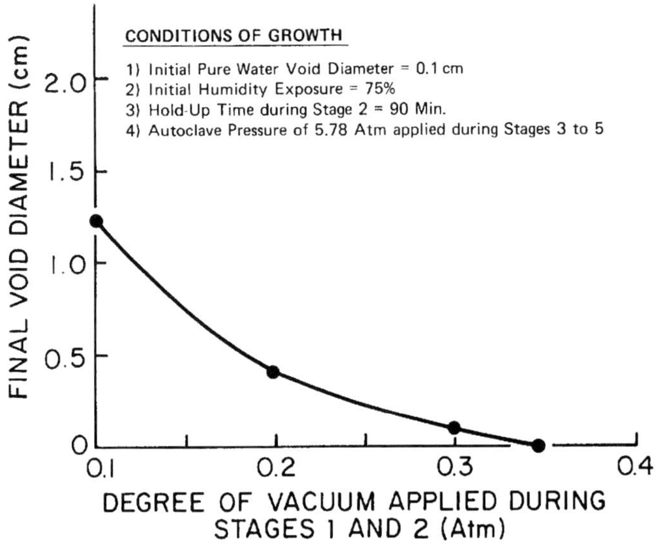

Figure 6.9 Effect of low matrix pressure (degree of vacuum) on the final void size for the process conditions shown

For a prepreg equilibrated with moisture at a particular relative humidity, in order to prevent the potential for pure water void growth by diffusion at all times and temperatures during the curing cycle, the pressure at all points of the prepreg must satisfy the following inequality:

$$p \geq 4.962 \times 10^3 \exp\left(\frac{-4892}{T}\right)(RH)_o \tag{6.32}$$

where

$(RH)_o$ (%) = initial relative humidity exposure of the prepreg
P (atm) = resin pressure in the prepreg at various times
T (K) = temperature during the curing cycle

Equation 6.32 was derived from the requirement that void growth by diffusion at any temperature cannot occur if the pressure within the void is greater than the saturated vapor pressure at that temperature (i.e., if $C_{\text{sat}} \geq C_\infty$).

Whereas Equation 6.32 holds exactly for a pure water void in any system, it would also hold well for small air–water voids after an initial growth period because it has already been shown in Figure 6.6 that the growth pattern of an air–water void is similar to that of a pure water void after an initial growth period.

A plot of Equation 6.32 for two relative humidities (50 and 100%) yields a void stability map, which is shown in Figure 6.10. It is evident from this map that vacuum can be applied without encouraging void growth if such application is coordinated with the temperature of the system. Brown and McKague [14] have experimentally observed that the void content is reduced significantly when pressure is applied to the prepreg early in the cycle, which is in accordance with the stability map.

The real question in light of Figure 6.10 is what exactly is the resin pressure throughout the product part? Is it equal to the gas pressure above the bag in the autoclave process, or the entry resin pressure in the RTM process? If not, how can the actual resin pressure be reliably predicted?

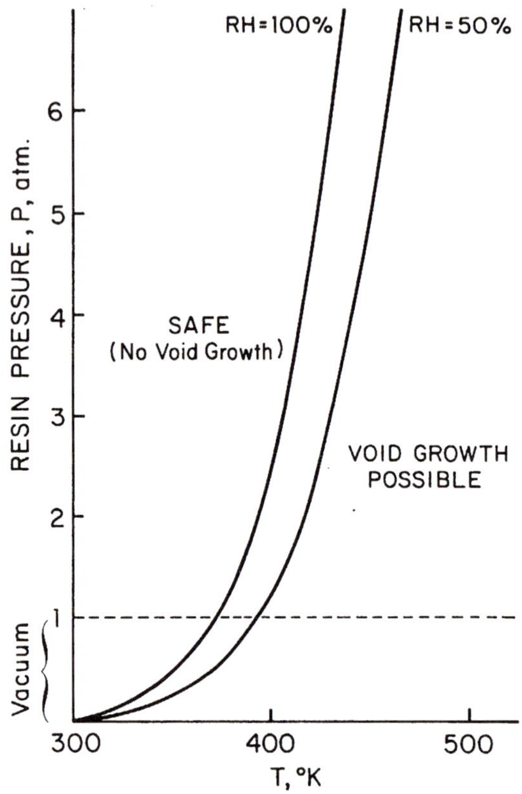

Figure 6.10 Void stability map for pure water void formation in epoxy matrixes. Note the significant effect of initial relative humidity exposure of the resin

6.4 Resin and Void Transport

For thick epoxy laminates processed in the autoclave, voids once formed and stabilized can only be removed by dissolution or by resin flow. Furthermore, resin gradients are deleterious to structural laminates. These two key phenomena make an understanding of resin transport vital to the development of any processing model.

In the past, various resin flow models have been proposed [2,15–19]. Two main approaches to predicting resin flow behavior in laminates have been suggested in the literature thus far. In the first case, Kardos et al. [2], Loos and Springer [15], Williams et al. [16], and Gutowski [17] assume that a pressure gradient develops in the laminate both in the vertical and horizontal directions. These approaches describe the resin flow in the laminate in terms of Darcy's Law for flow in porous media, which requires knowledge of the fiber network permeability and resin viscosity. Fiber network permeability is a function of fiber diameter, the porosity or void ratio of the porous medium, and the shape factor of the fibers. Viscosity of the resin is essentially a function of the extent of reaction and temperature. The second major approach is that of Lindt et al. [18] who use lubrication theory approximations to calculate the components of "squeezing' flow created by compaction of the plies. The first approach predicts consolidation of the plies from the top (bleeder surface) down, but the second assumes a plane of symmetry at the horizontal midplane of the laminate. Experimental evidence thus far [19] seems to support the Darcy's Law approach.

Among all the models mentioned earlier, only Loos and Springer [15] have considered flow in both the horizontal (laminate plate) and vertical directions. In their model, however, the two flows have been decoupled, which results in different resin pressure profiles along any horizontal line in the laminate for the vertical and the horizontal flows. They have modeled the horizontal flow as laminar flow between ply planes (based on a channel flow equation) with the implicit assumption that a linear pressure gradient exists in the horizontal direction. To apply this equation to the case of flow during consolidation would imply that there is an imaginary source at the vertical midplane of the laminate that is at a higher pressure than the pressure at the edges. Further, in their model, they have considered that the applied pressure on the laminate is borne solely by the pressure in the resin. The experimental evidence [20], however, indicates that the resin pressure at the center of the laminate below the ply nearest to the tool surface decreases with time during processing and does not remain constant under constant applied stress.

In overcoming the shortcomings of the earlier models, Davé et al. [21,22] proposed a comprehensive three-dimensional consolidation and resin flow model that can be used to predict the following parameters during cure: (1) the resin pressure and velocity profiles inside the composite as a function of position and time, (2) the consolidation profile of the laminate as a function of position and time, and (3) resin content profile as a function of position and time.

The Davé model considers a force balance on a porous medium (the fiber bed). The total force from the autoclave pressure acting on the medium is countered by both the force due to the springlike behavior of the fiber network and the hydrostatic force due to the liquid resin pressure within the porous fiber bed. Borrowing from consolidation theories developed for the compaction of soils [23,24], the Davé model describes one-dimensional consolidation with three-dimensional Darcy's Law flow. Numerical solutions were in excellent agreement with closed-form solutions for one- and two-dimensional resin flow cases in which the fiber bed permeabilities and compressibility, as well as the autoclave pressure, are all held constant [21].

Figure 6.11 depicts the numerical solutions for the time dependence of the resin pressure profile in the vertical direction for one-dimensional flow in the vertical direction (corresponding to an edge-dammed laminate). The laminate is 1.4-in thick (*z* direction) and is a unidirectional lay-up. Values for the specific permeability in the *z* direction, (k_z in.^{2}), the

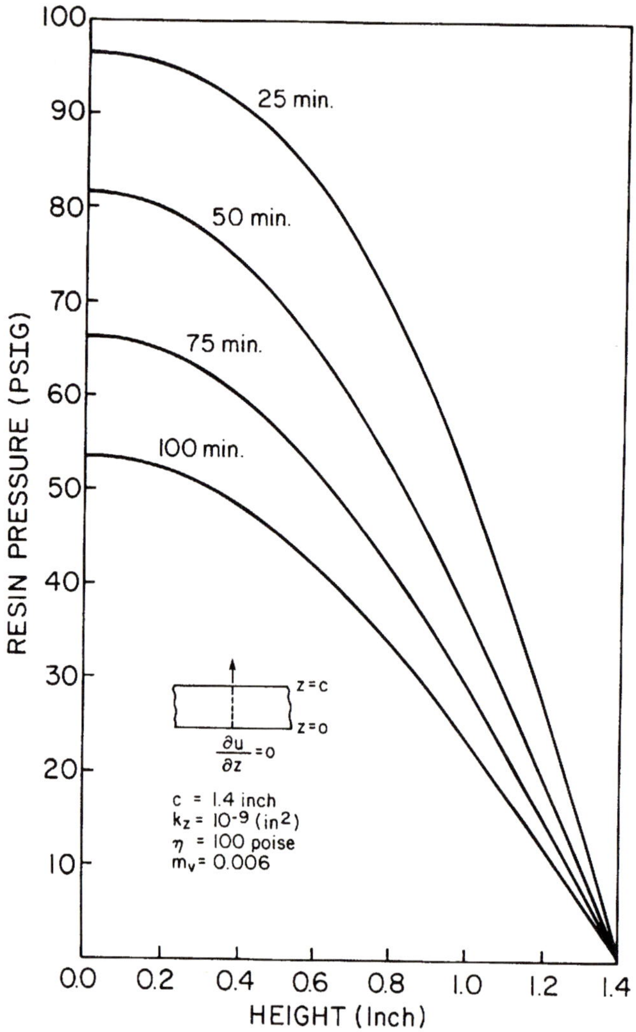

Figure 6.11 Resin pressure profiles in the laminate thickness direction (vertical) in a 1.4-in. thick unidirectional graphite–epoxy laminate for one-dimensional flow (edge-dammed) under conditions indicated in the figure

viscosity (η, poise), and the coefficient of volume change, m_V, are shown in the figure. The pressure gradient is clearly not linear as has been assumed in the past. Even more non-linearity results when one properly allows for changing k_z, η, and m_v.

This model can also provide resin pressure gradients, resin flow rates, consolidation profiles, and, when combined with the void model, void profiles at any point in the laminate.

The model also makes it very clear that when resin flow occurs, the autoclave pressure is almost never equal to the hydrostatic resin pressure anywhere in the laminate. Even in a net resin prepreg system where very little resin flow occurs, the fiber bed network can bear significant fractions of the autoclave pressure. When the resin pressure drops below that necessary to prevent void formation and growth, the part being autoclaved is in jeopardy. An example of such an occurrence is shown in Figure 6.12 from a study by Davé et al. [25] on an AS-4/3502 laminate having 48 plies. Profiles of the resin pressures monitored by the laminate, tool, and bleeder transducers are shown for both "dry" (exposed to 35% relative humidity for 72 h) and "wet" (exposed to 85% relative humidity for 72 h). There was no measurable difference between the autoclave pressure profiles for the "dry" and "wet" laminates. The profiles of time-temperature and corresponding time-P_{min} for the "dry" and "wet" laminates are also shown in the figure. It is evident that the resin pressure at various locations in the laminates is significantly below the autoclave pressure. It is also clear that at 130 min into the cycle (less than the 140-min resin gel time), all the transducer readings are decreasing below the minimum pressure required to prevent void formation in the "wet" laminate. Indeed, ultrasonic C-scans of the cured laminates showed the "wet" laminate to have large amounts of porosity, whereas the "dry" laminate was void free. The transducer responses were lost at 135 min, but it is not likely that the resin pressure dropped below P_{min} for the "dry" laminate before the resin gelled at 140 min.

Similar kinds of resin pressure calculations must also be made for other processes to examine the potential for void stability and growth. This has become a subject of considerable interest [26–30] as quality is sought in RTM parts.

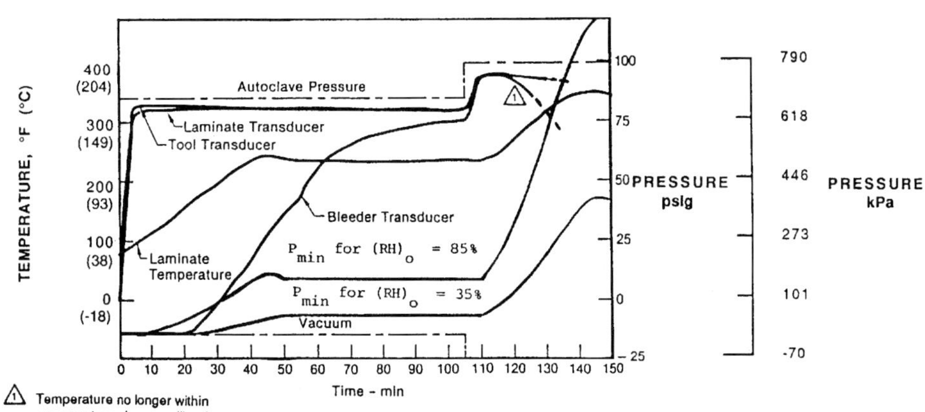

Figure 6.12 Profiles of typical resin pressures monitored by the laminate, tool, and bleeder transducers. The profiles of time–temperature and the corresponding time—P_{min} for 35% and 85% initial relative humidity exposure of the prepreg—are also shown. (*Source*: From [25])

6.5 Conclusions

Any on-line process control model used for computer-aided manufacturing of high-performance composite laminates must include a thorough treatment of void stability and growth as well as resin transport. These two key components, along with a heat transfer model and additional chemorheological information on kinetics and material properties, should permit optimized production of void-free, controlled-thickness parts. A number of advances have been made toward this goal.

Based strictly on equilibrium considerations, bounds can be set on the stability of voids as a function of temperature and pressure. Although this type of phase map does not depict the time dependency of an actual process, it does provide a limiting scenario toward which the actual process would be heading at any point in the curing cycle. It is surprising that high void pressures are possible if sufficient moisture is present in the resin.

A model and attendant computer code have been constructed to describe time-dependent void growth and stability for any processing cure cycle. Although the analysis is approximate, it does account for the moving void-resin boundary layer and its effect on the concentration profile and diffusion of water in the resin. It was found that for the duration of the cure cycle it makes little difference whether the initial small voids contain pure water or mixtures of air and water.

At the end of a cure cycle, the spread in the final void sizes is not as marked as the spread in the initial void sizes. For any particular cure cycle, the final diameter of the void increases almost linearly but very slowly with increasing initial diameter. Further, even at zero initial void diameter (i.e., void-free resin) the final void diameter may be relatively large due to rapid nucleation and growth. A particularly significant result is that application of vacuum during the initial stage of the cure cycle has the potential of creating large voids. Moreover, the initial moisture content of the prepreg is very important. The void diameter at the end of the cure cycle has a direct second-order dependency on the initial relative humidity exposure of the prepreg.

A pressure–temperature stability map can be constructed as a function of humidity exposure, which identifies the resin pressure values for each temperature below which void growth is possible and above which voids cannot grow but rather tend to collapse via dissolution.

A generalized three-dimensional resin flow model has been developed that employs soil mechanics consolidation theory to predict profiles of resin pressure, resin flow velocity, laminate consolidation, and resin content in a curing laminate.

The numerical solutions necessary to solve the practical three-dimensional problems agree well with the closed-form analytical solutions for simpler one- and two-dimensional cases with constant material properties. The resin pressure gradient in the thickness (vertical) direction for a well-dammed laminate (no horizontal flow) is nonlinear.

The resin pressure is almost never equal to the autoclave pressure. If the resin pressure drops due to resin flow, then it may become less than the minimum pressure necessary to prevent void stability and growth. In order to produce quality void-free laminates consistently, accurate resin pressure predictive software is a necessity.

Nomenclature

a	void radius, cm
C	concentration, gm/cm^3
C_∞	bulk concentration of water or solvent in resin, gm/cm^3
C_{sat}	concentration of water or solvent in void at void–resin interface, gm/cm^3
d_v, d_B	void diameter, cm
d_{B_o}	initial void diameter, cm
D	diffusion coefficient, cm^2/hr
D_o	pre-exponential constant, cm^2/hr
E_n	activation energy for diffusion, J/mol
ΔF^*	maximum free energy barrier for nucleation, ergs
ΔF_v	free energy change for phase transition, ergs/cm^3
ΔH_v	heat of transition, ergs/cm^3
k	Boltzmann constant, ergs/molecule K
k_1	proportionality constant
M	molecular weight, gm/mol
M_{H_2O}	molecular weight of water, gm/mol
M_{air}	molecular weight of air, gm/mol
m_{LV}	void volume-to-surface area ratio, cm
n	number of molecules per unit volume in nucleus
p	total pressure or resin pressure, atm
p_a	partial pressure of air, Pa
p_o	initial resin pressure, atm
p_{H_2O}	partial pressure of water, atm
p_1	component partial pressure in void gas mixture, Pa (psia)
P_{o_1}	pure component vapor pressure, Pa (psia)
P_c	resin pressure, dynes/cm^2
P_g	void pressure, dynes/cm^2
P^*	water vapor pressure, air pressure, or total mixture pressure, dynes/cm^2
$p^*_{H_2O}$	vapor pressure of water, atm
$p^*_{H_2O_o}$	vapor pressure of water at boiling (1 atm)
R	ideal gas constant, cal/mol K or J/mol K
R_T	heating rate, K/min
$(RH)_o$	relative humidity, %
r	radial coordinate, cm
r^*	critical nucleus radius, cm
S_o	solubility, wt% water per unit weight prepreg, %
t	time, min
t_{BEGIN}	time increment from start of cure cycle to time when $C_{sat} = C_\infty$, min
T	absolute temperature, K
T_o	equilibrium transition temperature or initial resin temperature, K
T_{CR}	temperature at which $C_{sat} = C_\infty$, K
T_{IN}	temperature at start of cure cycle, K

W_R	weight fraction of resin in prepreg
v_g, v_l	specific volumes of gas and liquid, respectively, cm^3/gm
x_1	mole fraction water in resin–water solution
y_{air}	mole fraction dry air in void
β	growth driving force, dimensionless
γ_{LV}	surface energy for resin–void surface, erg/cm^2
γ_{SV}	surface energy for void–substrate surface, erg/cm^2
θ	contact angle between void and substrate, degrees
ρ	void gas density, gm/cm^3
ρ_g	gas density, gm/cm^3
ρ_R	resin density, gm/cm^3

Acknowledgments

We wish to acknowledge support for this work by the Air Force Wright Aeronautical Laboratories, Materials Laboratory, Wright Patterson AFB, under Contract F33615-80-C-5021 to General Dynamics Corporation (L. McKague-Program Manager) and under Contract F33615-83-C-5088 to McDonnell Aircraft Company, McDonnell Douglas Corporation (F. Campbell, Program Manager). Figures 6.1, 6.3, 6.4, and 6.5 and some discussion thereof are reproduced under permission from American Society for Testing and Materials, Philadelphia, Penn.

References

1. Hinrichs, R.J. "Processing Science of Epoxy Resin Composites" (1982) Industry Contract Review F33615-80-C5021, General Dynamics Convair Division, San Diego, Calif. Jan. 19
2. Kardos, J.L., Dudukoviç, M.P., McKague, E.L., Lehman, M.W. Composite Materials: Quality Assurance and Processing (1983) ASTM STP 797, C.E. Browning (Ed.). American Society for Testing and Materials, Philadelphia, p. 96–109
3. May, C.L. (Ed.), *Chemorheology of Thermosetting Polymers* (1983) ACS Symposium Series, No. 227, American Chemical Society, Washington, DC
4. Kingery, W.D. *Introduction to Ceramics* (1960) Wiley, New York, p. 291
5. Epstein, P.S., Plesset, M.S. *J. Chem. Phys.* (1950) 18, 1505
6. Scriven, L.E. *Chem. Eng. Sci.* (1959) 10, 1
7. Bankoff, S.G. *Advances in Chemical Engineering*, 6 (1966) Academic Press, New York
8. Ready, D.E., Cooper, A.R., Jr. *Chem. Eng. Sci.* (1966) 21, 916
9. Duda, J.L., Vrentas, J.S. *AIChE J.* (1969) 15, 351
10. Subramanian, R.S., Weinberg, M.C. *J. Chem. Phys.* (1980) 72, 6811
11. Tao, L.N. *J. Chem. Phys.* (1978) 69, 4189
12. Tao, L.N. *J. Chem. Phys.* (1979) 71, 3455
13. Interim Report, "Processing Science of Epoxy Resin Composites" (1981) Contract No. F33615-80-C-5021, 9/15/80–10/15/81, Air Force Materials Laboratory, Weright Patterson AFB, Ohio

14. Brown, G.G. and McKague, E.L. "Processingi Science of Epoxy Resin Composites" (1982) 8th Quarterly Technical Report, Contract No. F33615-80-C-5021, Air Force Materials Laboratory, Wright-Patterson AFB, Ohio
15. Loos, A.C., Springer, G.S. *J. Comp. Mats.* (1983) 17, 135
16. Williams, J., Donnellan, T., Trabocco, R. (1984) Paper presented at 16th National SAMPE Tech. Conf., Albuquerque, NM
17. Gutowski, T.G. *SAMPE Q.* (1985) July p. 58
18. Lindt, J.T. *SAMPE Q.* (1982) Oct. p. 14
19. Brand, R.A., McKague, E.L. "Processing Science of Epoxy Resin Composites" (1983) Tenth Quarterly Report, Contract No. F33615-80-C-5021, Air Force Materials Lab., Wright-Patterson AFB, Ohio
20. Computer Aided Curing of Composites (1985) McDonnell Douglas Corp., 4th Interim Report, Contract No. F33615-C-5088, Air Force Materials Laboratory, Wright-Patterson AFB, Ohio
21. Davé, R., Kardos, J.L., Dudukoviç, M.P. *Polym. Comp.* (1987) 8, p. 29
22. Davé, R., Kardos, J.L., Dudukoviç, M.P. *Polymer Comp.* (1987) 8, p. 123
23. Terzaghi, K. *Theoretical Soil Mechanics* (1943) New York, John Wiley and Sons
24. Taylor, D.W. *Fundamentals of Soil Mechanics* (1948) New York, John Wiley and Sons
25. Davé, R., Mallow, A., Kardos, J.L., Dudukoviç, M.P. (1990) *SAMPE J.* 26(3), p. 31
26. Parnas, R.S., Phelan, F.R., Jr., *SAMPE Q.* (1991) 22(2), p. 53
27. Patel, N., Lee, L.J., *Polym. Comp.* (1996) 17, p. 96
28. Patel, N., Lee, L.J. *Polym. Comp.* (1996) 17, p. 104
29. Rohatgi, V., Patel, N., Lee, L.J. *Polym. Comp.* (1996) 17, 161
30. Lundström, T.S. *Polym. Comp.* (1996) 17, 770

7 Consolidation during Thermoplastic Composite Processing

Alfred C. Loos and Min-Chung Li

To overcome the shortcomings in thermosetting resins there is considerable interest in the use of thermoplastic resins as matrix materials for fiber-reinforced composites. Thermoplastic resins are high toughness materials; subsequently, they can improve the damage tolerance of composites. Fabrication costs of thermoplastic composites are potentially lower than are those for thermosetting composites due to reduced processing times and high-speed manufacturing methods. It has been established that thermoplastic composites consolidate by autohesive bonding of interply interfaces. The resulting autohesive bond strength is a function of the processing cycle to which the interply interfaces are subjected. This chapter reviews developments in modeling and analysis of thermoplastic composite consolidation. The effects of the processing cycle on mechanisms of intimate contact formation and polymer healing that occur during autohesive bonding are discussed.

7.1 Introduction

The use of advanced composites has increased significantly in the last decade. The properties of high-specific strength and stiffness make composites ideal for many aerospace, automotive, and infrastructure applications. Fiber-reinforced composites, which commonly use thermosetting resins such as epoxies as the matrix material, have some inherent deficiencies. These include the need for multistep processing, limited shelf-life, low toughness, sensitivity to moisture, and the inability to reprocess or reform the material [1].

Considerable attention has been focused on the use of tough, high-temperature, solvent-resistant thermoplastic polymers as matrix materials for fiber-reinforced composites, as an alternative to the previously mentioned thermosetting resin. Thermoplastic resin systems have shown potential for reducing manufacturing costs [2]. The increased toughness of thermoplastic resins can subsequently improve the damage tolerance of composites [1]. The mechanisms occurring during consolidation of thermoplastic matrix composites, however, are quite different than the mechanisms observed during curing of thermosetting matrix composites. Extreme tow height nonuniformity and lack of flow make thermoplastic prepregs more difficult to process than thermosetting prepregs. Unlike thermosetting prepregs, which rely on low viscosity and high flow of the resin to coalesce the ply interfaces, thermoplastic matrix prepregs must be physically deformed to produce intimate contact and coalescence of the ply interfaces. Table 7.1 shows a comparison between thermosetting and thermoplastic matrix composites.

Techniques commonly used to manufacture advanced composites from prepregs, which are available in unidirectional fiber or fabric form, include compression molding, vacuum-bag autoclave molding, stamp molding, thermoforming, and on-line (in situ) consolidation [3]. In compression and vacuum-bag autoclave molding techniques, the prepreg is cut to the dimensions of the structure, and the fiber directions are oriented to obtain the required

Table 7.1 Comparison of Thermosetting and Thermoplastic Matrix Composites

	Thermosetting resin composites	Thermoplastic resin composites
Neat resin	Minimum viscosity is low (~10 poise)	Minimum viscosity is high (~10^4 poise)
	Flow of resin is high	Flow of resin is very low
Prepreg (or raw materials)	Uniform	Nonuniform
	Requires refrigerated storage	Ambient storage anywhere
Processing	Long processing time	Short processing time
	Chemical reaction during cure (irreversible)	No reaction during consolidation (reversible)
	Low processing temperature (120–180°C)	High processing temperature (260–400°C)
Laminates (or finished parts)	Cannot be reformed	Multiple reforming possible
	Absorb moisture	No moisture absorption (depending on materials)

mechanical properties of the finished product. The lay-up is then placed in a rigid mold (compression molding) or vacuum-bagged (autoclave molding) and processed at an elevated temperature and pressure for a known period of time. The specific combination of processing temperature, pressure, and time is usually called a *processing cycle*. A typical processing cycle for a thermoplastic composite is shown in Figure 7.1. The composite is heated at a constant rate from room temperature to an elevated temperature above the glass transition temperature (T_g) of an amorphous thermoplastic matrix resin or the melting temperature (T_m) of a semicrystalline thermoplastic matrix resin. The consolidation pressure is applied at some point during the processing cycle. The composite is held for a specified period of time under elevated temperature and pressure to allow for intimate contact, coalescence, and complete autohesive bonding of the ply interfaces. At this point the consolidated composite is cooled under pressure to below the T_g of the matrix.

In composite thermoforming, the prepreg is preconsolidated into a laminated sheet. The laminate is preheated to a temperature above the melting/softening point of the resin. The laminate is then rapidly transferred to a matched mold and formed into final shape under pressure. In vacuum thermoforming, the preheated laminate is clamped and sealed to the mold. Application of vacuum to the mold cavity causes the heat-softened laminate to conform to the female mold surface. The advantages of thermoforming are fast cycle times, low forming pressures, and inexpensive tooling [4].

On-line consolidation is a composite manufacturing process where the resin-impregnated fiber bundles ("towpreg" or "prepreg') are continuously oriented, laid down, and consolidated onto the tool surface in a single step. Once the surface of the designed structure has been covered and the thickness has been achieved, the part is finished. Secondary processing steps, such as autoclave or hot-press consolidation, are eliminated. The process can be fully automated when integrated with a computer-controlled system, which leads to further cost savings in fabrication by increasing productivity and reducing labor cost.

In addition to the reduced cost, on-line consolidation also offers benefits for design flexibility and performance. With localized heating, this process is inherently suitable for manufacturing parts with large surfaces and moderate curvatures, such as fuselage structures and deep submersibles [5]. Because the towpreg is fully consolidated and locked in the vicinity of the melting zone as it is placed onto the structure, there is no conceptual limitation

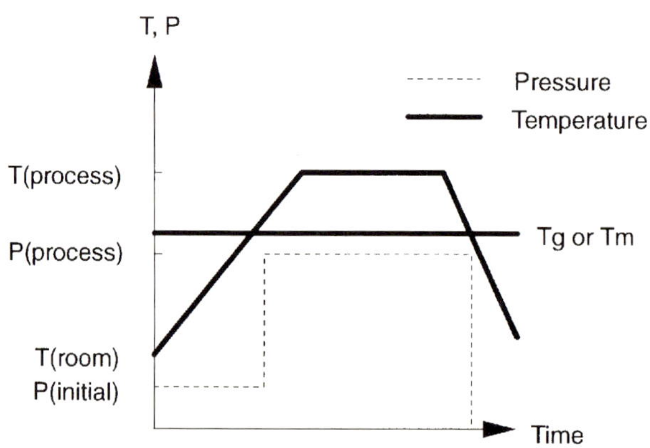

Figure 7.1 Typical processing cycle for thermoplastic composites

on producing parts with thick cross-sections and large surface areas [6]. Furthermore, complex, nongeodesic, and even concave winding paths are also achievable, thus allowing design flexibility [7].

On-line consolidation techniques are commonly used in manufacturing processes, such as filament winding and tape laying. Despite differences in the machines used to implement these two manufacturing processes, similar procedures are required to ensure complete bonding between the composite layers. The basic components of on-line consolidation process are illustrated in Figure 7.2.

A focused heat source is aimed at the interface between incoming towpreg and substrate to create a molten zone. The intensity of the heat source must be properly adjusted to ensure melting on both mating surfaces. Once the proper molten zone has been created by the heating system, adequate pressure must be applied via fiber tension and/or roller compaction to compress the uneven melting surfaces and to squeeze the matrix to fill the gaps. Intimate contact and diffusion bonding in the interface can subsequently be established. The roller pressure should be held constant until the temperature at the exit of the molten zone drops below the melting/softening point of the resin to prevent void formation by either volumetric shrinkage or the release of spring energy from the fiber network.

Continuous operation of the on-line consolidation process is achieved by either rotating the mandrel and moving the carriage during filament winding or, in the case of tape laying, by moving the lay-up head. The speed of on-line consolidation, however, is limited by the time required for intimate contact and diffusion bonding. In general, the speed of consolidation, the heat intensity, and the roller pressure are used to describe the processing window of on-line consolidation.

Because the towpreg must be heated to well above the melting/softening temperature in seconds, highly focused heaters must be used to guarantee melting of the mating surfaces. Excessive heating, however, may lead to matrix degradation by polymer chain scission and system oxidation, and consequently may affect the crystallization, adhesion of the fiber/matrix interface, and matrix properties [8]. Appropriate selection and arrangement of the heating sources, therefore, are the most important factors for a successful on-line consolidation.

It is clear from the description of the commonly used fabrication techniques that thermoplastic composite consolidation is a nonisothermal manufacturing process that involves continuous heating, compacting, and solidification. The major mechanisms,

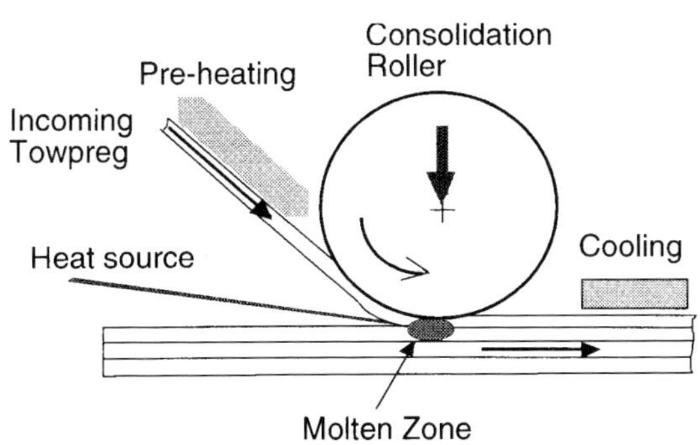

Figure 7.2 Schematic diagram of the on-line consolidation process

including intimate contact and diffusion bonding (healing), must be carefully investigated. It is believed that to form good contact and bonding, the previously mentioned mechanisms of consolidation must occur sequentially or simultaneously [8–11].

It is well recognized that the surfaces of prepregs or towpregs are uneven. When two surfaces are brought into contact without applying pressure, gaps exist at the interface. It is clear that polymer chains cannot diffuse without intimate contact. Providing a proper pressure field to compress the viscous matrix, fill the gaps, and form perfect contact are therefore essential for successful fusion bonding.

The time needed for polymer chains of amorphous thermoplastics above T_g and semicrystalline thermoplastics above T_m to diffuse across the interface and randomize is relatively short compared with the time needed for resin flow. It is believed, therefore, that diffusion bonding is completed immediately after the two molten surfaces merge, and that the microstructure of the contact zone is also assumed to be identical to that of the intraply sections [12,13].

The final step of thermoplastic composite consolidation is to cool and solidify the consolidated parts. It is well known that the physical and mechanical properties of composites are determined by the microstructure of the matrix in addition to the reinforcement, whereas the morphology of the thermoplastics is determined by its thermal history [14–17].

For semicrystalline thermoplastics, crystallization will occur during cooling. The equilibrium between crystalline and noncrystalline phases is delicate. Moreover, the presence of the reinforcing fibers, which provide a large surface for nucleation, further complicates the crystallization kinetics for a thermoplastic composite [8,14–16]. In order to thoroughly understand changes in the mechanical properties of thermoplastic composites processed under different conditions, we must study more than just the degree of crystallinity. Nucleation density, spherulite sizes, and the degree of crystalline perfection should also be taken into account [8]. Whereas the physical and mechanical properties of bulk thermoplastic composites are influenced by the entire processing cycle, the cooling rate has been identified as the most important influence on the morphology of the matrix and the degree of crystallinity. In general, slower cooling rates yield higher degrees of crystallinity, which correspond to an increase in tensile strength, compressive strength, and solvent resistance of the matrix [17].

From studies of thermoplastic composite processing, it has been recognized that individual prepreg plies consolidate by autohesive bonding of the ply interfaces [13,18]. The resulting bond strength is a function of the processing parameters (temperature, pressure, and time) to which the interface is subjected. It has also been established that there are two mechanisms governing the development of interply bonding: intimate contact formation and healing.

7.2 Intimate Contact

The mechanism by which a thermoplastic matrix composite consolidates to form a laminate was attributed to autohesive bond formation between plies [13,18]. Autohesion, however, can only occur after the two surfaces have coalesced (i.e., are physically in intimate contact). Macromolecules cannot diffuse across spatial gaps at the interface. The study in this section identifies the mechanisms by which the interfaces of a thermoplastic prepreg coalesce, which results in intimate contact. The effects of various processing parameters on the degree of intimate contact will be discussed. In the next section, models that have appeared in the literature that can be used to describe the intimate contact process during fabrication of thermoplastic matrix composites will be presented. These modeling efforts include intimate contact at the prepreg–towpreg ply interfaces during traditional molding processes such as compression molding and vacuum-bag–autoclave molding and during on-line (in situ) consolidation processes such as tape laying and filament winding.

7.2.1 Literature Review

Models of the intimate contact process that have appeared in the literature are commonly composed of three parts or submodels. The first submodel is used to describe the variation in the tow heights (surface waviness or roughness) across the width of the prepreg or towpreg. The second submodel, which is used to predict the elimination of spatial gaps and the establishment of intimate contact at the ply interfaces, relates the consolidation pressure to the rate of deformation of the resin impregnated fiber tow and resin flow at ply surface. Finally, the third submodel is the constitutive relationship for the resin or resin-saturated tow, which gives the shear viscosity as a function of temperature and shear rate.

Loos and Dara [13,18] developed an intimate contact model for composite laminates fabricated from graphite-polysulfone prepreg. The tow heights were measured with a micrometer across a 304.8 mm (12 in.)–wide prepreg sheet, perpendicular to the fibers. The measurements were made once every foot over a 5-ft length of the prepreg. A histogram of the tow height data was constructed, representing the percentage of tows within an interval of tow heights, and a two-parameter Weibull function was fit to the histogram. The resin-impregnated fiber tows were treated as a homogeneous fluid and the deformation of tow height was modeled by a viscometric shear-thinning flow in the transverse fiber direction. The fiber–matrix mixture viscosity was expressed by the power-law model and was obtained by multiplying the neat resin viscosity by a reinforcement–viscosity influence factor. During deformation, the tows are assumed to act independently concerning the disruption of flow, and in unison concerning the input of loading. The load was distributed evenly over only those tows that were in contact. The calculated degree of intimate contact was fit to the measured degree of intimate contact of $[0°/90°/0°]_T$ laminates by varying the power law parameters and the reinforcement/viscosity influence factor. The degree of intimate contact was measured using the ultrasonic C-scan technique on $[0°/90°/0°]_T$ panels. The threshold value of the relative signal attenuation was determined based on locating gaps in graphite/epoxy composites.

Lee and Springer [19] simplified the approach of Loos and Dara [13,18] in the development of an intimate contact model for graphite-PEEK (APC-2) prepreg. They represented the irregular ply surface (prepreg tow height variation) by a series of uniform rectangular elements of constant height, width and spacing. Note that the spacing between elements represents the spatial gaps and the areas of incomplete contact observed at the ply interfaces. Application of pressure will cause a decrease in element height with a corresponding increase in element width with time until the spatial gaps are eliminated and complete contact is achieved. The element geometric parameters (height, width, and spacing) were measured from photomicrographs of the cross-section of an uncompressed ply. The deformation of the rectangular elements was modeled as a laminar, one-dimensional "squeezing" flow between two parallel plates. The viscosity of the fiber–matrix mixture was obtained by matching the model to the experimental data of intimate contact versus time. The viscosity versus temperature relationship was described by an Arrhenius-type equation. In the verification of their model, Lee and Springer compressed a single prepreg ply in a mold under different temperature, pressure and time conditions. The prepreg ply was removed from the mold after processing, and the compressed area measured. The degree of intimate contact was determined by taking the ratio of the compressed area to the total area. The model matched the data well.

Mantell and Springer [20] modified the model developed by Lee and Springer [19] to simulate the manufacturing of thermoplastic composites by tape laying, filament winding, and compression molding. The geometric parameters describing the prepreg surface nonuniformity were combined into a single nondimensional parameter. This parameter was then determined by matching the model to a single data point. The fiber–matrix viscosity was assumed independent of shear rate and directly measured between two parallel plates in shear at constant temperature. Viscosity data at different temperatures were fit to an Arrhenius temperature dependence. For compression molding, Mantell and Springer modified the Lee-Springer model to predict intimate contact at the ply interfaces when the applied temperature and pressure vary with time during processing. For the tape-laying process, the Lee-Springer model was extended to predict the degree of intimate contact as a function of applied force, linear speed, and width of the compaction roller. By replacing the linear speed of the roller with the rotational speed of the mandrel and the outer radius of the cylinder, the degree of intimate contact for filament winding can be calculated.

The modified Mantell-Springer models for compression molding in a press and in tape laying were verified by fabricating lap shear and short-beam shear specimens made of graphite-PEEK (APC-2) unidirectional tape [21]. The degree of intimate contact was determined from C-scan measurements and defined as the area in contact divided by the total area. The procedure used to determine the C-scan threshold values, however, was not specified. Overall agreement between data and model results was very good.

Li and Loos [22,23] used a surface topology characterization machine to measure the waviness or roughness of the prepreg surfaces and directly measured the geometric parameters (i.e., height, width, and spacing of the rectangular elements representing the tow height variation across the width of the prepreg) required for solution of the Lee-Springer intimate contact model. As one would expect, the prepreg surface roughness depends on the fiber type, the matrix resin, and the prepregging process. By directly measuring the height, width, and spacing of the rectangular elements that represent the prepreg geometric nonuniformity and using a constitutive model based on the neat resin, the degree of intimate contact at the ply interfaces could be determined without the use of experimentally fitted parameters. In addition, Li and Loos used a two-step application of the Lee-Springer intimate contact model to predict intimate contact of a [0°, 90°] cross-ply interply interface.

The Li-Loos intimate contact model was verified for compression molded unidirectional graphite-polysulfone and graphite-PEEK (APC-2) laminae and graphite-PEEK (APC-2) cross-ply laminates. The degrees of intimate contact of the unidirectional and cross-ply specimens were measured by optical microscopy and scanning acoustic microscopy, respectively. The predicted degrees of intimate contact agreed well with the measured values for both the unidirectional and cross-ply specimens processed at different temperature and pressures.

The model of the intimate contact process developed for compression molding by Lee and Springer [19] and later modified for tape laying by Mantell and Springer [20] has been applied to in-situ consolidation of thermoplastic composites [24–26]. In these investigations, the intimate contact submodel was part of a comprehensive model of the automated tow-placement process. Pitchumani et al. [24], employed the Mantell-Springer intimate contact model, but used the actual temperature and pressure distributions in the nippoint beneath the compaction roller. The authors claimed that use of average pressure and a constant temperature at the nippoint could result in a 26 percent error. The pressure distribution under the roller was determined by considering bulk consolidation of the entire layer, whereas the temperature distribution was calculated by performing a heat transfer analysis of the tow placement head. The expression for degree of intimate contact was simplified for long contact times (i.e., initial contact area neglected), and the geometric parameters were combined into a single parameter denoted as the "roughness parameter" by Butler et al. [25]. Values of the roughness parameter were determined by experiment for various prepreg materials.

In the in situ consolidation model of Liu [26], the Lee-Springer intimate contact model was modified to account for the effects of shear rate–dependent viscosity of the non-Newtonian matrix resin and included a contact model to estimate the size of the contact area between the roller and the composite. The authors also considered lateral expansion of the composite tow, which can lead to gaps and/or laps between adjacent tows. For constant temperature and loading conditions, their analysis can be integrated exactly to give the expression developed by Wang and Gutowski [27]. In fact, the expression for lateral expansion was used to fit tow compression data to determine the temperature dependent non-Newtonian viscosity and the power law exponent of the fiber–matrix mixture.

Single-lap shear-test specimens were manufactured by in situ consolidation of graphite-PEEK unidirectional tow and used to verify the analysis for the interply bonding at the ply interfaces. The degree of bonding was defined as the ratio of the lap shear strength of the test interface to the lap shear strength of the specimens postprocessed in an autoclave. The predicted degrees of bonding compared reasonably well with the measured values for different torch temperatures and consolidation loads. The authors unfortunately did not discuss how they determined numerical values of the prepreg geometric parameters required for solution of the intimate contact model.

7.2.2 Intimate Contact Model

In the preceding section, a brief review of efforts to model the intimate contact process during consolidation of thermoplastic composites was given. Models based on the original work of

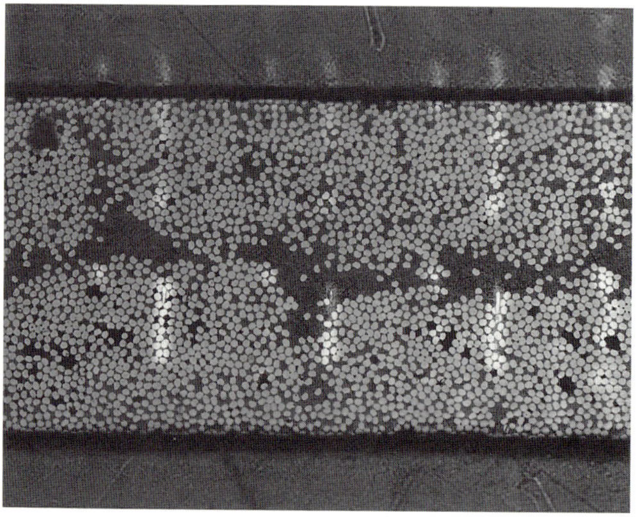

Figure 7.3 Photomicrograph of two prepreg plies after consolidation

Lee and Springer [19] and modified by Li and Loos [22,23] to include unidirectional and cross-ply lay-ups will be used to demonstrate the effects of various geometric, material, and processing parameters on the degree of intimate contact.

The uniformity of the prepreg microstructure and surface roughness are the critical parameters that govern selection of the processing pressure and time for consolidation. A photomicrograph of two prepreg plies during consolidation shown in Figure 7.3. A reasonable model of the prepreg is to assume that the prepreg consists of a fiber and resin core, possibly containing intraply voids, with resin-rich surfaces (Fig. 7.4). Spatial gaps due to prepreg surface roughness are eliminated during processing by localized deformation and flow at the resin rich surfaces.

7.2.2.1 Single Prepreg Ply Model

Lee and Springer [19] modeled the resin-rich prepreg surface roughness by a series of uniform rectangular elements, as shown in Figure 7.5. At the beginning of the consolidation

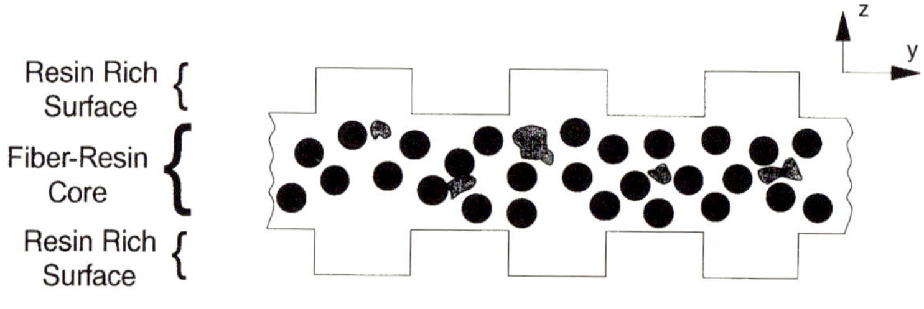

Figure 7.4 Prepreg model

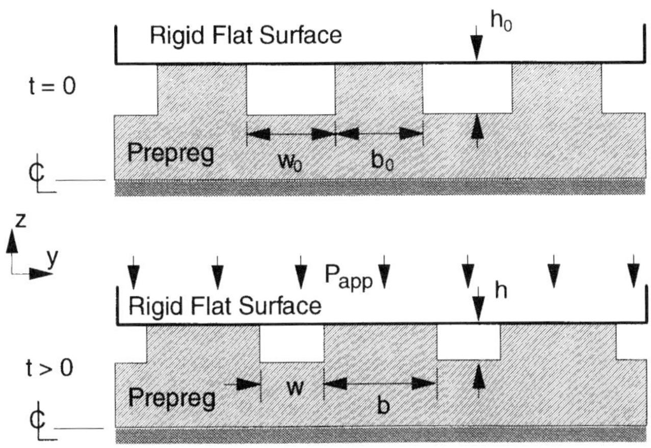

Figure 7.5 Deformation of the rectangular elements against a rigid flat surface

process (time, $t = 0$), the initial height, width, and spacing of the rectangular elements are given by the dimensions h_0, b_0, and w_0, respectively. Application of the consolidation pressure results in deformation of the rectangular elements, representing deformation and flow of the resin-rich prepreg surfaces (Fig. 7.5). Deformation and flow continues until the width of the elements is equal to the sum of the initial width and spacing of the elements. The surfaces are in intimate contact at this point.

The degree of intimate contact can therefore be defined as [19]:

$$\text{Dic} = \frac{b}{w_0 + b_0} \tag{7.1}$$

where b is the instantaneous width of the rectangular elements at the time t. The deformation and flow of the rectangular elements can be modeled as a squeezing flow between two rigid parallel plates that are of infinite length in the x-direction. Assuming that the resin viscosity is independent of shear rate, an expression for the degree of intimate contact can be written as [19]:

$$\text{Dic} = \frac{1}{1 + \dfrac{w_0}{b_0}} \left[1 + \frac{5P_{\text{app}}}{\eta_0} \left(1 + \frac{w_0}{b_0} \right) \left(\frac{h_0}{b_0} \right)^2 t \right]^{1/5} \tag{7.2}$$

where P_{app} is the nominal applied consolidation pressure and h_0 is the zero-shear-rate viscosity of the resin, which is a function of processing temperature. For a specified applied pressure P_{app}, Equation 7.2 can be used to calculate the degree of intimate contact as a function of time. If the initial element width and spacing are equal, Equation 7.2 can be simplified to:

$$\text{Dic} = \frac{1}{2} \left[1 + \frac{10P_{\text{app}}}{\eta_0} \left(\frac{h_0}{b_0} \right)^2 t \right]^{1/5} \tag{7.3}$$

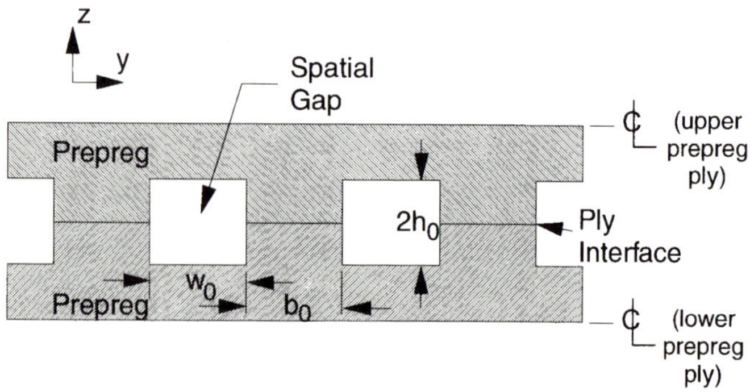

Figure 7.6 A schematic representation showing two unidirectional prepreg plies in contact at the beginning of consolidation

Equations 7.2 and 7.3 represent expressions for the degree of intimate contact of a single prepreg ply in contact with a smooth rigid surface (Fig. 7.5).

7.2.2.2 Unidirectional Interply Interface Model

A schematic representation of two unidirectional prepreg plies in contact at the beginning of consolidation is shown in Figure 7.6. For this case, it is assumed that there is no nesting of the rectangular elements and that adjacent plies are aligned to produce the largest spatial gaps (i.e., the rectangular elements of the lower ply match the rectangular elements of the upper ply). If interfacial wetting is instantaneous, then the deformation and flow of the stacked rectangular elements can be modeled as a squeezing flow between two rigid parallel plates which are of infinite length in the x-direction. The initial element height therefore becomes $2h_0$, and Equation 7.3 then becomes:

$$\mathrm{Dic} = \frac{1}{2}\left[1 + \frac{10P_{\mathrm{app}}}{\eta}\left(\frac{2h_0}{b_0}\right)^2 t\right]^{1/5} \tag{7.4}$$

7.2.2.3 Cross-Ply Interply Interface Model

Shown in Figure 7.7 is a cross-ply interply interface in which a 90 degree ply overlays a 0 degree ply. The shaded strips represent the rectangular elements. The strips in the y-direction belong to the upper prepreg ply surface, which faces down upon the lower prepreg ply surface represented by the strips in the x-direction. In Figure 7.7, therefore, the areas where two shaded strips intersect each other are the initial contacts. The remaining shaded areas are where the rectangular elements of one ply and the spatial gaps of the other ply are overlaid.

By closely examining the layout in Figure 7.7, a representative volume element can be selected for study, as shown in Figure 7.8. Region A, which comprises 25 percent of the area of the representative volume element, is in initial contact. When the consolidation pressure is applied, the deformation is initiated in region A, and the resin flows in the x–y plane to fill the gaps in regions B, C, and D.

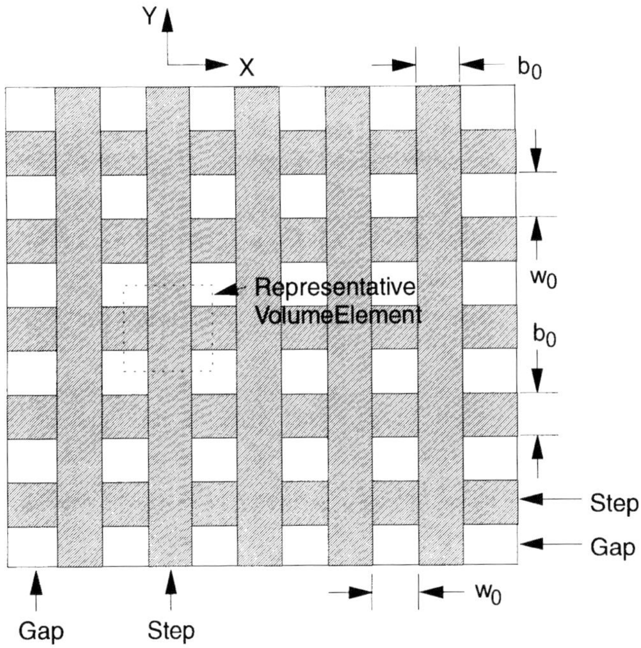

Figure 7.7 Cross-ply interply interface region

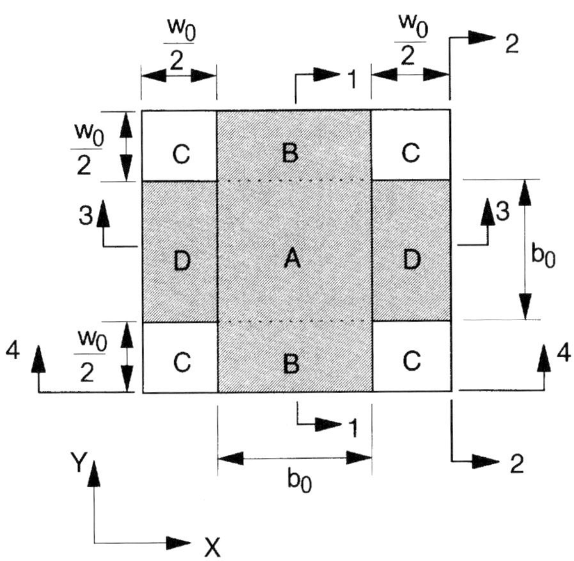

Figure 7.8 Representative volume element used in modeling the intimate contact achievement of a cross-ply interply interface

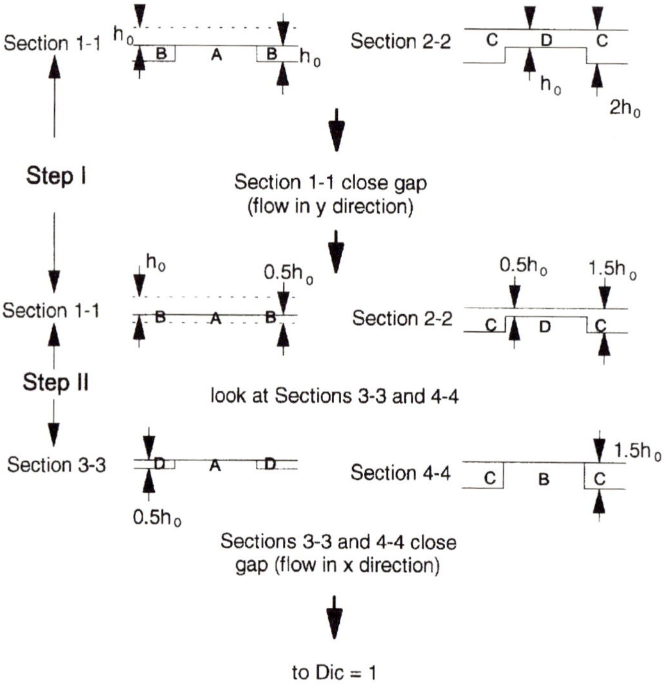

Figure 7.9 Representation of the two-step intimate contact process of a cross-ply interply interface

For simplicity, a two-step intimate contact process that describes the flow mechanisms at the resin-rich surfaces was developed. Figure 7.9 shows the two-step intimate contact process for the cross-ply interply interface. In step-I, flow is assumed to occur only in the y-direction to fill the gaps of h_0 in height in the 1–1 cross-section (regions B–A–B). No flow occurs in the 2–2 cross-section (region C–D–C) because no areas are in contact. After step I is completed, regions A and B are in contact; However, regions C and D are not yet in intimate contact. In step-II, localized resin flow occurs in the x-direction from region A and B to fill the gaps in regions D and C, $0.5h_0$ and $1.5h_0$ in height respectively, as shown in Figure 7.9. After step-II is accomplished, complete intimate contact at the cross-ply interply interface is achieved.

7.2.2.4 Surface Characterization of the Prepreg Plies

A critical parameter in the establishment of interply intimate contact is the surface roughness or waviness due to variations in the tow heights across the width of the prepreg. It is important, therefore, to be able to characterize the roughness of the prepreg surface. The parameters that influence the prepreg surface roughness include the fiber type, tow bundle size, matrix resin, and prepreg manufacturing method.

Techniques that have been used to measure the prepreg surface roughness were mentioned in Section 7.2.1 and include direct measurement with a micrometer and photo-micrographs of the prepreg cross section. Li and Loos [22,23] showed that a surface topology characterization machine (Talysurf 4) can be used to measure the waviness or roughness of the resin rich

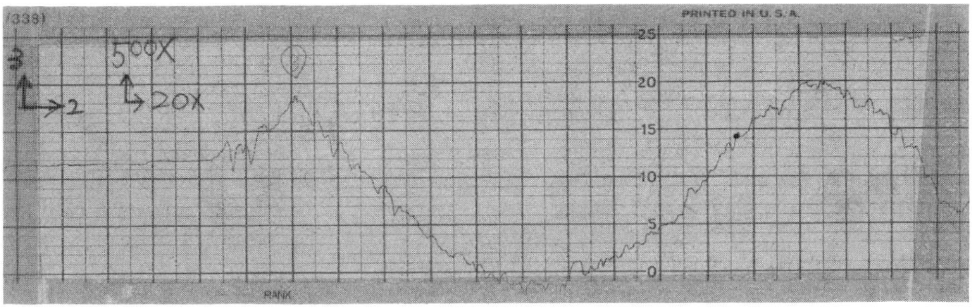

Figure 7.10 A typical result of the Talysurf 4 measurement on T300/P1700 prepreg

prepreg surfaces. Measurements are made across the width of the prepreg sheets, perpendicular to the fibers. Figure 7.10 is a typical result of the measurement for graphite-polysulfone prepreg. A typical measurement scanned about 9 mm along the width of the prepreg. Two or three measurements are made along a prepreg sample that was 25.4-mm wide. The measurements were magnified linearly by 20× in the transverse fiber direction (two-direction) and by 500× in the out-of-plane direction (three-direction). From the examination of the topology measurement, the surface waviness (roughness) is sinusoidal, and is represented by a dashed line in Figure 7.11.

For modeling purposes, the surface waviness is represented by a series of rectangular elements of height h_0, width b_0, and spacing w_0, shown as the solid line in Figure 7.11. The height of the rectangular elements (h_0) is equal to two times (twice) the amplitude of the sinusoidal wave, 2A. The width (b_0) and the spacing (w_0) of the rectangular elements are equal to one half of the wavelength of the sinusoidal wave, 0.5 P.

Prepreg surface characterizations for three different types of prepreg are shown in Table 7.2. The prepregs include T300/P1700 graphite–polysulfone prepreg, AS4/PES graphite–polyethersulfone prepreg, and (APC-2) AS4/PEEK graphite–polyetheretherketone prepreg. Among these three, T300/P1700 and AS4/PES are amorphous thermoplastic prepregs. Both T300/P1700 and AS4/PES prepregs were in the form of a 304.8 mm (12 in.)–wide sheet.

APC-2 is a semi-crystalline thermoplastic prepreg. The surface roughness characterization for two different batches of APC-2 prepreg are reported in Table 7.2. Batch I is a 152.4 mm (6 in.)–wide prepreg sheet, whereas batch II is 304.8-mm wide. In addition, the

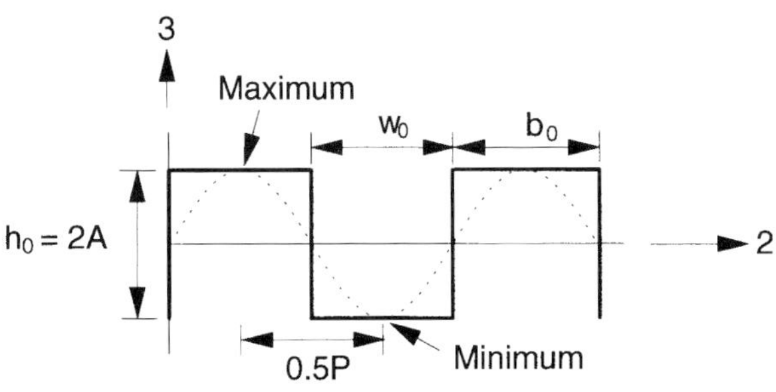

Figure 7.11 Prepreg surface roughness modeling

Table 7.2 Prepreg Surface Characterization Results

	Height (h_0) (mm) average ± standard deviation	Width and spacing (b_0, w_0)* average ± standard deviation
T300/P1700	0.05927 ± 0.02245	2.311 ± 0.655
AS4/PES	0.02152 ± 0.01084	2.2166 ± 1.6509
APC-2 Batch I	0.01263 ± 0.01084	1.3034 ± 0.8125
APC-2 Batch II Cross-Section 1	0.01512 ± 0.01676	1.2781 ± 0.666
APC-2 Batch II Cross-Section 2	0.02067 ± 0.01132	1.1482 ± 0.7212

[a] b_0 and w_0 are about one half of the fiber tow width.

topology measurements were made at two different sections of the APC-2 Batch II prepreg. The two cross-sections were about 1.83 m (6 ft) away from each other along the length of the prepreg.

From Table 7.2, we can see that T300/P1700 prepreg has both the greatest tow height and spacing of the rectangular elements. APC-2 Batch I prepreg has the smoothest surface. This data was supported by optical micrographs of the prepreg cross-sections [22]. In addition, the two batches of APC-2 prepreg have different surface roughness characteristics. Little change in the surface roughness characteristics was observed along the fiber direction of APC-2 Batch II prepreg. The Talysurf 4 surface topology characterization machine has proven to be a useful instrument for quickly measuring the surface roughness of a prepreg material. In Section 7.25 we will show the effects of prepreg surface roughness on the development of intimate contact.

7.2.3 Intimate Contact Measurements

Probably the two most commonly used techniques for measuring the overall quality of the composite consolidation are optical photomicrographs and through transmission C-scan. Both of these techniques can be readily adapted to measuring the degree of intimate contact at the ply interfaces.

An example of an optical photomicrograph of an interply interface was shown in Figure 7.3. This method is quite accurate because all of the ply interfaces can be directly examined. Each composite laminate, however, must be sectioned into small specimens, and each specimen must be potted and carefully polished. In addition, at high magnifications, each micrograph covers just a few millimeters of the length of the interply interface. The procedure is therefore very time consuming, so there is a practical limitation as to the number of specimens that can be examined. To reduce some of the work involved in measuring intimate contact, the micrographs can be digitized and image analysis software can be used to measure the void content (areas of incomplete contact) of the ply interfaces.

The ultrasonic C-scan technique is the most widely used nondestructive method of locating defects in the composite microstructure. The through transmission C-scan is easy to implement and a large composite panel can be scanned in a matter of minutes. The problem with this technique is that a C-scan cannot reveal the type of defect present. Hence, there is no way to determine if a flaw detected by the C-scan is due to incomplete contact of an interply interface or some other type of defect in the composite microstructure.

The most efficient approach to measuring intimate contact of a multiple ply composite laminate would probably be to first use the C-scan technique or some other nondestructive method to determine the location of any flaws in the panel. The cross-sections of the panel that contain the flaws can then be examined by preparing optical micrographs of those areas, and the interply interface examined for complete contact.

In the following example, scanning acoustic microscopy is combined with optical microscopy to measure intimate contact at the ply interfaces of a $[0^\circ/90^\circ/0^\circ]_T$ graphite-PEEK laminate. First, eight holes, 1.5 mm (0.059 in.) in diameter, were drilled into the composite specimen. These holes, shown in Figure 7.12, were used to locate where on the composite specimen the scanning acoustic microscope images were taken.

Scanning acoustic microscopy images of each of the two interply interfaces were obtained from the specimen. A typical image is shown in Figure 7.13 and encompasses an interface area of 20 mm × 15 mm. The black and white in the image corresponds to the "good" and "bad" interface regions, respectively. The holes that were drilled into the specimens to help locate defects in the image of the interply interface showed up as black circles in the acoustic microscope images. An apparent defect in each image was selected and the corresponding cross-section of the specimen was examined using optical microscopy. The defect was identified and its thickness was measured. The gray-scale value of the scanning acoustic microscopy image at the location of the defect was then determined using the image analysis

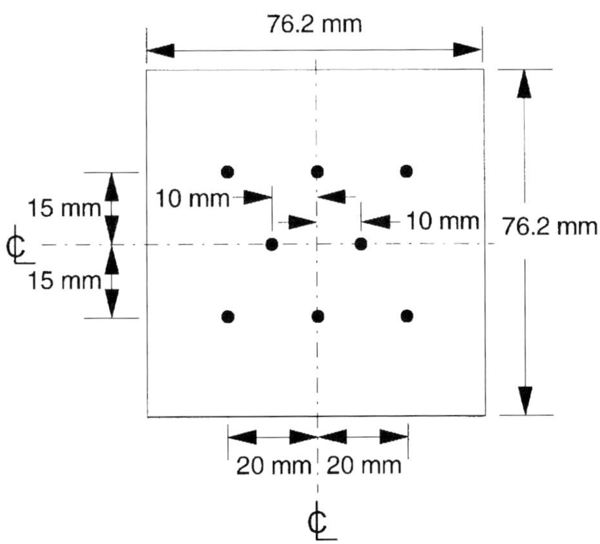

Figure 7.12 Locations of the holes drilled in the $[0/90/0]_T$ APC-2 specimens

Figure 7.13 Typical scanning acoustic microscopic image obtained from a $[0/90.0]_T$ APC-2 laminate processed at 370°C with 276 kPa for 50 s

system. This gray-scale value was recorded as the threshold value. The defect thickness was then correlated with the threshold gray-scale value. The gray-scale values across the whole scanning acoustic microscopy image were then measured and analyzed by the image analysis system. Any areas in the image where the gray-scale value was greater (paler) than the threshold value were determined to be voids. The remaining area with a lower gray-scale value (darker) than the threshold value was determined to be in intimate contact. The degree of intimate contact for each image was defined as the ratio of the intimate contact area to the total image area, excluding the holes. This procedure was used to obtain the intimate contact data from the $[0°/90°/0°]_T$ graphite-PEEK composites reported in the next section.

7.2.4 Model Verification

Composite specimens fabricated from graphite-polysulfone prepreg (T300/P1700) and graphite-PEEK (APC-2) prepreg were compression molded and the degrees of intimate contact of the consolidated panels were measured using the methods described in Section 7.2.3. The data are compared with the predicted degrees of intimate contact in order to assess the validity of the model for interply interfaces with different ply orientations.

7.2.4.1 Single Prepreg Ply

Intimate contact data reported by Lee and Springer [19] for a single prepreg ply against a rigid surface are compared with model predictions (see Eq. 7.3) in Figure 7.14. The prepreg material was APC-2. The surface characterization parameters for APC-2, Batch I prepreg in Table 7.2 and the zero-shear-rate viscosity for PEEK resin, reported in Reference 23, at the processing temperature were used as input in the intimate contact model. As can be seen in

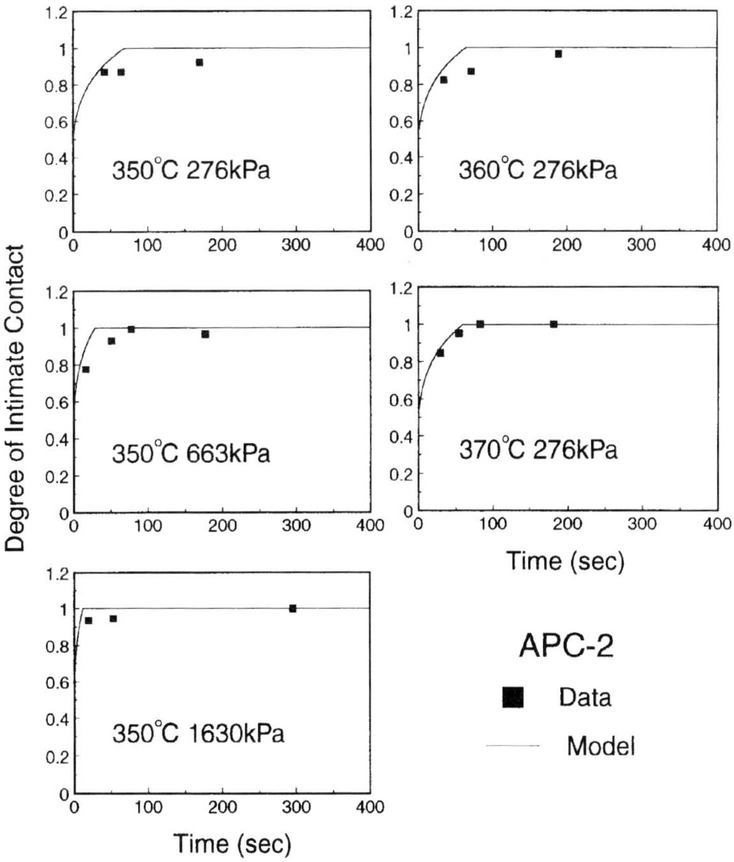

Figure 7.14 Degree of intimate contact (Dic) versus time for single-ply APC-2 samples

the figure, the calculated degree of intimate contact agrees well with the measured degree of intimate contact for samples consolidated at different temperatures and pressures.

7.2.4.2 Unidirectional Interply Interface

The data shown in Figure 7.15 were obtained from two-ply T300/P1700 unidirectional specimens that were compression molded in a 76.2 mm (3 in.) square steel mold. The cross-sections of the consolidated specimens were examined by optical microscopy and the degree of intimate contact was determined as the amount of the interply region that was in contact divided by the total area of the cross-section. Additional details of the experimental procedures are given in Reference 22.

The unidirectional interply interface model (see Eq. 7.4) was used to calculate the degree of intimate contact versus time for unidirectional samples consolidated at different temperatures and pressures. The surface characterization parameters for T300/P1700 and

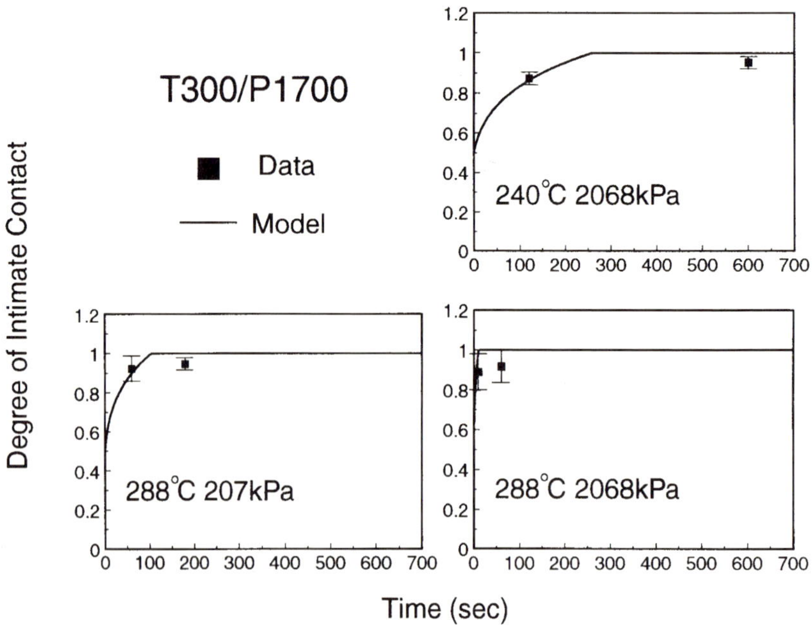

Figure 7.15 Degree of intimate contact (Dic) versus time for two-ply unidirectional T300/P1700 samples

the zero-shear-rate viscosity data of the polysulfone resin reported by Loos and Dara [13] were used as input for modeling intimate contact achievement of prepreg.

The results of the intimate contact model are compared with the experimental data in Figure 7.15. Again, there is good agreement between the calculated and the measured degree of intimate contact. The data show that for the processing conditions used in the experiments the maximum degree of intimate contact that can be achieved is about 97 percent. The 3 percent discrepancy may be due to voids that were trapped in the interply interface. The voids appeared as regions of incomplete contact in the photomicrographs. It is possible that higher consolidation pressures or longer times will be required to collapse the entrapped voids.

7.2.4.3 Cross-Ply Interply Interface

The intimate contact data shown in Figure 7.16 were obtained from three-ply, APC-2, $[0°/90°/0°]_T$ cross-ply laminates that were compression molded in a 76.2 mm (3 in.) square steel mold. The degree of intimate contact of the ply interfaces was measured using scanning acoustic microscopy and image analysis software (Section 7.4). The surface characterization parameters for APC-2 Batch II prepreg in Table 7.2 and the zero-shear-rate viscosity for PEEK resin were input into the intimate contact model for the cross-ply interface. Additional details of the experimental procedures and the viscosity data for PEEK resin are given in Reference 22.

Considering the complexities involved in the formation of the cross-ply interface during consolidation, the cross-ply interply interface model predicted degrees of intimate contact

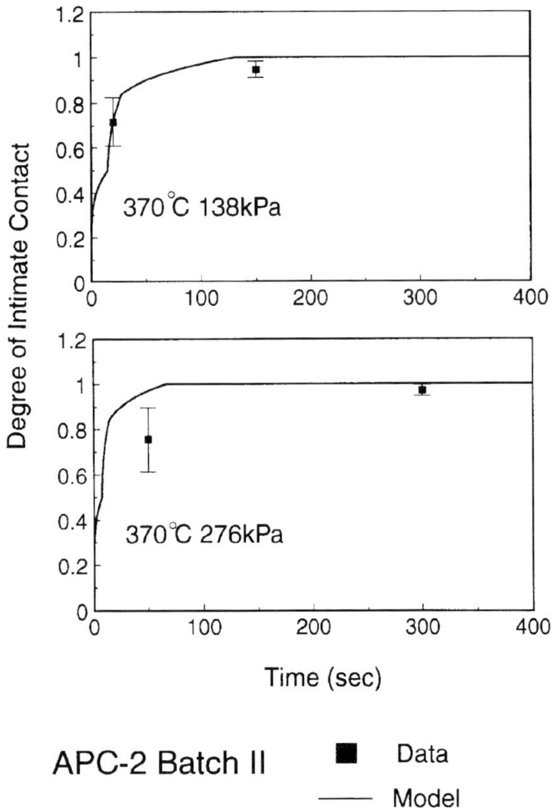

Figure 7.16 Degree of intimate contact (Dic) versus time for $[0/90/0]_T$ APC-2 Batch II samples at the same temperature

agree reasonably well with the measured values. Once again the maximum measured degree of intimate contact was between 0.95 and 0.97. The approximately 3–5 percent of the area not in contact was due to the presence of voids at the interply interface. The most likely cause of the voids are the low consolidation pressures and short processing times used to consolidate the laminates in the verification experiments. The experiments used a maximum pressure of 276 kPa, applied for 5 min. The manufacturer recommends a pressure of 1379 kPa, applied for about 5 min, which would most likely provide sufficient pressure to collapse any remaining interply or intraply voids. The low pressures were used in the experiments in order to obtain intimate contact data at the beginning of the consolidation process and generate experimental degree of intimate contact versus contact time curves shown in Figure 7.16.

7.2.4.4 Angle–Ply Interply Interface

Very few studies have appeared in the literature that specifically address the intimate contact of an angle ply $[+\theta/-\theta]$ interface. In a study reported by Li and Loos [22], APC-2 laminates,

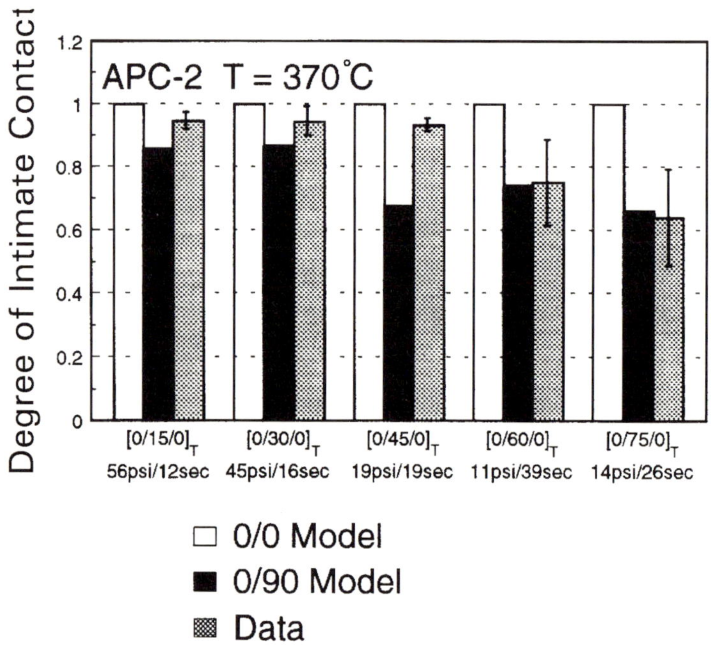

Figure 7.17 Comparison between models and the data for angle–ply interply interfaces

76.2 mm × 76.2 mm with the stacking sequence $[0°/\theta°/0°]_T$ were consolidated in a steel mold. The angles (θ) chosen for study were 15, 30, 45, 60 and 75 degrees. The experimental procedures for the cross-ply interface described in Section 7.4 were used to measure the degree of intimate contact of the angle ply interply interface. Comparisons between the unidirectional interply interface and the cross-ply interply interface models and the data are shown in Figure 7.17. It is observed that the unidirectional interply interface model can be used to estimate the degree of intimate contact for θ less than or equal to 45 degrees. For θ greater than 45 degrees, the cross-ply interply interface model seems to predict the degree of intimate contact very well.

7.2.5 Parametric Study

The intimate contact model was shown to give reasonable predictions of interply intimate contact development during thermoplastic composite consolidation. In this section, the model will be used to further examine the effects of the processing parameters, the surface roughness of the prepreg plies, and the ply orientation on the interply intimate contact achievement. Results from both the amorphous, graphite-polysulfone (T300/P1700) and the semicrystalline, graphite-PEEK (APC-2) prepregs will be reported. The consolidation time plotted in the figures, denoted t_c, represents the time required for the interply interface to achieve a degree of intimate contact of 0.97. Based on the results of the verification experiments, a degree of intimate contact of 0.97 can be readily obtained under all processing conditions.

Figure 7.18 shows the consolidation time versus consolidation pressure for unidirectional and cross-ply interply interfaces. As can be seen from both figures, the consolidation time decreases rapidly as pressure increases. Above 1000 kPa for T300/P1700 and 500 kPa for APC-2, increases in the consolidation pressure do not significantly decrease consolidation time.

Note that the differences in the processing characteristics between amorphous resin prepreg and semicrystalline resin prepreg can be clearly observed. The consolidation times are significantly longer for the amorphous thermoplastic prepreg than for the semicrystalline thermoplastic prepreg. This can be explained by the high melt viscosity of the amorphous resin and different surface characteristics of the T300/P1700 prepreg.

A comparison between the times required for consolidation of unidirectional and cross-ply lay-ups can be observed in Figure 7.18. The consolidation times required for the cross-ply lay-up are almost an order of magnitude higher than that required for the unidirectional lay-

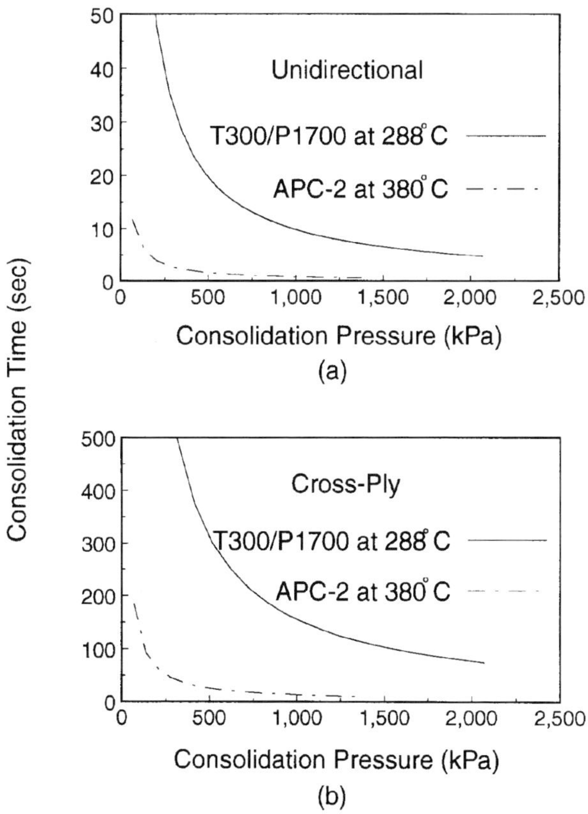

Figure 7.18 Consolidation time (t_c) versus consolidation pressure for T300/P1700 and APC-2 Batch II prepregs. (a) Unidirectional lay-up. (b) Cross-ply lay-up

up. This can be attributed to the increase in time required to achieve intimate contact of the 0-degree–90-degree interface.

The consolidation time versus processing temperature at two different consolidation pressures is shown in Figure 7.19 for T300/P1700 prepreg and in Figure 7.20 for APC-2, Batch II prepreg, respectively. The higher value of the pressure represents the manufacturer's recommended consolidation pressure for each prepreg. The consolidation time of the T300/P1700 prepreg decreases sharply as the processing temperature increases; however, the decrease in consolidation time at higher processing temperatures for APC-2 Batch II prepreg is minimal. This is because the zero-shear-rate viscosity (η_0) of P1700 polysulfone resin decreases significantly as the processing temperature rises above the glass transition temperature (T_g). Above the melt temperature of PEEK ($T_m = 343°C$), the zero-shear-rate viscosity remains relatively constant over the 30°C temperature range.

In Figure 7.21, the times required for consolidation of two different batches of APC-2 prepreg are presented. The influence of the prepreg surface roughness characteristics can be clearly observed. From the prepreg surface roughness characterization results, shown in Table 7.2, APC-2 Batch I prepreg and APC-2 Batch II prepreg have very close values of b_0 and w_0.

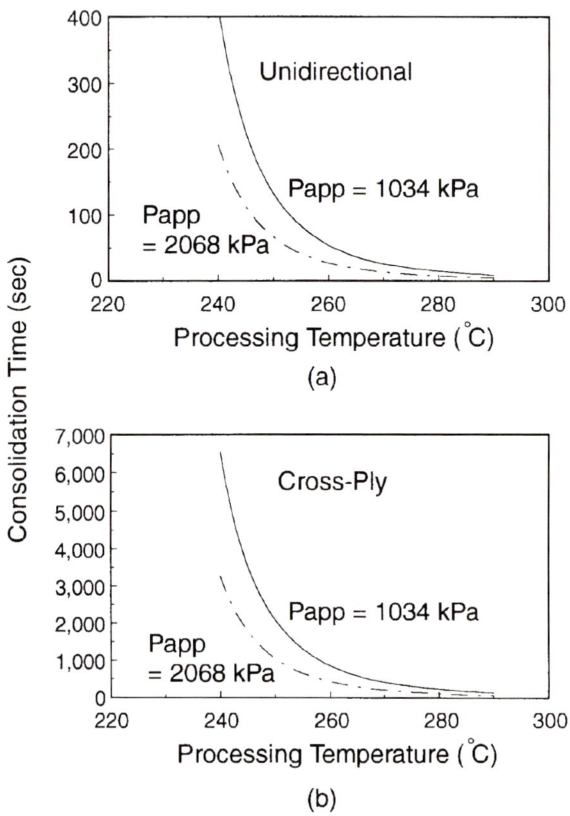

Figure 7.19 Consolidation time (t_c) versus processing temperature for T300/P1700 prepreg. (a) Unidirectional lay-up. (b) Cross-ply lay-up

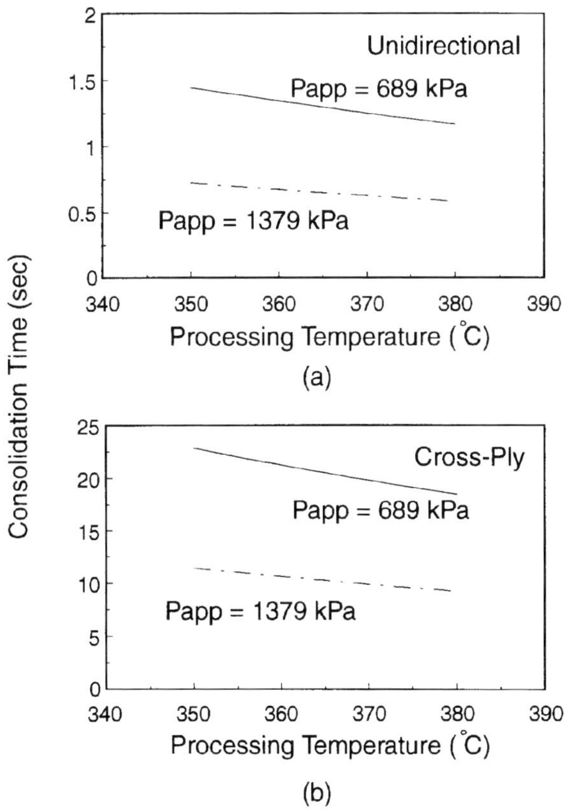

Figure 7.20 Consolidation time (t_c) versus processing temperature for APC-2 prepreg. (a) Unidirectional lay-up. (b) Cross-ply lay-up

APC-2 Batch II prepreg, however, has a larger value of h_0 than does APC-2 Batch I prepreg. As shown in Figure 7.21, the consolidation time needed for APC-2 Batch I prepreg is higher than that required for APC-2 Batch II prepreg. This is because the time to reach a specified degree of intimate contact depends on the ratio of b_0/h_0. APC Batch I has a higher ratio of b_0/h_0 then APC Batch II and hence requires longer times to reach complete contact.

7.3 Interply Bonding

The mechanism governing the formation of interply bonds has been established as autohesion or self-diffusion [28]. Autohesive bonding is controlled by two mechanisms: (1) intimate contact between the interfacial surfaces, and (2) diffusion of the macromolecules across the interface. Figure 7.22 shows the phenomenon of autohesion for an amorphous thermoplastic polymer. At time zero, the two surfaces are pressed together. Providing the temperature is

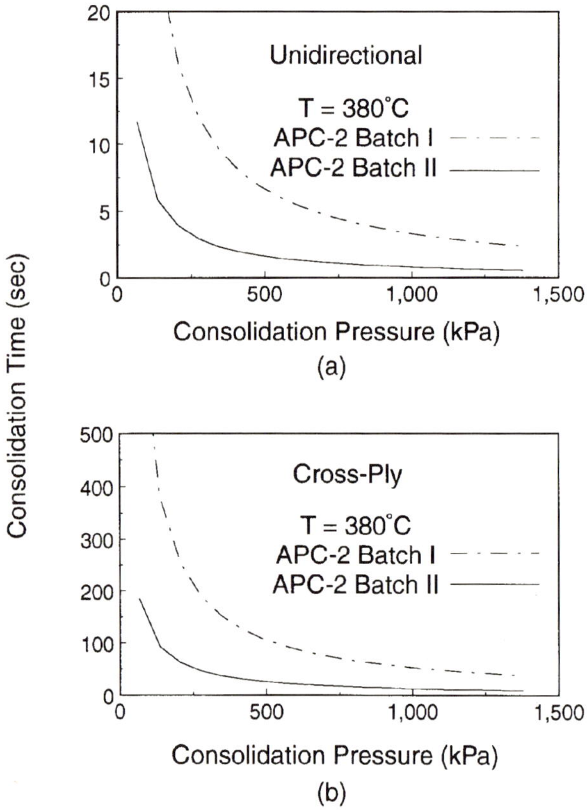

Figure 7.21 Consolidation time (t_c) versus pressure for APC-2 Batch I and Batch II prepregs at 380°C. (a) Unidirectional lay-up. (b) Cross-ply lay-up

high enough (normally above the glass transition temperature, T_g), the surfaces will deform viscoelastically, come into contact, and wet (Fig. 7.22a). The polymer chains will begin to diffuse across the interface due to random thermal motions. After time has passed, the chains will have partially diffused across the interface and entangled with molecular chains on the other side of the interface, thus giving the interface strength (Fig. 7.22b). Following a long period of time, the polymer chains will have penetrated and entangled into the adjacent interface so that the interface is no longer distinguishable from the bulk polymer. At this point, the interface is considered completely healed (Fig. 7.22c).

Either wetting or diffusion can account for significant portions of the interfacial strength. Diffusion is conditional upon the surfaces being in intimate contact, as the molecules cannot move across open space [29]. Theories describing polymer diffusion are based on De Gennes' reptation theory of molecular motion [30]. Wool [29,31], Wool and O'Connor [32-34], Prager and Tirrell [35], and Jud et al. [36] have developed theories explaining strength development of a polymer–polymer interface and crack healing in thermoplastic polymers. These

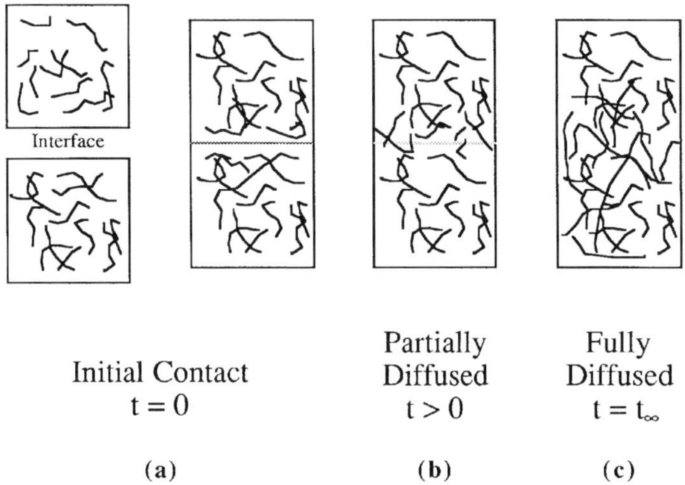

Figure 7.22 Schematic diagram of autohesive bond strength development across an interface

studies resulted in basic mathematical relationships between autohesive bond development, temperature, and contact time.

7.3.1 Healing Model

Destructive mechanical tests commonly are used to characterize autohesion of polymers. In the mechanical tests, two polymer surfaces normally are pressed together at a given temperature for a specified length of time. The fracture stress or fracture energy of the interface then is measured using the appropriate test.

If wetting is instantaneous and the instantaneous wetting load at initial time is negligible, then the autohesive bond fracture stress, σ, is proportional to the fourth root of contact time and the fracture energy, G_{IC}, is proportional to the square root of contact time as shown in the following equations:

$$\sigma \propto t^{1/4} \tag{7.5}$$

$$G_{IC} \propto t^{1/2} \tag{7.6}$$

These equations are valid for healing under isothermal conditions and the slope of the autohesive bond strength or fracture energy versus contact time curve is proportional to the temperature dependent self-diffusion coefficient. Hence, the relationships given in Equation 7.5 and 7.6 provide a means of relating temperature and contact time to interfacial strength development of thermoplastic resins. This approach has been followed in development of models to predict the degree of autohesion or degree of healing in both amorphous [12,13,18,37,38] and semicrystalline [19–22,24–26] thermoplastic polymers.

As an example, the compact tension (CT) fracture toughness test is commonly used to measure autohesive bond strength in thermoplastic polymers [36,38]. Precracked CT specimens are pressed together above the saturation pressure to ensure complete interfacial contact and wetting of the fractured surfaces. The specimen is then heated to the desired temperature (above T_g) and healed for the specified period of time. Following this procedure, Howes et al. [38] measured autohesive bond development in P1700 polysulfone resin. A nondimensional degree of healing D_h can be defined as follows [29,33,34,36]:

$$D_h = \frac{G_{IC}(t)}{G_{IC\infty}} = C(T)t^{1/2} \tag{7.7}$$

where $G_{IC}(t)$ is the critical strain energy release rate of a CT specimen healed for an amount of time t, $G_{IC\infty}$ is the critical strain energy release rate of a CT specimen healed for infinite time, $C(T)$ is a temperature-dependent parameter proportional to the polymer self-diffusion coefficient, and t is healing time.

Wool and O'Connor [33] stated that the self-diffusion coefficient should follow a Williams-Landel-Ferry (WLF) temperature dependence providing that the mode of failure remains the same between samples healed at different temperatures between T_g and $T_g + 100°\mathrm{C}$. Using a reference temperature of 196°C (469 K), the WLF relationship for P1700 polysulfone can be written as follows [38]:

$$\log a_T = \frac{-1.604(T-469)}{11.26+(T-469)} \tag{7.8}$$

where

$$a_T = \frac{1.94 \times 10^{-5}T}{C(T)} \tag{7.9}$$

Here, T is in absolute temperature. A comparison between the calculated and measured degree of healing for P1700 polysulfone is shown in Figure 7.23. Symbols represent the experimental data measured by Howes et al. [38]. Solid lines were calculated using Equation 7.7. As can be seen from the figure, there is good agreement between the calculated and measured degree of healing.

The interply bond strength for thermoplastic matrix composites has been shown to be dependent upon the processing parameters, pressure, temperature, and contact time. If the temperature distribution in the composite is nonuniform during processing, the ply interfaces will bond (or heal) at different rates. Thus, for a specified processing cycle, it is important to know precisely the temperature and degree of autohesive bonding at every point in the composite laminate in order to estimate the required process time.

Autohesive bond strength development during nonisothermal processing can be analyzed by using a heat transfer analysis to predict the instantaneous temperature distribution at the ply interfaces. The actual temperature versus time profile is then used to determine the degree of healing at all points along the ply interfaces. Following this approach, Loos and Li [37] used a numerical scheme to solve the degree of healing model for small time steps. In the study by Liu [26], the degree of healing equation was transformed into its time derivative and then integrated over time.

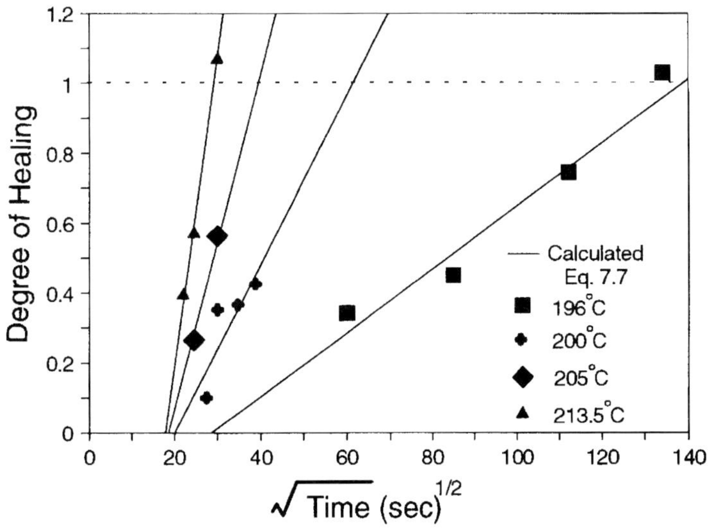

Figure 7.23 Degree of healing versus the square root of time. Symbols are data from Reference 38. Solid lines are predicted by healing model

7.3.2 Degree of Bonding

During processing, autohesive bonding of the ply interfaces occurs at the areas that are in intimate contact. The areas that are in intimate contact will begin to heal once the temperature exceeds the glass transition temperature for amorphous polymers or the melt temperature for semi-crystalline polymers. Hence, the degree of bonding, D_b, at the ply interfaces is a convolution integral of the degree of intimate contact and the degree of healing, and is given by the following equation [24–26]:

$$D_b(t) = D_h(t, 0)\mathrm{Dic}(0) + \int_0^t D_h(t, \tau)\frac{\partial \mathrm{Dic}}{\partial \tau}d\tau \tag{7.10}$$

In the equation, D_b, is the interfacial bond strength normalized with respect to the maximum bond strength in the composite and the integration variable, τ, represents the time an incremental area came into intimate contact.

The degree of bonding analysis has been verified for both compression molding and on-line consolidation of thermoplastic composites. In these studies, composite test specimens were consolidated under controlled processing conditions. The most common types of tests performed to measure the interply bond strength were the interlaminar (short beam) shear test [21,25] or the lap shear test [12,21,26].

7.4 Conclusions

The process by which a thermoplastic matrix composite consolidates to form a laminated structure has been attributed to autohesive bond formation at the ply interfaces. Autohesive bond formation is controlled by two mechanisms: (1) intimate contact at the ply interfaces, and (2) diffusion of the polymer chains across the interface (healing). The rate of autohesive bond formation and hence the speed of the composite consolidation process is directly related to the temperature–pressure–time processing cycle.

Healing or bond formation does not occur unless the ply interfaces are in intimate contact. Intimate contact achievement depends on the surface roughness of the prepreg, the viscosity of the matrix resin, the ply stacking sequence, and the processing cycle. In Section 7.2, models of the intimate contact process during fabrication of thermoplastic composites were presented. The models were used to determine the effects of material properties and processing parameters on the degree of intimate contact.

Once intimate contact is achieved, bonding of the ply interfaces can occur. The mathematical relationships between interply bond formation and processing temperature and time were discussed in Section 7.3. The analyses are based on the theories explaining strength development of a polymer–polymer interface and crack healing in polymers.

Nomenclature

a_T	Shift factor
$C(T)$	Temperature dependent parameter in Equation 7.7
b	Instantaneous width of the rectangular elements
b_0	Initial width of the rectangular elements
D_b	Degree of bonding
D_h	Degree of healing
Dic	Degree of intimate contact
G_{IC}	Critical strain energy release rate
$G_{IC\infty}$	Critical strain energy release rate at infinite time
h	Instantaneous height of the rectangular elements
h_0	Initial height of the rectangular elements
P_{app}	Applied consolidation pressure
t	Time
t_c	Consolidation time
T	Temperature
T_g	Glass transition temperature
T_m	Melt temperature
w	Instantaneous spacing of the rectangular elements
w_0	Initial spacing of the rectangular elements
x	Coordinate direction
y	Coordinate direction

z	Coordinate direction
η_0	Zero shear viscosity of the resin
θ	Ply stacking sequence angle
σ	Autohesive bond fracture stress
τ	Intimate contact time

Acknowledgments

This work was supported by the Virginia Institute for Material Systems (VIMS) and the NSF Science and Technology Center; High-Performance Polymeric Adhesives and Composites (NSF grant number DMR-912004). The authors would like to thank Dr. Po-Jen Shih and Mr. Todd Bullions for their assistance is preparing this chapter.

References

1. Newaz, G.M. *ASTM Standardization News* (1987) 15(10), p. 32
2. Wang, E.L., Gutowski, T.G. *SAMPE Journal* (1990) 26(6), p. 19
3. Lubin. G. *Handbook of Composite Materials* (1982) Van Norstrand Reinhold, New York
4. Brady, D.G., SME Technical Paper MF85-502 (1995).
5. Coffenberry, B.S., Hauber, D.E., Cirino, M., 38th International SAMPE Symposium and Exhibition, Anaheim, CA May 10–14, 1993, SAMPE, Covina, CA, p. 1640
6. Ghasemi Nejhad, M.N., Cope, R.D., Güçeri, S.I., *J Thermoplastic Comp Mat* (1991), 4, p. 20
7. Wells, G.M., McAnulty, K.F. *Proceedings, Sixth International Conference on Composite Materials, Second European Conference on Composite Materials, ICCM & ECCM*, (1987) 1:1.161 Imperial College of Science and Technology, London, UK 20–24 July 1987, Elsevier Applied Science, London
8. Denault, J., Vu-Khanh, T. *Polym. Comp.* (1992) 13(5) p. 361–371
9. Muzzy, J., Norpoth, L., Varughese, B., *SAMPE J.* (1989) 25, p. 23
10. Gutowski, T.G., *SAMPE Q.* (1985) 16(4), p. 58
11. Gutowski, T.G., Cai, Z., Kingery, J., Wineman, S.J. *SAMPE Q* (1986) 17(4), p. 54
12. Bastien, L.J., Gillespie, J.W., Jr., *Polym. Eng. Sci.* (1991) 31(24), p. 1720
13. Dara, P.H., Loos, A.C., Report CCMS-85-10, VPI-E-85-21, (1985) Virginia Polytechnic Institute and State University, Blacksburg, Virginia.
14. Velisaris, C.N., Seferis, J.C. *Poly. Eng.Sci.* (1986) 26(22), p. 1574
15. Velisaris, C.N., Seferis, J.C., *Polym. Eng. Sci.* (1988) 28(9), p. 583
16. Lee, Y., Porter, R.S. *Polym. Eng. Sci.* (1986) 26(8), p. 633
17. Talbott, M.F., Springer, G.S., Berglund, L.A. *J. Comp. Mat.* (1987) 21(11), p. 1056
18. Loos, A.C., Dara, P.H. In *Review of Progress in Quantitative Nondestructive Evaluation* (1987) Thompson, D.O., Chimenti, D.E. (Eds.). Plenum Press, New York
19. Lee, W.I., Springer, G.S. *J. Comp. Mat.* (1987) 21, p. 1017
20. Mantell, S.C., Springer, G.S. *J. Comp. Mat.* (1992) 26, p. 2348
21. Mantell, S.C., Wang, Q., Springer, G.S. *J. Comp. Mat.* (1992) 26, p. 2378
22. Li, M.C., Loos, A.C., Report CCMS-94-01, VPI-E-94-02 (1994) Virginia Polytechnic Institute and State University, Blacksburg, Virginia
23. Li, M.C., Loos, A.C. In *Proceedings of the Ninth International Conference on Composite Materials, Vol. 11 Cermanic Matrix Composites and Other Systems* July, 12–16 (1993) A. Miravete (Ed.). Woodhead Publishing Ltd.

24. Pitchumani, R., Ranganathan, S., Don, R.C., Gillespie, J.W., Jr., Lamontia, M.A., *Int. J. Heat Mass Transfer* (1996) 39(9), p. 1883
25. Butler, C.A., Pitchumani, R., Gillespie, J.W., Wedgewood, A.R. In *Proceedings of the Tenth Annual ASM/ESD Advanced Composites Conference* (1994) ASM International, Materials Park, Ohio
26. Liu, K.S., Ph.D. Thesis, "A Mathematical Model for In-Situ Consolidation of Thermoplastic Composites, (1995) Stanford University, Stanford, CA
27. Wang, E.L., Gutowski, T. G. *Comp. Manufact.* (1991) 2, p. 69
28. Voyutskii, S.S., *Polymer Reviews, Vol. 4* (1963) Interscience Publishers, New York
29. Wool, R.P. *Rubber Chem. Tech.* (1983) 57, p. 307
30. De Gennes, P.G., *Phys. Today* (June 1983) p. 33
31. Wool, R.P. *ACS Poly. Preprints*, 23(2), p. 62
32. Wool, R.P., O'Connor, K.M., *Polym. Eng. Sci.,* (1982) 21(14), p. 970
33. Wool, R.P., O'Connor, K.M. *J. Appl. Phys.* (1981) 52(10), p. 5953
34. Wool, R.P., O'Connor, K.M., *J. Polym.: Polym. Lett. Ed.* (1982) 20, p. 7
35. Prager, S., Tirrel, M. *J. Chem. Phys.* (1981) 75(10), p. 5194
36. Jud, K., Kausch, H.H., Williams, J.G, *J. Mat. Sci.* (1981) 16, p. 204
37. Loos, A.C., Li, M.C. *J. of Thermoplastic Comp.* (1994) 7, p. 311
38. Howes, J.C., Loos, A.C., Hinkley, J.A. In *Advances in Thermoplastic Matrix Composite Materials, ASTM STP 1044* (1989) G.M. Newaz (Ed.). ASTM, Philadelphia

8 Processing-Induced Residual Stresses in Composites

Scott R. White

Residual stresses are inherent to composite materials. They arise because the two (or more) materials that constitute the composite behave differently when subjected to a nonmechanical load (e.g., temperature). Processing-induced residual stresses occur as a result of nonmechanical loading during cure. Residual stresses induced during processing can be traced to one of two causes: chemical shrinkage strains or thermal expansion/contraction strains. A thermosetting matrix typically undergoes 5 percent volumetric shrinkage during crosslinking. Thermal expansion coefficients of polymer matrices are usually an order of magnitude higher than the reinforcement. The effect of processing-induced residual stresses can be dramatic. In some cases they are high enough to cause matrix cracking even before mechanical loading. In all cases a preloading of the fibers occurs. This chapter begins with a review of process models used to predict residual stresses in polymer matrix composites. A process model is then developed using phenomenological cure kinetics and cure-dependent mechanical properties. Both elastic and viscoelastic predictions are examined. Finally, the effect of processing conditions on residual stresses is discussed and ways of reducing them are explored.

8.1 Introduction

Composite materials inherently develop residual stresses during processing. This happens because the two (or more) phases that constitute the composite behave differently when subjected to nonmechanical loading. For example, consider a reinforcing phase that has low thermal expansion characteristics embedded in a matrix phase with high thermal expansion characteristics. If the material is initially stress free and the temperature is decreased, then the matrix will try to shrink more than the reinforcement. This places the reinforcement in a state of compression (i.e. a compressive residual stress). If the phases are well bonded, then models can be developed to predict the residual stress field that is induced during processing.

The standard process cycle for polymer matrix composites is a two-step cure cycle, as seen in Figure 8.1. In such cycles the temperature of the material is increased from room temperature to the first dwell temperature and this temperature is held constant for the first dwell period (~ 1 hour). Afterward, the temperature is increased again to the second dwell temperature and held constant for the second dwell period (2–8 hours). After the second dwell, the part is cooled down to room temperature at a constant rate. Because there are two dwell periods, this type of cure cycle is referred to as a two-step cure cycle. The purpose of the first dwell is to allow gases (e.g., entrapped air, water vapor, or volatiles) to escape and to allow the matrix to flow, which leads to compaction of the part. Thus, the viscosity of the matrix must be low during the first dwell. Typical viscosity versus temperature profiles of polymer matrices show that as the temperature is increased, the viscosity of the polymer decreases until a minimum viscosity is reached. As the temperature is increased further, the polymer begins to cure rapidly and the viscosity increases dramatically. The first dwell temperature must be chosen judiciously so that the viscosity of the resin is low while the cure is kept to a minimum.

The purpose of the second dwell is to allow crosslinking of the matrix to take place. It is during the second dwell when the strength and related mechanical properties of the composite are developed. To characterize the exothermic crosslinking reaction of a thermosetting polymer matrix, a thermal cure monitor technique such as Differential Scanning Calorimetry

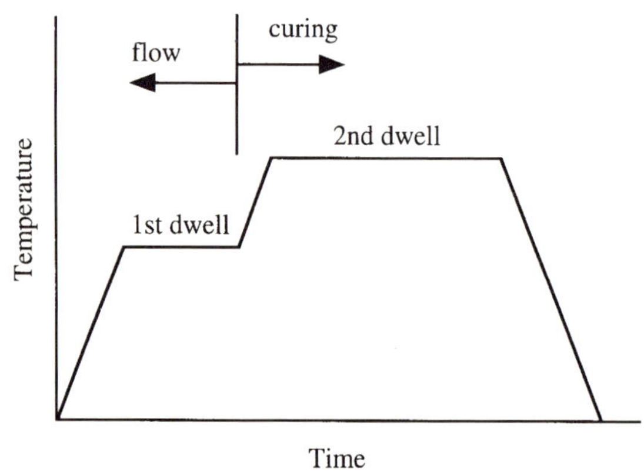

Figure 8.1 Typical two-step cure cycle

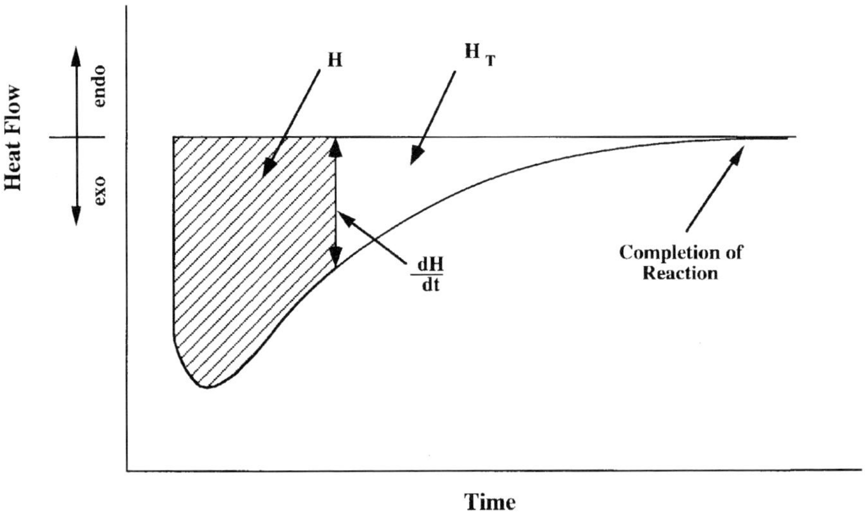

Figure 8.2 Typical isothermal DSC thermogram for a polymer matrix composite

(DSC) is commonly used. A typical DSC trace for a thermosetting polymer is shown in Figure 8.2. The DSC trace shows the amount of energy released during the cure. As the DSC trace approaches a flat line, the crosslinking reaction is nearing completion. If the cure temperature is increased, then the reaction rate increases and the time to complete the reaction decreases. Two competing priorities take place in the choice of the second dwell temperature. First, a low temperature is desirable for ease of manufacturing and to reduce residual stresses arising from the effects of thermal mismatch between fiber and matrix. Second, the processing time should be as short as possible for economic considerations. Because low temperatures require longer dwell times, these two concerns must be compromised.

One of the most critical processing parameters is the second dwell temperature. Its choice is largely material dependent. A certain minimum temperature must be reached before the crosslinking reaction begins. Economics dictate that a short process cycle is desirable, and this can be accomplished by processing at high temperature. However, the thermal mismatch between fiber and matrix means that higher processing temperatures lead to higher residual stresses after processing. Residual stresses are problematic for high-temperature resins like bismaleimide (BMI) because they are cured at higher temperatures than are conventional epoxy–matrix composites.

Residual stresses are not solely thermally induced. Chemical shrinkage can also play an important role during processing. Chemically, the reinforcing fibers are affected very little during the process cycle while the matrix contracts during crosslinking. For epoxies the volumetric chemical shrinkage can be as much as 6 percent [1]. For thermoplastics the chemical shrinkage due to crystallization can be much higher.

The effect of processing-induced residual stresses can be dramatic. In some cases they are high enough to cause cracking within the matrix even before mechanical loading [2]. Microcracking of the matrix can expose the fibers to degradation by chemical attack and strength is adversely affected because a "pre-loading" has been introduced [3].

The ability to predict residual stresses during processing is critical for the design of composite structures; unfortunately, the problem is particularly difficult. The material behavior during cure spans the entire range from fluid to solid which necessitates using a viscoelastic constitutive law. The nonmechanical deformations that are induced during cure are transient and can only be predicted by solving coupled thermochemical energy balance relations.

This chapter presents an overview of the development of process models for residual stresses. It is largely based on a series of three papers published in the composites literature [4–6] that reported on the development of a viscoelastic model accounting for thermally and chemically induced residual stresses in thermosetting polymer matrix composites. In the first section a discussion of process modeling is given covering cure kinetics, thermochemical modeling, and residual stresses. The chapter concludes with a discussion of the effects of processing conditions (i.e., cure temperature, postcure, intermediate dwells) on residual stresses for a graphite–bismaleimide composite material.

8.2 Process Modeling

8.2.1 Cure Kinetics

A cure kinetics model relates chemical composition with time and temperature during chemical reaction in the form of a reaction rate expression. Kinetic models may be phenomenological or mechanistic. A phenomenological model captures the main features of the reaction kinetics ignoring the details of how individual species react with each other. Mechanistic models, on the other hand, are obtained from balances of species involved in the reaction; hence, they are better for prediction and interpretation of composition. Due to the complexity of thermosetting reactions, however, phenomenological models are the most common.

Phenomenological reaction rate expressions have the general formula [7],

$$\frac{d\alpha}{dt} = Kf(\alpha) \tag{8.1}$$

where α is the fractional conversion of the reactive group (degree of cure), t is the reaction time, K is a reaction rate constant, and $f(\alpha)$ is some function of the reactive group conversion. The rate constant, K, is temperature-dependent according to the Arrhenius equation,

$$K = F\exp\left(\frac{-\Delta E}{g_c T}\right) \tag{8.2}$$

where F is the frequency factor, g_c is the universal gas constant, ΔE is the activation energy, and T is absolute temperature. As the degree of cure approaches unity, the reaction rate should decrease to zero. In order to satisfy this condition, the function $f(\alpha)$ may be expressed as,

$$f(\alpha) = (1 - \alpha)^n g(\alpha) \tag{8.3}$$

Substituting Equation 8.2 and Equation 8.3 into Equation 8.1, the general expression for the reaction rate of curing polymers is obtained,

$$\frac{d\alpha}{dt} = F \exp\left(\frac{-\Delta E}{RT}\right)(1-\alpha)^n g(\alpha) \tag{8.4}$$

This equation includes the particular cases of an nth order reaction model with $g(\alpha) = 1$, and the autocatalytic model with $g(\alpha) = 1 + I\alpha$, where I is the autocatalysis intensity. In general, the form of $g(\alpha)$ must be determined from experimental data.

Equation 8.4 is integrated to obtain the degree of cure at any given time,

$$\alpha(t) = \int_o^t \left(\frac{d\alpha}{dt}\right) dt \tag{8.5}$$

Equation 8.5 is an implicit integral equation because the rate of cure depends upon the current degree of cure. Whereas closed form solutions exist for the nth order and autocatalytic reaction models, a numerical integration technique, such as Runga-Kutta, is often used to solve Equation 8.5.

Several researchers have modeled the cure kinetics of thermosetting resins in the past, including an unsaturated polyester resin [8], epoxies [9–11], and bismaleimide [5]. As an example, a graphite/BMI material, IM6/3100, was modeled in [5] using

$$\frac{d\alpha}{dt} = K(1-\alpha)^n \alpha^m \tag{8.6}$$

To obtain the cure kinetic parameters K, m, and n, cure rate and cure state must be measured simultaneously. This is most commonly accomplished by thermal analysis techniques such as DSC. In isothermal DSC testing several different isothermal cures are analyzed to develop the temperature dependence of the kinetic parameters. With the temperature dependence of the kinetic parameters known, the degree of cure can be predicted for any temperature history by integration of Equation 8.5.

In an isothermal DSC test the heat flow to or from the sample is monitored over time (Fig. 8.2). As the crosslinking reaction commences the DSC monitors the release of energy and the heat flow occurs in the exothermic region. After a short period of time the heat flow reaches its maximum and begins to decrease gradually until reaching the baseline (zero heat release) upon completion of the reaction. The total area under the exotherm is the heat of reaction at the isothermal temperature, H_T. After completion of the isothermal cure, the sample is dynamically scanned from room temperature to the highest cure temperature, and the resulting exotherm is measured to obtain the residual heat of reaction, H_R. The summation of H_T and H_R is the ultimate heat of reaction, H_U. If the isothermal cure temperature is high, then the residual heat of reaction will be small and H_T is approximately equal to H_U. The final degree of cure, α_f, at the end of the isothermal cure is found from,

$$\alpha_f = \frac{H_T}{H_U} \tag{8.7}$$

The area under the exotherm up to time t is $H(t)$ and the degree of cure at that time is

$$\alpha(t) = \frac{H(t)}{H_U} \tag{8.8}$$

By differentiating Equation 8.8 the cure rate is obtained as,

$$\frac{d\alpha(t)}{dt} = \frac{dH(t)/dt}{H_U} \tag{8.9}$$

where $dH(t)/dt$ is just the height of the exotherm from the baseline.

For isothermal DSC testing Equations 8.8 and 8.9 are used in data reduction to produce degree of cure and cure rate histories. With these histories, the solution of Equations 8.2 and 8.6 yields the kinetic parameters of interest.

In Figure 8.3 the degree of cure histories for six different isothermal cures of IM6/3100 are presented. The cure rate is initially high but it decreases as the cure reaction progresses. Each of the curves shows the same type of behavior: the degree of cure exponentially increases until reaching an asymptote. In some cases the reaction is still progressing significantly after 120 minutes. For the high-temperature cures, the reaction reaches completion within this time frame. For example, at 200°C the reaction appears to be completed after 60 minutes. During manufacture of typical composite structures temperature variations of 10–20°C are not uncommon. For IM6/3100 such a temperature variation could lead to significant variation in the degree of cure after the cure cycle is completed.

Figure 8.4 shows the model predictions and experimental data for degree of cure during the manufacturer's recommended cure (MRC) cycle for IM6/3100. The MRC cycle is a two-step cure with the second dwell at 182°C. Note that most of the reaction occurs during the second dwell period. Full cure is reached during the latter half of the second dwell.

The ability to model the cure kinetics of the matrix accurately is critically important to achieving full and uniform cure state after processing. If significant temperature variations

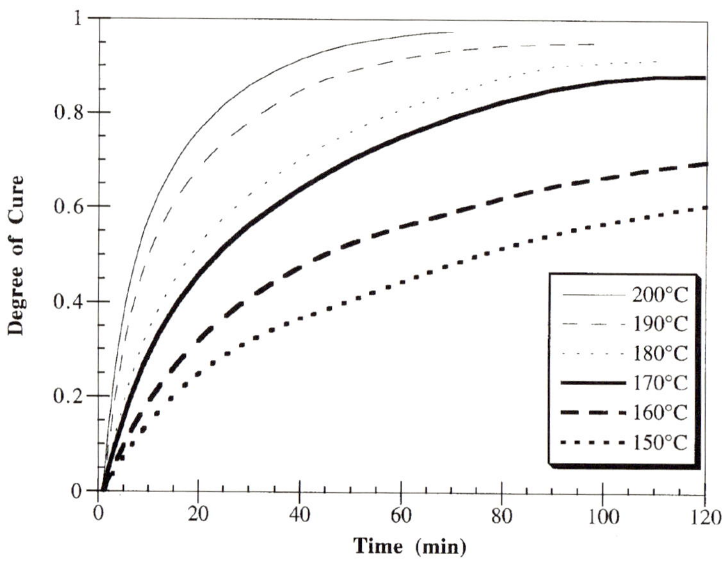

Figure 8.3 Degree of cure advancement during isothermal curing of a graphite/BMI [IM6/3100] composite

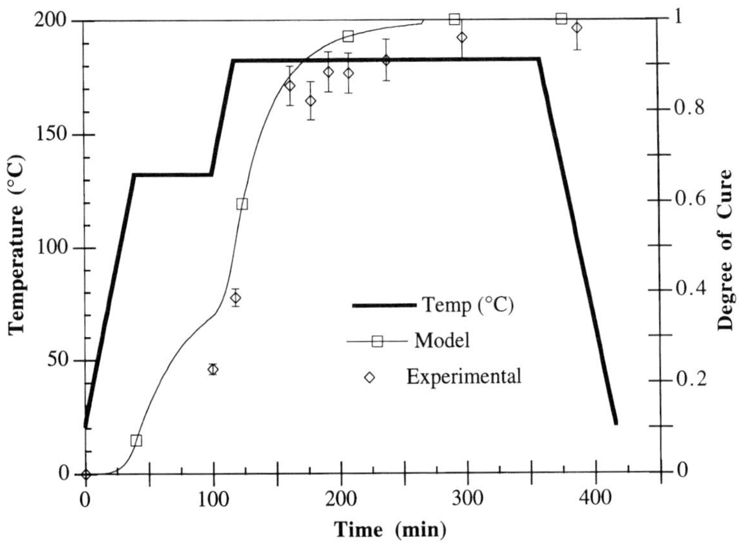

Figure 8.4 Degree of cure during MRC cycle [IM6/3100]

exist within the material, on-line modifications in the cure cycle may be necessary. New research and development in this area has provided manufacturers with the ability to incorporate automated control schemes with on-line temperature and cure state feedback.

8.2.2 Thermochemical Modeling

Before developing the necessary equations to predict the temperature within a composite part, it is enlightening to consider the thermal diffusion problem in polymer matrix composites. Thermal diffusivity is defined as

$$D_{th} = \left(\frac{k}{\rho C_p}\right) \tag{8.10}$$

where k is the thermal conductivity, ρ is the density, C_p is the heat capacity, and D_{th} is the thermal diffusivity. Table 8.1 lists the thermal diffusivity for several common materials, including graphite and glass-reinforced composites. The low thermal diffusivities for these composites is the cause of several problems in their manufacture. Materials with low thermal diffusivity take longer to heat or cool. This lengthens the cure cycle and effectively limits the maximum heating and cooling rates. In addition, thick composites are difficult to process because the outer edges will reach the cure temperature first, whereas the interior temperature of the part lags behind. This nonuniformity of temperature history can lead to incomplete curing, residual stresses, or incomplete consolidation.

Table 8.1 Thermal Diffusivity of Several Engineering Materials

Material	Thermal diffusivity [$\times 10^{-5}$ m^2/sec]
Polymer matrix composites [transverse]	
• Graphite fiber-reinforced	0.044
• Glass fiber-reinforced	0.035
Wood	0.009–0.013
Glass	0.020
Copper	11.2
Tool steel	1.48
Iron	2.03
Nickel	2.27
Plaster	0.040
High-temperature epoxy	0.013
Aluminum	8.42

To develop the governing equations for thermochemical modeling, consider the material volume element in Figure 8.5. Performing an energy balance over this volume while neglecting convective processes yields

$$\rho C_p \frac{\partial T}{\partial t} = \frac{-\partial q_i^{th}}{\partial x_i} + \rho \dot{r} \tag{8.11}$$

where $\dot{r}$ is the rate of internal heat generation, x_i are the coordinate directions, and q_i^{th} are the components of the heat flux vector. Here the thermal properties have been assumed to remain

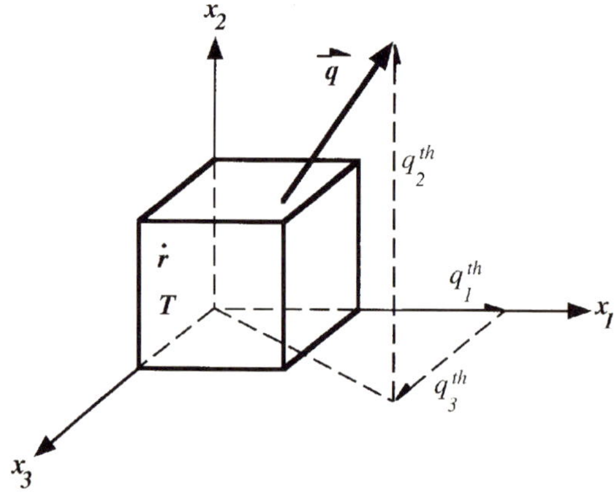

Figure 8.5 Thermochemical model material volume element

constant throughout the cure cycle. Fourier's Law can be used to relate the heat flux to temperature gradients,

$$q_i^{th} = -k_{ij} \frac{\partial T}{\partial x_j} \tag{8.12}$$

where k_{ij} is the thermal conductivity tensor. For orthotropic materials with the three principal axes oriented in the (x_1, x_2, x_3) axes directions, the thermal conductivity tensor has three components.

$$k_{ij} = \begin{bmatrix} k_{11} & 0 & 0 \\ 0 & k_{22} & 0 \\ 0 & 0 & k_{33} \end{bmatrix} \tag{8.13}$$

If the material is transversely isotropic then $k_{22} = k_{33}$. For other orientations the thermal conductivity tensor can be transformed according to normal tensor transformation relations. Combining Equations 8.11–8.13 gives the three-dimensional energy balance equation.

$$\rho C_p \frac{\partial T}{\partial t} = \frac{\partial}{\partial x_1}\left(k_{11} \frac{\partial T}{\partial x_1}\right) + \frac{\partial}{\partial x_2}\left(k_{22} \frac{\partial T}{\partial x_2}\right) + \frac{\partial}{\partial x_3}\left(k_{33} \frac{\partial T}{\partial x_3}\right) + \rho \dot{r} \tag{8.14}$$

In most cases, for composite structures the planar dimensions are sufficiently large compared with the thickness that one-dimensional heat transfer is assumed. In this case,

$$\rho C_p \frac{\partial T}{\partial t} = k_{33} \frac{\partial^2 T}{\partial x_3^2} + \rho \dot{r} \tag{8.15}$$

where the thermal conductivity k_{33} is assumed to be independent of x_3. The rate of internal heat generation can be found from the cure rate as,

$$\dot{r} = H_U \frac{d\alpha}{dt} \tag{8.16}$$

where H_U is the ultimate heat of reaction. The reaction rate in Equation 8.16 is given by the generalized expression in Equation 8.4. Combining these equations yields the governing equations for one-dimensional thermochemical modeling,

$$\rho C_p \frac{\partial T}{\partial t} = k_{33} \frac{\partial^2 T}{\partial x_3^2} + \rho H_U \frac{d\alpha}{dt} \tag{8.17}$$

This equation, along with Equation 8.4, constitutes a coupled set of a differential equations governing the flow of thermal energy in a composite part during cure. Two boundary conditions (for temperature) and two initial conditions (for temperature and degree of cure) are required. An analytic solution to these equations is usually not possible. Numerical techniques such as finite difference or finite element are commonly used.

The relative importance of internal heating during cure is demonstrated in Figures 8.6–8.8. In the first figure a 6 mm thick glass/polyester laminate is modeled during cure. The bottom surface temperature ($z = 0$) is held constant at 177°C and thermal diffusion transfers heat through the laminate. The top surface ($z = h$) is modeled as a convective boundary. The boundary conditions are representative of autoclave processing in which the bottom surface is next to the mold (fixed temperature) and the top surface is exposed to the autoclave environment (convective boundary). Two cases are shown in Figure 8.6. The first case is the model predictions for combined thermal diffusion and internal heating, and the second is the prediction when the internal heating from the cure reaction is ignored ($H_U = 0$). There is little difference in the results until 200 secs. Steady state is reached for the thermal diffusion case at this point and the temperature distribution is linear from the top to the bottom surfaces. The combined case shows elevated temperatures are reached throughout the laminate. The difference is relatively small for this thin laminate (< 15°C) As the thickness increases, however, this effect becomes more pronounced. Figure 8.7 shows the center line temperature prediction for a 25 mm (200 ply) AS4/3501-6 laminate during cure. The autoclave temperature cycle is overlaid on the figure. In this case both the upper and lower surface temperatures were assumed to be fixed and equal to the autoclave temperature. There is a small thermal lag during heat-up to the first dwell temperature. The center line temperature overshoots the autoclave temperature at the first dwell by about 10°C. There is another thermal lag during heat-up to the second dwell, but then the centerline temperature rapidly increases and overshoots the autoclave temperature significantly during the second well period. The thermal spike here is about 26°C.

As the thickness of the laminate increases, the strength of this thermal spike and the degree of thermal lag during heat-up increases. Figure 8.8 shows the results for a 62.5-mm (500 ply) laminate of the same material. Now the center-line temperature never reaches the autoclave temperature during the first dwell, and the thermal spike during the second dwell is nearly 135°C. The thermal spike is directly related to the release of internal heat during cure. The thermal lag is a manifestation of the low thermal diffusivity of polymer matrix composites.

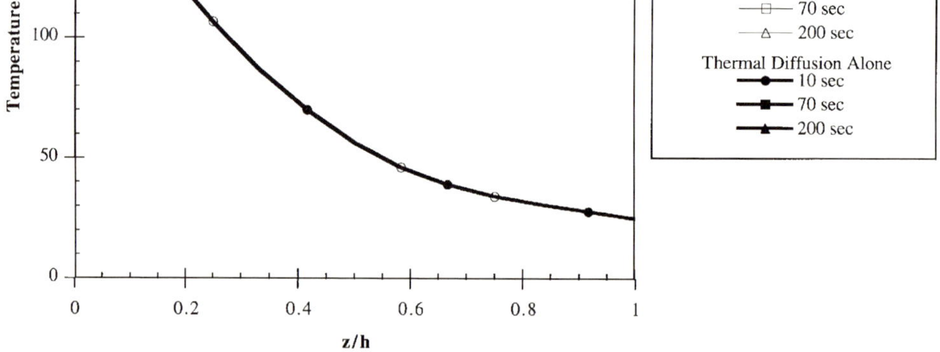

Figure 8.6 Thermal diffusion through a 6-mm-thick glass/polyester laminate with and without cure reaction

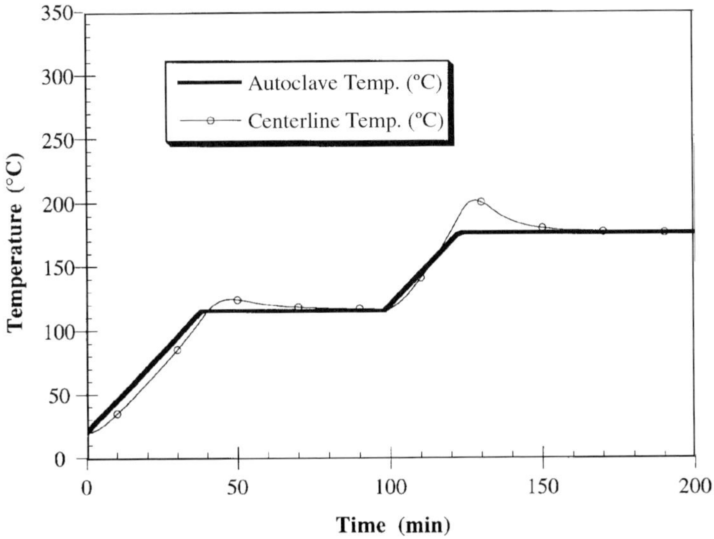

Figure 8.7 Centerline and autoclave temperature histories during cure of a 25-mm (200 ply)-thick AS4/3501-6 laminate

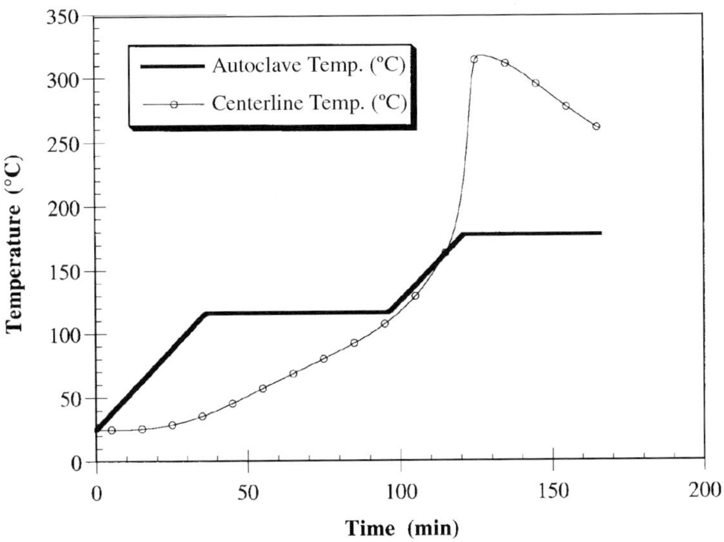

Figure 8.8 Centerline and autoclave temperature histories during cure of a 62.5-mm (500 ply)-thick AS4/3501-6 laminate

Once the temperature and degree of cure can be predicted throughout the material during cure, the nonmechanical loadings (thermal expansion and chemical shrinkage) can be obtained. With this knowledge in hand the residual stresses can then be analyzed.

8.2.3 Residual Stress Modeling

As with any stress, residual stresses are not directly measurable; instead, the *effects* of residual stresses are measured. These measurements are then used, with an appropriate model, to back-calculate the level of stresses necessary to produce the given effect. Residual stresses have been experimentally investigated in the past [12–17] by measuring the curvature development in unsymmetric cross-ply laminates. Unsymmetric laminates develop curvature after processing as shown in Figure 8.9 and the curvature can be directly related to the level of residual stresses in the laminate. For the purpose of providing experimental correlation with the model output, the curvature development in unsymmetric cross-ply strips will be analyzed in the following sections. First, an elastic analysis is developed. A viscoelastic model is subsequently formulated based on the elastic solution using the elastic–viscoelastic analogy.

8.2.3.1 Elastic Analysis

The elastic stress–strain relations for an orthotropic lamina under plane stress conditions are

$$\sigma_i = Q_{ij}(\varepsilon_j - e_j) \tag{8.18}$$

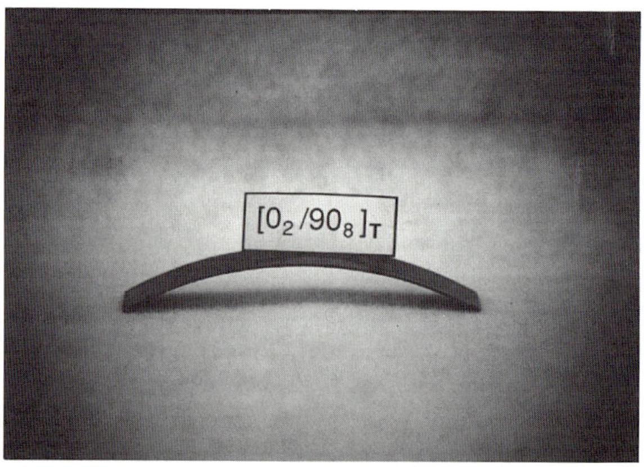

Figure 8.9 Warpage induced by processing for an IM6/3100 $[0_2/90_8]_T$ unsymmetric cross-ply specimen

where σ_i are the stresses, ε_j are the total strains, e_j are the nonmechanical strains, and Q_{ij} are the lamina stiffnesses. The nonmechanical strains consist of chemical strains, e_j^{ch}, and thermal strains, e_j^{th}

$$e_j = e_j^{ch} + e_j^{th} \tag{8.19}$$

The longitudinal chemical strains are neglected because the fibers do not experience chemical strains during cure and they are much stiffer in this direction in comparison to the matrix. The transverse chemical strains are dominated by the matrix shrinkage associated with the cross-linking reaction. The matrix chemical shrinkage strains have been modeled in the past assuming a linear dependence on degree of cure [18]. Experimental results for IM6/3100 [5] indicate that a more accurate representation for this composite system is obtained from

$$\begin{aligned} e_2^{ch} &= c_1 10^{(c_2\alpha)} \quad && \alpha \le \alpha^{ch}, \\ e_2^{ch} &= e_2^{ch_f} && \alpha > \alpha^{ch}. \end{aligned} \tag{8.20}$$

The final transverse chemical shrinkage strain is e_2^{ch} and α^{ch} is the degree of cure when chemical shrinkage is complete. The parameters c_1 and c_2 are empirical constants obtained from chemical shrinkage characterization tests. In most cases α^{ch} can be approximated as the degree of cure at gelation.

The thermal strains can be modeled using the longitudinal and transverse thermal expansion coefficients. From experimental testing of IM6/3100 [5] these coefficients were not found to be significantly dependent on degree of cure . In this case the thermal strains are

$$e_i^{th} = \alpha_i(T - T_0) \tag{8.21}$$

where α_i are the thermal expansion coefficients and T_0 is the initial stress-free temperature. The nonmechanical strain history during curing can be calculated knowing the temperature and degree of cure histories and utilizing Equations 8.20 and 8.21.

Consider the freely warping $[0_n/90_n]_T$ laminate as shown in Figure 8.10. The total strains ε_j at position z from the midplane can be found from the midplane strains, ε_j^o, and curvatures, κ_j, by

$$\varepsilon_j = \varepsilon_j^o + z\kappa_j \tag{8.22}$$

The longitudinal and transverse stresses in the individual plies are given by the laminated plate theory as [19]

$$\begin{aligned} \sigma_1^{(0^\circ)} &= Q_{11}\varepsilon_1 + Q_{12}\varepsilon_2 - (Q_{11}e_1 + Q_12e_2) \\ \sigma_2^{(0^\circ)} &= Q_{12}\varepsilon_1 + Q_{22}\varepsilon_2 - (Q_{12}e_1 + Q_22e_2) \\ \sigma_1^{(90^\circ)} &= Q_{22}\varepsilon_1 + Q_{12}\varepsilon_2 - (Q_{22}e_1 + Q_12e_2) \\ \sigma_1^{(90^\circ)} &= Q_{12}\varepsilon_1 + Q_{11}\varepsilon_2 - (Q_{12}e_1 + Q_{11}e_2) \end{aligned} \tag{8.23}$$

In Equation 8.23 the superscripts refer to the 0 and 90-degree plies and the subscripts 1 and 2 refer to the axis directions as depicted in Figure 8.10. Note that the laminate stiffnesses Q_{ij} in Equation 8.23 are cure dependent [i.e. $Q_{ij}(\alpha)$].

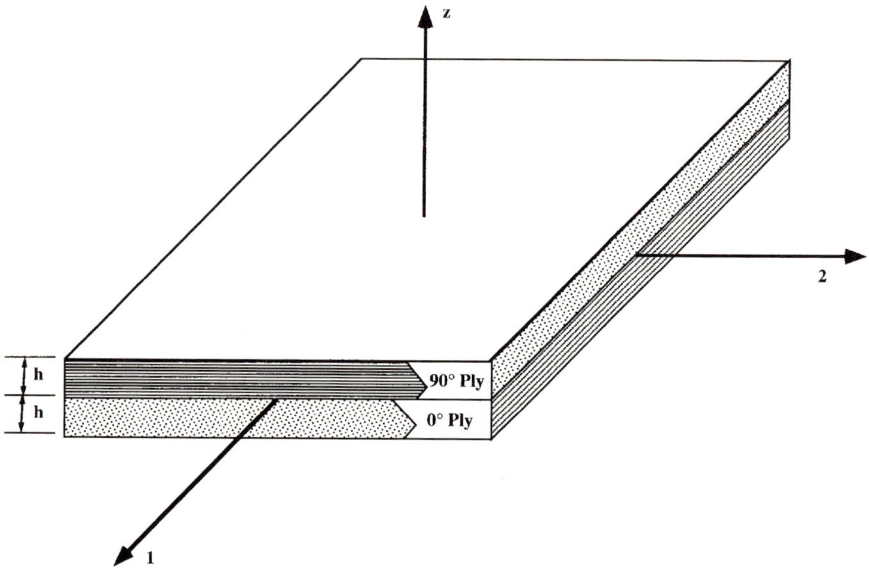

Figure 8.10 Geometry of a cross-ply laminated plate

For the freely warping laminate of Figure 8.10, the midplane strains and curvatures are calculated by requiring the stress resultants (N_1, N_2) and the bending moments (M_1, M_2) to be zero. The resulting deformation is anticlastic with the anticlastic curvature, κ_a, given by

$$\kappa_a = \kappa_1 = -\kappa_2 = \frac{12}{h} Y_a E(e_1 - e_2) \tag{8.24}$$

where

$$Y_a = \frac{(1 - \nu_{12}^2 E)}{1 + 14E + (1 - 16\nu_{12}^2)E^2} \tag{8.25}$$

The major Poisson's ratio is ν_{12} and $E = E_2/E_1$ is the ratio of the transverse to longitudinal modulus. Equation 8.24 gives the induced curvature for anticlastic deformation of an unsymmetric cross-ply laminate. The curvature is dependent on the thermal and chemical strain mismatch $(e_1 - e_2)$, lamina mechanical properties (ν_{12}, E) and the half-thickness, h.

Equation 8.24 is valid when the overall dimensions of the laminate are small so that the deflections are small. For large laminates, however, the deflections may become too large to satisfy the assumptions of linear theory. In this case a large-deflection plate theory must be used [15]. For long, narrow strips, the small-deformation laminated plate theory can still be used with the modified condition $\kappa_2 = 0$ in place of $M_2 = 0$. This results in a cylindrical curvature, κ_c, given by

$$\kappa_c = \kappa_1 = \frac{12}{h} Y_c E(e_1 - e_2) \tag{8.26}$$

where,

$$Y_c = \left\{ \frac{1 + (1 - 2\nu_{12} - \nu_{12}^2)E + (1 + 2\nu_{12})\nu_{12}^2 E^2}{1 + 15E + (15 - 16\nu_{12}^2)E^2 + (1 - 16\nu_{12}^2)E^3} \right\} \tag{8.27}$$

To account for small thickness variations between different specimens, the dimensionless curvature $\eta = 2h \cdot \kappa_c$ is used. Note that η depends directly on material properties through Y_c and E and on the nonmechanical loading through $(e_1 - e_2)$.

The midplane strains can be calculated similarly from laminated plate theory. For the cross-ply case they are found from

$$\varepsilon_1^0 = \varepsilon_2^0 = \frac{[1 + (7 - \nu_{12})E + \nu_{12}(1 - \nu_{12})E^2]e_1 + [\nu_{12}E + (1 - \nu_{12}(1 - 8\nu_{12}))E^2]e_2}{1 + 14E + (1 - 16_{12}^2)E^2} \tag{8.28}$$

Note that if $e_1 = e_2$ then the curvatures are zero from Equations 8.24 and 8.26 and the midplane strain is just e_1 from equation (8.28). In this case the residual stresses are identically zero.

During the processing of composite materials in a hot press or an autoclave, the laminate is usually kept flat until cure is complete. If the platen surfaces are assumed frictionless, the effect of the constraints is to require that the curvatures κ_1 and κ_2 be zero throughout cure. To develop the elastic solution under these constrained conditions, the laminated plate theory may be used with conditions of $N_i = 0$ and $\kappa_i = 0$. The resulting midplane strains are given by

$$\varepsilon_j^0 = \frac{(Q_{11} + Q_{12})e_1 + (Q_{12} + Q_{22})e_2}{Q_{11} + Q_{22} + 2Q_{12}} \tag{8.29}$$

With these strains determined and realizing that the curvatures are zero, the moment resultants can be obtained from

$$M_1 = -M_2 = \frac{h^2(Q_{12}^2 + Q_{11}Q_{22})(e_1 - e_2)}{Q_{11} + Q_{22} + 2Q_{12}} \tag{8.30}$$

Once the platens (or mold) are opened, the laminate freely warps. The curvature that develops can be found by applying the moments found earlier in Equation 8.30 to the laminate.

What remains is to examine the cure-dependence of the mechanical properties found in the preceding equations. Matrix-dominated properties will be significantly affected by the change in degree of cure during processing. Fiber-dominated properties are also affected, but to a lesser extent. For IM6/3100, experimental testing has shown that the longitudinal modulus and major Poisson's ratio are linearly dependent on degree of cure [5]. In this case they can be modeled according to

$$E_1(\alpha) = E_1^i + (E_1^f - E_1^i)\alpha \tag{8.31}$$

$$\nu_{12}(\alpha) = \nu_{12}^i + (\nu_{12}^f - \nu_{12}^i)\alpha \tag{8.32}$$

where the superscripts i and f indicate initial and final values. The transverse modulus increases quickly after a certain degree of cure, α^*, before asymptotically approaching a fully cured value. For IM6/3100 the transverse modulus can be modeled as

$$\begin{aligned} E_2^{el}(\alpha) &= E_2^i & 0 \le \alpha < \alpha^* \\ E_2^{el}(\alpha) &= c_3 + c_4\alpha + c_5\alpha^2 & \alpha^* \le \alpha \end{aligned} \tag{8.33}$$

Extensive experimental testing was performed on IM6/3100 to obtain longitudinal and transverse stiffnesses and major Poisson's ratio as a function of degree of cure. A detailed explanation of the test procedure can be found in [5]. The model predictions and experimental results for this material system are shown in Figures 8.11–8.13. From these relations the lamina stiffnesses Q_{ij} can be obtained from

$$Q_{11} = \frac{E_1}{1 - \nu_{12}(E_2/E_1)}, \tag{8.34}$$

$$Q_{12} = \frac{\nu_{12}E_2}{1 - \nu_{12}(E_2/E_1)} \tag{8.35}$$

$$Q_{22} = \frac{E_2}{1 - \nu_{12}(E_2/E_1)} \tag{8.36}$$

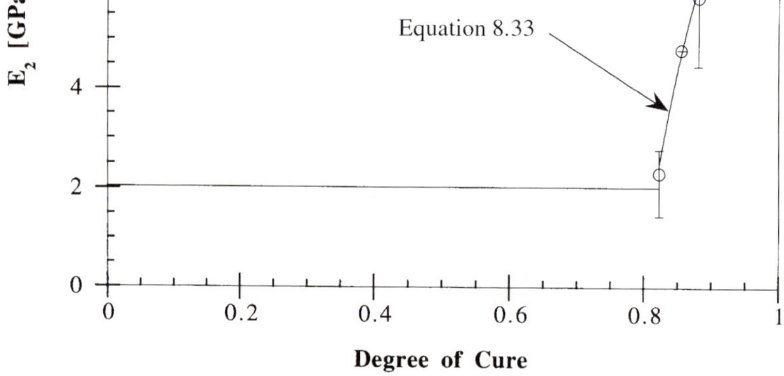

Figure 8.11 Transverse modulus development during cure for IM6/3100

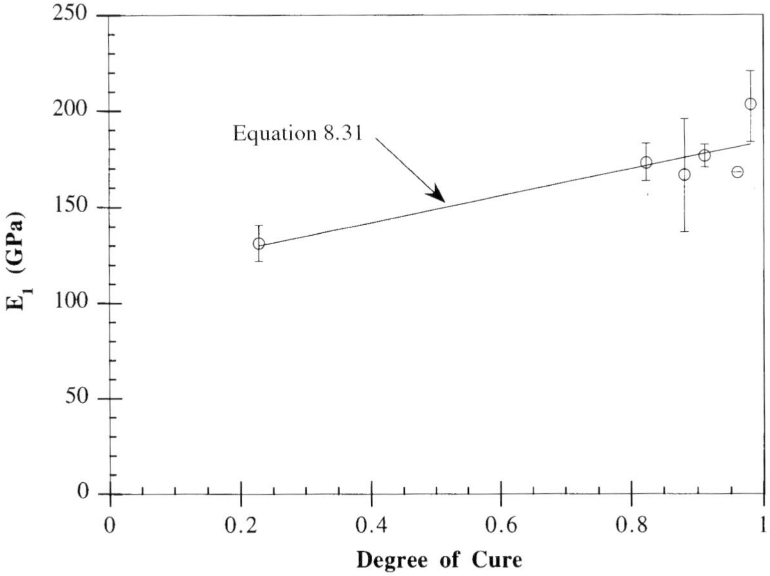

Figure 8.12 Longitudinal modulus development during cure for IM6/3100

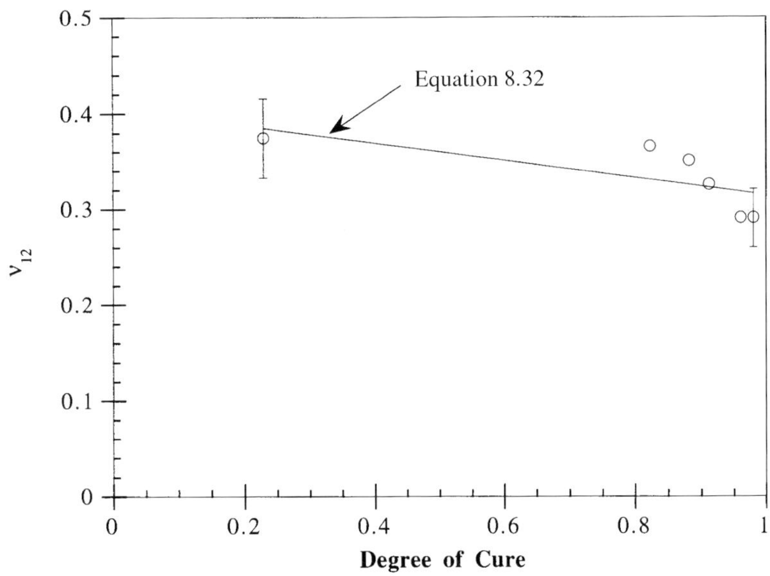

Figure 8.13 Major Poisson's ratio development during cure for IM6/3100

8.2.3.2 Viscoelastic Analysis

Polymeric materials are known to exhibit time-dependent mechanical behavior, especially at high temperature and incomplete cure, which are two conditions that are present during processing. The elastic analysis previously developed will be shown subsequently to be quite useful for determining final residual stresses in thin laminates if chemical shrinkage is neglected. The stresses *during* cure, however, *are not* predictable from an elastic model. In this case, a viscoelastic formulation must be used. An elastic model works well for thin laminates when chemical shrinkage is neglected because the chemical shrinkage strains occur at high temperature and at partial cure states when the matrix is highly viscoelastic. Any stresses that are developed as a result of chemical shrinkage are immediately relaxed. Thus, at the end of curing the laminate is essentially stress free and residual stresses are primarily a result of thermal shrinkage during cool down from the cure temperature.

For linear thermorheologically simple materials a single temperature-dependent shift factor, $a_T(T)$, can be used to predict the transient thermal response [20]. The mechanical response is history dependent and involves the use of reduced times, $\xi(t)$ and $\xi(\tau)$, which can be found from the shift factor as

$$\xi(t) = \int_0^t \frac{ds}{a_T[T(s)]}, \qquad \xi(\tau) = \int_0^\tau \frac{ds}{a_T[T(s)]} \tag{8.37}$$

The shift factor is modeled either as a modified Williams-Landel-Ferry (WLF) equation, or as a best fit to the general form of the Equation [20–25]

$$a_T(T) = \exp\left(\frac{c_6}{T} + c_7\right) \tag{8.38}$$

In Equation 8.38, c_6 and c_7 are constants to be determined from experimental testing.

The quasielastic method as developed by Schapery [26] is used in the development of the viscoelastic residual stress model. The use of the quasielastic method is motivated by the fact that the relaxation moduli are required in the viscoelastic analysis of residual stresses, whereas the experimental characterization of composite materials is usually in terms of the creep compliances. An excellent account of the development of the quasielastic method is given in [27]. The underlying restriction in the application of the quasielastic method is that the compliance response of the material shows little curvature when plotted versus log time [28]. Harper [27] shows excellent agreement between the quasielastic method and direct inversion for AS4/3510-6 graphite/epoxy composite. For most graphite/thermoset systems, the restrictions imposed by the quasielastic method are satisfied.

When the quasi-elastic method is used, the viscoelastic resultant moment can be approximated by substituting the time-dependent stiffnesses for elastic stiffnesses in Equation 8.30 and making use of the convolution integral. The resulting moments are

$$M_1(t) = -M_2(t) = h^2 \int_0^t \left\{ \Gamma(\alpha(\tau), \xi(t) - \xi(\tau)) \frac{d[\Delta e(\tau)]}{d\tau} \right\} d\tau \tag{8.39}$$

where

$$\Gamma(\alpha, t) = \frac{Q_{12}^2(\alpha, t) + Q_{11}(\alpha, t)Q_{22}(\alpha, t)}{Q_{11}(\alpha, t) + Q_{22}(\alpha, t) + 2Q_{12}(\alpha, t)} \tag{8.40}$$

Because Equation 8.39 is a history-dependent integral, the degree of cure is represented as a function of τ and not of the current time, t. Equation 8.39 can be solved by discretizing the time domain into N equal portions Δt with the initial time $t_1 = 0$ and the current time $t_N = t$ once the time-dependent mechanical properties in $\Gamma(\alpha, t)$ have been determined.

For cross-ply laminates only three in-plane stiffnesses are required to determine the residual stresses. These are the longitudinal stiffness Q_{11}, the transverse stiffness Q_{22}, and the Poisson's stiffness term Q_{12}. On the other hand, the corresponding compliances may be found and the stiffness determined through inversion of the compliances. Experimental testing has shown that for graphite/polymer composites the only compliance term of these three which exhibits significant time-dependent behavior is the transverse compliance, $S_{22}(t)$ [12,21,22,29]. As such, the longitudinal compliance S_{11} and the Poisson's compliance term S_{12} are regarded as perfectly elastic. The elastic compliances can now be found from

$$S_{22}^{el}(\alpha) = \frac{1}{E_2^{el}(\alpha)} \tag{8.41}$$

$$S_{11}(\alpha) = \frac{1}{E_1(\alpha)} \tag{8.42}$$

$$S_{12}(\alpha) = \frac{-\nu_{12}(\alpha)}{E_1(\alpha)} \tag{8.43}$$

The transverse compliance can be modeled by a power law equation of the form [12,21,22]

$$S_{22}(\alpha, t) = S_{22}^{el}(\alpha) + J(\alpha)\left(\frac{t}{a_T(\alpha, T)}\right)^{\gamma(\alpha)} \tag{8.44}$$

The creep parameters J and γ are obtained through transverse creep experiments. The initial compliance is the elastic response of the material (Equation 8.41). In general, the creep parameters J and γ, and the shift factor a_T may all be dependent on the cure state of the material. For the current process model the shift factor is assumed to be separable and, as such, is only temperature dependent. As a first approximation the creep parameters are represented as linear functions of the degree of cure.

$$J(\alpha) = J_i + (J_f - J_i)\alpha \tag{8.45}$$

$$\gamma(\alpha) = \gamma_i + (\gamma_f - \gamma_i)\alpha \tag{8.46}$$

Finally, the time-dependent stiffnesses in Equation 8.40 can then be obtained through inversion of the compliance matrix following the quasielastic approximation

$$Q_{11}(\alpha, t) = \frac{S_{22}(\alpha, t)}{S_{11}(\alpha)S_{22}(\alpha, t) - S_{12}^2(\alpha)} \tag{8.47}$$

$$Q_{22}(\alpha, t) = \frac{S_{11}(\alpha)}{S_{11}(\alpha)S_{22}(\alpha, t) - S_{12}^2(\alpha)} \tag{8.48}$$

$$Q_{12}(\alpha, t) = \frac{-S_{12}(\alpha)}{S_{11}(\alpha)S_{22}(\alpha, t) - S_{12}^2(\alpha)} \tag{8.49}$$

8.2.3.3 Curvature-Moment Relations

Once the elastic residual moment from Equation 8.30 or the viscoelastic residual moment history from Equation 8.39 have been determined, the final value may be applied elastically as platens open (hot pressing) or as the applied pressure is released (autoclave processing) to find the resulting cross-ply curvature. This curvature can then be correlated to experimental data.

The elastic midplane strains and curvatures are related to the resultant forces (N_i) and moments (M_i) through the laminate compliance matrix

$$\begin{Bmatrix} \varepsilon^o \\ \kappa \end{Bmatrix} = \begin{bmatrix} \alpha^m & \beta \\ \beta^T & \delta \end{bmatrix} \begin{Bmatrix} N \\ \hline M \end{Bmatrix} \tag{8.50}$$

Because $N_i = 0$ and $M_1 = -M_2$ during cure, the pertinent compliances for cross-ply laminates are δ_{11} and δ_{12}. The δ_{ij} compliance matrix can be found from [19]

$$\delta_{ij} = (D_{ij} - B_{ik}a_{kl}B_{lj})^{-1} \tag{8.51}$$

where $a_{ij} = A_{ij}^{-1}$ is the inverse of the laminate in-plane stiffness matrix, D_{ij} is the laminate flexural stiffness matrix, and B_{ij} is the laminate coupling stiffness matrix. With knowledge of the a_{ij}, B_{ij}, and D_{ij} matrices after processing, residual curvatures can be determined by applying the final residual moments from Equations 8.30 or 8.39 in Equation 8.50.

8.3 Experimental Results

Thin laminated strips 25.4 mm wide × 152.4 mm long of IM6/3100 graphite/BMI $[0_4/90_4]_T$ cross-ply construction were used to measure residual stresses. These specimens exhibit cylindrical curvature after processing. The curvature was measured by placing specimens on a 127-mm base plate (with 12.5 mm of overhang on each end to preclude end effects) and measuring the deflection above the plate. Knowing the deflection, Δ, and chord length, L, and measuring the specimen thickness, $2h$, the dimensionless curvature, η, can be calculated by

$$\eta = 2h \cdot \kappa_c = \frac{4h\Delta}{(L/4)^2 + \Delta^2} \tag{8.52}$$

The dimensionless curvature is used to preclude the influence of small thickness variations between specimens.

An intermittent cure technique was selected as a means to investigate how residual stresses develop during cure. The MRC cycle was interrupted at eight predetermined points and the specimens cooled down to room temperature at a constant rate of 5.6°C/min. The resulting specimens are incompletely cured and they provide information on the development of residual stresses during the cure cycle. Figure 8.14 shows the dimensionless curvature versus degree of cure. The curvature is initially quite small and begins to increase rapidly around $\alpha = 0.6$. It asymptotically reaches a fully cured value of 30.8×10^{-4} at full cure.

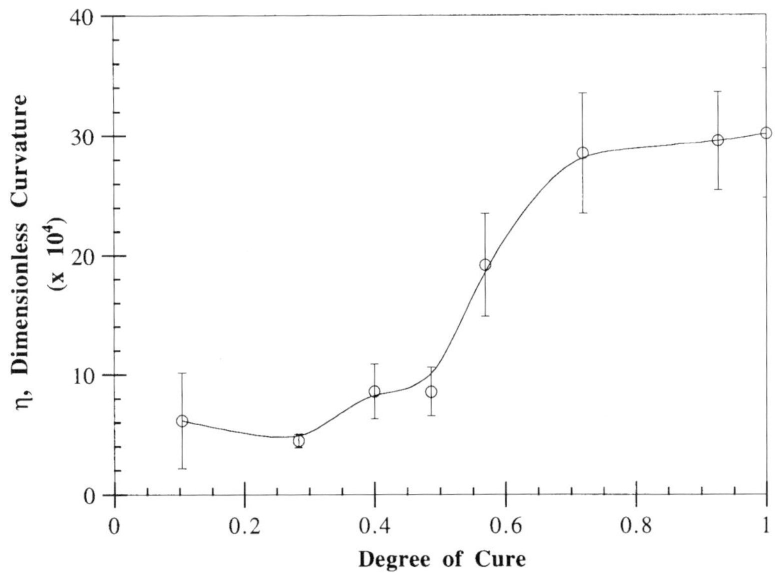

Figure 8.14 Dimensionless curvature development during cure for IM6/3100

8.3.1 Elastic Model Correlation

Elastic predictions using the constrained model of Equation 8.30 together with the curvature–moment relations of Equation 8.50 were applied to the intermittent cure specimens. Table 8.2 summarizes the input data for the elastic model for these specimens. The dimensionless curvature from the elastic model is plotted versus the experimental data in Figure 8.15. Within the experimental scatter, the elastic model predicts the curvature induced during cure except for one group of data in which the measured curvature was significantly less than the predicted value. This group of specimens was analyzed for evidence of transverse cracking, which was believed to have relieved some of the residual stresses and lowered the measured curvature. These specimens showed significant transverse cracking, and the crack densities were measured to be between 10 and 12 cracks/cm. These specimens experienced transverse cracking because of the lag in transverse strength development compared with the transverse modulus during the cure cycle. The degree of cure of these specimens was between 0.86 and 0.89. The transverse strength is quite low at this cure state, whereas the transverse modulus is nearly fully developed. Thus, during cool down the residual stresses build up until the transverse strength of the laminate is exceeded and transverse cracks are induced. Specimens cured longer than these two groups have transverse strengths that are greater than the residual stresses induced; therefore, no transverse cracking occurs. Specimens that are cured less than these groups show no signs of transverse cracking because the modulus is quite low and, thus, the residual stresses are low as well.

Table 8.2 Input Parameters (Experimental) for Elastic Model

Degree of cure	ν_{12}	E	Δe [$\mu\varepsilon$]
0.23	0.37	0.009	2510
0.82	0.35	0.013	4540
0.86	0.37	0.026	4540
0.88	0.33	0.032	4540
0.89	0.34	0.038	4540
0.91	0.29	0.042	4540
0.96	0.32	0.041	4540
0.98	0.29	0.048	4540

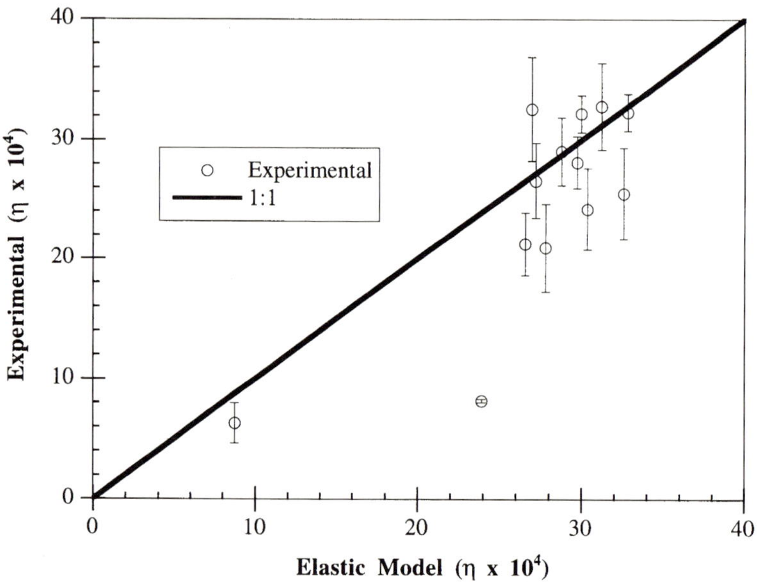

Figure 8.15 Comparison of dimensionless curvature (η) predicted by elastic model and experimental data

8.3.2 Viscoelastic Model Correlation

Based on the results of the characterization studies, the viscoelastic residual moments of the cross-ply specimens were predicted by the viscoelastic analysis using Equation 8.39 together with the curvature–moment relations in Equation 8.50 for the intermittent cure specimens.

The residual moment histories during cure for each specimen are very similar. Figure 8.16 shows the residual moment history for one particular cycle in which the final degree of cure is 0.75. There is an initial positive moment induced upon heat-up from room temperature. The moment induced is opposite in sign to the final moment since the heat-up generates expansional strains in the matrix. During cool down and during chemical shrinkage the strains are of the opposite sign. During heat-up to the second dwell temperature there is a smaller positive moment induced, but it is quickly overcome by the chemical shrinkage strain that is beginning to occur. In addition, the absolute temperature difference is smaller in comparison to heat-up from room temperature. In Figure 8.16 the rate of increase in moment is seen to decrease as the temperature approaches the first dwell temperature. This is indicative of stress relaxation occurring because the degree of cure is low and the temperature is elevated. During the first dwell period the residual moment is nearly fully relaxed as stress relaxation occurs.

In the early stages of the second dwell period the chemical shrinkage prevails and a negative moment is induced. The rate at which this negative moment is generated (the slope of the curve) is nearly equal to that during the final stages of cool down when the material behaves elastically. As the degree of cure increases, the rate of chemical shrinkage decreases and the slope of the moment curve begins to decrease. The slope of the moment curve begins to flatten as viscoelastic stress relaxation begins to become more prominent when the chemical shrinkage strains are nearing completion. As the temperature decreases further, the stress relaxation effects are reduced, no more chemical shrinkage occurs, and the material behaves elastically. The cool down curve shows a slight curvature that becomes steeper as the temperature decreases. The viscoelastic effects at high temperatures are again evidenced here. Perfectly elastic behavior would show a linear cool down curve. The final residual moment is -1.1 Nm/m.

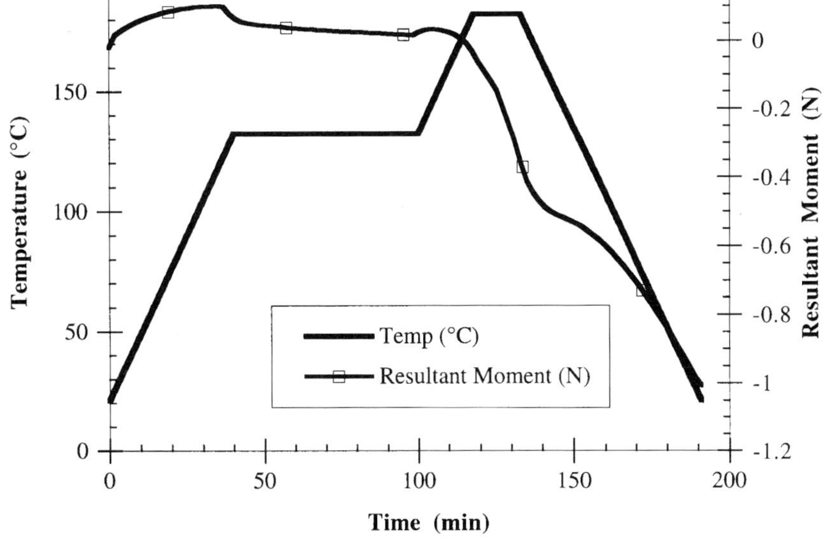

Figure 8.16 Residual resultant moment development during cure for IM6/3100

In Figure 8.17 the predicted dimensionless curvature at the end of curing and the experimental data are plotted versus degree of cure for the intermittent cure study. Most of the changes occur in the region above $\alpha = 0.8$. Up to $\alpha = 0.95$ and at $\alpha = 1$, the predictions are very good. Between $\alpha = 0.95$ and 1, however, the model underpredicts the dimensionless curvature by about 40 percent. This could be the result of various assumptions introduced in the material property models.

The contribution of chemical shrinkage to the development of residual stresses can be analyzed by taking the moment value immediately before cool down and comparing this value to the final moment. The residual moment at the beginning of cool down is primarily due to chemical shrinkage during the second dwell period. For the MRC cycle the moment immediately before cool down is -0.367 Nm/m, and the final moment is -10.0 Nm/m. Thus, the contribution by chemical shrinkage represents 3.7 percent of the final moment. The dimensionless curvature values are 1.09 for chemical shrinkage and 29.7 at the end of cool down. Neglecting the chemical shrinkage for the MRC cycle, therefore leads to an underprediction of the resulting curvature development of less than 4 percent.

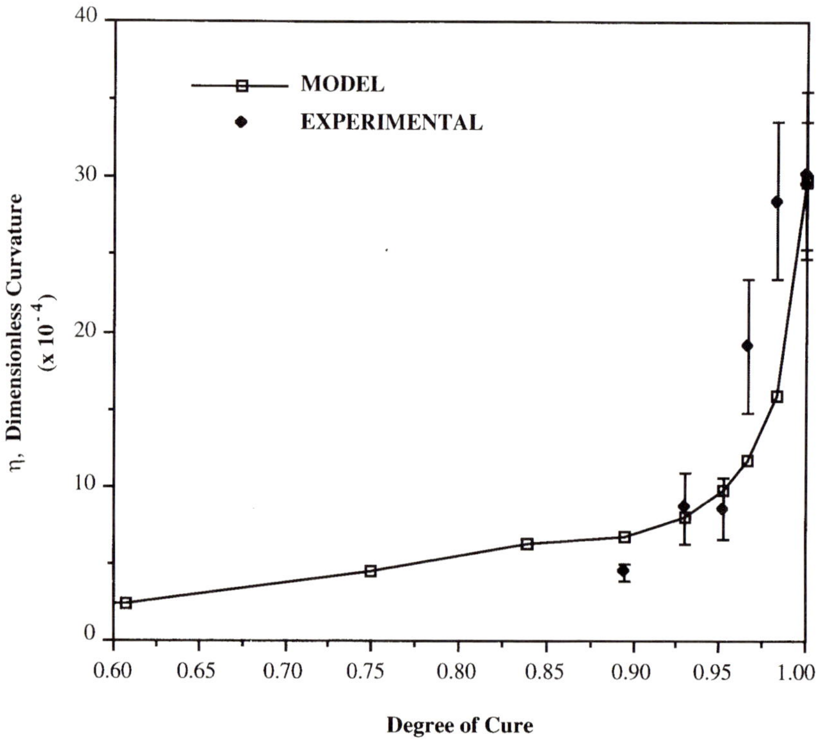

Figure 8.17 Dimensionless curvature development during MRC cycle

8.4 Processing Effects on Residual Stresses

The modeling of residual stress development during cure can be used to optimize the processing conditions to reduce or control residual stresses. The current process model is used next to assess the effects of several processing conditions on residual stresses. Reduced cure temperature, longer dwell times, slower cool down rate, and the use of novel cure cycles are all feasible for the reduction of residual stresses.

8.4.1 Cure Temperature

The MRC cycle calls for a 182°C cure temperature. The effect of cure temperature on residual stress was investigated by curing specimens at four other cure temperatures (171, 165, 160, and 149°C) while holding the dwell time (4 hours) constant. In Figure 8.18 the dimensionless curvature for these specimens is plotted versus the cure temperature. The curvature is reduced as the cure temperature is decreased with significant reduction in curvature obtained for dwell temperatures of 165°C or less. The final curvature as predicted by the viscoelastic process model is overlaid with the experimental data in Figure 8.18 and is shown to capture the trend.

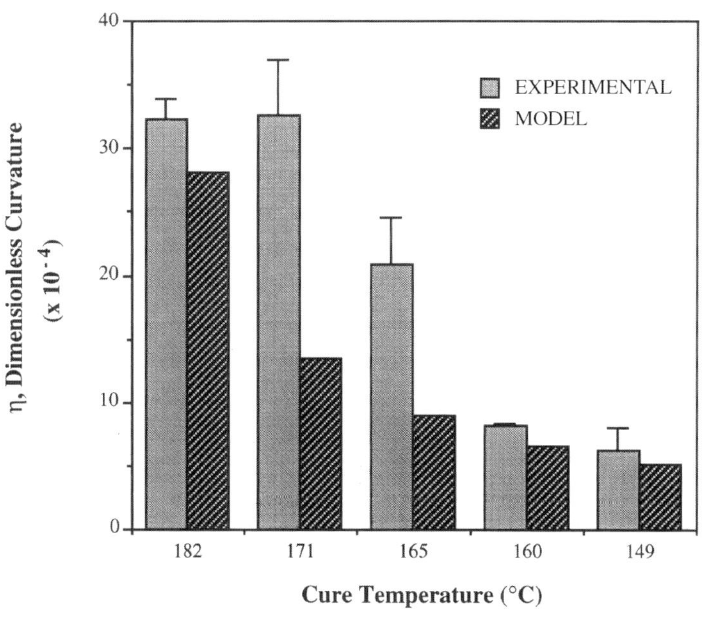

Figure 8.18 Effect of cure temperature on dimensionless curvature [model and experimental results]

Table 8.3 Effect of Cure Temperature on Degree of Cure and Transverse Mechanical Properties

Cure temperature (°C)	α Degree of cure	E_T (GPa) transverse modulus [standard deviation]	X_T (MPa) transverse strength [standard deviation]
182 (MRC)	1.0	10.2 [0.46]	25.8 [5.08]
171	0.976	8.07 [0.10]	20.0 [3.50]
165	0.947	9.04 [0.47]	15.7 [4.98]
160	0.902	—	3.45 [1.72]
149	0.784	—	2.45 [0.09]

The degree of cure for these specimens is shown in Table 8.3. Relatively complete curing (i.e. $\alpha \geq 0.95$), is obtained at 165, 171, and 182°C cure temperatures. Specimens cured at 160°C and 149°C are undercured with degrees of cure of 0.90 and 0.78, respectively. Incomplete curing leads to a sacrifice in mechanical properties, as shown in Table 8.3. The transverse strength and modulus are listed for the various cure temperatures investigated. For the 149°C cure the strength and modulus reductions are quite significant. The transverse modulus is shown to have developed sufficiently for cure temperatures greater than 165°C.

8.4.2 Postcure

The effect of postcure on the residual stress state was investigated by subjecting partially cured specimens to the standard postcure cycle, which calls for a 4-hour dwell at 221°C and ambient pressure. Dimensionless curvature and weight loss results for postcured specimens are shown in Figure 8.19. The horizontal axis is plotted as initial degree of cure (i.e., the degree of cure before postcure). The curvature is seen to increase after postcure for all specimens. A number of factors could contribute to an increase in curvature after postcure. For example, any additional chemical shrinkage that takes place during postcure will contribute to increased residual stresses. In addition, the loss of moisture during postcure reduces the moisture-induced swelling strains in the laminate that act to relieve residual stresses. In addition, an increase in the transverse modulus after postcure will result in increased curvature through elastic effects.

From Figure 8.19 the weight loss is seen to be especially prevalent in the undercured specimens. For example the weight loss for $\alpha = 0.23$ specimens was 0.54 percent after postcure. For the fully cured specimens the weight loss was only 0.14 percent.

To ascertain the effect of increased transverse modulus after postcure, nine $[90]_8$ 254 mm × 152.4 mm postcured specimens were mechanically tested to obtain the transverse modulus after postcure. Once this data was obtained the increase in curvature due

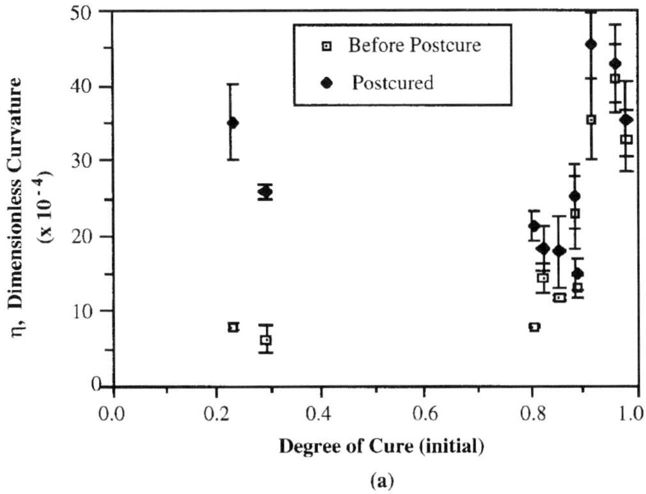

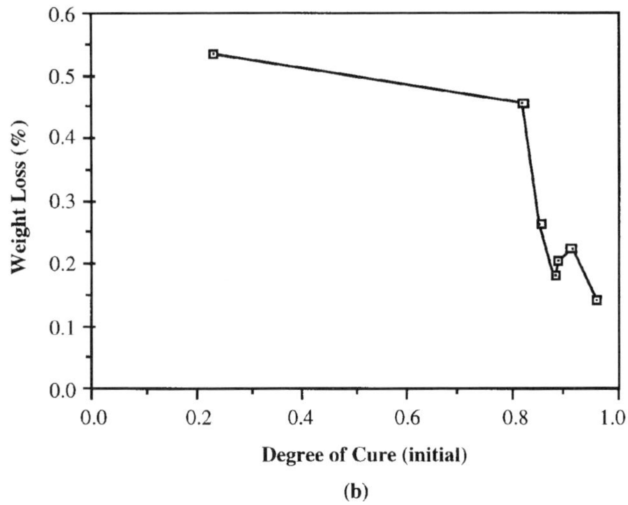

Figure 8.19 Dimensionless curvature (a) and weight loss (b) after postcure of partially cured specimens

to the change in transverse modulus was predicted using the process model. Table 8.4 shows the results of this testing and analysis. The change in curvature after postcure can be predicted by accounting for the increase in transverse modulus. The data at $\alpha = 0.80$ showed some transverse cracking before postcure that reduced the experimental curvature from the prediction.

Table 8.4 Change in Curvature Predicted after Posture

Degree of cure	ΔE_2 (GPa)	$\Delta\eta$ (Model)	$\Delta\eta$ (Experimental)
0.80	4.92	15.2	13.3
0.91	1.66	2.96	3.60
0.98	1.08	2.06	3.09

8.4.3 Three-Step Cure Cycles

Under normal processing conditions the cure temperature is a good estimate of the stress-free temperature [3]. Thus, curing at lower temperatures will lower the stress-free temperature, whereas curing at higher temperatures will raise the stress-free temperature. This effect is evident in the postcure cycle results. For the MRC specimens the curvature after postcure was shown to increase slightly, but this increase could be accounted for by the increase in the transverse modulus. Thus, the stress-free temperature remained constant and equal to the cure temperature (182°C) even though the postcure cycle occurs at 221°C. It should be possible to use this phenomenon during the primary cure in an effort to reduce the stress-free temperature.

Three-step cure cycles were investigated to mimic the behavior of the standard two-step cure cycle followed by a postcure cycle. The second step in the cure cycle allows significant curing to occur at a reduced temperature followed by a third step at higher temperature which assures completion of the cure reaction. The cure cycles investigated are shown graphically in Figure 8.20. Second dwell temperatures of 149 and 160°C were held for 2 hours, followed by a third dwell of 2 hours at 182°C. If sufficient cure occurs at the second dwell, then the stress-free temperature should be reduced from 182°C and fall between the second and third dwell temperatures. Figure 8.21 shows the dimensionless curvatures for these specimens according to the secondary dwell temperature. Both cure cycles showed a reduction in curvature of more than 20 percent from the MRC cycle. The lack of further reduction in curvature between 160 and 149°C secondary dwell cycles indicates that the dwell time at 149°C was too short. A longer dwell at 149°C would allow further curing to occur at the reduced temperature, ultimately leading to a further reduction in the stress-free temperature. Figure 8.22 shows the degree of cure development during these cure cycles. During the second dwell period the 149°C dwell has a slower cure rate and reaches a degree of cure of 0.69. In contrast the 160°C dwell reaches a degree of cure of 0.81 after 2 hours. During the final dwell at 182°C both cure cycles reach a degree of cure of nearly 1.

Mechanical testing of the three-step cure specimens indicated that no sacrifice in properties resulted from the modification of the process cycle. The retainment of mechanical properties (transverse strength and modulus) coupled with the reduction in dimensionless curvature for the three-step cure cycles investigated provides another suitable cure cycle modification for reduction of residual stresses in composite materials. Overall processing time has not been increased beyond that specified in the MRC cycle. Thus, with no increase in process time and comparable mechanical properties, the residual stresses have been reduced by more than 20 percent in comparison to the MRC cycle baseline data.

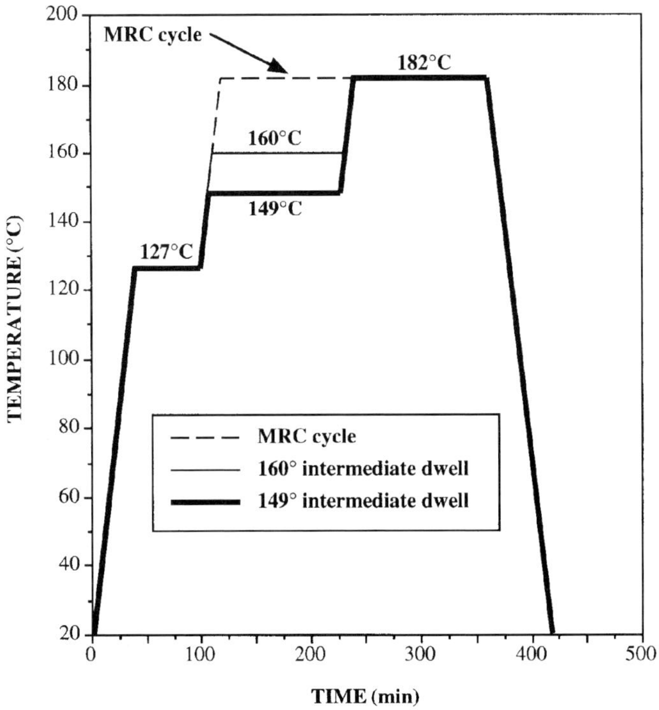

Figure 8.20 Three-step cure cycles and MRC cycle for IM6/3100

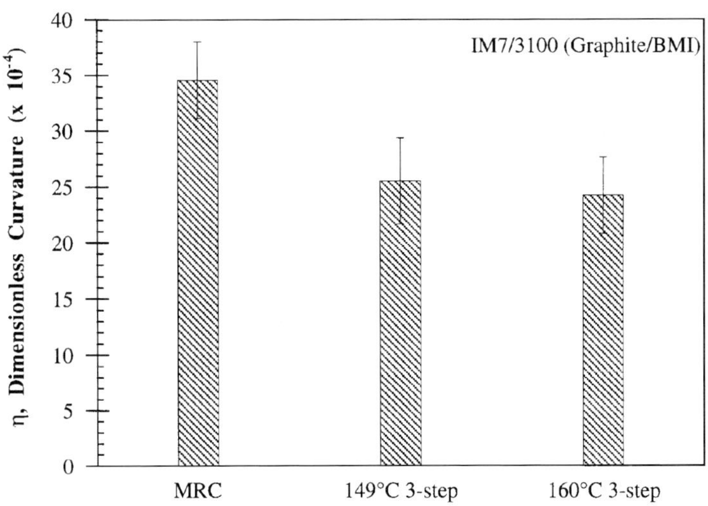

Figure 8.21 Dimensionless curvature (η) for MRC and three-step cure cycles

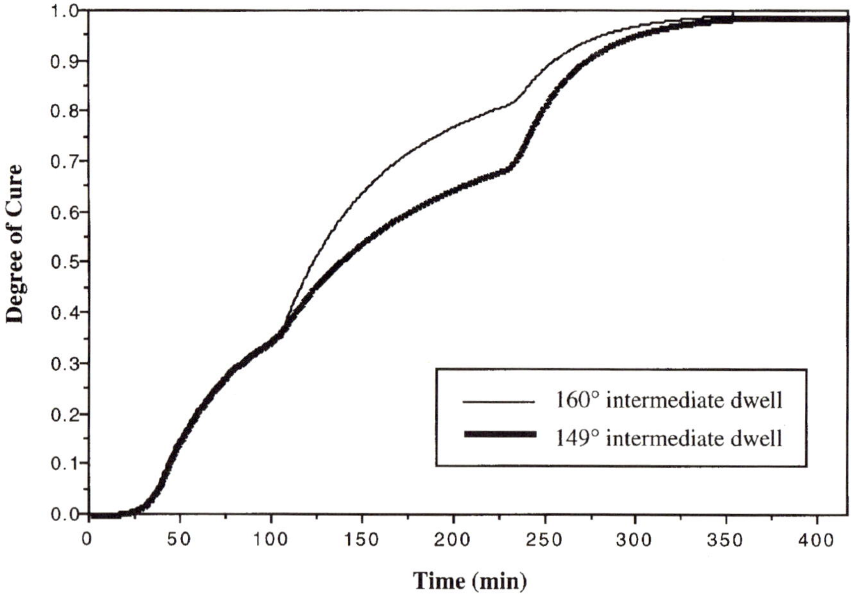

Figure 8.22 Degree of cure history for three-step cure cycles

8.5 Conclusions

Composite materials develop residual stresses because two or more phases are bonded together at elevated temperature and then cooled down to room temperature. These residual stresses develop as a result of chemical shrinkage associated with the crosslinking reaction and thermal expansion mismatch between constituent phases. Process models can be constructed to investigate how residual stresses develop during cure and how they can be controlled or reduced. In this chapter both an elastic and a viscoelastic process model were developed for unsymmetric cross-ply laminates of a graphite/BMI composite system. Thermochemical modeling yields important information characterizing the cure state of the material during the cure cycle. From this, mechanical properties and nonmechanical strains can be predicted throughout the cure cycle. The laminate curvature or residual moment can then be obtained using the elastic prediction (Eq. 8.30) or the viscoelastic formulation (Eq. 8.39). Once these are obtained, the final residual stresses for each ply can be obtained using laminated plate theory.

The methodology presented is generic and can be applied to other types of materials, laminate constructions, and processing methods. In all cases a characterization of the cure state is needed first. As a result, models must be developed to predict the mechanical response as a function of the cure state. Mechanics models can then be developed for the specific application of interest.

Nomenclature

a_{ij}	laminate in-plane compliances
a_T	shift factor
B_{ij}	laminate coupling stiffnesses
c_1–c_7	experimental constants
C_p	specific heat
D_{ij}	laminate flexural stiffnesses
D_{th}	thermal diffusivity
e_j	nonmechanical laminate strains
e_j^{ch}	laminate chemical strains
e_j^{th}	laminate thermal strains
$e_2^{ch_f}$	final transverse chemical shrinkage strain
E_1	longitudinal modulus
E_1^f	fully cured longitudinal modulus
E_1^i	uncured longitudinal modulus
E_2^{el}	elastic transverse modulus
E_2^i	initial elastic transverse modulus
F	frequency factor
g_c	universal gas constant
h	laminate half thickness
H	heat of reaction
H_R	residual heat of reaction
H_T	heat of reaction at an isothermal temperature, T
H_U	ultimate heat of reaction
J	transverse creep compliance coefficient
J_i	initial transverse creep compliance coefficient
J_f	final transverse creep compliance coefficient
k	thermal conductivity
k_{ij}	thermal conductivity tensor
K	Arrhenius kinetic constant
L	chord length
m	cure kinetics exponents
M_i	bending moments
n	cure kinetics exponents
N_i	stress resultants
q_i^{th}	heat flux vector
Q_{ij}	in-plane stiffnesses
$\dot{r}$	rate of internal heat generation
S_{ij}	in-plane compliances
S_{22}^{el}	elastic transverse compliance
t	time
T	temperature
T_0	initial stress-free temperature

x_i	Cartesian coordinates
Y_a	mechanical property constant (for anticlastic curvature)
Y_c	mechanical property constant (for cylindrical curvature)
α	degree of cure
α_{ij}^m	generalized in-plane compliances
α^*	degree of cure at initial transverse modulus development
α^{ch}	degree of cure when chemical shrinkage is complete
α_i	thermal expansion coefficients
β_{ij}	generalized coupling compliances
γ	creep exponent
γ_i	initial creep exponent
γ_f	final creep exponent
$\Gamma(\alpha, t)$	time and cure dependent mechanical properties in calculation of residual moments
δ_{ij}	generalized flexural compliances
Δ	deflection
Δe	nonmechanical strain difference (e_1–e_2)
ΔE	activation energy
ε_j	laminate strains
ε_j^0	laminate midplane strains
ξ	reduced time
η	dimensionless curvature
κ_a	anticlastic curvature
κ_c	cylindrical curvature
κ_j	laminate curvatures
ν_{12}	major Poisson's ratio
ν_{12}^i	uncured major Poisson's ratio
ν_{12}^f	fully cured major Poisson's ratio
ρ	density
σ_i	laminate stresses
τ	time

References

1. Yates, B., McCalla, B.A., Phillips, L.N., Kingston-Lee, D.M., Rogers, K.F. *J. Mat. Sci.* (1979) 14, p. 1207
2. Doner, D.R., Novak, R.C. (1969) Twenty-Fourth Annual Technical Conference, SPI, Inc., Paper No. 2-d
3. Hahn, H.T. *J. Astronautical Sci.* (1984) 32, p. 253
4. White, S.R., Hahn, H.T. *J. Comp. Mat.* (1992) 26(16), p. 2402–2422
5. White, S.R., Hahn, H.T. *J. Comp. Mat.* (1992) 26(16), p. 2423–2453
6. White, S.R., Hahn, H.T. *Polymer Engineering and Science* (1993) 27(14), p. 1352–1378
7. Gonzalez-Romero, V.M., Casillas, N. *Polym. Eng. Sci.* (1989) 29, p. 295
8. Bogetti, T.A., Gillespie, J.W. *J. Comp. Mat.* (1991) 25, p. 239
9. Loos, A.C., Springer, G.S. *J. Comp. Mat.* (1983) 17, p. 135
10. Han, C.D., Lem, K.-W. *J. Appl. Polym. Sci.* (1983) 28, p. 3155

11. Dusi, M.R., Lee, W.I., Ciriscioli, P.R., Springer, G.S. *J. Comp. Mat.* (1987) 21, p. 243
12. Harper, B.D., Weitsman, Y. AIAA/ASME/ASCE/AHS 22nd Structures, Structural Dynamics and Materials Conference, Atlanta, Georgia, April 6–8 (1981) AIAA-81-0580, p. 325–332
13. Harper, B.D., Peret, D., Weitsman, Y AIAA/ASME/ASCE/AHS Twenty-fourth Structures, Structural Dynamics and Materials Conference, Lake Tahoe, Nevada, May 2–4, 1983, AIAA-83-0799-CP.
14. Hyer, M.W. *J. Comp. Mat.* (1981) 15, 175
15. Hyer, M.W. *J. Comp. Mat.* (1981) 15, p. 296
16. Hyer, M.W. *J. Comp. Mat.* (1982) 16, p. 318
17. White, S.R. Hahn, H.T. *Polym. Eng. Sci.* (1990) 30(22), p. 1465–1473
18. Bogetti, T.A., Gillespie, J.W., Jr. *J. Comp. Mat.* (1992) 26(5), p. 626
19. Tsai, S.W., Hahn, H.T. *Introduction to Composite Materials* (1980) Technomic Publishing Co., Lancaster, Penn.
20. Harper, B.D., Weitsman, Y. *Int. J. Solids Structures* (1985) 21 p. 907
21. Weitsman, Y. *J. Appl. Mech.* (1979) 46, p. 563
22. Weitsman, Y. *J. Appl. Mech.* (1980) 47, p. 35
23. Yeow, Y.T., Morris, D.H., Brison, H.F. *Composite Materials: Testing and Design (Fifth Conference)*, (1979) ASTM STP 674, S.W. Tsai (Ed.). American Society for Testing and Materials, Philadelphia, p. 263–281.
24. Wang, A.S.D., McQuillen, E.J., Ahmadi, A.S. U.S. Naval Air Systems Command Tech. Report. (1975).
25. Plazek, D.J., Choy, I.-C. *J. Polym. Sci.: Part B: Polym. Phys.* (1989) 27, p. 307
26. Schapery, R.A., *Journal of the Franklin Institute* (1965) 279, p. 268–289
27. Harper, B.D. 1983, PhD Thesis, Texas A&M University, College Station, August, 1983
28. Schapery, R.A. *Mechanics of Composite Materials*, Vol. 2 (1974) G.P. Sendeckj (Ed.) Academic Press, New York, p. 85–168
29. Amijima, S. and Adachi, T. *Progress in Science and Engineering of Composites* (1982) T. Hayashi, K. Kawata, S. Umekawa (Eds.). ICCM-IV, Tokyo, p. 811

9 Intelligent Control of Product Quality in Composite Manufacturing

Babu Joseph and Matthew M. Thomas

Quality control is an important issue in composite manufacturing because of the high cost associated with the processing steps. The lack of accurate processing models, inadequate sensors to monitor quality on-line, and the batch or sequential nature of processing all make quality control difficult to achieve. Advances in computer hardware, artificial intelligence, model predictive control, and sensor technology have enabled the implementation of so-called intelligent or smart control strategies in composites processing. Such strategies attempt to combine prior experiential knowledge and available processing models to predict product quality so that any deviations from the quality specifications can be detected and corrected before the processing is completed. This chapter examines the evolution of these intelligent processing strategies and the current research in this field.

9.1 Introduction

Quality control is defined as the ability to adjust processing conditions in response to raw material variations (e.g., resin content, fiber properties, resin properties, impurities present, etc.), product requirements (e.g., shape, size, strength, etc.), and unexpected situations (e.g., sensor bias, sensor failure, changes in the equipment processing environment, etc.). The end result of such adjustment must be a product—a composite part, in this case—that meets product quality specifications (e.g., target thickness, minimal void content, maximum structural and strength properties, etc.). Due to the complex nature of composite manufacturing, automatic control of the process must be realized through experience from past operational history of the process, as well as from an understanding of the physical and chemical events that take place during the processing cycle.

Automation and on-line quality control in the composite manufacturing industry have been stimulated by developments in computer technology (e.g., artificial intelligence, neural network modeling, increased microprocessor speed, etc.), inferential control, control of nonlinear systems, and new sensor technology (e.g., dielectric cure sensors, fiberoptic sensors, noncontact sensors, etc.). Synergism of these technologies is the driving force behind the development of intelligent process control strategies. By intelligent control, we mean control that uses prior experiential knowledge and available process models to manipulate the processing conditions in response to available on-line sensor data.

In this chapter, we will review the current practice using traditional SPC/SQC techniques, knowledge based control strategies, model predictive control strategies, and models for on-line control. A novel control strategy called Shrinking Horizon Model Predictive Control (SHMPC), which uses nonlinear process models, secondary process measurements, and on-line optimization to minimize quality variations, will be presented in detail. Results from autoclave curing fiberglass–epoxy composite laminates using such a control strategy will also be given. The chapter concludes with a discussion of future trends in this field.

Whereas intelligent control comprises the aforementioned computer technology, control methods, and sensors, the control methods and advances in them are the focus of this chapter. Sensors parallel control methods in their importance to intelligent control of product quality, but a discussion of them is left to other works [1–8]. The control methods discussed in this chapter are on-line supervisory control methods, as opposed to regulatory control methods. The latter are commonly used, for example, to keep the operation of an autoclave in control: When a ramp in temperature is called for, regulatory control methods govern the autoclave in achieving that ramp; however, the supervisory control determines the ramp rate in order to influence product quality.

Section 9.2 will review traditional statistical process control/statistical quality control (SPC/SQC) techniques used in quality control. Section 9.3 will follow this review with a discussion of techniques based primarily on an experiential rule base and expert system technology. Section 9.4 will discuss control strategies that use an on-line process model; a variety of models can be used in such model predictive control. Section 9.5 will discuss this variety of models. Section 9.6 will summarize this chapter and discusses future trends in the field.

9.2 Traditional Approaches Using SPC/SQC

The traditional approach to quality control is to generate charts of various kinds to monitor the performance of a production unit. At a superficial level, *statistical process control* (SPC) and *statistical quality control* (SQC) [9] are terms used interchangeably to describe traditional process monitoring techniques. Processes lacking physical models and providing an abundance of data are well suited for SPC application. With heightened interest in Total Quality Management concepts and in the ISO 9000 standards, the first tool that quality seekers have turned to is SPC. SPC techniques rely upon various types of control charts to determine if a process is under control. Charts that are commonly used include Shewhart charts, Cumulative SUM (CUSUM) charts, and Exponentially Weighted Moving Average (EWMA) charts.

The Shewhart chart graphically tests the hypothesis that a process measurement is not different from the desired target value of that measurement. If one assumes that the measurement, y_k, follow a normal distribution about the target value, y_t, assuming that the target value y_t represents the true mean y_m for the given measurement, and assuming that the standard deviation σ for the process is known, then

$$\Pr[(y_t - 3\sigma) < y_k < (y_t + 3\sigma)] = 0.9973 \tag{9.1}$$

(i.e., the probability that y_k will be within $\pm 3\sigma$ of y_t is 0.9973). Only 27 out of 10,000 measurements, are expected to fall outside of this range. The Shewhart chart plots measurement y_k versus discrete time, with a center line $y = y_t$ representing the target for that measurement, a line at $y = y_t + 3\sigma$ which represents an upper control limit (UCL), and a line at $y = y_t - 3\sigma$ which represents a lower control limit (LCL). If a value of y_k should fall above the UCL or below the LCL, there is a 0.9973 probability that a significant event has occurred. When such an event occurs, those responsible for keeping the process in control (i.e., keeping LCL $< y_k <$ UCL) must determine what physical cause triggered that event. The Shewhart chart is of no help in making this determination.

Shewhart charts are adept at detecting mean value shifts on the order of 3σ or higher. To detect more subtle shifts in the mean value, the CUSUM chart has been developed [10,11]. The cumulative sum is defined as:

$$S_k = \sum_{i=1}^{k} (y_i - y_t) \tag{9.2}$$

where y_i is the ith measurement in the process, y_t is the target value for that measurement, and k is the number of time steps into the process. The CUSUM chart, then, is a plot of S_k versus time index, k. The difference in visual representation between Shewhart and CUSUM charts is that the latter indicates both that a significant event has occurred as well as when it occurred (i.e., when S_k no longer falls randomly about zero, but begins to show steady, sustained growth or decay from zero). In contrast to Shewhart charts, CUSUM charts do not feature explicit UCL and LCL lines. As with the Shewhart chart, however, the CUSUM chart cannot tell the process operator why a significant event has occurred.

Of a class of moving average techniques that fall between Shewhart and CUSUM, the most popular is the EWMA method [12,13]. The EWMA introduces $y_{\mathrm{EWMA},k}$, where

$$y_{\mathrm{EWMA},k} = r y_k + (1 - r) y_{\mathrm{EWMA},k-1} \tag{9.3}$$

where $0 < r \leq 1$. The variable $y_{\mathrm{EWMA},k}$ denotes the geometric moving average or exponentially weighted moving average of measurements, y. Weights on each measurement are a function of r, and they decrease as a geometric progression from the most recent measurement to the first. Just as a Shewhart chart plots y_k versus discrete time an EWMA chart plots $y_{\mathrm{EWMA},k}$ versus discrete time, the EWMA chart can be more sensitive to significant events than its Shewhart precursor; however, it is, as easy to use as a Shewhart chart; EWMA and Shewhart charts are often used in tandem as part of an SPC strategy. If $r = 1$, the two charts are the same. As with Shewhart and CUSUM charts, though, an EWMA chart cannot reveal the physical meaning behind a significant event.

Unlike SPC techniques, standard feedback control methods such as PID-control, do exert control upon a process, in an effort to minimize $|y_t - y_k|$. Control in Statistical Process Control is as such not regulatory control, but a semantic means of relating SPC to quality control — a means that often leads to the hybrid term SQC. Ogunnaike and Ray [14, Sec. 28.4] offer advice on when to use SPC and when to use standard feedback control methods: When the sampling interval is much greater than the process response time, when zero-mean Gaussian measurement noise dominates process disturbances, and when the cost of regulatory control action is considerable, SPC is preferred.

As can be seen, these SPC/SQC methods provide a means of determining if a process is under control, but they provide little help in the way of taking corrective action should a process not be under control. For the latter, the process engineer usually must determine the quality deviation cause, and determine what — if any — action is necessary to return the process to normal. Prior experience is of considerable value in making this decision. This value led to the investigation of knowledge-based control strategies that relied on a set of control rules derived from prior experience.

9.3 Knowledge-Based (Expert System) Control

The knowledge-based or rule-based expert system (KBES) arguably is the most visible technology to emerge from the artificial intelligence (AI) heyday of the late 1980s. Expert systems are based upon heuristic rules developed through a process known as knowledge acquisition: A KBES comprises a "knowledge base" of such rules, and an "inference engine" that determines (1) when to "fire" (implement) a rule and (2) what rule to fire in case of a conflict among rules. This knowledge acquisition process, through which KBES rules are developed and encoded for a physical system, has been considered the most important — and most problematic — aspect of expert system development [15]. Inference engines are primarily based upon first-order predicate logic; advances upon such logic (e.g., nonmonotonic logic and defeasible reasoning) result in more sophisticated inference engines. The primary languages for KBES development have been LISP, PROLOG, C, and C++; "shells" written in these languages allow users to develop a KBES without knowing the underlying language.

Autoclave curing parameters have traditionally been set on the basis of previous curing cycles. The ramp rates, hold durations, hold temperatures, and time and magnitude of pressure application are fixed in advance of the curing, and the cure cycle is run without deviation from this fixed "recipe." No on-line modifications are made to this recipe based upon temperatures recorded during the run (industrial autoclaves traditionally operate with but one thermocouple in, on, or near the stacked prepreg). Early on-line curing control efforts at the regulatory level relied heavily upon dielectric sensor feedback to characterize resin viscosity [16–19]. The seminal efforts at on-line curing control above the regulatory level came through KBES application, based upon the "Naïve Physics Manifesto" [20] and Qualitative Process Theory [21]. The most significant of these efforts was Qualitative Process Automation (QPA).

Evolving from efforts [22] to use the best features of trial-and-error, process model, expert system, and expert model approaches, QPA [23–25] combines KBES traits with on-line dielectric, pressure, and temperature data to implement autoclave curing control. QPA combines extensive sensor data with KBES rules to determine control actions. These rules determine curing progress based upon process feedback, and implement control action. QPA adjusts production parameters on-line; as such — within the limits of its heuristics — QPA can accommodate batch-to-batch prepreg variations.

The heuristics in QPA, though extracted from process data, do not ignore first (physical) principles. An example of such is a rule designed to prevent a runaway exothermic reaction within the prepreg resin during its cure. Given a flowing resin ready for curing: If prepreg stack temperature is rising and the rate of temperature increase is also rising, QPA assumes that a runaway reaction is underway and cools the autoclave. Here is the rule that QPA fires:

$$\text{If} \quad \frac{dt}{dt} > 0 \quad \text{and} \quad \frac{d^2T}{dt^2} > 0 \quad \text{then} \quad \text{COOL} \quad \text{else} \quad \text{HEAT}$$

Because it considers no analytical model, QPA does not make explicit use of heat transfer dynamics. Nonetheless, QPA does reduce the autoclave curing cycle durations in several experimental autoclave curing runs.

A rule-based expert system developed by Springer and coworkers [26,27] also governs autoclave temperature and pressure in fiber-reinforced thermoset resin composite curing. The intent of that KBES is to produce a fully compacted and cured laminate with minimal void content and minimal residual (curing) stresses. Inputs to this KBES are instantaneous autoclave temperatures and pressures, prepreg stack midpoint and surface temperatures, instantaneous composite thickness, and ionic conductivity. Many autoclaves cannot be instrumented with the means for on-line thickness and ionic conductivity measurements; however, such measurements when available are useful to any controller. Given these inputs, the KBES outputs are autoclave heater, cooler, and pressure settings.

The objective of the Springer KBES is twofold: To ensure a high-quality part in the shortest autoclave curing cycle duration. This KBES is similar to QPA in that sensor outputs are combined with heuristics not with an analytical curing model. The rules for compaction dictate that dielectrically measured resin viscosity be held Constant during the First temperature hold in the autoclave curing run. The autoclave temperature is made to oscillate about the target hold temperature in an attempt to attain constant viscosity. Full pressure is applied from the cure cycle start.

Rules in this KBES govern void content (through manipulating saturation pressure), degree of cure, and residual stresses (through manipulating maximum autoclave temperature). All of these rules must be ready to fire concurrently; the KBES has to perform resolution among conflicting rules. This KBES has been used to cure several laminates of varying prepregs, stack thickness, and stack surface area: Results in cure duration, mechanical properties, and void content are said to be competitive with those from standard curing recipes.

Perry and Lee [28,29] offer an enhancement of QPA, based upon use of dual heat flux sensors and additional thermocouples in autoclave curing. This enhancement entails determining heat transfer properties during the cure, then using these properties in conjunction with PID regulatory control of autoclave temperature. Using the additional sensors, Perry and Lee employ an on-line Damköhler number in lieu of the second time-derivative of temperature to avoid exothermic thermal runaway within the prepreg stack thermoset resin. The Damköhler number is defined as:

$$Da = \frac{L^2(\Delta H_{r \times n}) r_{\text{REF}}}{k(T_1 - T_0)} = \frac{\text{rate of heat generation}}{\text{rate of heat conduction}} \tag{9.4}$$

and L is prepreg stack thickness (cm), r_{REF} is a reference rate of curing (gmol/cm_3-sec), and T_1 and T_0 are the maximum and minimum autoclave temperatures (K) during the cure. Perry and Lee use a discretized version of Da, estimated and updated on-line using the additional sensors, to determine three particular curing states: $Da > 0$ (onset of curing); $Da > 1$ (acceleration of curing); $Da = 0$ (end of cure). That the on-line Da number indicates these states was confirmed by differential scanning calorimetry. Da proves a more accurate indicator (than the noise-ridden second temperature derivative) of these states.

Pillai et al. [30] used an approach said to differ from those involving heuristics, but still involving a collection of previously used temperature recipes, to minimize the required autoclave curing duration. An expert system is still used, but the role of heuristics is downplayed in favor of Process Trend Analysis (PTA). The critical autoclave-curing decision is when to begin the second linear temperature ramp and how steep that ramp should be. According to this approach, the second ramp should begin 7 min after an inflection point in the temperature at the prepreg stack center, or 3 min after a temperature inflection point on the prepreg stack surface, depending upon which occurs first. A heuristic, extracted from previous runs, governs that setting, but the rest of the cure is run according to an optimal temperature profile compiled from previously used profiles. This profile is updated on-line by the expert system, which uses an inverted autoclave response model to modify the temperature set point every 30 seconds. This approach was used in an experimental cure of a 6-in. by 6-in. prepreg stack 1-in. (42 plies)-thick; degree of cure and gel point were computed for that curing run.

Wu and Joseph [31,32] presented a KBES for real-time regulatory control of chemical process systems, and demonstrated this KBES on a simulated autoclave curing process. Inputs to the KBES were thermoset resin-type, fiber-type, fraction of resin and of fiber in the precured prepreg stack, and initial autoclave temperature. An initial operating plan that best matches the current initial conditions is retrieved from an existing data file of past curing runs. During the cure, KBES rules classify process variables; these classifications are then sent to a "blackboard," which coordinates control actions. This approach combines heuristics with a one-dimensional process model: It aims to apply pressure to the prepreg stack while the thermoset resin viscosity is lowest. Its primary goal is to minimize void size and deviation from target thickness within the cured laminate. Wu and Joseph also studied the use of fuzzy logic to incorporate uncertainty into the measurements and the rules.

Fuzzy logic represents a quantification of the first-order predicate logic that is the backbone of KBES. Given that quantification, fuzzy logic improves the KBES' capability to solve problems such as car steering [33] and parallel parking. For the latter, the ill-parked car has either one or two tires up on the curb, or is so far away from the curb that it impedes traffic. The well-parked car, however, falls between these extremes and is equally well-parked, be it 2, 3, 4, 5, 6, or 7 cm from the curb — the quantitative measurement is irrelevant. Given the irrelevance of such quantitative measurements, KBES and fuzzy logic approaches are well suited for such problems. In the outdated definition of quality, quantitative measurements of controlled variables — as long as they fell between the UCL and LCL — were just as irrelevant: Hitting the target was no better than falling just above the LCL or just below the UCL.

The modern definition of *quality* [34], which is at the core of high-volume industrial production, has an impact on the increasing commercialization of composites manufacturing. In this modern definition, quality minima still occur outside the lower and upper control limits. The quality maximum still occurs when the controlled variable perfectly matches its target value, but a quality decrease now occurs *within* the control limits as a function of controlled variable deviation from the target value. Proponents of qualitative control methods accurately noted the shortcomings of SPC [24] for [modern] quality control. As it happens, however, KBES approaches (even fuzzy-logic enhanced) to modern quality control also have shortcomings: It is the outdated quality definition to which these approaches are best suited. The solutions to those shortcomings paradoxically lie in *quantitative*, not *qualitative*, approaches to maximizing *quality* (as now defined). We will examine these quantitative approaches next.

9.4 Model-Based (Model-Predictive) Control

For purposes of this chapter, model-predictive control can be discussed in the context of continuous and batch process applications. The former constitutes traditional applications; the latter describes processes such as autoclave curing. We will discuss the former first, to put the latter into context.

9.4.1 Model-Predictive Control of Continuous Processes

When thermodynamics or physics relates secondary measurements to product quality, it is easy to use secondary measurements to infer the effects of process disturbances upon product quality. When such a relation does not exist, however, one needs a solid knowledge of process operation to infer product quality from secondary measurements. This knowledge can be codified as a process model relating secondary to primary measurements. These strategies are within the domain of model-based control: Dynamic Matrix Control (DMC), Model Algorithmic Control (MAC), Internal Model Control (IMC), and Model Predictive Control (MPC — perhaps the broadest of model-based control strategies).

DMC has its origins in industry. Cutler and Ramaker [35] outlined the DMC strategy; Prett and Gillette [36] showed its application to a fluid catalytic cracking unit. The most popular version of DMC is now Quadratic Dynamic Matrix Control (QDMC) [37,38] which

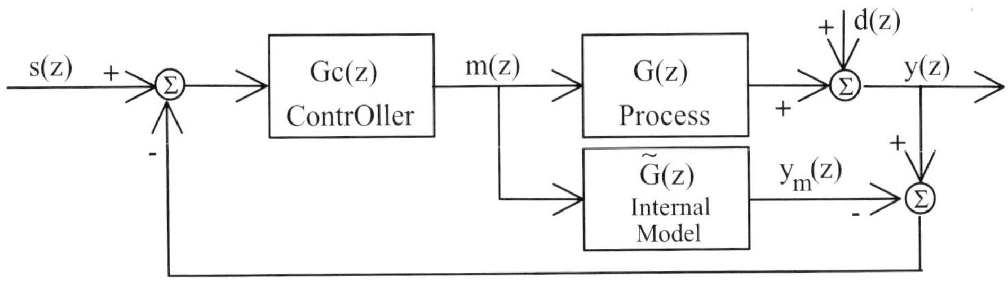

Figure 9.1 Basic internal model control (IMC) structure

formulates DMC as a quadratic expression to be minimized — subject to linear process inequality constraints—through quadratic programming approaches.

Model Algorithm Control (MAC), introduced as model-predictive heuristic control [39], is strongly similar to DMC. Rouhani and Mehra [40] provided the basic theoretical properties of MAC through a detailed treatment of the single-input–single-output case. IDCOM (IDentification and COMmand) is a commercial package that implements the model-based MAC strategy; Froisy and Richalet [41] surveyed industrial applications of IDCOM.

The similarity between DMC and MAC has been formalized; that formalism has been labeled the IMC structure. Garcia and Morari [42] produced the seminal work on IMC, summarizing DMC- and MAC-based results, developing the structure illustrated in Figure 9.1, and acknowledging that inferential control concepts [43] motivated the IMC structure.

The MPC strategy can be summarized as follows. A dynamic process model (usually linear) is used to predict the expected behavior of the controlled output variable over a finite horizon into the future. On-line measurement of the output is used to make corrections to this predicted output trajectory, and hence provide a feedback correction. The moves of the manipulated variable required in the near future are computed to bring the predicted output as close to the desired target as possible without violating the constraints. The procedure is repeated each time a new output measurement becomes available.

MPC and other versions of model-based control have gained widespread industrial acceptance, according to surveys [44,45] of MPC in industry. With widespread acceptance in industry, and with research in the area continuing, model-based control in general and MPC in particular are common strategies for controlling continuous processes — processes with no well-defined start and end points; processes with no discrete stages or phases — which can be modeled reasonably well using linear input–output models.

9.4.2 Model Predictive Control of Batch Processes (SHMPC)

Model predictive control, as put forward in Section 9.4.1, is commonly known in its linear form as *receding* horizon model predictive control and has been used primarily in the control of continuous processes. Work with receding horizon model predictive control generally assumes linear systems. Nonlinear model predictive control is needed [14] for batch and semi-batch processes — processes that are carried out in their entirety in transient mode; processes that have no steady state. Two practical issues have made difficult the use of a nonlinear model in an MPC context: the rarity of good nonlinear process models and nonlinear model inversion difficulty. Biegler and Rawlings [46] survey optimization approaches to nonlinear MPC; these approaches often entail solving a nonlinear programming (NLP) problem. A survey of approaches to MPC for both constrained linear and constrained nonlinear plants is presented by Rawlings et al. [47].

In direct contrast to traditional MPC, in which the control horizon recedes or moves, Shrinking Horizon Model Predictive Control (SHMPC) [48–50] features a control horizon that narrows or "shrinks" with time. SHMPC is geared toward batch chemical processes — processes with well-defined start and (especially) end points; processes for which the window of control opportunity decreases as the batch run progresses. The SHMPC strategy is not limited to using linear process models; indeed, it cannot be if it is to be used on inherently nonlinear and dynamic batch processes, such as autoclave curing. Pardee et al. [51] discussed pyrolysis optimization applications that use secondary measurements, recognize linear control theory limitations, and consider "observation" (control) and "planning" (prediction) horizons: As Brosilow [43] linked inferential control to IMC, so do Pardee et al. appear to link qualitative approaches to SHMPC.

A batch chemical process comprises discrete stages. In autoclave curing of composites, for example, industry generally accepts (though salient exceptions do exist) that there are four well-defined stages: an initial temperature ramp, a fixed temperature hold, a second temperature ramp, and a second fixed temperature hold. SHMPC differs from traditional MPC, in that SHMPC is a quality control strategy operating at the supervisory or optimization level in a control hierarchy, whereas the other approaches operate at the regulatory control level of that hierarchy.

In SHMPC, m denotes the number of phases during which manipulated inputs can be altered (i.e., during which control action can be taken), and p denotes the number of discrete phases in the batch process. Both horizons shrink as the batch run progresses. The prediction horizon shrinks from p (initially) to p-i (after phase i); similarly, the control horizon shrinks from m to m-i. The inequality $m < p$ describes those instances when the batch run continues after the control horizon shrinks to zero. A phase changes with each measurement or each manipulated input setting made during the batch run; however, measurements for the batch process are not made continuously, but are instead recorded at discrete times during the run. For application of SHMPC, these measurements must be richer in process information than measurements used in simple inferential control of continuous processes. As is the case with KBES approaches, a localized temperature ramp rate (for example) is much more likely to be useful than a temperature reading every 10 s in SHMPC. There should be a physical relation between these measurements and predicted final quality variable values; in the absence of such a relationship, there must be an inherent understanding of the batch process and its models. Details behind the measurements for applying SHMPC to autoclave curing are presented by Thomas [52].

Figure 9.2 illustrates the inner workings of SHMPC during a given phase i. In this figure, let $\mathbf{x}_0$ be the initial state, $\mathbf{d}_i$ be a vector of measured disturbances available after phase i, $\mathbf{y}_i$ be a

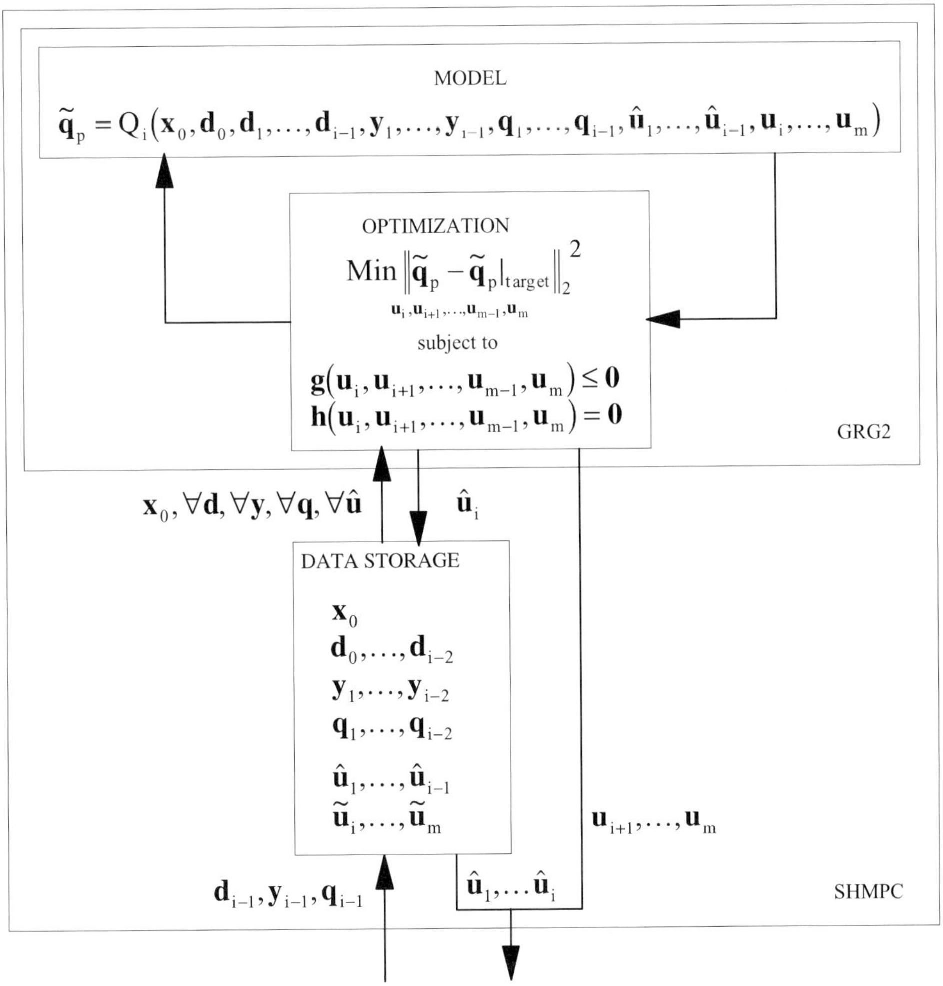

Figure 9.2 Shrinking horizon model predictive control (SHMPC)

vector of secondary measurements available after that phase, $\mathbf{q}_i$ be postphase i primary measurements (quality variable values), $\hat{\mathbf{u}}_i$ be postphase i manipulated inputs fixed by the SHMPC scheme, and $\mathbf{u}_i$ be postphase i manipulated inputs not yet fixed. The premise of SHMPC is to use all information available, at each phase of the batch process, to predict final quality variable values, and to fix manipulated inputs to maximize the likelihood of realizing those final values. The predicted final quality variable values are denoted by $\tilde{\mathbf{q}}_p$. During each phase i, $\tilde{\mathbf{q}}_p$ is predicted by linear or nonlinear model Q_i.

In SHMPC, newly available values of primary and secondary measurements are combined with previously obtained values, as well as with manipulated input settings previously fixed by the SHMPC scheme. All of these values are then fed to the optimization component of the SHMPC scheme. That component addresses the problem:

$$\min_{\mathbf{u}_i, \mathbf{u}_{i+1}, \ldots, \mathbf{u}_{m-1}, \mathbf{u}_m} \|\tilde{\mathbf{q}}_p - \tilde{\mathbf{q}}_p|_{\text{target}}\|_2^2$$

subject to

$$\mathbf{g}(\mathbf{u}_i, \mathbf{u}_{i+1}, \ldots, \mathbf{u}_{m-1}, \mathbf{u}_m) \le \mathbf{0}$$

and

$$\mathbf{h}(\mathbf{u}_i, \mathbf{u}_{i+1}, \ldots, \mathbf{u}_{m-1}, \mathbf{u}_m) = \mathbf{0}$$

by interacting with the model component of SHMPC. This model component comprises the linear or nonlinear model:

$$\tilde{\mathbf{q}}_p = Q_i(\mathbf{x}_0, \mathbf{d}_0, \mathbf{d}_1, \ldots, \mathbf{d}_{i-1}, \mathbf{y}_1, \ldots, \mathbf{y}_{i-1}, \mathbf{q}_1, \ldots, \mathbf{q}_{i-1}, \hat{\mathbf{u}}_1, \ldots, \hat{\mathbf{u}}_{i-1}, \mathbf{u}_i, \ldots, \mathbf{u}_m) \tag{9.5}$$

A nonlinear optimization strategy, fast enough to work on-line, is required to combine the model and optimization components of SHMPC. A common application of this strategy has been provided by the GRG2 software [53], which combines these components within a generalized reduced gradient algorithm for solving NLP problems. During a given phase, free manipulated input settings are chosen so as to minimize deviation of $\tilde{\mathbf{q}}_p$ from its target value; previously fixed manipulated input settings, which have been implemented and are no longer adjustable, serve only as model inputs. The process of choosing free manipulated input settings is a numerical process rather than an analytical one. It is governed by the inherently quantitative nonlinear optimization strategy not by qualitative rules. As manipulated inputs are fixed, they no longer play an active role in SHMPC; rather, they become other model inputs, much as the primary, secondary, and disturbance measurements are other model inputs.

Figure 9.2 illustrates the broadest possible application of SHMPC, which is an application in which vectors **d**, **y**, and **q** are available at each phase. The applications where all vectors are present in each phase are rare. Often, especially in autoclave curing, only a secondary measurement or a measured disturbance will be available in a given phase. In the nomenclature of Figure 9.2, most of the vectors in the batch process phases will be null vectors. This general lacking of available information leads to a sparseness or "parsimony" in the SHMPC models, which is a favorable condition in that the effects of fewer variables upon $\tilde{\mathbf{q}}_p$ need to be gauged. For autoclave curing, quality variable vector **q** is a null vector until after the final phase p because real-time primary measurements cannot be made.

In this section, we examined the model predictive control strategies that can be applied to composite manufacturing. We did not, however, discuss the *models* that are useful in such strategies. That discussion will be presented in the following section.

9.5 Models for On-Line Control

Batch processes, such as autoclave curing, are inherently nonlinear and dynamic. For on-line quality control, the model must predict the outcome of the batch (i.e., product quality) in terms of the input and processing variables. The variables associated with the process are:

1. Measured disturbances, **d**: Uncontrollable variations in raw material (prepreg) properties and variations in environmental variables such as prepreg storage humidity, initial weight fraction of resin in the prepreg, amount of impurity in the prepreg, and prepreg age.
2. Manipulated inputs, **u**: Variables such as pressure, time of pressure application, first and second hold temperatures, first and second temperature hold durations, and temperature ramp rates. These variables represent changes in regulatory controller set points with time, the time sequence of process events, and corrective actions that can be taken during the process.
3. Intermediate secondary measurements, **y**: Dependent variables whose values are recorded by sensors used in the process. These variables are indirect indicators of final product quality; as such, they should be useful in predicting autoclave curing outcomes.
4. Output quality variables, **q**: End-process measurements of product laminate quality.

A model must have the following properties for on-line quality control purposes

- It must relate product quality to the manipulated input variables.
- It must take into account the effect of measured disturbances.
- It must make use of intermediate secondary measurements.

For use in a strategy such as SHMPC, a model must also be relatively small because it must provide real-time output in an on-line application. In a batch process, such as autoclave curing, model output must be available before the process moves into its next phase (i.e., before the next measurement is recorded). This requirement for real-time models in an on-line application limits the types of models that can serve in SHMPC.

9.5.1 Categories of Models

Quantitative models for predicting quality can be classified into two categories: (1) *fundamental process models*, which are based on physical and chemical events that occur in the autoclave, and (2) *regression-type models*, which are based on a statistical fit of the observed product quality to the input raw material properties and the process conditions used.

When available, fundamental process models are preferred. For many complex processes such as composite manufacturing in general and autoclave curing in particular, however, these models are often not available. This lack of availability is due to an inadequate understanding of the complex events that take place during the process. A fundamental process model is occasionally available, but it is still unsuitable for on-linc model predictive control application due to the extensive computing time required to solve the model's equations. This lack of suitability is especially the case when non-Newtonian flow through complex geometric shapes is involved. Fundamental process models for autoclave curing have long been under development [54,55]; however, these models are not yet in a form suitable for on-line implementation. Another difficulty for fundamental process models is accommodating available on-line process measurements.

For continuous process systems, empirical models are used most often for control system development and implementation. Model predictive control strategies often make use of linear input–output models, developed through empirical identification steps conducted on the actual plant. Linear input–output models are obtained from a fit to input–output data from this plant. For batch processes such as autoclave curing, however, the time-dependent nature of these processes — and the extreme state variations that occur during them — prevent use of these models. Hence, one must use a nonlinear process model, obtained through a nonlinear regression technique for fitting data from many batch runs.

Regression-type models rely upon data collected from test and production runs. Because these models represent a statistical fit, they can predict product quality accurately only over a narrow operating window. Operations outside this window require collection of new data for updating the model. Though regression-type models are not fundamental process models, fundamental process knowledge should be used as much as possible to set the input variables for regression-type models. Doing so eliminates meaningless inputs, thus preventing the model from growing too large and thus too slow. Doing so also lends fundamental process meaning to the regression-type model inputs, thus lowering the chances of model regression abuse that arises from spurious inputs [56].

Traditional regression-type models have been linear and quadratic regression models. Linear and quadratic regression models unfortunately impose further constraints upon the nature of the process nonlinearity; as such, these models are limited in the range of their applicability. A relatively new nonlinear regression-type model — the Artificial Neural Network (ANN) — is not as limited, and is worthy of additional discussion.

9.5.2 ANNs as On-Line Quality Models for SHMPC

Artificial neural networks arose from efforts to model the functioning of the mammalian brain. The most popular ANN — the *feedforward* ANN — has deeper roots in statistics than in neurobiology, though. A form of ANN (a Probability Neural Network) has been used within a QPA context to improve sensor data reliability, but not as an on-line quality model [57]. The best way to represent a feedforward ANN as an on-line quality model for SHMPC is

$$\mathbf{q} = S(\mathbf{B} \cdot \mathrm{S}(\mathbf{A} \cdot \mathbf{x} + \mathbf{a}_{\text{bias}}) + \mathbf{b}_{\text{bias}}) \tag{9.6}$$

where $\mathbf{x}$ is a vector of inputs (measured disturbances, manipulated inputs, and intermediate secondary measurements), $\mathbf{q}$ is a vector of output quality variables, and matrixes $\mathbf{A}$ and $\mathbf{B}$ and vectors $\mathbf{a}_{\text{bias}}$ and $\mathbf{b}_{\text{bias}}$ contain fitted parameters ("weights" in ANN parlance). Operator $\sum(\,)$ is a vector operator defined by

$$S(\mathbf{w}) = [(1+e^{-w_1})^{-1}(1+e^{-w_2})^{-1}\cdots(1+e^{-w_{n-1}})^{-1}(1+e^{-w_n})^{-1}]^T \tag{9.7}$$

and each element in this equation is what is called an *activation* function. The *sigmoid* is among the most commonly used activation functions, and is the activation function represented in Equation 9.7.

Let $\mathbf{A}$ be an $\mathbf{m} \times \mathbf{p}$ matrix, $\mathbf{a}_{\text{bias}}$ be an $\mathbf{m} \times 1$ vector, $\mathbf{B}$ be an $\mathbf{n} \times \mathbf{m}$ matrix, and $\mathbf{b}_{\text{bias}}$ be an $\mathbf{n} \times 1$ vector. Vectors $\mathbf{x}$ and $\mathbf{q}$ are then $\mathbf{p} \times 1$ and $\mathbf{n} \times 1$ vectors, respectively, and $\mathbf{m}$ becomes an arbitrarily determined number equal to the rows in $\mathbf{A}$, the length of $\mathbf{a}_{\text{bias}}$, and the columns in $\mathbf{B}$. The total fitted parameters in Equation 9.6 is then $\mathbf{m}(\mathbf{p} + \mathbf{n} + 1) + \mathbf{n}$; as such, an upper limit for $\mathbf{m}$) is set by the number of $\mathbf{x}$, $\mathbf{q}$ data records available for parameter fitting. A corresponding lower limit for $\mathbf{m}$ is determined empirically: That lower limit is reached when is so low that Equation 9.6 can no longer map $\mathbf{x}$ to $\mathbf{q}$ with the required degree of accuracy.

Backpropagation is currently the most popular method of fitting the $\mathbf{m}(\mathbf{p} + \mathbf{n} + 1) + \mathbf{n}$ parameters in Equation 9.6. Backpropagation is a gradient descent-based procedure that addresses the problem

$$\underset{A, B, a_{\text{bias}}, b_{\text{bias}}}{\text{Min}} \quad \|\mathbf{q}_{\text{data records}} - S(\mathbf{B} \cdot S(\mathbf{A} \cdot \mathbf{x} + \mathbf{a}_{\text{bias}}) + \mathbf{b}_{\text{bias}})\|_2^2$$

subject to constraints upon the magnitudes of $\mathbf{A}$, $\mathbf{B}$, $\mathbf{a}_{\text{bias}}$, and $\mathbf{b}_{\text{bias}}$. This procedure entails presenting $\mathbf{x}$, $\mathbf{q}$ data records repeatedly to the ANN, and iteratively adjusting the fitted parameters. The error between the value of $\mathbf{q}$ predicted by the ANN and the value of $\mathbf{q}$ in the data records is minimized in a least-squares sense. Once this parameter-fitting procedure is complete, the ANN is ready to serve as an on-line quality model: Given $\mathbf{x}$, it predicts $\mathbf{q}$.

The details behind backpropagation, as well as early discussions of feedforward ANNs, were provided by Rumelhart et al. [58]. Hecht-Nielsen [59] and Hertz et al. [60] offered an instructive examination of neural networks in general. Sarle [61] placed the ANN into the context of the regression-type model. The ANN clearly fit best into the category of regression-type models, not of fundamental process models. Unlike linear and quadratic regression models, however, ANNs impose less severe constraints upon the nature of process nonlinearity. Given the sophisticated interrelationship among its fitted parameters (see Eq. 9.6), yet given the relative speed with which it can map $\mathbf{x}$ to $\mathbf{q}$, the ANN is well-suited for on-line quality modeling in the SHMPC strategy. ANN applications in this role follow.

9.5.3 Applications to Autoclave Curing

The usefulness of applying SHMPC to autoclave curing, using ANNs as the on-line models, has been verified on both simulated and actual (experimental) autoclave curing. The application to simulated curing differed slightly from the application to actual curing. We will discuss herein each application, as well as a hybrid application using both simulated and actual autoclave curing results.

Application to simulated curing was made using an autoclave curing simulator [31]. There were two steps to the simulated curing application: Assessment of ANN on-line models compared with quadratic regression on-line models, and use of SHMPC for on-line product quality control. For the assessment, four measured disturbances $\mathbf{d}$ (initial resin weight fraction, prepreg impurity, initial void size, and initial prepreg temperature) and three manipulated inputs $\mathbf{u}$ (pressure and the first and second hold temperatures) were the seven inputs **x** to the regression-type models. No intermediate secondary measurements **y** were considered. Output quality variables **q** were cured composite laminate thickness and maximum laminate void size. The autoclave simulator was then used to generate 198 **x**), **q** data records, for use in fitting the parameters of two 7-input, 1-output ANNs (one for thickness; one for void size), and for use in fitting the parameters of two similar quadratic regression models. After parameter fitting, 99 additional **x**, **q** data records were generated by the simulator. The ANNs and the quadratic models were then applied to those 99 additional data records, and a correlation coefficient R^2 and an average of model error absolute values $\bar{\varepsilon}$ were computed for each of these four models. Values R^2 and $\bar{\varepsilon}$ for each model appear in Table 9.1.

The results in Table 9.1, as well as a more detailed study [62], indicate that the ANN models outperform their quadratic regression counterparts for predicting both thickness and maximum void size. For predicting maximum void size, ANN model performance could be improvised by optimizing the number of rows in ANN parameter matrix **A**—no such optimization is possible with quadratic regression models. These and other results indicate more importantly that the ANN can model nonlinear relationships without any a priori knowledge of the type of nonlinearity present in the system. As such, process models can be rapidly built using ANNs, whose execution times are also rapid enough to allow for on-line applications.

For use of SHMPC to achieve quality control of the simulated autoclave curing process, three intermediate secondary measurements **y** (laminate top temperature near the end of the first autoclave temperature ramp, maximum laminate temperature during the first autoclave temperature hold, and laminate top temperature near the end of the second autoclave temperature ramp) were added to the seven inputs comprised by **x**. In contrast to the on-line model assessment, which used separate single-output ANNs for each quality variable in **q**, a single ANN with 10 inputs and two outputs was used as the model in SHMPC. In addition, the length of **d** was shortened to three, as initial void size was removed; the time of pressure application was added to **u** to take its place in **m**. Recall that **x** comprises the contents of **d**, **u**, and **y**. The autoclave curing simulator was again used to generate **x**, **q** data records for this 10-input, 2-output ANN.

After ANN parameter fitting was complete, 37 more simulations were run with the SHMPC strategy in place, using the fitted ANN as its model. The goal for SHMPC was to minimize both laminate thickness and void size. Detailed results from this SHMPC application are presented by Joseph and Wang Hanratty [48,63]. Of the 37 simulated curing runs, all but three met the minimum void size goal, and all but one met the minimum thickness goal. It was found, more importantly, that omitting **y** values from the ANN input adversely affected SHMPC performance. When **y** values were excluded from SHMPC, the

Table 9.1 Comparing ANNs to Quadratic Regression Models

Independent variable	Measure of fit	Quadratic regression	ANN model
Product	R^2	0.9186	0.9999
Thickness	$\bar{\varepsilon}$	0.0168	0.0105
Maximum	R^2	0.4962	0.6567
Void Size	$\bar{\varepsilon}$	0.0119	0.0077

number of deviant void size values stayed roughly the same, but the number of deviant thickness values rose from 1 to 6—a considerable increase. These results attest to the value of **y**, used by SHMPC (through the ANN model) to update **u** in order to optimize quality variables **q**.

In an experimental effort that complements this simulation-based work, 112 stacks (each 7.62 cm by 7.62 cm and 64 plies per stack) of Hercules 120-glass/8551-7A resin prepreg underwent autoclave curing. For these 112 experimental curing runs, measured disturbances **d** were prepreg age and the delay in beginning the run after the prepreg stack had been assembled. Manipulated inputs **u** were pressure, time of pressure application, duration of first temperature hold, and first hold temperature. Intermediate secondary measurements **y** were two instantaneous top laminate temperature rates of increase, recorded near the middle of and near the end of the first autoclave temperature increase. Output-quality variables **q** were laminate thickness (as measured using a micrometer accurate to 0.001 in. = 0.00254 cm) and laminate void content (as measured using ASTM-standard ignition loss, specific gravity, and void content tests). Design-Of-Experiments–based orthogonal arrays [34] were used to set **u** levels. Another work [49] offered more detail on how those 112 experiments were performed.

In this effort, note that hold duration is a constraint, whereas thickness and void content are quality variables to be optimized. In much of the KBES-based work, curing duration is to be minimized, whereas nominal thickness (occasionally) and minimal void content (often) represent constraints.

Results from these experimental runs were used as **x**, **q** data records to fit the parameters of six ANNs. In the experimental effort, a different feedforward ANN was used after each intermediate secondary measurement was obtained; in the simulation-based effort, only one ANN accommodates all secondary measurements, and averaged "dummy" inputs are used for those secondary measurements not yet obtained. In addition in the experimental effort, a different ANN was used for final thickness and final void content predictions; in the simulation-based effort, one ANN was used to predict both final thickness and final void content. The advantage of using one ANN to predict all values of **q** is that the parameters of only one ANN need be fitted. Fitting the parameters of an ANN for each variable in **q** is much more time-consuming. The disadvantage, however, is that the parameters **A** and $\mathbf{a}_{\text{bias}}$ are the same for each variable in **q** when just one ANN is used as an on-line model. When a different ANN is used for each variable in **q**, the parameters in **A** and $\mathbf{a}_{\text{bias}}$ are unique for each of those output variables, which results in increased on-line prediction accuracy. Similar speed-versus-accuracy arguments apply to the choice of one ANN for all secondary measurements versus an ANN for each secondary measurement.

Final laminate thickness values from these 112 experimental runs ranged from 0.493 cm through 0.564 cm. SHMPC was then applied to the actual autoclave curing process. The targets for SHMPC were five laminates at 0.516-cm thickness, five laminates at a more demanding 0.541-cm thickness, and all 10 controlled laminates at minimum void content. The same type of prepreg stack used in the 112 experimental runs (64 plies per stack, Hercules 120-glass/8551-7A resin, 7.62 cm by 7.62 cm) was used for these 10 controlled runs. Detailed results are presented by Thomas [52]. Figure 9.3 compares void content to thickness for all 112 experimental runs and all 10 SHMPC-controlled runs. For a given thickness, the void content of an SHMPC laminate tends to be less than that of its non-SHMPC counterpart. For the lower target thickness, void content for all five laminates is well below 1 percent, whereas no actual thickness strays from the 0.516-cm target by more than

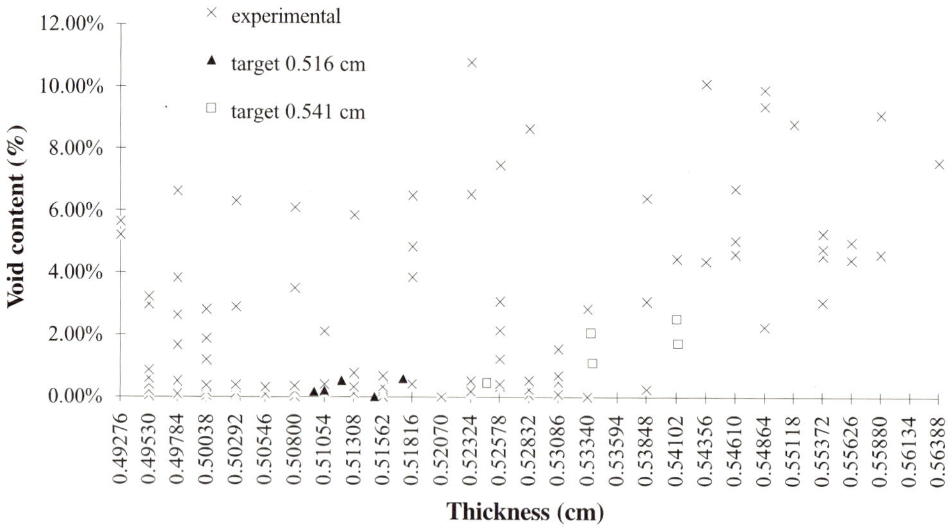

Figure 9.3 Void content variation with laminate thickness

0.008 cm. For the more challenging target thickness, void content is higher (but still quite low compared to other void percentages in the area of a 0.541 cm thickness), and where the maximum thickness error is 0.017 cm, the average thickness error is only 0.008 cm. The average thickness error for all 10 SHMPC laminates is 0.005 cm. As such, the applicability of SHMPC to both simulated and actual autoclave curing is confirmed. [Note: Five digits are used for the *x*-axis, given the 1 in. = 2.54 cm conversion.]

For industrial practice, a *hybrid* SHMPC approach should combine the speed of simulated autoclave curing with the freedom from simulator error offered by actual autoclave curing. Such a hybrid approach has been applied to the emulsion polymerization of vinyl acetate in a latex reactor [50]. This strategy uses a simulation model to generate additional data for fitting the ANN parameters through interpolation among available experimental data points. Given an acceptably accurate composite-manufacturing simulator, and given the opportunity to do experimental composite manufacturing runs on the process to be controlled, one can apply this hybrid SHMPC approach to composite manufacturing as well.

9.6 Summary and Future Trends

Composite material processing in general, and autoclave curing in particular, pose control problems more difficult than those from traditional chemical processing. As is decidedly not the case in traditional chemical processing, quality variables in industrial autoclave curing (e.g., laminate thickness, void content, structural and strength properties) are impractical at best and impossible at worst to measure before the end of the cure and the opening of the autoclave. This lack of access to the quality variables during the cure is what makes autoclave curing control so difficult. In the absence of such control, however, the rejection rate of cured composite parts has been unacceptably high. A strong economic incentive thus exists to develop on-line control of product quality in composite manufacturing.

Traditional quality control approaches offered by SPC/SQC methods cannot dictate the corrective action to be taken if the process is not under control. It is "intelligent" product quality control that must be applied to composite manufacturing. Initial attempts at applying intelligent control were through knowledge-based expert systems, a highly visible result of advances in artificial intelligence. Expert systems, however, are often as difficult to develop and maintain as rigorous physical models for batch processes such as autoclave curing. They also lack a statistical basis for control decision making. Surmounting those problems is a synergism of technologies, such as neural network modeling, inferential control, control of nonlinear systems, and SHMPC. Whereas an early effort [64] labeled premature a qualitative-quantitative methodology comparison, that effort did not consider what at that time was the MPC state-of-the-art [35–44]: Synergism-based strategies were derived from MPC.

Future trends will bring the inevitable advances in sensor, control, and computer technologies. To complement and perhaps supplant dielectric and fiber optic sensors (which themselves are just now gaining acceptance in industrial applications), nondestructive composite sensors are now in development. The SHMPC algorithm is ideally suited to use these advanced sensors in the intelligent control of product quality. Designed for controlling batch processes in general, and not just for autoclave curing control in particular, the SHMPC strategy with ANN models shall undergo refinement and maturation. With microprocessors inexorably increasing in speed, the complexity of on-line models in composite manufacturing control will likewise be able to increase. The end result of these future trends will be to make a complicated quality control problem much less of a challenge than what it now is.

Nomenclature

$\mathbf{A}$	matrix of ANN "weights" used in Equation 9.6
$\mathbf{a}_{\text{bias}}$	vector of ANN "bias weights" used in Equation 9.6
AI	Artificial Intelligence
ANN	Artificial Neural Network
$\mathbf{B}$	matrix of ANN "weights" used in Equation 9.6
$\mathbf{b}_{\text{bias}}$	vector of ANN "bias weights" used in Equation 9.6
CUSUM	CUmulative SUM
$\mathbf{d}_i$	measured disturbances after SHMPC phase i
Da	Damköhler number
DMC	Dynamic Matrix Control
EWMA	Exponentially Weighted Moving Average
$\mathbf{g}(\,)$	inequality vector constraint function for nonlinear optimization
$\mathbf{h}(\,)$	equality vector constraint function for nonlinear optimization
$\Delta H_{r \times n}$	heat of resin curing reaction
IDCOM	IDentification and COMmand
IMC	Internal Model Control

k	time index; resin thermal conductivity
KBES	Knowledge (Rule) Based Expert System
L	prepreg stack thickness
LCL	Lower Control Limit
m	control horizon in MPC and SHMPC
m	arbitrary number of ANN "weights" in Equation 9.6
MAC	Model Algorithmic Control
MPC	Model Predictive Control
n	output vector length used in Equation 9.6
NLP	Nonlinear Programming
p	prediction horizon in MPC and SHMPC
p	input vector length used in Equation 9.6
Pr[ ]	probability
PTA	Process Trend Analysis
$\mathbf{q}$	vector of ANN output (predicted quality) variables
$\mathbf{q}_i$	primary measurements after SHMPC phase i
Q_i	linear–nonlinear model used during SHMPC phase i
$\tilde{\mathbf{q}}_p$	predicted final quality variable values (SHMPC)
QDMC	Quadratic Dynamic Matrix Control
QPA	Qualitative Process Automation
r	fractional weighting factor in EWMA
R^2	regression correlation coefficient
r_{REF}	reference rate of resin curing
S_k	sum of deviations of k measurements from target (CUSUM)
SHMPC	Shrinking Horizon Model Predictive Control
SPC	Statistical Process Control
SQC	Statistical Quality Control
T	temperature
t	time
T_0	minimum autoclave temperature in Da definition
T_1	maximum autoclave temperature in Da definition
$\mathbf{u}_i$	manipulated inputs not yet fixed after SHMPC phase i
$\hat{\mathbf{u}}_i$	manipulated inputs fixed by SHMPC after SHMPC phase i
UCL	upper control limit
$\mathbf{x}$	vector of ANN inputs in Equation 9.6
$\mathbf{x}_0$	initial state (SHMPC)
$y_{\mathrm{EWMA},k}$	exponentially weighted moving average of measurements 1 through k
$\mathbf{y}_i$	secondary measurements after SHMPC phase i
y_k	measured value in generic process
y_m	true mean for measured value y_k
y_t	target value in generic process
$\bar{\varepsilon}$	average of model error absolute values
σ	standard deviation of process whose True mean is y_m

References

1. Baumgartner, W.E., Ricker, T. *SAMPE J.* (1983) 19(4), p. 6
2. Sentura, S.D., Sheppard, N.F. Jr. *Adv. Polym. Sci.* (1986) 80, p. 1
3. Kranbuehl, D., Hoff, M., Haverty, P., Loos, A., Freeman, T. In *Procedings of the 33rd International SAMPE Symposium* (1988) Anaheim, p. 1276
4. Walsh, S.M. In *Proceedings of the 35th International SAMPE Symposium* (1990) Anaheim, p. 1280
5. Day, D.R., Shepard, D.D., Craven, K.J. In *Proceedings of the 36th International SAMPE Symposium* (1991) San Diego, p. 571
6. Buczek, M.B., Lee, C.W., In *Proceedings of the 40th International SAMPE Symposium* (1995) Anaheim, 696
7. Lee, C.W., Rice, B., Buczek, M. In *Proceedings of the 41st International SAMPE Symposium* (1996) Anaheim, p. 1511
8. Rice, B.P., Lee, C.W., In *Proceedings of the 41st International SAMPE Symposium* (1996) Anaheim p. 1518
9. Juran, J.M. (Ed.) *Juran's Quality Control Handbook, 4th ed.* (1988) McGraw-Hill Inc., New York
10. Page, E.S. *Biometrika* (1954) 41, p. 100
11. Page, E.S. *Technometrics* (1961) 3, p. 1
12. Roberts, S.W., *Technometrics* (1959) 1, p. 239
13. Hunter, J.S. *J. Qual. Tech.* (1986) 18, 203
14. Ogunnaike, B.A.,, Ray, W.H., *Process Dynamics, Modeling, and Control* (1994) Oxford University Press, New York
15. McGraw, K.L., Harbison-Briggs, K. *Knowledge Acquisition: Principles and Guidelines* (1989) Prentice-Hall Inc., Englewood Cliffs, New Jersey
16. Zsolnay, A., Perkins, K.M., Blad, L.H. *U.S. Pat. 4,399,100* (1983)
17. Hinrichs, R.J., Thuen, J.M. *U.S. Pat. 4,455,268* (1984)
18. Zsolnay, A., Perkins, K.M., Blad, L.H. *U.S. Pat. 4,496,697* (1985)
19. Hinrichs, R.J., Thuen, J. *U.S. Pat. 4,515,545* (1985).
20. Hayes, P.J. In *Expert Systems in the Micro-Electronic Age* (1979) Michie, D. (Ed.). Edinburgh University Press, Edinburgh, p. 242
21. Forbus, K.D. *Artif. Intell.* (1984) 24, p. 85
22. Servais, R.A., Lee, C.W., Browning, C.E. *SAMPE J.* (1986) 22(5), p. 14
23. Abrams, F.L., Garrett, P., Lagnese, T., LeClair, S., Lee, C.W., Park, J., AFWAL-TR-87-4083, Materials Laboratory, Air force Wright Aeronautical Laboratories (1987) Wright-Patterson AFB, OH
24. LeClair, S.R., Abrams, F.L., Matejka, R.F. *Artif. Intell. Des. Anal. Manuf.* (1989) 3, p. 125
25. Lee, C.W., Park, J.B., LeClair, S.R., Abrams, F.L., Garrett, P.H., Servais, R.A., *U.S. Pat. 5,032,525* (1991)
26. Ciriscioli, P.R., Springer, G.S. *Smart Autoclave Cure of Composites* (1990) Technomic Publishing Co. Inc., Lancaster, Penn
27. Ciriscioli, P.R., Springer, G.S., Lee, W.I. *J. Comp. Mat.* (1991) 25, p. 1542
28. Kalra, L., Perry, M.J., Lee, L.J. *J. Comp. Mat.* (1992) 25, p. 2567
29. Perry, M.J., Lee, L.J., Lee, C.W. *J. Comp. Mat.* (1992) 26, p. 274
30. Pillai, V., Beris, A., Dhurjati, P. *Advances in Instrumentation and Control: International Conference and Exhibition*, (1993) 48, p. 1431
31. Wu, H.-T. Joseph, B. *SAMPE J.* (1990) 26(6), p. 39
32. Wu, H.-T. Joseph, B. *Chem. Engng. Sci.* (1992) 47, p. 917
33. Zadeh, L.A. *IEEE Spectrum* (1984) 21(8), p. 26
34. Taguchi, G. *System of Experimental Design* (1987) American Supplier Institute, Dearborn, Mich.
35. Cutler, C.R., Ramaker, B.L. "Dynamic Marix Control — A Computer Control Algorithm" (1979) AIChE 1979 National Meeting, Houston

36. Prett, D.M., Gillette, R.D. "Optimization and Constrained Multivariable Control of a Catalytic Cracking Unit" (1979) AIChE 1979 National Meeting, Houston
37. Garcia, C.E. "Quadratic/Dynamic Matrix Control of Nonlinear Processes: An Application to a Batch Reaction Process," (1984) AIChE 1984 National Meeting, San Francisco
38. Garcia, C.E., Morshedi, A.M. *Chem. Engng. Comm.* (1986) 46, p. 73
39. Richalet, J., Rault, A., Testud, J.L., Papon, J. *Automatica* (1978) 14, p. 413
40. Rouhani, R., Mehra, R.K. *Automatica* (1982) 18, p. 401
41. Froisy, J.B., Richalet, J. In *Proceedings of the Third International Conference on Chemical Process Control (CPC III)* (1986) M. Morari, T.J. McAvoy (Eds.). Elsevier Science Publishers Inc., New York, p. 233
42. Garcia, C.E., Morari, M. *Ind. Engng. Chem. Process Des. Dev.* (1982) 21, p. 308
43. Brosilow, C.B. "The Structure and Design of Smith Predictors from the Viewpoint of Inferential Control" (1979) Joint Automatic Control Conference, Denver
44. Garcia, C.E., Prett, D.M. In *Proceedings of the Third International Conference on Chemical Process Control (CPC III)* (1986) Morari, M., McAvoy, T.J. (Eds.). Elsevier Science Publishers Inc., New York, p. 245
45. Richalet, J. *Automatica* (1993) 29, p. 1251
46. Biegler, L.T., Rawlings, J.B. In *Proceedings of the Fourth International Conference on Chemical Process Control (CPC-IV)* (1991) Y. Arkun, W.H. Ray, (Eds.), CACHE, Austin, Texas, p. 543
47. Rawlings, J.B., Meadows, E.S., Muske, K.R. *IFAC Symposium ADCHEM '94: Advanced Control of Chemical Processes* (1994) Kyato p. 203
48. Joseph, B., Wang Hanratty, F. *Ind. Engng. Chem. Res.* (1993) 32, 1951
49. Thomas, M.M., Kardos, J.L., Joseph, B. In *Proceedings of the 1994 American Control Conference* (1994) Baltimore, p. 505
50. Tsen, A.Y.-D., Jang, S.-S., Wong, D.S.-H., Joseph, B. *AIChE J.* (1996) 42, p. 455
51. Pardee, W.J., Shaff, M.A., Hayes-Roth, B. *Artif. Intell. Eng. Des. Anal. Manuf.* (1989) 4, p. 55
52. Thomas, M.M. D.Sc. Thesis (1955) Washington University, St. Louis, Missouri
53. Lasdon, L.S., Waren, A.D., Jain, A., Ratner, M. *ACM Transactions on Mathematical Software* (1978) 4, p. 34
54. Loos, A.C., Springer, G.S. *J. Comp. Mat.* (1983) 17, p. 135
55. Kardos, J.L., Dudukovic', M.P., Davé, R. *Adv. Polym. Sci.* (1986) 80, p. 101
56. Box, G.E.P., *Technometrics* (1966) 8, p. 625
57. Chen, C.T., LeClair, S.R. *J. Reinf. Plast. Comp.* (1991) (10), pp. 379–390
58. Rumelhart, D.E., Hinton, G.E., Williams, R.J. In *Parallel Distributed Processing: Explorations in the Microstructure of Cognition. Volume 1: Foundations* (1986) D.E. Rumelhart, J.L. McClelland and the PDP Research Group (Eds.). MIT Press, Cambridge, Mass. p. 318
59. Hecht-Nielsen, R. *Neurocomputing* (1990) Addision-Wesley, Reading, Mass.
60. Hertz, J., Krogh, A.S., Palmer, R.G., *Introduction to the Theory of Neural Computation* (1991) Addison-Wesley, Redwood City, Calif.
61. Sarle, W.S. In *Proceedings of the Nineteenth Annual SAS Users Group International Conference* (1994) Cary, NC. Available as /pub/sugi19/neural/neural1.ps via anonymous ftp from ftp.sas.com
62. Joseph, B., Wang, F.H., Shieh, D.S.-S. *Comp. Chem. Eng.* (1992) 16, p. 413
63. Joseph, B., Wang Hanratty, F., Kardos, J.L. *J. Comp. Mat.* (1995) 29, p. 1000
64. Garrett, P., Lee, C.W., LeClair, S.R., In *Proceedings of the 1987 American Control Conference* (1987) Minneapolis, p. 1368

Part II
Process

10 Autoclave Processing

Andrew R. Mallow and Flake C. Campbell

Autoclave processing remains the mainstay for fabricating continuous fiber-reinforced thermoset composite parts. During autoclave processing, several complex and interrelated phenomena occur, including heat transfer, resin viscosity changes, resin flow, chemical crosslinking, and void formation, growth, and transport. Numerous factors influence the cure cycle, but the ultimate goal remains the same: to produce a high-quality composite part free of voids and porosity. To achieve this goal, researchers have developed a fundamental understanding of many aspects of autoclave processing. In some cases this understanding is a mathematical representation of one aspect of the cure process. These models allow process engineers to conduct off-line interrogations of the autoclave process. Numerous models are available for simulating heat transfer and resin flow, viscosity and kinetic changes, and void formation, growth, and transport. For other cases, this fundamental understanding is a general principle. These general principles address the difficult to model, but more influential factors that affect autoclave processing, including bagging method, incoming material quality, prepreg moisture content, and debulking procedures. This fundamental understanding has been further explored for making off-line process decisions, for real-time process control, for processing of high temperature condensation polymers, and for extensions to nonautoclave composite processes. This chapter provides an overview and reference guide for this fundamental understanding of autoclave processing.

10.1 Introduction

Autoclaves lend considerable versatility to the manufacturing process. They can accompany a single large composite part, as illustrated in Figure 10.1 for an AV-8B wing skin, or numerous smaller parts loaded onto racks and cured as a batch. Whereas autoclave processing is not the most significant cost driver for total part cost, it does represent a culmination of all the previously performed manufacturing operations because final part quality (per ply thickness, degree of crosslinking, and void and porosity content) is determined during this operation. Manufacturing steps prior to autoclave processing include raw material production, prepreg manufacturing, material cutting, kitting, storage, hand or automated collation, and bagging.

Figure 10.1 AV-8B wing skin preparing for autoclave processing

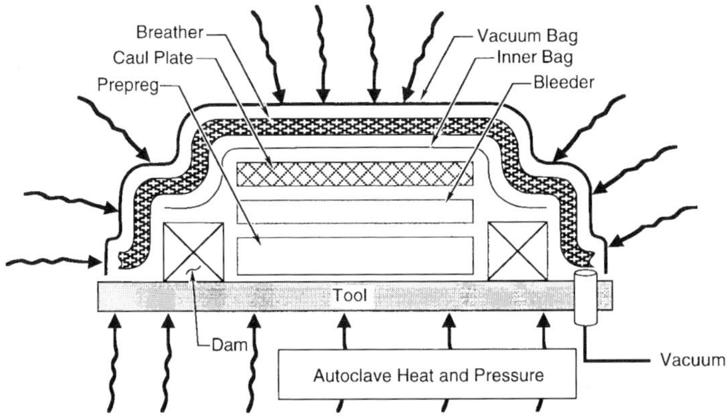

Figure 10.2 Typical composite bagging sequence

A bagging schematic for a typical autoclave bagging sequence is shown in Figure 10.2. The prepreg laminate or part is collated on a hard tool. Typical tooling materials are aluminum, steel, nickel, invar, or composite. Solid or cork dams are placed around the perimeter of the laminate to control resin flow from the ply edges (horizontal flow). Many parts utilize a release ply on top of the laminate to allow for easy removal of adjacent bagging materials. For laminates that require resin bleed, a bleeder cloth is placed on top of the release ply. A bleeder pack is not used for net resin content prepregs. A caul plate typically follows the bleeder cloth or release ply. A caul plate improves part surface finish and evenly distributes the autoclave pressure. Following the caul sheet is an inner bag. The inner bag is sealed to the tool surface or perimeter dams to control vertical resin flow. Following the inner bag is the breather and vacuum bag, which establish the driving force behind consolidation by maintaining the pressure differential between the autoclave pressure and vacuum pressure underneath the bag. This combination of bagging and tooling materials, combined with the actual part geometry, yields complex heat transfer and resin flow behavior during autoclave processing.

Beginning in the early 1980s a number of government-, industry-, and university-sponsored research programs sought to develop a scientific understanding of the autoclave cure process. These programs have expanded our knowledge of the effect of processing variables on final part quality. This chapter focuses on the key issues and complexities of autoclave processing, especially variables that affect final part quality.

10.2 Autoclave Processing Description

10.2.1 The Cure Cycle

A typical cure cycle used to process many thermoset epoxy composite parts is shown in Figure 10.3. It contains two ramps and two isothermal hold periods. The first ramp and

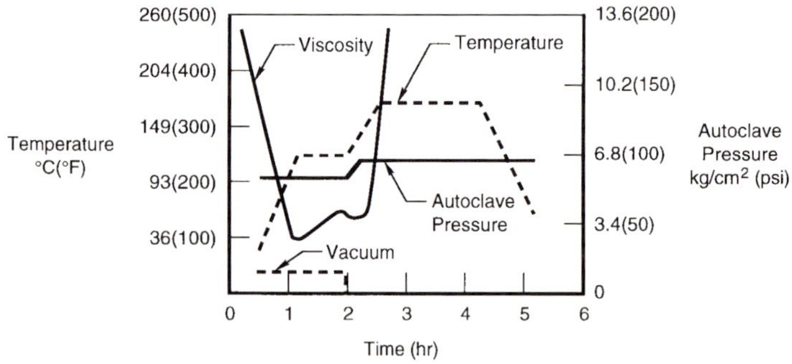

Figure 10.3 Typical composite cure cycle

isothermal hold is used to allow the resin to flow and volatiles to escape. The applied autoclave pressure is typically 5.8 kg/cm^2 (85 psi), and a full vacuum is applied underneath the bag. The imposed viscosity curve on the figure shows the dramatic drop in viscosity during this first ramp as the epoxy resin melts and flows.

The second ramp and hold is the polymerization portion of the cure cycle. During this stage, the viscosity rises dramatically to gel and additional crosslinking occurs. Thermal integrity and structural strength are developed during this portion of the cycle. The vacuum is vented to atmosphere and autoclave pressure is increased to 6.8 kg/cm^2 (100 psi) immediately preceding this stage. The vacuum is usually removed prior to this second ramp because vacuum, even though it promotes devolatilization, can also lead to increased porosity if outgassing is still occurring during resin gellation.

The second ramp portion of this cure cycle is critical from a void nucleation and growth standpoint. During this ramp, the temperature is high, the resin pressure can be near its minimum, and the volatile vapor pressure is high and rising with temperature. These are the ideal conditions for void formation and growth.

10.2.2 Resin Viscosity and Kinetic Models

The resin viscosity cycle is also shown with the temperature, pressure, and vacuum segments. A knowledge of resin viscosity behavior is an important part of understanding a materials processing behavior. High-flow resin systems (i.e., resins with low minimum viscosities) are more sensitive to bagging and tooling methods that may lead to overbleeding. Low flow resin systems (i.e., resins with high minimum viscosities) tend to be easier to process, but are more sensitive to room temperature out-time.

Kinetic models determine the minimum time required to cure the resin (i.e., guarantee sufficient physical and mechanical properties). They also determine the heat of reaction of the resin for use by heat transfer models and the degree of crosslinking for use in viscosity submodels. The exothermic cure reaction for the transformation of the epoxy resin to the cured matrix polymer can be expressed as:

$$[\text{RESIN}] \xrightarrow{k} [\text{CROSS-LINKED POLYMER}] + \Delta H \quad (10.1)$$

where k = rate constant and ΔH = heat generated. Kinetic models of this expression are typically developed from experimental measurements of the rate of heat generation as a function of time and temperature.

Several viscosity and kinetic models, and experimental procedures for developing these models, are available for a number of commercially available resin systems [1–5]. These models allow insight into autoclave process decisions based on changes in resin viscosity and kinetic behavior and can be used to determine hold temperatures and durations that allow sufficient resin flow and cross-linking to avoid over bleeding, exotherms, and void formation.

10.2.3 Resin Hydrostatic Pressure and Flow

High pressures are commonly used during autoclave processing to provide ply compaction and suppress void formation. Autoclave gas pressure is transferred to the laminate due to the pressure differential between the autoclave environment and the vacuum bag interior. Translation of the autoclave pressure to the resin depends on several factors, including the fiber content, laminate configuration, and the amount of bleeder used.

An appreciation of the importance of hydrostatic resin pressure must be developed to understand void growth fully. Because of the load-carrying capability of the fiber bed in a composite layup, the hydrostatic resin pressure needed to suppress void formation and growth is typically only a fraction of the applied autoclave pressure. The hydrostatic resin pressure is critical because it is the pressure that helps to keep volatiles dissolved in solution. If the resin pressure drops below the volatile vapor pressure, then the volatiles will come out of solution and form voids.

In the early stages of the cure cycle, the hydrostatic resin pressure should be equal to the applied autoclave pressure. As resin flow occurs, the resin pressure drops. If a laminate is severely overbled, then the resin pressure could drop low enough to allow void formation. Thus, the hydrostatic resin pressure is directly dependent on the amount of resin bleeding that occurs. As the amount of bleeding increases, the fiber volume increases, resulting in an increase in the load carrying capability of the fiber bed.

To help illustrate the interactions of the resin flow process, hydrostatic resin pressure, and the load-carrying capability of the fiber bed, a mechanical analogy is presented in Figure 10.4. In this analogy, a laminate undergoing cure is simulated as a piston–spring–valve set-up. The spring represents the fiber bed and thus has a load carrying capability. Just like a spring, the fiber bed supports larger loads as it undergoes compression. The liquid contained in the piston represents the ungelled resin. Finally, the valve is the means by which the liquid resin can leave the system.

The rate of bleeding is dependent on several factors, including the permeability of the fiber bed, both vertically and horizontally, and the viscosity of the liquid resin. The permeability of the fiber bed will depend on the weave of the fabric, the fiber diameter, and the fiber volume fraction. The resin viscosity is determined by the chemistry of the resin and the thermal profile of the cure cycle. The cure cycle greatly affects resin viscosity and the flow process, both directly through the pressure application and indirectly through the effect of the thermal profile on resin viscosity.

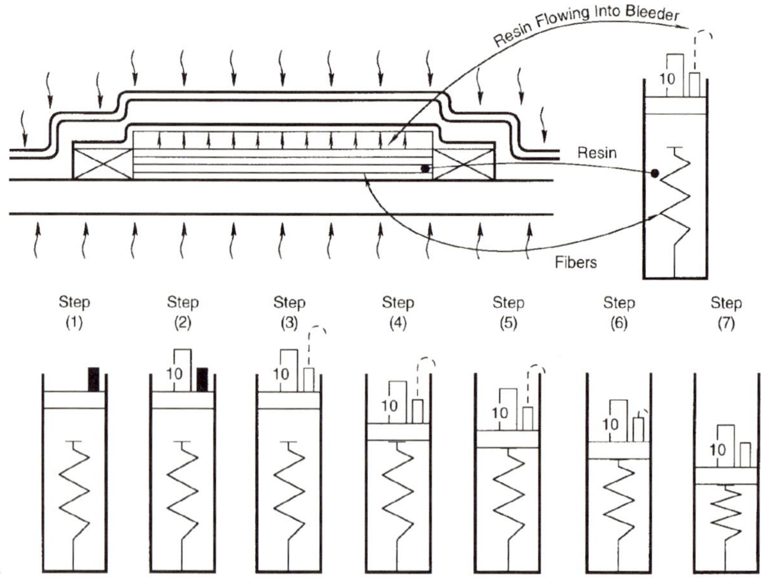

$P_{applied}$	0	10	10	10	10	10	10
$P_{hydrostatic}$	0	10	10	7.5	5	2.5	0
P_{spring}	0	0	0	2.5	5	7.5	10

Legend:

▮ Closed Valve (No Bleeding) ▯ Open Valve (Bleeding)

Step 1 There is initially No Load on the system. The Liquid Hydrostatic Pressure and the Load Carried by the Fiber Bed is Zero.

Step 2 A 10 kg/cm^2 Force is Applied to the System, But No Liquid Has Escaped (Closed Valve). The Liquid Carries the Entire Load and the Load on the Fiber Bed Is Zero. Note That the Downward Force (in This Case 10 kg/cm^2) Is Equal to the Upward Force (Again 10 kg/cm^2). This Upward Force Is the Sum of the Load Carried by the Liquid (10 kg/cm^2) and the Springlike Fiber Bed (0 kg/cm^2).

Step 3 The Valve Is Now Opened Allowing Resin to Escape(e.g., Resin Bleeding). At This Point, However, the Resin Still Carries the entire 10 kg/cm^2 Load.

Step 4 Liquid Continues to Escape, but at a Decreasing Rate Due to a Portion of the Load that Is Now Carried by the Spring (2.5 kg/cm^2). This Is Analogous to Bleeding in a Laminate Occurring Rapidly Until the Fiber Bed Starts Supporting a Portion of the Applied Autoclave Pressure.

Steps 5 and 6 Liquid Continues to Escape, But at an Ever-Decreasing Rate as a Greater Portion of the Load Is Being Born by the Spring. In an Actual Laminate, the Rate of Bleeding Would Be Retarded Both by the Increasing Load Carrying Capability and the Reduced Permeability of the Fiber Bed as Its Compacted.

Step 7 No Further Bleeding Occurs Because the Pressure on the Resin Has Now Dropped to Zero and the Entire Load (10 kg/cm^2) Is Being Carried by the Spring. If This Condition Occurs During Actual Autoclave Processing Before the Resin Gels(Solidifies), It Would Be Quite Easy for Dissolved Volatiles to Vaporize Out of Solution and Form Voids.

Figure 10.4 Resin flow analogy

10.2.4 Resin Flow Models

Resin flow and pressure distributions are among the most challenging aspects of the autoclave process cycle to model. Accounting for tooling and bagging variations, such as inner bag perforations, mismatched tooling details, and gaps between the edge of a part and dams requires models of significant detail for accurate predictions. Finite element models representative of these conditions require two and three dimensional models consisting of hundreds of elements. This in turn requires considerable computational time and hours of analysis. It is advantageous in many instances to represent a complex model with several simplified one- and two-dimensional models. These smaller models are computationally faster and the results are easier to interpret [6].

Resin flow models are capable of determining the flow of resin through a porous medium (prepreg and bleeder), accounting for both vertical and horizontal flow. Flow models treat a number of variables, including fiber compaction, resin viscosity, resin pressure, number and orientation of plies, ply drop-off effects, and part size and shape. An important flow model output is the resin hydrostatic pressure, which is critical for determining void formation and growth.

Over the past decade, significant progress has been made in improving the accuracy of these models by developing a better understanding of the role the fiber bed plays in the flow process. The fiber bed supports a portion of the applied autoclave pressure, and the permeability of the fiber bed changes depending on fiber type, fiber volume fraction, and layup orientation. Researchers [7,8] have strived to further our understanding of these relationships to predict the resin flow process more accurately.

Experimental studies, such as those to be discussed in Section 10.2.5, have been key to demonstrating the validity of these flow models. Many of the more sophisticated models [3,6,8] have shown to follow the resin flow process accurately. These models are extremely useful for choosing material systems and for determining optimum pressure cycles and bagging procedures.

10.2.5 Experimental Studies

To investigate the potential pressure gradients that exist within a laminate during autoclave processing, miniature pressure transducers (Fig. 10.5), which are capable of measuring the hydrostatic resin pressure, were embedded at multiple locations within several laminates to study the effects of vertical and horizontal pressure gradients [9].

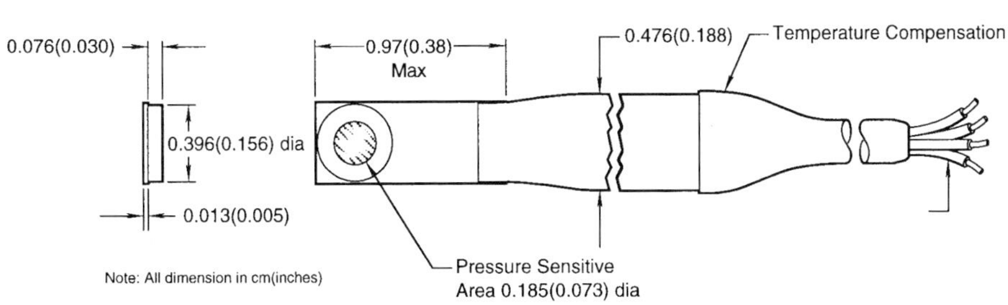

Figure 10.5 Miniature pressure transducer design

A-60 ply thick laminate was collated to allow vertical flow only. Miniature pressure transducers were embedded through the thickness within both the laminate and the bleeder. Three tool-mounted transducers were also used to monitor the resin pressure at several locations along the surface of the tool. The pressure curves from this vertical flow test (Fig. 10.6) confirmed that a significant pressure gradient can exist within a laminate. These results also illustrate the vertical compaction process. The resin pressure within the entire laminate is initially near the applied autoclave pressure. As resin bleeding occurs, the hydrostatic resin pressure at the top of the laminate drops (Transducer 3). At this point, resin begins to bleed from the middle of the laminate toward the top, and the pressure drops here also (Transducer 2), but remains above the pressure at the top. The opposite process occurs in the bleeder. As resin fills the bleeder, the pressure in the bleeder rises. Resin contents and photomicrographs were taken to confirm the resin pressure results. Resin content specimens were split into three pieces, one each from the top, middle, and bottom 20 plies. The resin content at the top was 24.6 percent, the middle was 31.0 percent, and the bottom was 33.0 percent, which confirms the resin pressure results. Photomicrographs showed the same results (i.e., the top plies were compacted considerably more than those at the laminate bottom near the tool surface).

Just as the vertical flow laminate experienced a pressure gradient, a horizontal pressure gradient also exists. Two additional laminates were processed to investigate horizontal pressure gradients. The first had cork dams located as close as possible to the laminate edges, but the second had a large gap (0.5 in.) between the laminate edges and the dams. Both laminates (Fig. 10.7) contained nonporous release film against the top surface to prevent vertical flow.

The pressure curves for both laminates (Fig. 10.8) showed the existence of a horizontal pressure gradient and that the magnitude of the gradient depends on the amount of horizontal flow (i.e., the larger gap distance between the laminate edges and the dams resulted in more horizontal flow).

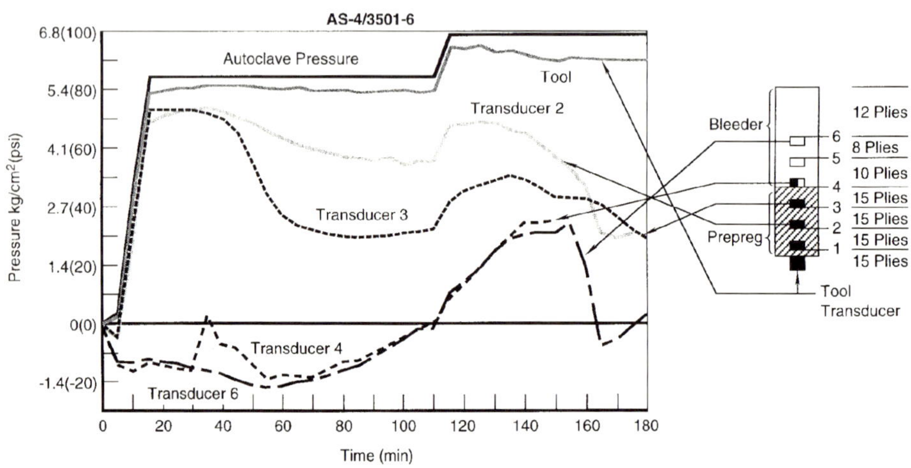

Figure 10.6 Vertical pressure gradient

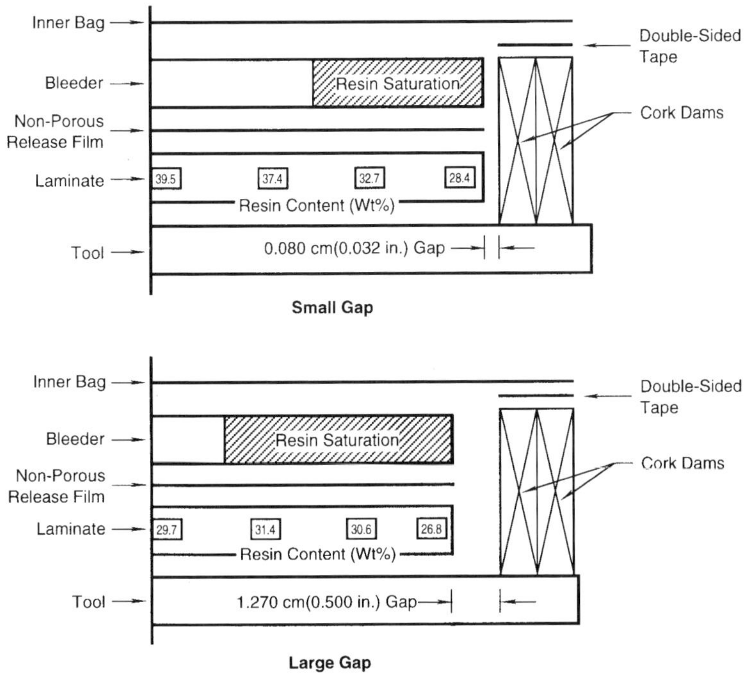

Figure 10.7 Horizontal flow bagging arrangement

The pressure curves also illustrate the horizontal flow process. The resin pressure initially approaches the applied autoclave pressure and then decreases as bleeding occurs. The opposite occurs in the bleeder. The applied vacuum is measured initially, and the pressure increases as resin begins to fill the bleeder. Note that the horizontal pressure gradient is very small for a majority of the laminate but becomes large near the edges.

Resin content results confirmed the resin pressure results, showing a large resin content gradient (Fig. 10.7) existed at the edge of the laminate. The figure also illustrates that resin bled further into the bleeder when a large gap was used between the laminate edges and the dams.

10.2.6 Caul Plates and Pressure Intensifiers

As illustrated in Figure 10.2, many composite applications utilize a caul plate or pressure intensifier. Caul plates result in a greatly improved part surface finish compared with a bag surface, improved dimensional control, and improved radius quality. Caul plates are also used to reduce ply movement during processing of honeycomb reinforced parts. Caul plates may be semi-rigid or rigid in nature. Semi-rigid caul plates, which are the most common type, are typically constructed of thin metal, composite, or rubber materials so they are flexible in

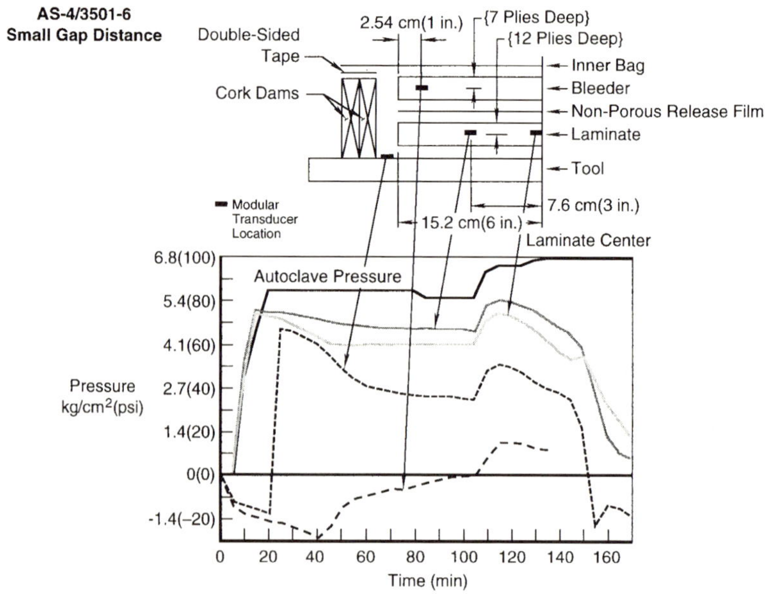

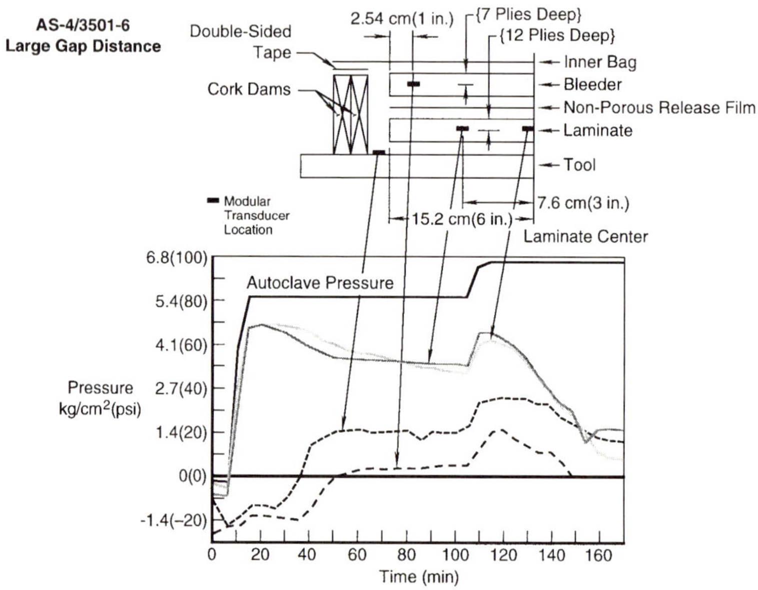

Figure 10.8 Horizontal flow laminates

nature. Pressure intensifiers are used at part corners to reduce ply thinning at male radii and ply bridging at female radii.

Rigid caul plates are typically constructed of thick metal or composite materials. Thick caul plates are used on very complex part applications or cocured parts where dimensional control is critical. Many rigid caul plates result in a matched die configuration similar to compression or resin transfer molding. Parts processed in this manner are extremely challenging because resin pressure is much more dependant on tool accuracy and the difference in thermal expansion between the tool and the part. Tool accuracy is critical to ensure no pinch points are encountered that would inhibit a tool from forming to the net shape of the part.

Both semirigid and rigid caul plates dramatically influence resin flow, resin pressure distribution, and final part quality. Semi-rigid caul sheets and pressure intensifiers reduce vacuum bag bridging around tight corners, resulting in a more uniform pressure distribution; however, rigid caul plates completely alter resin pressure distributions by minimizing laminate thinning. A rigid caul plate applies pressure only at laminate high points. The flow process illustrated in Figure 10.8 would be greatly altered by a rigid caul plate. The laminate would no longer be thick at the center and thin at the edge, but would be uniform across the part. If a rigid caul plate is trapped on a high spot or mating tool surface, then the resin pressure can drop extremely low in areas, resulting in gross porosity. Many times when rigid caul plates are utilized, a portion of the autoclave pressure is applied underneath the vacuum bag (internally pressurized bag) to ensure a minimum resin pressure is maintained.

10.2.7 Net Resin and Low Flow Resin Systems

Prepregs introduced in the 1970s were considered moderate flow systems that facilitated the intentional bleeding of resin during autoclave processing. As part applications became larger and more complex through the 1980s, the use of net resin and low (controlled) flow resin prepregs became more prevalent. Carbon–epoxy prepregs for resin bleed processing are specified at resin contents ranging from 37 to 42 percent by weight resin, while the desired or nominal resin content for these materials are 31 percent resin content for unidirectional prepregs and 35 percent resin content for woven cloth prepregs. A 57–60 percent fiber final fiber volume is typically desired to balance both mechanical performance and part quality. For net resin systems, it is critical to ensure that little or no resin bleed occurs to maintain resin pressure during autoclave processing. Many net resin systems utilize controlled flow resins that have higher minimum viscosities and lower flow numbers than moderate flow systems. Net resin, controlled-flow prepregs offer improved dimensional control, reduced debulking requirements, reduced comsumables (bagging materials) and reduced ply movement during processing over bleed resin systems. In addition, resin flow into honeycomb core, tooling details and bagging regions are reduced.

10.3 Voids and Porosity

10.3.1 Theory of Void Formation

Void formation and growth in addition curing composite laminates is primarily due to entrapped volatiles. Higher temperatures result in higher volatile pressures. Void growth will potentially occur if the void pressure (i.e., the volatile vapor pressure) exceeds the actual pressure on the resin (i.e., the hydrostatic resin pressure) while the resin is a liquid (Fig. 10.9). The prevailing relationship, therefore, is:

$$\text{If } P_{\text{void}} > P_{\text{hydrostatic}} \rightarrow \text{Then void formation and growth can occur} \qquad (10.2)$$

When the resin viscosity dramatically increases, or gellation occurs, the voids are locked into the resin matrix. Note that the applied pressure on the laminate is not necessarily a factor. As shown previously, the hydrostatic resin pressure can be low due to resin flow even though the applied autoclave pressure is high, leading to void formation and growth. The void problem unfortunately cannot be resolved simply by maintaining the hydrostatic resin pressure above the potential void pressure of the volatiles (although this is a good start). During collation or ply layup, air can become entrapped in the layup between the prepreg plies. The amount of air entrapped depends on many variables: the prepreg tack, the resin viscosity at room temperature, the degree of impregnation of the prepreg and its surface smoothness, the number of intermediate debulk cycles used during collation, and geometrical factors such as ply drop-offs and radii. An obvious place where entrapped air pockets form is at the terminations of the internal ply drop-offs. In addition, air can be entrained in the resin itself during the mixing and prepregging operations. This entrained air can also lead to voids, or at least serve as nucleation sites.

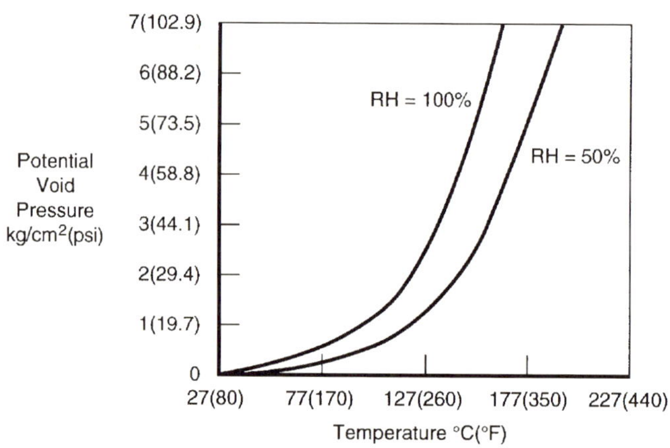

Figure 10.9 Potential void formation

10.3.2 Void Models

Void minimization remains the key to generating optimum cure cycles. However, the void nucleation, growth, and transport process is not only difficult to model, but is also difficult to verify. The complexities of the void process are immense. This effort is further compounded by the lack of good in situ measurement techniques for monitoring the void process; however, a few simple models [10] have been used extensively for helping optimize process cycles by avoiding situations that may result in void formation and growth. Further research is required in modeling voids before void models become as useful as the existing kinetic, viscosity, resin flow, and heat transfer models. Experimental studies continue to be the primary guide for processing decisions concerned with void management.

10.3.3 Resin and Prepreg Variables

The resin mixing and prepregging operations influence the processability of the final prepreg. During normal mixing operations, air can be easily mixed into the resin. This entrained air can later serve as nucleation sites for voids and porosity. Many mixing vessels, however, are equipped with seals that allow vacuum degassing during the mixing operation, a practice that has been found to be effective in removing entrained air and may be beneficial in producing superior quality laminates.

In one study [11], several laminates were fabricated from carbon–epoxy prepreg produced with vacuum degassed resins. Standard prepreg (i.e., not degassed) was also used for making baseline laminates. One layup condition studied was moisturizing the prepreg to a 1.0 percent moisture content prior to ply collation. Of all the laminates produced under this "wet" layup condition, the laminate made from the prepreg using vacuum degassed resin produced the highest quality laminate. This again this could be due to the reduced number of nucleation sites for void growth in the vacuum degassed resin. It should be pointed out that both the vacuum degassed and nondegassed resin produced acceptable laminates when moisture was not a factor.

Prepreg physical properties can also influence final laminate quality. Prepreg tack is one such property. Prepreg tack is a measure of the stickiness or self-adhesive nature of the prepreg plies. Prepregs with a high tack level have often resulted in laminates with severe voids and porosity. This could be due to the potential difficulty of removing entrapped air pockets during collation with tacky prepreg. Moisture can again be a factor. Prepregs with a high moisture content have been found to be inherently tackier than low-moisture content material. Previous work [10,12] has indicated a possible correlation between prepreg tack and resin viscosity (i.e., prepregs that are extremely tacky also have high initial resin viscosities). Resins with such high viscosities will be less likely to cold flow and eliminate voids at ply terminations.

Prepreg physical quality can greatly influence final laminate quality. Three prepreg conditions are summarized in the highly idealized schematic shown in Figure 10.10. Prepreg that appears "good" (i.e., smooth and well impregnated) ironically may not necessarily

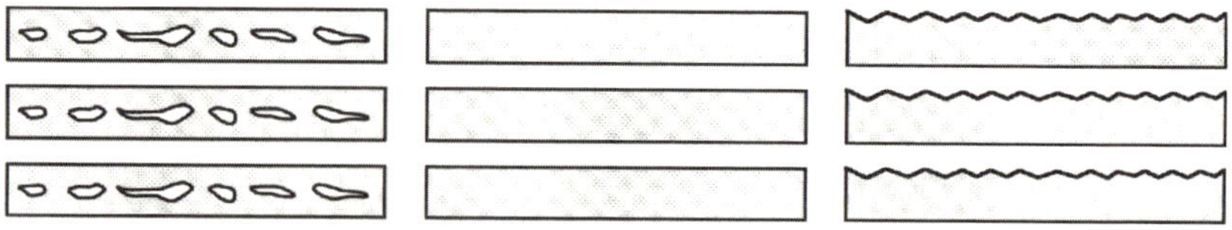

Figure 10.10 Effect of prepreg physical quality

produce the best laminates. Several material suppliers have determined that only partially impregnating the fibers during prepregging results in a prepreg that consistently yields high quality parts, whereas a "good" (i.e., smooth and well impregnated) prepreg can result in laminates with voids and porosity. Partially impregnated prepregs have the same resin content and fiber areal weight as do the fully impregnated material. The only difference is the placement of the resin with respect to the fibers. The partial impregnation process provides an evacuation path for air and low temperature volatiles entrapped in the layup. As the resin melts and flows, full impregnation occurs during autoclave processing.

One series of studies [13,14] showed that partially impregnated prepregs can yield high-quality laminates even with moisturized prepreg. They successfully applied partial impregnation to five different resin systems and eight different fibers.

A closely related phenomena relates to the surface condition of the prepreg. Fully impregnated prepregs will not cause a problem if a surface has the impressions of the fibers (sometimes called the corduroy texture), again providing an evacuation path.

10.3.4 Debulking Operations

Debulking is the process of removing air and compacting the plies during the layup operation. A typical debulking operation might consist of pulling a vacuum on the layup for 10–15 min. Debulking operations usually depend on the geometry of the part and the thickness; however, a frequency of about every three to five plies is fairly typical. Although considerable controversy exists over the merits of debulking, a general guideline is that more is better because a more compacted laminate will have fewer void nucleation sites. This must be weighed against the additional cost of interrupting the collation process to conduct the debulk operation.

Although debulking is routinely conducted at room temperature using vacuum pressure, hot debulking in an autoclave is sometimes required to sufficiently reduce the bulk factor when close tolerance matched die tools are used (i.e., the tool will not fit together unless the bulk factor is reduced to near final dimensions). It is unfortunately a nuisance to debulk every part every several plies and it can dramatically increase the collation costs. A trade-off between part quality and production cost must once again be made in a logical manner.

10.3.5 Debulking Studies

To investigate the combined effects of initial debulking, prepreg moisture content, and laminate configuration on final laminate quality, four laminates were processed according to the procedures shown in Figure 10.11 [15]. The laminates were fabricated from a standard carbon–epoxy prepreg. To evaluate different laminate configurations, a cross-plied laminate with internal ply drop-offs was used. This laminate configuration contained a thin area (36 plies), a thick area (48 plies), and a tapered area with internal ply terminations.

All four laminates were ultrasonically inspected prior to final autoclave processing. The ultrasonic results shown in Figure 10.12 illustrate that essentially no sound was transmitted through the laminates debulked only once, but the hot precompacted laminates transmitted a much larger percentage of the original signal. It is also worth noting that the ply drop-off and thick areas of the laminate were of much poorer quality.

Of the two hot precompacted laminates (Laminates 3 and 4), the wet exposed laminate was of significantly poorer quality. The attenuation ranges for Laminate 3 was from 30 to 62 dB, whereas that for Laminate 4 was 10 to 42 dB. Because the hot precompaction temperature was only 66°C (150°F), the additional porosity should not be due to voids caused by the vaporization of absorbed moisture, but is more likely due to voids caused by air trapped during collation. The wet exposed prepreg exhibited a noticeable increase in tack that could contribute to increased air entrapment during collation.

All four laminates were autoclave processed using the standard cure cycle discussed previously. After processing, all four laminates were ultrasonically inspected. The results are presented in Figure 10.13. As expected, the dry-hot precompacted laminate (Laminate 4) was

Laminate 1	***Laminate 2***
Moisturized prepreg • Poor compaction	• Dry prepreg • Poor compaction
Laminate 3	***Laminate 4***
• Moisturized prepreg • Good compaction	• Dry prepreg • Good compaction

Where:

Poor compaction = No vacuum debulks

Good compaction = Vacuum debulk every 5 plies during collation,
Followed by 66°C (150°F) @ 6.8 kg/cm^2 (100 psi)
Autoclave debulk for 2 hours (Bagged for No Resin Bleed)

Moisturized prepeg = Plies exposed to 32°C (90°F), 85% RH for 72 hours
Prior to collation

Dry prepeg = Plies exposed to 24°C (75°F), 35% RH for 72 hours
Prior to collation

Figure 10.11 Laminate collation procedures

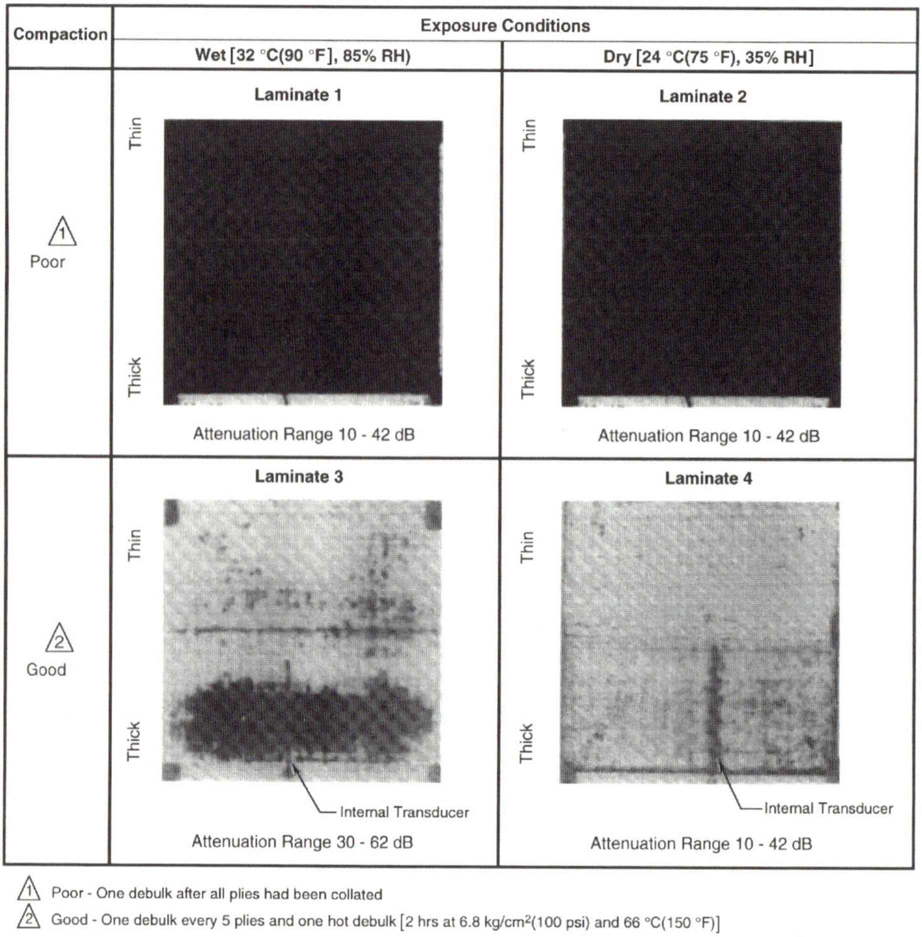

Figure 10.12 Precure NDT results

of the highest quality. It was also of interest that the extent of compaction did not significantly affect laminate quality for dry prepreg. Good initial compaction, however, did significantly reduce the amount of porosity in the wet exposed laminates. The results of this limited study showed the significant impact of absorbed prepreg moisture on final laminate quality. Although initial laminate precompaction helped to reduce the amount of porosity when moisture was present, considerable porosity was still present after curing, particularly in the thick section and at internal ply terminations. In addition to being a volatile that can cause void formation, absorbed moisture can also dramatically increase the prepreg tack level, which, in turn, increases the amount of entrapped air during collation. This further enhances the opportunity for additional void formation during autoclave processing.

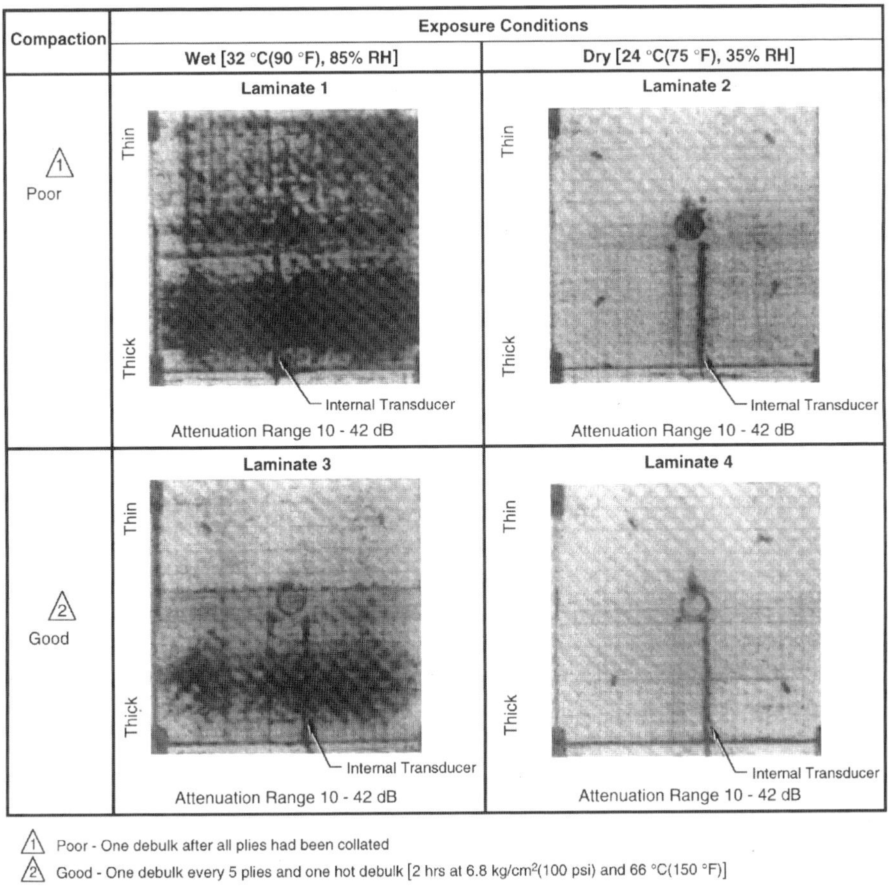

Figure 10.13 After cure NDT results

10.4 Tooling

10.4.1 Part Thermal Response

Part heat-up rate during autoclave processing can dramatically influence final part quality. At least three variables can affect the autoclave heat-up rate for composite parts: (1) tool material and design, (2) the actual placement of the tool within the autoclave, and (3) the autoclave cure cycle used. Recommendations for the design of an individual tool are fairly obvious and well understood in industry (e.g., thin tools heat faster than thick tools; materials with a high thermal conductivity heat faster than those with lower thermal conductivity; and tools with well-designed gas flow paths heat-up faster than those with restricted flow paths [e.g., tools with open egg-crate support structures heat-up faster than those without open support structures]).

The location of tooling within the autoclave can also affect the heat-up rate. Gas flow studies conducted previously [16] showed that for large commercial autoclaves, higher heat-up rates are experienced near the door due to the turbulence of the gas flow bouncing off the door and then decreases as it flows toward the rear. This phenomena is dependent on the actual design of the autoclave and its gas flow characteristics.

The previous discussion of autoclave heat-up rate is important because a number of commercial cure cycles specify using only vacuum pressure during the initial portion of the cure cycle to allow for volatile removal. The part is typically heated to the first isothermal hold with only vacuum pressure, and full autoclave pressure is then applied at the end of the hold. This can create two problems. The first problem occurs if a number of parts with different tool heat-up rates are to be cured in the same autoclave run. The dilemma facing an autoclave operator is shown in Figure 10.14. It is not clear when the hold period should start or what is the proper point to vent the bag and apply full autoclave pressure. If the resin gels during this first isothermal hold with only vacuum pressure applied to the laminate, then the probability of gross porosity is very high.

The second potential problem is illustrated in Figure 10.15. With only vacuum pressure applied during the initial part of the cure cycle, the hydrostatic pressure on the resin can be extremely low, even negative. This is an ideal condition for void formation and growth if allowed to persist to high enough temperatures.

To circumvent both of these problems in a production environment, a significant portion of the autoclave pressure should be applied immediately before initiating the heat-up cycle (Refer to Fig. 10.3). For standard epoxy systems, a full vacuum and 5.8 kg/cm^2 (85 psi)

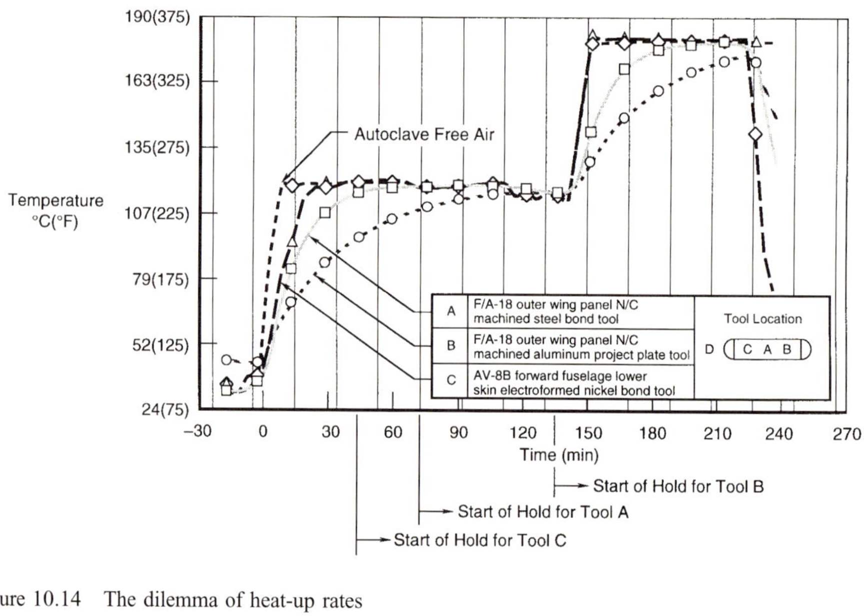

Figure 10.14 The dilemma of heat-up rates

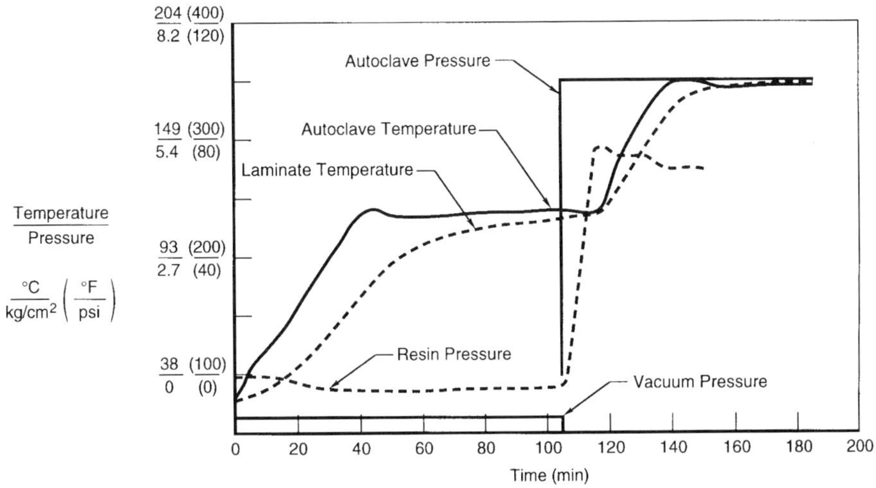

Figure 10.15 Vacuum only can create negative resin pressure

autoclave pressure can be applied through the first hold, and then the bag vented to atmosphere and 6.8 kg/cm^2 (100 psi) autoclave pressure applied before ramping up to the final cure temperature.

10.4.2 Heat Transfer Models

Heat transfer models of the autoclave process are the most accurate and well understood of all the process models. Much of this understanding is because the models are so easily verified through thermocouple measurements. Thermocouples are the most common part-sensing technique used in production. The challenging aspects are the incorporation of the affects of resin flow, resin kinetics, and autoclave position on heat transfer properties. The importance of incorporating resin kinetic models is to properly predict conditions that may lead to exotherms, especially for thick laminates [17].

Heat transfer models are a powerful tool for developing autoclave process cycles. They are especially useful in aiding tool designers in choosing tooling materials, thicknesses, and thermocouple locations. Models can also be used to determine if a tooling concept would be detrimental in a specific position in the autoclave and the types of tools that should be processed together to optimize the cure cycle.

Changes in convective heat transfer coefficients due to autoclave position and loading scheme are difficult to model. These coefficients are more easily correlated from experimental data. This correlation can be determined from monitoring thermocouples attached to tools or by correlating air flow based on autoclave position. These coefficients are crucial for determining the rate of heat transfer from the autoclave environment into the part. Heat transfer coefficients are also a function of autoclave pressure; however, the adjustment for autoclave pressure is straightforward after the autoclave position affect is known. This understanding is essential for simultaneously modeling an entire autoclave load. This capability in-turn is an extremely powerful tool, because it allows the process engineer to develop an understanding on how changes in a process cycle for one part may have a negative impact on another part in the autoclave.

10.5 Conclusions

Autoclave processing remains the mainstay for processing continuous fiber-reinforced thermoset composite parts. Considerable research has been conducted to establish a scientific understanding of the many complex and interrelated phenomena that occur during autoclave processing. This research has yielded tremendous insight into the fundamental principals that guide heat transfer, resin viscosity, resin flow, chemical crosslinking, and void formation, growth, and transport during processing. Several researchers have developed mathematical models that are powerful tools for understanding how changes in autoclave processing parameters (i.e., temperature, pressure, and vacuum), materials, tooling, and bagging can affect final part quality.

Much of this chapter focused on final part quality through control of resin pressure and void management. Maintaining the resin hydrostatic pressure above the potential void pressure is key to minimizing void formation. The resin pressure, however, is typically lower than the autoclave pressure due to resin flow, bagging, tooling concept, and support materials such as honeycomb core. Other variables influence void formation, such as material surface texture, entrained air in the resin, prepreg moisture content, and prepreg tack. Whereas these variables are much less understood, their impact is significant. In many instances, the resin pressure can be high throughout the entire cure cycle, but voids will remain that are present before, or created during, the collation operation.

Several advances have been made in autoclave process optimization. Both off-line [18] and real-time [19,20] optimization approaches have been successfully demonstrated. Off-line studies allow the ease of simulation models to be joined with optimization approaches to determine efficient autoclave positioning and process cycles, while maintaining part quality. On the other hand, real-time approaches allow for the adjustment of autoclave parameters during processing to account for changes in resin pressure, resin viscosity, part temperatures, or cure kinetics. Real-time approaches, unfortunately, require accurate and interpretable sensor information. Low-cost, reliable sensors to provide all the necessary process information are not yet available. Although neither approach is perfect, both provide tremendous insight into developing efficient processing, tooling, and bagging practices and are critical assets to process engineers. Both approaches are now being actively explored for extensions to other composite processes, such as resin transfer molding, vacuum assisted resin transfer molding, forming, and condensation reaction material systems.

Nomenclature

k	Rate constant
ΔH	Heat of reaction
P_x	Pressure (x = void, hydrostatic, applied, or spring)
dB	Decibel

References

1. Carpenter, J. "Processing Science for AS/3501-6 Carbon/Epoxy Composites," Technical Report N00019-81-C-0814, NASC, April 1983
2. Souror, S., Kamal, M. *Thermochem. Acta* (1976) 14, p. 41–59
3. Loos, A., Springer, G. "Curing of Graphite/Epoxy Composites," Air Force Contract AFWAL-TR-83-4040, Interim Report for June 1982–March 1983
4. Mijovi', J., Kim, J., Salby, J. *J Appl. Polym. Sci.* (1984) 29, p.1449–1462
5. Chung, T., *J. Appl. Polym. Sci.* (1984) 29, p. 4403–4406
6. Mallow, A., Muncaster, F., Campbell, F. "Science Based Cure Model for Composites," American Society for Composites, First Technical Conference, Dayton, OH, Oct. 1986
7. Gutowski, T. *SAMPE Q.* (1985) 16(4) p. 58–64
8. Dave, R., Kardos, J., Dudukovic, M. "Process Modeling of Thermosetting Matrix Composites: A Guide For Autoclave Cure Cycle Selection," American Society for Composites, First Technical Conference, Dayton, OH, Oct. 1986
9. Feldmeier, W., et al. "Composite Curing Process Nondestructive Evaluation," Air Force Contract No. F33615-85-C-5024, Second Interim Technical Report, 1 April–30 September 1986
10. Brand, R.A., Brown, G.G., McKague, E.L., "Processing Science of Epoxy Resin Composites," Air Force Contract No. F33615-80-C-5021, Final Report for August 1980–December 1983
11. Campbell, F.C., Mallow, A.R., Carpenter, J.F. "Chemical Composition and Processing of Carbon/Epoxy Composites," American Society of Composites, Second Technical Conference, September 1987
12. Campbell, F.C., Mallow, A.R, Muncaster, F.R., Boman, B.L., Blase, G.R. "Computer-Aided Curing of Composites," Air Force Contract No. F33615-83-C-5088, Interim Report for April 1984-April 1986, AFWAL-TR-85-4060
13. Thorfinnson, B., Biermann, T.F. "Production of Void Free Composite Parts Without Debulking," Thirty-First International SAMPE Symposium and Exposition, April 1986
14. Thorfinnson, B., Biermann, T.F. "Measurement and Control of Prepreg Impregnation for Elimination of Porosity in Composite Parts," Society of Manufacturing Engineers, Fabricating Composites '88, Anaheim, September 1988
15. Browning, C.E., Campbell, F.C., Mallow, A.R., "Effect of Precompaction on Carbon/Epoxy Laminate Quality," AIChE Conference on Emerging Materials, August 1987
16. Griffith, J.M., Campbell, F.C., Mallow, A.R., "Effect of Tool Design on Autoclave Heat-Up Rates," Society of Manufacturing Engineers, Composites in Manufacturing 7 Conference and Exposition, 1987. Anaheim, CA
17. Kays, A.O., "Exploratory Development on Processing Science of Thick-Section Composites," Air Force Contract No. F33615-82-C-5059, Final Report for September 1982 to May 1985, AFWAL-TR-85-4090
18. Thomas, J., et al. "Computer-Aided Curing of Composites," Air Force Contract No. F33615-83-C-5088, Final Report for Period April 1984–March 1989

19. West, B., Kardos, J. "Advanced Composite Processing Technology Development," Air Force Contract No. F33615-88-C-5455, Sixteenth Quarterly Interim Technical Report for February 1993–April 1993
20. Abrams, F., et al. "Qualitative Process Automation for Autoclave Curing of Composites," AFWAL-TR-87-4083, Interim Report for the Period 15 October–15 May 1987

11 Pultrusion

B. Tomas Åström

The pultrusion process is briefly introduced in terms of machinery, materials, markets, process characteristics, and technology issues, whereupon modeling of matrix-flow–related issues is discussed in some detail. Despite the fact that pultrusion in its thermoset incarnation is a very economical and widely used technique to manufacture polymer-matrix composites, and that its thermoplastic counterpart has received considerable interest from both industry and academia, they are still—from a scientific point of view—relatively poorly understood processes. Although most of the early modeling work on the thermoset process focused on heat transfer and crosslinking kinetics, it has gradually been recognized that the issue of matrix flow is of great importance in terms of impregnation, consolidation, void migration, and the like. To shed light on the relevance of flow phenomena in both thermoset and thermoplastic pultrusion, models to predict pressure distributions within the composite during molding and models to predict the pulling resistance of the die are presented. Experimental verifications of the models are discussed and the usefulness of the models in their dimensionless forms is illustrated with numerical examples.

11.1 Introduction

Already in 1951 the process of pulling fibrous reinforcement impregnated with a polymeric fluid through a die was patented as a method to manufacture fishing rods [1]. Since its inception pultrusion has developed into the most economical of all composites manufacturing processes, and even though rapid technology improvements have widened the range of applications to enable manufacturing of complex components, it is still inexpensive and uncomplicated mass products that dominate. The amount of pultruded composites sold, and consequently the number of companies involved in pultrusion, is steadily growing. Pultrusion was conceived in the United States and most production and product innovation still take place there; nevertheless, pultruders in Europe, Japan, and the rest of the world are doing their best to close the technology and market gaps.

This chapter commences with a description of pultrusion equipment, technology, and market. While reading this section, the uninitiated reader may get the impression that the pultrusion process is uncomplicated; however, its abundant intricacies are revealed upon closer scrutiny. On a worldwide basis pultruders tend to be small enterprises relying on the no-doubt significant processing experience of their employees, however, the technological approach is often "if it works, don't fix it." Since there is often a significant lack of fundamental understanding for the governing mechanisms of the process, a fair amount of effort has been spent trying to increase the understanding through process modeling, which is the subject of the bulk of this chapter. The chapter will be rounded off with an outlook section.

11.2 Process Description

In essence, *pultrusion* is a process where continuous fibrous reinforcement is impregnated with a matrix and then is continuously consolidated into a solid composite. Whereas there are several different ways to achieve impregnation and consolidation, Figure 11.1 illustrates the different processing steps in a basic pultrusion process for thermoset–matrix composites. The reinforcement is pulled from packages in a creel stand and is then gradually brought together and pulled into an open resin bath, within which the reinforcement is impregnated with the liquid resin. When the impregnated reinforcement emerges from the impregnation bath, it may require some additional guidance or shaping before entering the die. The die has a constant–cross-section cavity throughout most of its length; the exception is the tapered entrance, where excess resin is squeezed out of the reinforcement. The die is heated and the heat transferred to the (liquid) reinforcement–resin mass initiates the crosslinking reaction; the resin thus gradually solidifies from the perimeter of the composite toward the center. Although the die initially heats the reinforcement-resin mass, the exotherm of the crosslinking reaction causes the temperature of the newly solidified composite to exceed that of the die toward its end, thus cooling the composite. When the process is correctly run the temperature peak caused by the exotherm will take place within the confines of the die and the contraction of the resin due to crosslinking will cause the composite to shrink away from the die. The composite thus emerges from the die as a hot solid and is allowed to cool off before being pulled by the pulling mechanism, which is followed by a saw to cut the composite to desired lengths continuously.

11.2.1 Equipment

The preceding short process description is based on the most basic—but also most common—pultrusion setup. There are several variations to the theme, however, and many details were intentionally left out. In the following, the process is described in some detail and different processing and machinery options are briefly discussed. The variety in machinery solutions reflect the fact that many, perhaps most, pultrusion machines used to be designed and built inhouse by the pultruder rather than by a dedicated machinery manufacturer. There

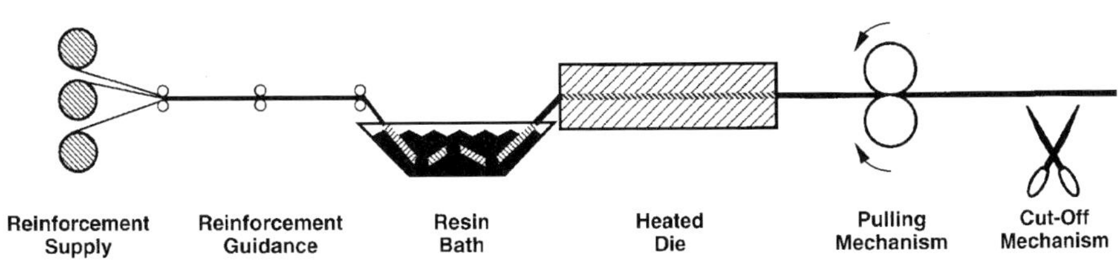

Figure 11.1 Schematic of basic pultrusion operation for thermosets

are now several commercial equipment manufacturers, however, that supply the bulk of the new machines to the marketplace.

11.2.1.1 Reinforcement Supply

Due to the nature of the process the reinforcement must be continuous, either in the form of roving packages or rolls of fabrics or mats. The reinforcement is stored in a creel stand that is of the simplest possible rack construction because no significant loads, except the weight of the reinforcement, normally need to be supported. The entire creel stand is often placed on wheels to allow off-line preparation of reinforcement supply and rapid reconfiguration of a pultrusion line.

11.2.1.2 Pre-Impregnation Reinforcement Guidance

The overall consideration of preimpregnation reinforcement guidance is that the reinforcement is fragile and, in the case of glass, also very abrasive. The dry rovings are often guided by ceramic eyelets to reduce friction and wear on both the fibers and the guidance device, although steel rings also suffice in many applications. Fabrics, mats, veils, and also rovings may be guided by plastic or metal sheets with machined slots or holes. Several sets of consecutive guides are employed to gradually shape and position the different reinforcement layers prior to impregnation. To satisfy especially high torsional requirements, roving winding units that overwrap the predominantly axial reinforcement with an angle pattern of choice may be used. This technique is referred to as *pull-winding*.

11.2.1.3 Reinforcement Impregnation

Four possible impregnation options, the first three of which are in common practice, will be described below. The most common impregnation method is the one illustrated in Figure 11.1 (i.e., the reinforcement is guided *down* into an open resin-filled bath). The impregnation is accomplished by capillary forces and the fact that the reinforcement is guided over and under rods located below the resin surface. The major advantage of this impregnation method is simplicity and good impregnation results, whereas the major disadvantage is significant volatile (usually styrene) emissions.

The second impregnation option also employs an open resin-filled bath, but the reinforcement is not guided down into the bath; rather, it horizontally enters and exits the bath through holes and/or slots in the resin container and thus maintains its horizontal travel. The major advantage of this method, which often is used in production of hollow composites, is that the reinforcement does not need to be bent, which otherwise may exclude the use, or at least wrinkle or rip, of vertically oriented fabrics, mats, and veils. This impregnation method also has the disadvantage of significant volatile emissions.

The third impregnation method, which is often referred to as injection or reaction–injection molding (RIM) pultrusion, employs a different kind of die. Unimpregnated reinforcement is guided into a narrow opening in the die, which widens into a cavity further "downstream" where the resin is injected under pressure (see Fig. 11.2). The cavity, which is maintained at a temperature that ensures that the resin does not start crosslinking prematurely,

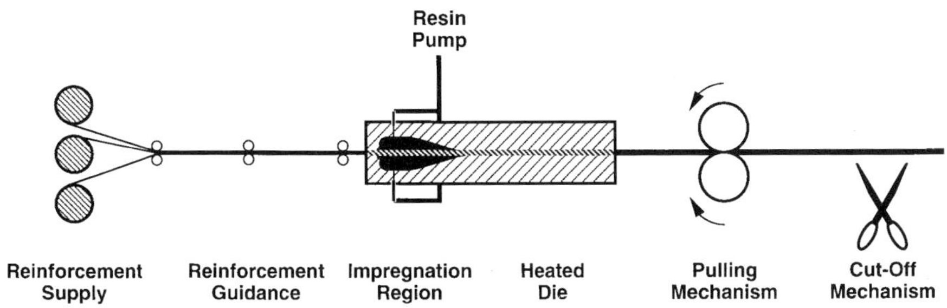

Figure 11.2 Schematic of injection pultrusion die for thermosets

is tapered toward the latter section of the die, which has a geometry similar to a traditional pultrusion die. The advantages of this impregnation method are low resin loss; no stagnant resin; that more reactive resin systems may be used (hence RIM pultrusion), thus expanding the range of resin candidates; and that the work environment is greatly improved because the impregnation takes place in a closed die resulting in very low monomer levels. The major disadvantages are complex and expensive dies and potential problems in successfully impregnating large amounts of axially oriented reinforcement.

A fourth, but rather unusual, impregnation alternative involves the use of preimpregnated reinforcement (prepregs). The reason for its limited use is the high cost of prepregs, whereas the reason for it ever having been considered is that improved composite properties may be achieved.

The first two impregnation methods clearly dominate in North America and, probably, most of the world, whereas injection pultrusion is favored in several European countries where the permissible volatile levels in the workplace tend to be the lowest. Although injection pultrusion may be the simplest way of solving the monomer emission problem, it is clearly possible to reduce this disadvantage of the first two impregnation methods through proper ventilation.

11.2.1.4 Postimpregnation Reinforcement Guidance

Once the reinforcement has been impregnated it is much less sensitive to friction than it is when it is dry and thus may be guided by for example sheet-metal guides. When manufacturing composites with complex geometries (e.g., multicavity, thin-walled sections) or with complex reinforcement orientations (e.g., high degree of off-axis reinforcement), the gradual forming of the impregnated reinforcement before it enters the die is one of the most critical and most difficult aspects of pultrusion. With injection pultrusion, all positioning of the reinforcement obviously must be accomplished before it enters the die, thereby potentially making it a more complex issue.

11.2.1.5 Preheating

To prevent cracking and excessive, exotherm-induced residual stresses in thick composites, as well as to allow increased pulling speed, preheating may be used to heat either the reinforcement prior to impregnation or to heat the liquid reinforcement–resin mass through the thickness. Common heating principles include radio and induction frequency heating, depending on whether the reinforcement is conductive (i.e. carbon) or not. Alternatively, the resin may be heated prior to impregnation to lower its viscosity and thus facilitate reinforcement impregnation; however, due to pot-life concerns, this is most commonly used with injection pultrusion.

11.2.1.6 Consolidating Die

A pultrusion die is usually machined from tool steel and typically has a length of 600–1,500 mm. With the exception of the tapered entrance, the die normally has a constant–cross-section cavity with extremely smooth surfaces that are chrome-plated to lower friction and decrease wear. The die usually is sectioned to facilitate machining, inspection, cleaning, and so on. Most of the die is heated, and typically employs two or more independently controlled temperature zones, whereas the final section of the die may need active cooling. Because excess resin is squeezed out of the reinforcement at the tapered entrance to the die, also this section may require active cooling to prevent premature gelation of the resin (i.e., before it reaches the constant–cross-section cavity).

For pultrusion of hollow cross-sections, cantilever mandrels that are mounted in front of the die (i.e., "upstream") are used. In the transverse direction, the mandrel (which may be heated) is oriented solely by the amount of reinforcement around it. The fact that it is difficult to produce perfectly concentric pipes and hollow sections with well-defined wall thicknesses underscores the significance of predie reinforcement guidance and positioning.

As eluded to in the impregnation section, a die for injection pultrusion is of a different design. One may look upon such a die as a conventional die with an added impregnation section (see Fig. 11.2). Such a die, therefore, is longer, more complex, and more expensive than its more conventional counterpart.

To increase throughput when pultruding reasonably simple components, it is common practice to mount several dies in parallel in the same pultrusion machine. This procedure is referred to as *multistream* or *multicavity pultrusion*.

11.2.1.7 Pulling Mechanism

The die is separated from the pulling mechanism by a long section to ensure that the composite cools off enough to be gripped by the time it reaches the pulling mechanism, which may be of four different kinds. The simplest one employs wheels with rubber surfaces that grip the composite in pairs; normally, several consecutive pairs are used. This kind of pulling mechanism is usually limited to laboratory-scale equipment. For commercial machinery, conventional belt pullers (e.g., akin to those used in extrusion) may be used, although it is more common to use caterpillar belt pullers, where successive rubber pads are mounted on the belt. The most common pulling concept, however, involves hydraulic clamp pullers with rubber pads. Either one or two of these may be used; with one pulling mechanism the pulling is intermittent, while with two coordinated reciprocating pullers the pulling is continuous. Except for the simplest of composites, the rubber wheels or pads that grip the composite are tailored to fit the specific geometry of the composite, thus reducing the lateral pressure needed to grip the composite, thereby reducing the risk of crushing a hollow composite. A run-of-the-mill pulling mechanism may have a pulling capacity of 50–100 kN, although larger machines may have a capacity of several hundred kN.

11.2.1.8 Cut-Off Mechanism

Because pultrusion is a truly continuous manufacturing method, composites may be produced in any length that can be handled; small, flexible cross-section composites may even be wound onto drums "endlessly." Nevertheless, the composite is normally cut, in the simplest case with a hand-held hack saw, but more commonly to predetermined lengths with an automatic saw that follows the moving composite during cutting.

11.2.2 Materials

Virtually any reinforcement–matrix combination feasible in any other composites application also may be used in pultrusion; however, glass heavily dominates as reinforcement with 95 percent in the United States and 98 percent in Europe (see Table 11.1), whereas polyester resin dominates as matrix material with 79 percent in the United States and 66 percent in Europe [2].

Most of the reinforcement used in pultrusion is in continuous roving form, although it is standard practice to interleave rovings with chopped or continuous strand mats (CSMs) to improve transverse properties, and surfacing mats, veils, and nonwovens to create a resin-rich, therefore cosmetically appealing and environmentally resistant surface. For applications that require higher transverse or torsional properties, fabrics and braids, which may be stitched together, are used. Pull-winding of rovings is used for the highest torsional requirements. Rovings are delivered on (nonrotating) inside-pull (glass) or (rotating) outside-pull (glass, carbon, and aramid) packages, whereas fabrics, CSMs, veils, and the like, are stored on outside-pull packages. Whereas fabrics may be woven or braided to the desired

Table 11.1 Material Consumption in the United States and Europe in 1991 [2]

Material combination	United States (%)	Europe (%)
Glass/polyester	79	66
Glass/vinylester	8	8
Glass/acrylics	1	16
Glass/phenolics	3	2
Glass/epoxy	4	6
Carbon/epoxy	2	2
Other	3	—
Total	100	100

Table 11.2 Characteristics of Theremoset Matrixes Commonly Used in Pultrusion [3]

System	Polyester	Vinylester	Epoxy
Resin type	Orthophthalic, isophthalic, low profile, halogenated	Epoxy Novolak, methacrylate esters of bisphenol epoxy resins	Epichlorohydrin/ bisphenol-A
Initiator type	Organic peroxide	Organic peroxide	Amines, acid anhydrides
Polymerization characteristics	Gelation occurs prior to exotherm	Gelation occurs prior to exotherm	Exotherm occurs prior to gelation
Volume shrinkage	7–9%	7–9%	1–4%
Interfacial adhesion	Low	Medium	High
Relative line speed	Normal	Slower	Very slow
Relative pultrusion difficulty	Nominal	More difficult	Very difficult
Die temperature	100–150°C	125–160°C	200°C
Posttreatment	None	None	Post cure

widths, CSMs, veils, and the like are typically cut to the desired width by the user. At the end of a roving package, a new roving is spliced to the end of the former one with an air-splicing device or a simple knot may be tied in less critical applications. Fabrics, CSMs, veils, and the like, are usually hand-sewn together. In more demanding applications, where a splice may be objectionable, the operator can easily track the composite section that contains the splice through the process to ensure that it is discarded.

The characteristics of the three most common thermoset resin systems used in pultrusion are compiled in Table 11.2 [3]. It is noteworthy that unreinforced polyesters and vinylesters shrink 7–9% upon crosslinking, whereas epoxies shrink much less and tend to adhere to the die. These epoxy characteristics translate into processing difficulties, reduced processing speed, and inferior component surface finish. It is normal practice to use resin additives to improve processability, mechanical properties, electrical properties, shrinkage, environmental resistance, temperature tolerance, fire tolerance, color, cost, and volatile evaporation. It is normally the resin, or rather its reactivity, that determines the pulling speed. Typical pulling speeds for polyesters tend to be on the order of 10–20 mm/s, whereas speeds may exceed 100 mm/s under certain circumstances. Apart from the resins characterized in Table 11.2, several other thermosets, such as phenolics, acrylics, and polyurethanes, have been tried, as have several thermoplastics (as will be discussed in Sec. 11.2.6).

11.2.3 Market

The applications of pultruded products range from sporting goods to electrical and construction. Figure 11.3 illustrates the application distribution in the United States, Japan, and Europe [2]. Some applications in the respective categories of Figure 11.3 include:

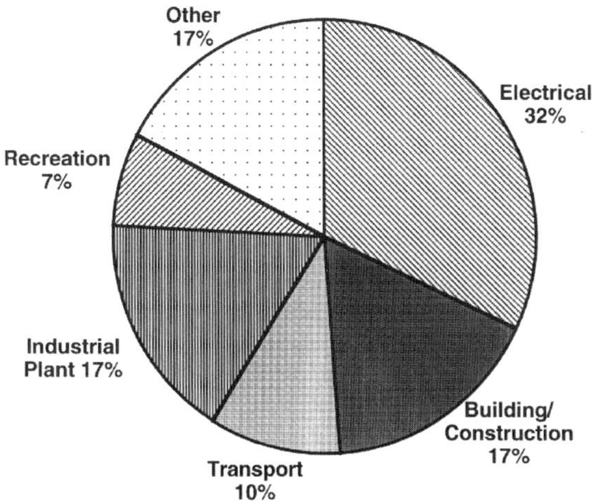

Figure 11.3 Applications of pultruded composites in the United States, Japan, and Europe in 1991 [2]

Electrical—transformer spacers, insulators, ladders, booms for electrical bucket trucks; Building/construction—construction beams and panels; Transport—cable trays in tunnels, exterior vehicle panels, automobile bumper beams, third rail cover boards for commuter trains; Industrial—grating, hand rails, pipes, cable trays; Recreation—ski poles, golf clubs, and golf bags. Some outstanding examples of the development of pultrusion technology have resulted in several footbridges, a drawbridge for vehicle traffic, a 48-m-long airfoil-shaped windmill blade, and many others.

The most spectacular application example to date is perhaps the Aberfeldy footbridge over the river Tay in Scotland (see Fig. 11.4). This bridge is 113-m long, has a deck width of 2.2 m, and a main span of 113 m [4]. The entire deck structure, hand rails, and A-frame towers are pultruded composites, and the cable stays are Kevlar ropes. The deck structure is assembled from a modular system of pultruded 6-m long, hollow components, which consist of 70 percent by weight of E glass and 30 percent pigmented isophtalic polyester resin.

World production and supply structure are shown in Table 11.3 [2,5]. The table illustrates the dominance of the United States pultrusion industry and the fact that pultrusion companies in other regions tend to be significantly smaller. One also notes the high value-to-weight ratio of Japanese and to some extent European pultrusions and the comparatively low value-to-weight ratio of pultrusions in the United States and the rest of the world, which indicate different types of products and market conditions.

11.2.4 Process Characteristics

The predominant advantage of pultrusion as a manufacturing method for thermoset composites is the low production cost, which is closely related to the reasonably high

Figure 11.4 The all-composite Aberfeldy footbridge over the river Tay in Scotland. Photograph published courtesy of Maunsell Structural Plastics Ltd

Table 11.3 World Production and Supply Structure in 1991 [2,5]

Region	Production		Number of companies	Number of machines
	10^6 kg	10^6 US$		
United States	57	240	50+	300
Europe	20	100	62	220
Japan	8	60	28	40
Rest of World	15	60	56	210
Total	100	460	200	770

processing rate and the inherent continuous nature of the process. On the list of advantages one also finds low raw material scrap, inexpensive and uncomplicated machinery, and a high degree of automation. Further, almost any constant–cross-section part can be manufactured, including components for demanding structural applications. There are limitations with pultrusion, such as the fact that only constant–cross-section components can be produced (with rare exceptions), components may have void contents too high for some applications, the bulk of the reinforcement tends to be aligned with the pulling direction, the resin must have a long pot life and a low viscosity, thus reducing available resin candidates, and there are worker health problems from volatiles, usually styrene, emitted from the impregnation bath (the last two disadvantages do not apply to injection pultrusion).

Despite the excellent stiffness- and strength-to-weight ratios obtainable in pultruded composites, it is interesting to note that more than half of all pultruded composites were probably sold for nonstructural applications (cf. Fig. 11.3).

11.2.5 Key Technology Issues

As pointed out in the beginning of the chapter—and perhaps also noticeable from the process description earlier—the pultrusion process at first appears rather simple and straightforward; however, several intricacies arise on closer inspection. The key technology issues of pultrusion of thermoset–matrix composites are usually considered to be resin formulation, temperature control, material guidance, and die design (not in order of importance). It is typically the skill in these areas that distinguishes a successful pultruder from his hapless competitor.

For open-bath impregnation, the resin must have a long pot life because (some of) it will remain in the impregnation bath for a long time and cannot be allowed to gel. In contrast, highly reactive resins with short pot lives may be used with injection pultrusion. In either case the resin viscosity must be low enough (on the order of 1 Pa·s) to enable satisfactory reinforcement impregnation. The low initial resin viscosity, however, is often partially offset by additives and fillers, which are used to alter properties and processability. It is possible to lower the resin viscosity to a certain degree by raising the temperature, although heating simultaneously decreases pot life because it also increases resin reactivity. The key issue is to tailor the resin formulation to achieve the required combination of properties, pot life, and crosslinking characteristics. To this end several different crosslinking agents, inhibitors, and accelerators that are active in different temperature regimes may be required.

The resin should preferably crosslink throughout the composite cross-section before it exits the die: The faster it crosslinks while still keeping the maximum exotherm temperature within reasonable limits, the faster and more economical the process will be. To ensure that the resin crosslinks with an exotherm temperature high enough to achieve complete conversion, but not so high as to cause excessive thermally induced stresses or even cracks in the composite, precise multizone temperature control is essential. Whereas precise control of the temperature of the reinforcement–resin mass and later the composite would ultimately be desirable, one must resort to controlling the die temperature at a few discrete points (where thermocouples are embedded). To get some feedback on how to achieve the desired composite temperature profile as a function of location in the die, it is common to embed bare thermocouples within the reinforcement–resin mass prior to entry into the die; thus, the thermocouple monitors the temperature development within the composite throughout impregnation and consolidation.

In pultrusion of composites with all the reinforcement axially aligned, the pre- and post-impregnation guidance is quite straightforward. In contrast, to place mats, fabrics, and veils corrrectly to form an intricate, thin-walled, and hollow cross-section, such as a window profile, considerable ingenuity, skill, and experience are required. From a composite strength point of view, it is of course of utmost importance that the reinforcement ends up where intended; however, if the cross-section is hollow, incorrectly placed reinforcement may force the cantilever mandrel from its intended (transverse) position, thus also changing the composite geometry, (e.g. the concentricity of a pipe).

Because the reinforcement–resin mass solidifies within the confines of the die, its geometry determines the composite geometry. Although the cross-sectional geometry of most of the die thus is determined by the desired composite geometry, the design of the die taper, the length of the die, the cavity surface finish, and location of heating/cooling zones also affect composite properties and processability. The design of the composite cross-section geometry preferably should take into account that it is to be pultruded to avoid a die geometry (i.e., corners with very small radii, large thickness variations, etc.) that is likely to result in unnecessary processing difficulties. With injection pultrusion, the die design issue is much more complex than it is with bath impregnation because the impregnation portion of the die—which makes the die longer—must be carefully designed to obtain satisfactory impregnation.

11.2.6 Pultrusion of Thermoplastic–Matrix Composites

Although still quite rare and representing only a small fraction of all pultruded composites, the use of thermoplastic matrixes in pultrusion has long been predicted to have a promising future. The interest in thermoplastics in pultrusion applications started to become obvious in the second half of the 1980s [6–8]. The reasons for this interest lie partly in the inherent advantages of thermoplastic composites in comparison to their thermoset brethren, but, perhaps more important, thermoplastics offer the possibility of greatly increasing the processing rate because no chemical reactions need to take place to consolidate the composite. Despite considerable efforts, pultrusion with thermoplastics must still be considered in the early stages of development and the commercial availability of pultruded thermoplastic composites is severely limited. Part of the reason for this is the fact that, to date, the pulling speeds reported are quite low. Although pulling speeds up to 100 mm/s have been reported [9], lower speeds are more common [7,10,11]. With the exception of the production of injection molding raw material and prepreg tapes (where relatively small fiber bundles are melt-impregnated and then pultruded through a small die), pultrusion of thermoplastic composites currently involves the use of prepregs as feedstock. "Prepreg" is here taken to include solvent-impregnated, melt-impregnated, powder-impregnated, and commingled materials: The most commonly used forms are melt-impregnated and commingled materials. Among the commonly used matrixes are polypropylene (PP), poly(phenylene sulfide) (PPS), and poly(ether ether ketone) (PEEK), although several others have been tried as well. Reinforcements used include glass and carbon.

The reinforcement ideally should be impregnated with the matrix in line with the pultrusion operation (in analogy with the thermoset process) in order to improve process economy. Although work on including on-line impregnation units is under way, this concept still must be considered under development due to the difficulties encountered in impregnation with the highly viscous thermoplastic matrices. The use of prepregs as feedstock

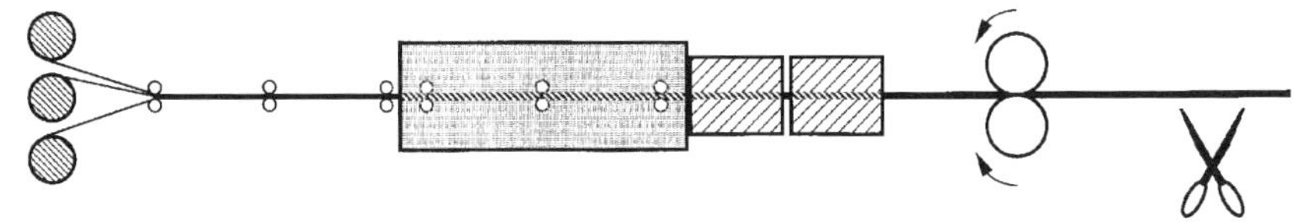

Figure 11.5 Schematic of basic pultrusion operation for thermoplastics

together with the modest production rates so far achieved probably are the greatest reasons for thermoplastic pultrusion not yet having seen a commercial breakthrough.

The machinery used in thermoplastic pultrusion is somewhat different from that used with thermosets. Whereas reinforcement supply, guidance devices, preheater, pulling mechanism, and saw are very similar or identical to the equipment used for thermosets, the feedstock is consolidated in different types of dies. Figure 11.5 illustrates the processing steps of a basic facility for pultrusion of thermoplastic–matrix composites. As with conventional, thermoset pultrusion, the preheater is used to increase the pulling speed, but the temperature required to melt the thermoplastic matrix is considerably higher than that commonly achieved with the preheater of a thermoset pultrusion process. Although one-piece dies have been tried in thermoplastic pultrusion, it is more common that at least two separate dies are used, the last of which is cooled. The heated die(s) is/are tapered to some degree, but the cooled die has a constant–cross-section cavity; typically, the dies used are much shorter than their thermoset counterparts. In a typical processing scenario, the preheater heats the feedstock to a temperature near or in excess of the softening point of the matrix before entering the heated die(s), where the material is heated further. In the taper(s), the composite is gradually shaped before it is consolidated in the cooled die.

A close relation of thermoplastic pultrusion is rollforming, where the dies have been replaced by several consecutive pairs of contoured rollers—normally four or more—that gradually shape and consolidate a stack of preheated and melted prepregs to the desired shape. The rollers, which are driven, are normally unheated and they consequently gradually cool the component. Rollforming is potentially capable of manufacturing any constant cross-section geometry and the products may be curved if desired; nevertheless, thermoplastic rollforming has been used so far to manufacture hat and Z shapes only. An important advantage of rollforming is a potentially high pulling speed—up to 83 mm/s has been reported [12]. Compared with thermoplastic pultrusion little work on rollforming appears to be ongoing despite the feasibility of the technique having been proven.

11.3 Process Modeling

Processing and design problems related to the technology issues discussed in the previous sections are usually addressed through trial-and-error type experiments, no doubt with great aid from the considerable processing experience of machine operators. The fundamental process understanding, however, is often lacking and what goes on before and inside the die is not fully understood. Typical processing problems result in parts that have scaly and cracked surfaces, contain porosities and internal cracks, or are warped and discolored. Many of these problems result in an increased pulling resistance, which thus becomes an important indicator of the state of the process. Although compilations of common processing problems and likely remedies are available (e.g., Ref. 13), much would be gained if these problems never occurred in the first place. These problems can be reduced through clever part design, die design, resin formulation and process parameter selection, which ideally should be based on modeling of the different processing mechanisms and numerical simulation of hypothetical processing situations. Thus, by modeling crosslinking, heat transfer, matrix flow, and pulling resistance a genuine and scientifically based process understanding is within reach, and one can better appreciate the importance of changes in matrix formulation, temperature setpoints, pulling speeds, die design, and so on. The next section attempts to give some examples of how a modeling capability perhaps could be of use to a pultruder.

11.3.1 How Can Modeling Help?

As previously described, it is standard procedure to feed an occasional bare thermocouple into the die together with the reinforcement–resin mass to measure the temperature distribution at a specific location within the composite cross-section as a function of (axial) position in the facility. The pultruder knows from experience at what axial position in the die the "peak exotherm" (the maximum composite temperature) should occur to produce a well-consolidated part and uses "moving" thermocouples and experience to ensure that this is achieved. The desire to maintain the peak exotherm at a specific position is in conflict with the desire to improve process economy through an increased pulling speed, because an increased pulling speed moves the peak exotherm closer to the die exit or even out of the die. With an experimentally verified model of the crosslinking reaction and the heat transfer, the need for most of the moving thermocouples may be reduced or even eliminated because the position of the peak exotherm as a function of resin formulation and pulling speed then can be predicted. Further, the maximum processing speed as a function of die temperatures may be calculated for a specific resin formulation and die length without the need for experiments. If the die is not yet designed, one may consider manufacturing a longer die than originally intended to allow a higher pulling speed while still keeping the peak exotherm within the die; however, a pulling resistance model will then reveal that a longer die (as well as increased pulling speed) results in a higher pulling resistance which may be beyond the capacity of the pulling mechanism. Further, a pressure model may reveal whether the desired consolidation pressure is achieved or not and may help determine what taper geometry best removes voids in the resin upon entry into the die.

11.3.2 Previous Modeling Work

During the 1980s an increased interest in mathematically modeling the pultrusion process resulted in numerous studies. The majority of the studies considered pultrusion of thermoset composites and concentrated on crosslinking kinetics and heat transfer, whereas some also considered matrix flow relative to the fibers and contributions to the pulling resistance. During the 1990s the interest in modeling thermoplastic pultrusion resulted in several studies of heat transfer, matrix flow, and calculations of the pulling resistance. Reference 14 provides a compilation of references on previous modeling efforts.

In thermoset pultrusion the most crucial aspects to model no doubt are heat transfer and crosslinking kinetics. In thermoplastic pultrusion heat transfer and crystallization kinetics similarly are of paramount importance. In both of these cases, modeling of the heat transfer situation is reasonably straightforward, whereas modeling of the crosslinking and crystallization kinetics, respectively, are complicated tasks that may involve a different approach for each matrix system. Most previous modeling efforts have concentrated on heat-transfer–related aspects, and these have been well documented in the literature (cf. Ref. 14), but comparatively little work has been dedicated to flow-related issues, and the bulk of the remainder of this chapter is dedicated to the latter, including issues such as pressure distributions within the reinforcement–resin mass and contributions to the pulling resistance from a pultrusion die.

Matrix flow relative to the reinforcing fibers is caused by thermal expansion of the fiber–matrix mass within the confines of the die and by the geometrical constriction of the die taper. Once the matrix flow distribution is known, the matrix pressure distribution may be determined using a flow rate–pressure drop relationship. One-dimensional flow models of thermoset pultrusion have been reasonably well verified qualitatively [15–17]. A one-dimensional flow model of thermoplastic pultrusion [14,18] has similarly been compared with experimental data and the correlation found to be encouraging [19].

The pulling resistance of a generic, tapered pultrusion die is caused by shearing of a lubricating matrix boundary layer between the outermost layer of fibers and the die wall, matrix flow relative to the fibers due to compaction, work required to compact the fiber bed, and solid-to-solid friction. To calculate the contributions to the pulling resistance when the matrix is still liquid, the matrix pressure distribution in the die taper and the distribution of the fiber compaction force throughout the die are required. To calculate the contributions to the pulling resistance when the matrix of the composite surface is solid, the normal-stress distribution between composite surface and die is required. Attempts at experimentally verifying pulling resistance models have been reasonably successful for both thermoset [16,20] and thermoplastic [19] pultrusion.

Whereas the analyses of the following sections concentrate on one specific approach to pultrusion process modeling, careful evaluation of these and other existing model approaches reveals that, with few exceptions, the other models are special cases of the more generic models presented herein. The following models specifically focus on pultrusion with shear-thinning matrixes (i.e., thermoplastics); however, by choosing the correct viscosity descriptors, the model expressions reduce to Newtonian relationships, thereby also applying to thermosets.

The exceptions mentioned in the previous paragraph are a couple of models of thermoplastic pultrusion of melt-impregnated prepregs that assume that no significant matrix flow relative to the fibers occurs [21,22]. In Reference 21, it is assumed that the lack of matrix flow may be due either to perfect metering of the feedstock or to compression of the matrix. In the former case there will be no pressure within the composite (except possibly through thermal expansion); in the latter case the matrix compresses rather than flows, thus creating a pressure increase. In Reference 22, the pressure in the composite within the die is assumed to be due to compliance and transverse deformation of the uneven prepreg plies. The only flow in this case is that of the transverse deformation of the reinforcement–matrix mass to fill the gaps between the plies, and no matrix flow relative to the fibers is considered. Another model [23] develops this concept further: The feedstock is assumed to be melt-impregnated prepregs with a matrix-rich surface on one side of each ply. This model assumes that the matrix flow *within* the fiber bundles that constitute the plies is negligible compared with the matrix flow in the matrix-rich gaps *between* the bundles. This model also assumes that the feedstock is more or less precisely metered to fit through the die and that the occurring matrix flow is caused by the axial dimension variations of the feedstock only. Attempts at experimentally verifying this flow model in terms of the predicted pulling force proved inconclusive [23].

The major difference between these models and the main approach of this chapter lies in the assumption of these models that the amount of feedstock is so precisely metered that all material fits through the die, translating into no matrix flow [21,22] or local flow only [23], whereas the models elaborated on herein assume that to achieve low void content and good consolidation a forced and constant matrix backflow—as in thermoset pultrusion—is required. The flow assumption of this chapter is based on experimental experience [19], which suggests that precise metering is not likely to be successful without paying the price of high void content and poor consolidation. At least it assumes that it would be very difficult, especially because the variations in the current commercially available feedstock is considerable [23]. With a flexible die or rollforming, however a no-flow assumption certainly may be realistic.

In contrast to the flow—or no-flow—situation, there appears to be a consensus on what mechanisms contribute to the pulling resistance of a die; although some models choose to neglect certain contributions, the following model includes all contributions discussed in the literature.

11.4 Matrix Flow Modeling

For the analysis of the matrix flow in pultrusion, the linearly tapered, constant-width die of Figure 11.6 is considered. The treatment herein applies to an idealized process at steady state; it is assumed that all fibers in the oversaturated and void-free bundle that enters this heated die are parallel to one another and to the pulling direction throughout the die, and that matrix flows parallel to the fibers only. The tapered section of a die is normally followed by a

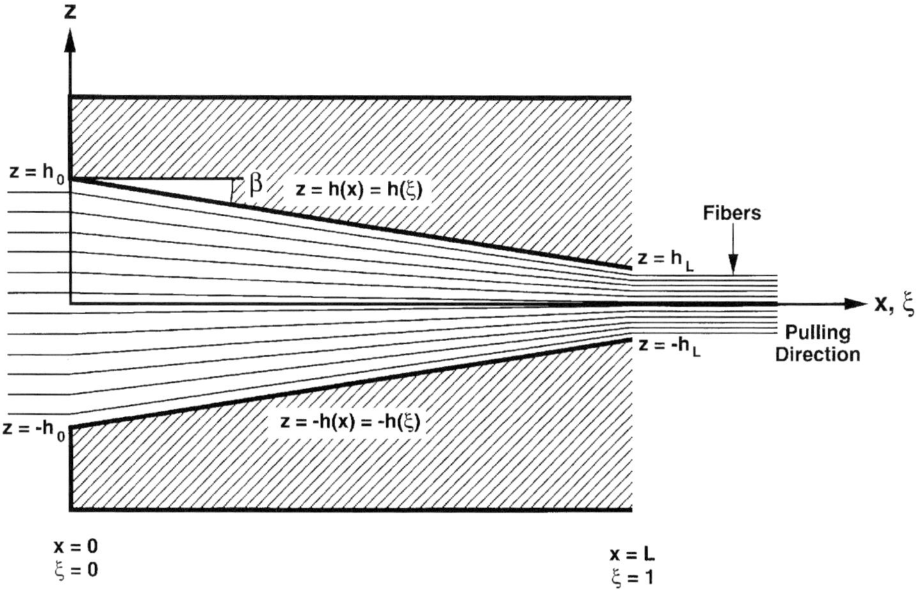

Figure 11.6 Linearly tapered die considered in pressure and pulling resistance analyses

constant–cross-section cavity, but it is assumed that no flow relative to the fibers occurs in this section. With the exception of matrix viscosity and density, the properties of fibers, matrix, and composite are assumed to be constant (i.e., unaffected by changes in temperature, pressure, and fiber volume fraction). In the model developed, the matrix viscosity changes with shear rate only and the matrix density with temperature only. Throughout the taper the fiber distribution is assumed transversely isotropic. The matrix flow is also assumed to be independent of transverse location; therefore, only the average flow of a cross section is considered.

The fiber continuity requirement for the taper follows from the assumption that all fibers are infinitely long and their number is constant. A mass balance gives the fiber–volume–fraction dependency along the x-axis as

$$v_f(x) = \frac{v_L h_L}{h(x)} \tag{11.1}$$

where $v_L = v_f(L)$, $h(x)$ is half the height of the die, and $h_L = h(L)$ (see Fig. 11.6). For a linear taper, the fiber continuity requirement may be written as

$$v_f(x) = \frac{v_L}{\dfrac{v_L}{v_0} + \left(1 - \dfrac{v_L}{v_0}\right)\dfrac{x}{L}} \tag{11.2}$$

where $v_0 = v_f(0)$. Whereas the assumption of a linear taper simplifies the subsequent analyses, it is later illustrated that the formulation is not limited to this geometry.

The matrix continuity requirement is obtained from consideration of the matrix flux at two control surfaces in the taper—one at an arbitrary coordinate x and one at $x = L$ (see Fig. 11.6). At x the matrix flux is unknown, and at the end of the taper the matrix travels with the same velocity as the fibers due to the "plug-flow" assumption. Neglecting the difference between the flow component in the x-direction and the component along the fibers, the matrix flux through the respective control surfaces is obtained as

$$q(x)h(x) = U(1 - v_L)h_L \tag{11.3}$$

where q is the matrix flow rate per unit area in a die-fixed coordinate system and U is the (constant) pulling speed. It should be noted that due to the assumption that the difference between the matrix flow in the x-direction and the matrix flow parallel to the fibers is negligible, this equation is valid for "small" angles only.

Introducing Equation 11.1 into Equation 11.3, the matrix flux through a cross-section at a coordinate x is obtained as

$$q(x) = U\left(\frac{v_f(x)}{v_L} - v_f(x)\right) \tag{11.4}$$

for an isothermal situation. The pertinent matrix flux, however, is that occurring relative to the fibers

$$\hat{q}(x) = q(x) - U(1 - v_f(x)) = U\left(\frac{v_f(x)}{v_L} - 1\right) \tag{11.5}$$

where $\hat{q}$ is the matrix flow rate per unit area in a fiber-fixed coordinate system. Equation 11.5 was first derived in Reference 20. If the temperature of the composite is increased within the taper, then the matrix flow relative to the fibers will be greater than what is indicated by Equation 11.5. Hence, because it has been assumed that the matrix density varies with temperature only, the thermal-expansion–induced flow may be included to yield

$$\hat{q}(x) = U\left(\frac{v_f(x)}{v_L} - 1\right) - U\alpha_{TE,Z}(T_{AV}(L) - T_{AV}(x)) \tag{11.6}$$

where $\alpha_{TE,Z}$ is the transverse thermal expansion coefficient of the composite and T_{AV} is the average temperature of a composite cross section, which may be obtained from a suitable heat transfer model. Equation 11.2 may be introduced into Equation 11.5 or Equation 11.6 to obtain the flow rate in a linear taper explicitly in terms of the x-coordinate. Figure 11.7 shows the dimensionless matrix flow relative to the fibers, $\hat{\theta}(\xi) = \hat{q}(\xi)/U$, according to Equation 11.5, (i.e., for an isothermal situation), as a function of the dimensionless length coordinate, $\xi = x/L$, in a linearly tapered die (cf. Fig. 11.6). In this example, the final fiber volume fraction is held constant at 0.60 while the initial fiber volume fraction is varied. The fact that the matrix flow is negative indicates a backflow (opposite direction to the pulling motion), which ceases at the end of the taper due to the plug-flow assumption. It should be noted that, because the fiber volume fraction does not linearly depend on the length coordinate in a linear taper (cf. Eq. 11.2), neither does the flow distribution (see Fig. 11.7).

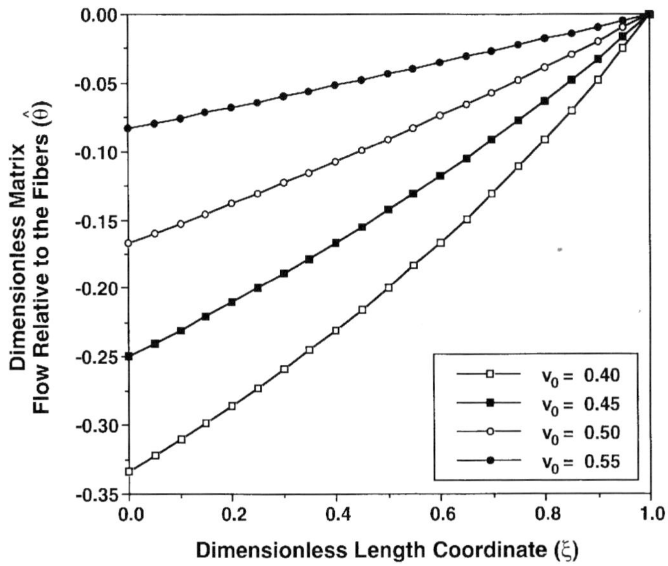

Figure 11.7 Dimensionless matrix flow relative to the fibers in a linearly tapered die as a function of initial fiber volume fraction (thermoset pultrusion)

11.5 Pressure Modeling

11.5.1 Flow Rate–Pressure Drop Relationships

To obtain the pressure distribution in the tapered die from the flow distribution, a flow rate–pressure drop relationship is required. One approach to obtain such a relationship is through an extension of the empirically derived Darcy's law for steady-state flow of Newtonian fluids through macroscopically isotropic beds [24]. It has been suggested that an equivalent expression for a bed of aligned fibers may be obtained through application of the hydraulic-radius concept, thus yielding [25]

$$\frac{\Delta P}{L} = \frac{4K_0K_1^2\mu\hat{q}}{r_f^2}\frac{v_f^2}{(1-v_f)^3} = \frac{\mu\hat{q}}{S} \tag{11.7}$$

where ΔP is the fluid pressure drop over the bed thickness, L; K_0 is a shape factor depending on the cross-sectional shape of the supposed capillaries of the porous medium; K_1 is the tortuosity of the capillaries; μ is the Newtonian fluid viscosity; r_f is the fiber radius; v_f is the fiber volume fraction; and S is the permeability of the porous medium.

A bed of aligned fibers is approximately transversely isotropic and flow parallel to the fibers is greatly favored over perpendicular flow. One could thus argue that Equation 11.7 should be generalized to incorporate a permeability tensor. Due to its lack of scientific foundation, however, Equation 11.7 has been the subject of significant criticism. It has proven to yield less than satisfactory correlation with experimental data, particularly for perpendicular flow. Thus, in an attempt to avoid some of the criticism associated with Equation 11.7, this chapter considers one-dimensional flow parallel to aligned fibers only, where the flow channels are really capillaries.

The scientific literature commonly refers to the product $K_0K_1^2$ of Equation 11.7 as the *Kozeny constant*, which has been found to equal approximately 5 for macroscopically isotropic beds for a range of conditions; in contrast, it has been found experimentally to vary with flow direction, fiber volume fraction, fiber, fluid, and pressure gradient for aligned fiber beds [25]. Nevertheless, there appears to be no realistic alternative to Equation 11.7 if one requires a first-order analytical flow rate–pressure drop relationship for aligned fiber beds, so despite the earlier reservations the following analyses will employ this flow rate–pressure drop relationship.

Virtually all previous investigators of flow in pultrusion have employed a Newtonian flow rate–pressure drop relationship, (i.e., Eq. 11.7), which is unquestionably a reasonable assumption when modeling thermoset processes because most uncured thermoset polymers exhibit near-Newtonian behavior (i.e., shear-rate–independent viscosities). Most thermoplastic polymer melts, however, are shear thinning (i.e., their viscosities decrease with increasing shear rate) [26]. Thus, it would be incorrect to assume that molten thermoplastics have constant viscosities, especially as shear rates increase. By using the concept of the generalized Newtonian fluid, which incorporates a shear-rate–dependent viscosity [26], shear-rate–dependent flow rate–pressure drop relationships may be derived. One such shear-rate–dependent fluid representation is the *Carreau model* [26], which allows one to express the viscosity of a thermoplastic melt in terms of three material parameters as

$$\eta = \eta_0[1 + (\lambda\dot{\gamma})^2]^{(n-1)/2} \tag{11.8}$$

where η_0 is the zero shear-rate viscosity; λ is a time constant that indicates when shear-thinning effects become noticeable; $\dot{\gamma}$ is the shear rate, which is equal to the magnitude of the rate-of-strain tensor; and the so-called power-law index n describes the degree of shear thinning. In this form, the Carreau model correctly represents the constant viscosity at low shear rates and the shear-thinning behavior at high shear rates. The viscosity descriptors (n, η_0, λ) in Equation 11.8 may refer to either nominal (uncorrected) or real (corrected) viscosity data, where the correction refers to the nonparabolic velocity profile of the non-Newtonian fluid. In flow-through-porous-media applications, it has proved particularly convenient [27] to calculate the non-Newtonian viscosity data from experimental measurements as if the fluid were Newtonian (i.e. had a parabolic velocity profile), thus obtaining nominal viscosity data. Using such nominal data, the flow rate–pressure drop relationship for the Carreau model is obtained as [25]:

$$\frac{\Delta P}{L} = \frac{4K_0K_1^2\hat{q}\eta_0}{r_f^2}\frac{v_f^2}{(1-v_f)^3}\left[1 + \left(\frac{2K_0K_1\hat{q}\lambda}{r_f}\frac{v_f}{(1-v_f)^2}\right)^2\right]^{(n-1)/2} \tag{11.9}$$

An explicit expression for the flow rate cannot be obtained because $\hat{q}$ appears twice in this equation, but, given the flow rate, the pressure drop can be calculated. Setting n equal to unity in Equation 11.9, one recovers the Newtonian expression (Eq. 11.7). To this author's knowledge there have been no attempts at experimentally verifying Equation 11.9 for flow through fiber beds. Because the Newtonian version of this flow rate–pressure drop relationship (Eq 11.7) has already been the subject of criticism, results obtained with both Equations 11.7 and 11.9 should be treated with caution. It is nevertheless assumed that Equations 11.7 and 11.9 yield qualitatively correct results useful in parametric studies.

11.5.2 Pressure Distributions

By substituting the matrix-flow distribution (Eqs. 11.5 or 11.6) into the relevant flow rate–pressure drop relationship (Eqs. 11.7 or 11.9), the pressure-gradient distribution in the tapered die is obtained. Inserting Equation 11.5 into Equation 11.7 yields a pressure-gradient expression that is analytically integrable. The resulting pressure distribution is obtained as

$$P(x) = \frac{4K_0K_1^2UL\mu}{r_f^2}\frac{1}{\frac{\upsilon_L}{\upsilon_0}-1}\left(\frac{1}{1-\upsilon_f(x)}-\frac{1-\upsilon_L}{2(1-\upsilon_f(x))^2}-\frac{1}{1-\upsilon_0}+\frac{1-\upsilon_L}{2(1-\upsilon_0)^2}\right) \quad (11.10)$$

as first shown in Reference 28. This pressure distribution applies to isothermal thermoset pultrusion. The isothermal assumption may be justified by the fact that the tapered part of a die for thermoset pultrusion is usually unheated or even cooled to prevent premature gelation (see Sec.11.2).

Whereas the assumption of a Newtonian fluid behavior is acceptable in thermoset pultrusion, this is not the case with its thermoplastic counterpart unless the shear rates encountered are very low. Thus, replacing the Newtonian flow rate–pressure drop relationship (Eq. 11.7) with the shear rate–dependent relationship (Eq. 11.9), a more realistic expression is obtained. Although the pressure-gradient expression thus obtained is not analytically integrable, numerical integration is straightforward. Also in thermoplastic pultrusion the isothermal assumption may be acceptable under certain conditions (e.g., when the preheater heats the feedstock to the temperature of the heated die).

If the thermoplastic-based feedstock is so precisely metered that no backflow occurs in the taper of the die, then the pressure distributions derived in either one of References 21–23 should be considered instead of the one suggested earlier, because it is not applicable if there is no matrix flow relative to the fibers.

11.5.3 Comparison Between Model Predictions and Experiments

By choosing an appropriate value of the Kozeny constant, Equation (11.10) has been qualitatively verified in terms of the matrix pressure right after the taper [15,16]. Similar experimental verifications have been presented for prediction of the maximum injection pressure to avoid "upstream" leakage in injection pultrusion applications [17] and the qualitative correlation is very good. Moreover, predicted pressure-gradient distributions in the taper based on the Carreau model have been compared with experimental data from a thermoplastic pultrusion process [19]. The experimental data were obtained with a custom-built thermoplastic pultrusion facility (see Fig. 11.8), having one heated and one cooled die preceded by a preheater (see Fig. 11.9, where the insulation on one side of the preheater and the heated die has been removed for visibility) [11]. The feedstock used in the experiments was unidirectionally reinforced melt-impregnated prepregs. The pressure distributions where obtained by feeding force-sensing resistors onto the preheater and the dies together with the prepregs [11,19] in analogy with the conventional use of moving thermocouples as previously described. Although a fair quantitative agreement between predictions and experimental data was obtained [19], this is probably fortuitous because the value of the shape factor and the tortuosity were not experimentally determined for the pertinent flow situation; rather, it was based upon a literature study on flow of Newtonian fluids through aligned fiber beds [25]. In addition, the accuracy of the pressure sensors was considered unsatisfactory; hence, a purely qualitative comparison between predictions and experimental data was performed [19]. Figure 11.10 thus shows experimental data and predictions for pultrusion of glass/PP composites normalized with respect to the maximum pressures as functions of location in the taper, where $\xi = 1$ represents the end of the taper (cf. Fig. 11.6). Two different pulling speeds and two different taper angles were experimentally evaluated. The developed pressure model reveals that changes in temperature do not affect the predicted *normalized* pressure distribution at all, and changes in pulling speed do not affect it noticeably on the scale of the figure [19]. One finds by studying Figure 11.10 that the qualitative correlation between experimental data and predictions is encouraging for changes in taper angle (curves 1 and

Figure 11.8 Experimental pultrusion facility used to verify model predictions

Figure 11.9 Preheater and die assembly of experimental pultrusion facility. Ceramic insulation on one side of the preheater and the heated die removed for visibility

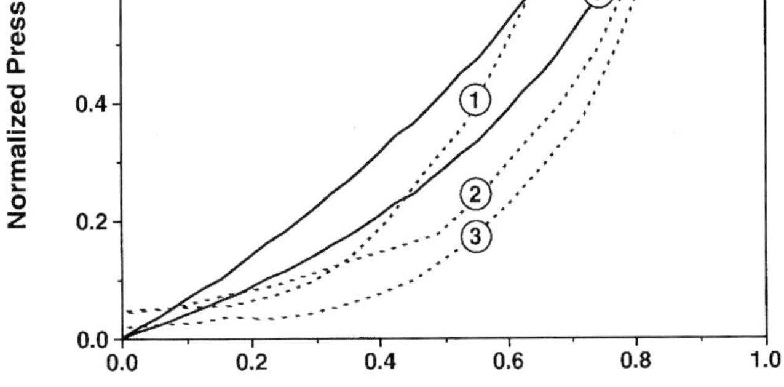

Figure 11.10 Experimentally determined and predicted pressure distributions for pultrusion of glass/PP composites normalized with respect to peak pressure

2/3, vs. 4 and 5, respectively). In addition curves 2 and 3 are in approximate agreement, as the pressure model predicts. It is thus concluded that the pressure model presented earlier yields at least qualitatively accurate results. For details on the experiments and the parameters used in the predictions, the reader is referred to References 11 and 19, respectively.

11.5.4 Sample Model Applications

To illustrate the differences between the pressure predictions based on the two fluid models for nominal viscosity data discussed earlier, a basic parametric study may be undertaken. By introducing the dimensionless pulling speed

$$\alpha_1 = \frac{K_0 K_1 U \lambda}{r_f} \tag{11.11}$$

the dimensionless pressure, Π, may be written as a function of the dimensionless length coordinate as

$$\Pi(\xi) = \frac{P(\xi) r_f}{4K_1 L}\frac{\lambda}{\eta_0} = \int_0^{\xi} \alpha_1 g_1(v_f(\bar{\xi}))[1 + (2\alpha_1 g_2(v_f(\bar{\xi}))^2]^{(n-1)/2} d\bar{\xi} \tag{11.12}$$

for the Carreau model, where

$$g_1(v_f(\xi)) = \left(1 - \frac{v_f(\xi)}{v_1}\right)\frac{v_f^2(\xi)}{(1 - v_f(\xi))^3} \tag{11.13}$$

$$g_2(v_f(\xi)) = \left(1 - \frac{v_f(\xi)}{v_1}\right)\frac{v_f(\xi)}{(1 - v_f(\xi))^2} \tag{11.14}$$

The equivalent Newtonian expression is obtained by setting the power-law index n equal to unity and $\eta_0 = \mu$. For pultrusion of carbon/PEEK, realistic numerical values for the parameters needed in the following analyses are shown in Table 11.4 [14]. Figure 11.11 shows the dimensionless pressure in the taper according to Equation 11.12 for $\alpha_1 = 10$. The predictions illustrate that $\alpha_1 = 10$ corresponds to a flow situation in the shear-thinning domain, because the Newtonian model predicts a significantly higher pressure than the Carreau model. It should be noted that $\alpha_1 = 10$ corresponds to a speed on the order of 1 mm/s for carbon/PEEK (depending on temperature) [19]. Because one of the potential advantages of thermoplastics in pultrusion is to enable pulling speeds at least two orders of magnitude greater than this, the need for inclusion of the shear-thinning behavior in process modeling is underscored. It should be noted that choosing a value for α_1 much less than unity (i.e., well into the Newtonian domain) would cause the predictions for the Newtonian and the Carreau models to coincide exactly.

It has been argued that among the requirements on a "good taper" for a thermoset pultrusion process are smooth fiber lines, easy machining, lack of dead spots, and a large pressure gradient [16]. It has also been illustrated that Equation 11.10 may be utilized to evaluate the importance of different taper shapes upon the pressure distribution [16]. As a result, three taper shapes will be considered herein: linear, parabolic, and round (circular).

Table 11.4 Dimensionless Parameters Used in Predictions Resulting in Figures 11.11–11.14

	Figure			
Parameter	11.11	11.12	11.13	11.14
α_1	10	1	1	10
v_0	0.55	0.50	Varied	0.55
v_1	0.60	0.75	0.75	0.60
n	0.8	(1.0)	(1.0)	0.8

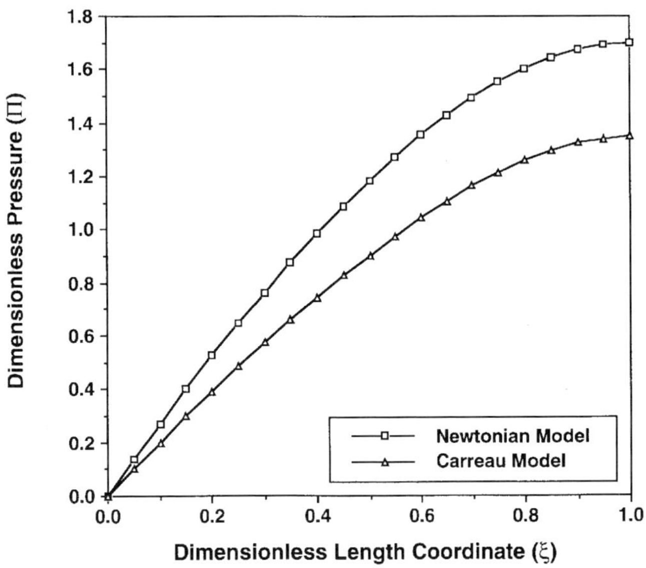

Figure 11.11 Dimensionless pressure distributions for Newtonian and Carreau fluid models (thermoplastic pultrusion)

Using the Newtonian fluid model to calculate the respective dimensionless pressure distributions at a dimensionless pulling speed of $\alpha_1 = 1$ and a fiber-volume-fraction increase realistic in thermoset pultrusion, $v_0 = 0.50$ and $v_1 = 0.75$ [16], results in Figure 11.12, which is identical in concept to Figure 10.7 of Reference 16. In this example one notes that the effect of taper shape upon the final pressure is modest and—as noted earlier—one should bear in mind that for a Newtonian fluid the pulling speed does not qualitatively influence the pressure distribution. However, by changing the initial fiber volume fraction while keeping the final fiber volume fraction constant, one finds that the final (maximum) pressure may be highly dependent upon the taper shape for certain compaction ratios (the ratio of the initial and the final fiber volume fractions), see Figure 11.13, which is similar in concept to Fig. 10.8 of Ref. 16.

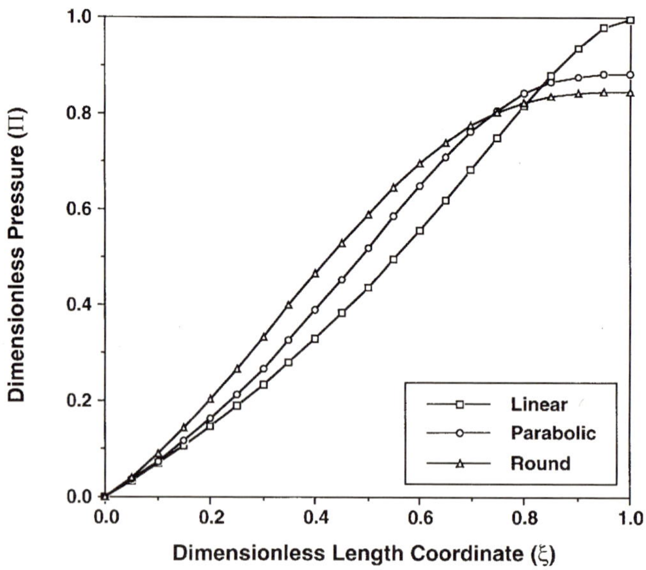

Figure 11.12 Dimensionless pressure distributions for three different taper shapes (thermoset pultrusion)

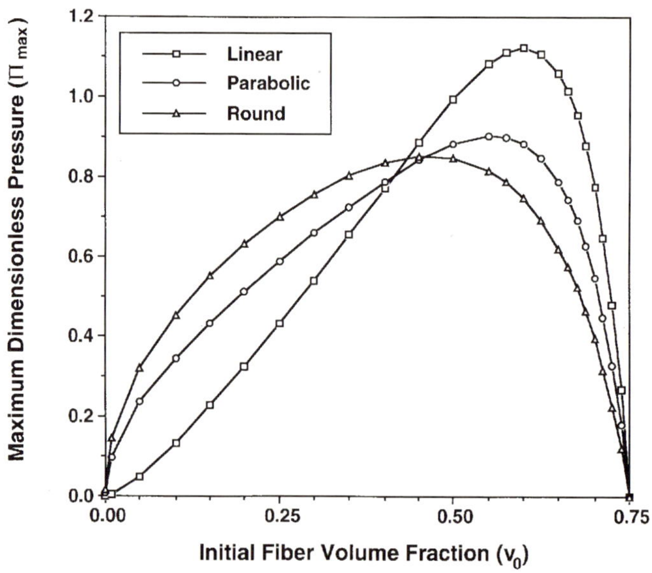

Figure 11.13 Dimensionless maximum pressure for three different taper shapes as a function of initial fiber volume fraction

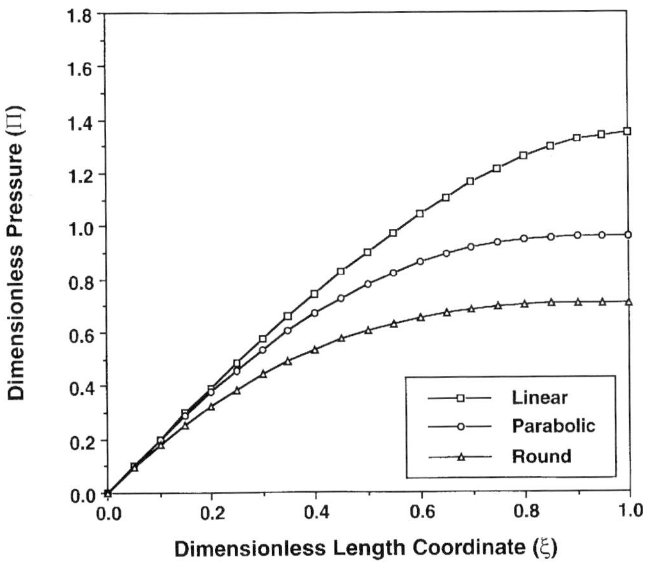

Figure 11.14 Dimensionless pressure distributions for three different taper shapes (thermoplastic pultrusion)

For thermoplastic pultrusion the first three requirements on the thermoset-pultrusion taper still apply, but the pressure must be kept within reasonable limits. The fact that the viscosity of thermoplastics is orders of magnitude larger than that for thermosets already results in much larger pressures, and the pressure contributes significantly to the pulling resistance of the die, which quickly may become the rate-limiting feature of the process [11]. Moreover, the results of Figures 11.12 and 11.13 are not directly applicable to thermoplastic pultrusion because the fiber-volume-fraction increase is generally much lower. Using the Carreau fluid model and parameters realistic in thermoplastic pultrusion (see Table 11.4), one thus obtains Figure 11.14. One finds that the taper shape that compresses the fiber–matrix mass most in the beginning of the taper (round) results in the lowest pressures, although it should be borne in mind that this depends on the compaction ratio (initial and the final fiber volume fractions) (cf. Fig. 11.13).

11.6 Pulling Resistance Modeling

The pulling resistance in a pultrusion process, which is easily monitored continuously during processing, is a crucial parameter that is often used as a "composite" indicator of the state of the process. It is thus of great interest to be able to interpret the importance of changes in the required pulling force or to predict the maximum pulling forces to be encountered in a pultrusion facility already in the design stage. The predictions of a pressure model—whether based upon matrix flow relative to the fibers or not—may be used as input to a model to predict the pulling resistance of a generic die with a cavity of local taper angle $\beta(x)$, $0 \leq \beta(x) < 90°$ (see Fig. 11.6). The total force required to pull a composite through a die is

$$F = \int_A f_{tot}(x) dA(x) \tag{11.15}$$

where f_{tot} is the total resisting force per unit area of the composite in contact with the die and A the internal surface area of the die. The resistance of the die is assumed to have three contributing factors: viscous, compaction, and friction resistance.

11.6.1 Viscous Resistance

It is assumed that a neat matrix boundary layer acts as a lubricant between the outermost layer of fibers and the die wall with no-slip conditions at both interfaces, which makes the situation analogous to wire coating. The force per unit area required to pull the composite at a constant speed is equal to the shear stress at the outermost layer of fibers. Assuming that the shear rate is constant throughout the thickness of the boundary layer, the viscous resistance for a Carreau fluid equals

$$f_v(x) = \frac{U\eta_0}{\delta(x)} \left[1 + \left(\frac{U\lambda}{\delta(x)} \right)^2 \right]^{(n-1)/2} \cos(\beta(x)) \tag{11.16}$$

where the cosine term yields the component of the resistance in the pulling direction. The equivalent Newtonian expression is obtained by setting the power-law index n equal to unity and $\eta_0 = \mu$. For this expression to be useful, an estimate of the thickness of the boundary layer, δ, is required. Several different estimates may be considered plausible. First, the thickness of the boundary layer may be of the same magnitude as the shortest distance between fibers within the composite for square or equilateral triangular arrangement. Second, fibers in square or triangular arrangement may be in direct contact with the die wall, in which case the thickness of the matrix layer may be averaged over the width of a unit cell. Third, it has been suggested that a more accurate expression for the boundary layer thickness may be the square root of the permeability [29]. In the following, the estimate based on triangularly arranged fibers in direct contact with the die wall is used, resulting in

$$\delta(x) = r_f \left(1 - \frac{1}{2} \sqrt{\frac{\sqrt{3}\pi v_f(x)}{2}} \right) \tag{11.17}$$

11.6.2 Compaction Resistance

The compaction resistance consists of the combined effects of the matrix pressure and the pressure load carried by the fiber bed, where the latter is defined as [28]

$$P_f(x) = C\left(\sqrt{\frac{v_f(x)}{v_0}} - 1\right) \cdot \left(\sqrt{\frac{v_\infty}{v_f(x)}} - 1\right)^{-4} \tag{11.18}$$

where C is a spring constant, v_0 the fiber volume fraction of the uncompressed fiber bed, and v_∞ the maximum fiber volume fraction obtainable. Through geometrical considerations one finds that the resistance from compaction equals

$$f_c(x) = [P(x) + P_f(x)\cos(\beta(x))]\sin(\beta(x)) \tag{11.19}$$

where $P(\mathrm{x})$ is given by Equation 11.10 in the isothermal Newtonian case, and by integrating Equation 11.7 or 11.9 with Equation 11.5 or 11.6 inserted, in other situations. On the other hand, if no significant matrix flow occurs, then $P(\mathrm{x})$ may be obtained from the relationships derived in References 21–23. The compaction contribution vanishes when the taper angle is zero, because for a constant–cross-section cavity there is no further compaction of the composite.

11.6.3 Friction Resistance

Two different types of friction resistance are considered: The friction that occurs where bare fibers touch the die without a lubricating matrix boundary layer in between, and the friction that occurs between solid composite surface and die before shrinkage from crosslinking and/or thermal contraction eliminates contact. In the former case the friction resistance may be written as

$$f_f(x) = P_f(x)\mu_f\cos^2(\beta(x)) \tag{11.20}$$

where μ_f is the friction coefficient. When the surface of the composite is solid, $P_f(x)$ is replaced by the normal stress distribution between composite and die.

11.6.4 Total Pulling Resistance

Assuming that the width of the die cavity shown in Figure 11.6 is constant and equal to $2w$, the total force required to pull a composite through the die may be expressed as

$$\begin{aligned} F_{tot} = {} & 4 \cdot \int_0^L (1 - \mathscr{F}(x))f_v(x)[w + h(x)]dx + 4 \cdot \int_0^L f_c(x)w\,dx \\ & + 4 \cdot \int_0^L \mathscr{F}(x)f_f(x)[w + h(x)]dx \end{aligned} \tag{11.21}$$

where the three integrals represent the contributions from viscous, compaction, and friction resistance, respectively, and $\mathscr{F}$ denotes the fraction of the composite perimeter that is subject to friction. By the same reasoning, the fraction that is subject to viscous resistance must be $1 - \mathscr{F}$. $\mathscr{F}$ may vary from zero when the fiber–matrix mass is completely surrounded by a neat matrix boundary layer, to unity when the composite surface is solid. It should be noted that because the die is assumed to be of constant width the fiber–matrix mass is compacted in the z-direction only (see Fig. 11.6), and that the compaction thus is integrated over the width of the die only (the second integral of Equation 11.21), whereas the other contributions are also integrated over the height of the die. It should further be noted that to calculate the resisting contributions of a heated die with part of the cavity untapered, the distribution of the pressure load carried by the fiber bed throughout the die is required (Eq. 11.18), whereas the matrix pressure distribution in the taper is sufficient, because the matrix pressure contributes to the pulling resistance solely in compaction.

11.6.5 Comparison Between Model Predictions and Experiments

Attempts at verifying a pulling resistance model similar in concept to the model presented earlier have proved successful for isothermal "pultrusion" of a thermoset resin without initiator [16], and the qualitative correlation is also quite good for (real, nonisothermal) pultrusion of a resin with initiator added [16,20]. In the case of the nonreactive resin, the friction resistance was found to be roughly twice the viscous resistance [16], whereas for the reactive resin the friction resistance has clearly been found to dominate over the other resistance mechanisms [16,20,21]. In thermoset pultrusion the friction resistance increases linearly until the resin at the die wall gels, at which point the friction resistance abruptly increases, whereupon it equally quickly vanishes as the solidifying composite shrinks away from the die [16,21].

Attempts at experimentally verifying the pulling resistance model presented earlier for a thermoplastic pultrusion process have provided encouraging results [19]. The experimental data were obtained with a pultrusion facility (see Fig. 11.8), which employs one heated and one cooled die preceded by a preheater (see Fig. 11.9). Figures 11.15 and 11.16 show experimentally determined pulling forces together with predictions from the pulling resistance model [19]. The figures show the pulling force for pultrusion of 3.2×6.4-mm carbon/PEEK composites as functions of pulling speed and common temperature of preheater and heated die, respectively; error bars show standard deviation. For details on the experiments and the parameters used in the predictions, see References 11 and 19, respectively. For both figures both preheater and heated die were set to the same temperature to eliminate thermal-expansion contributions to the pulling resistance.

The lower set of data (unfilled symbols) of Figure 11.15 is from an experiment where the heated die was just slightly overfilled and very little material was accumulated in the taper, whereas the upper set of data (filled symbols) is from an experiment where the taper was almost entirely filled with material. (The degree of die filling refers to the amount of material that is fed into the die in relation to the die cavity cross-section. If a die is perfectly filled, then all material "fits" into the die and no material is squeezed back (i.e. no flow occurs as

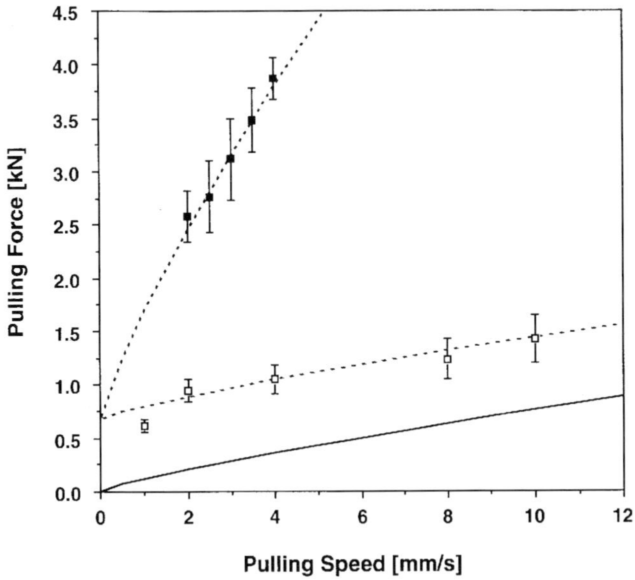

Figure 11.15 Experimentally determined and predicted pulling force as a function of pulling speed for pultrusion of carbon/PEEK composites

assumed in References [21–23]. If a die is overfilled, then not all material fits and some material is squeezed back toward the entrance of the die [i.e. flow occurs]).

In both sets of experiments the fiber volume fraction was increased from 0.55 to 0.60 from prepreg to composite, so the pressures and thus the pulling resistances predicted by the models are the same in both cases. The solid line at the bottom of the figure represents the predictions of the pulling resistance model based on the fiber-volume-fraction increase only. As the line suggests, the model predicts that even an infinitesimal pulling force would start the process, which is not consistent with experimental experiences. Because there will be significant friction between prepreg tapes and between fibers right before and in the taper when the 23 tapes used are being brought together, it seems likely that the system would have a minimum force required to start the process. Assuming that this is the case, the original predictions may be shifted to coincide with the data from the slightly overfilled die (unfilled symbols). Based on the 4-mm/s data point, the minimum pulling force required to start the process is 680 N.

From the pulling resistance predictions based on the experimentally determined fiber-volume-fraction increase from 0.55 to 0.60, one finds that the contributions from friction and compaction are negligible. Hence, virtually all the resistance is of viscous nature (because the aforementioned friction between tapes and between fibers at the entrance of the heated die is not included in the model). It is obvious that any friction resistance from fiber-to-die contact within the heated die must be very small because the fiber-volume-fraction increase is so small (cf. Eq. 11.18). In addition, such a small fiber-volume-fraction increase translates into very little matrix backflow in the taper; therefore, the compaction contribution to the pulling force is also negligible.

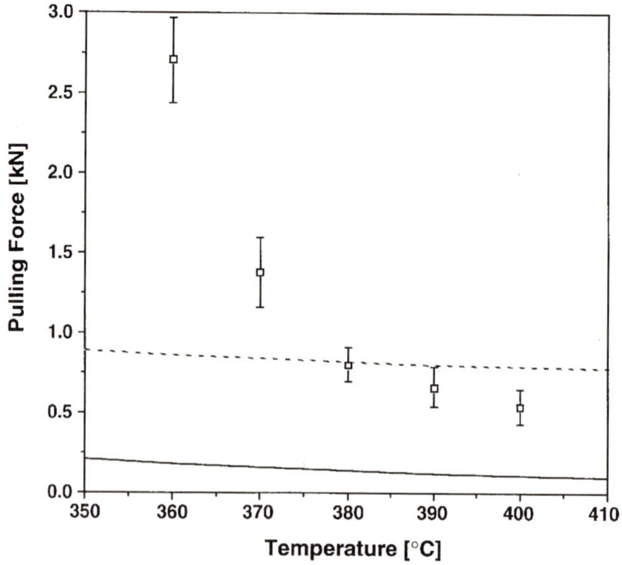

Figure 11.16 Experimentally determined and predicted pulling force as a function of common temperature of preheater and heated die for pultrusion of carbon/PEEK composites

On the other hand, in the case where the taper was filled with material, there obviously must have been significant matrix backflow over a greater length of the taper than indicated solely by the fiber-volume-fraction increase. Hence, by using the same minimum force required to start the process as for the data from the slightly overfilled die, one finds, through fitting, a quite realistic *effective* taper length—slightly less than the real taper length—resulting in the dashed line correlating the filled symbols (see Fig. 11.15). By using this taper length in the predictions, one finds that in this case the compaction is the clearly dominant contribution to the pulling resistance, which explains the different slopes of the two sets of predictions and experimental data.

For Figure 11.16, the pulling speed and the temperature of the cooled die were held constant, while the common temperature setpoint of the preheater and the heated die was varied. The fiber volume fraction was increased from 0.55 to 0.60 from prepreg to composite. The only aspect of the pulling resistance predictions that changes with composite temperature is the matrix viscosity, which decreases with increasing temperature, thus decreasing the pressures. The solid line represents the predictions based on the fiber-volume-fraction increase only, while the dashed line includes the minimum force required to start the process determined from the data of Figure 11.15. All data of Figure 11.15 were obtained with the temperature of the preheater and the heated die set to 380°C, which explains why the correlation with the 380°C data point in Figure 11.16 is so good. Nevertheless, one clearly notes that the experimentally determined temperature dependency is much stronger than indicated by the model predictions, especially at low temperatures. Because the concept of a minimum force required to start the process relates to the deformability of prepreg tapes and collimated fiber bundles before and in the taper of the heated die, it would seem appropriate to assume that this minimum force would be temperature dependent, because the higher the temperature of the tapes, the more pliable they are. Through relatively rough pulling-speed–variation experiments at different temperatures it was determined that the minimum force required to start the process is indeed temperature dependent; however, a more thorough investigation is necessary to quantitatively capture this temperature dependency.

Figures 11.15 and 11.16 illustrate that the proposed pulling resistance model does not appear to include all relevant mechanisms contributing to the resistance. Based on the preceding discussion, however as well as further attempts at experimentally verifying the pulling resistance model [16,19,20], it is concluded that the model yields results that are at least qualitatively correct, while at the same time acknowledging that further improvements are necessary.

11.6.6 Sample Model Applications

Because the pulling force of a pultrusion process is such a convenient indicator of the state of the process, it is of great interest to study the influence of process parameters and die geometry on the pulling resistance of a die. This is conveniently accomplished by defining the dimensionless pulling resistance

$$\Phi(\xi) = \frac{F(\xi) r_f}{4 K_1 L^2 w} \frac{\lambda}{\eta_0} \tag{11.22}$$

the dimensionless pulling speed, α_1 (Eq (11.11)), and the following dimensionless parameters

$$\alpha_2 = K_0 K_1 \tag{11.23}$$

$$\alpha_3 = \frac{C r_f}{K_1 L} \frac{\lambda}{\eta_0} \tag{11.24}$$

$$\alpha_4 = \frac{r_f}{K_1 L} \tag{11.25}$$

$$\alpha_5 = \frac{h_L}{w} \tag{11.26}$$

Assuming a Carreau fluid and using these five dimensionless groups to rewrite Equation 11.21, the accumulated pulling resistance at any given axial location within a die may be written as

$$\begin{aligned}\Phi(\xi) = &\int_0^{\xi} (1-\mathscr{F}(\bar{\xi}))\left(\frac{\alpha_1\alpha_4}{\alpha_2 g_3(v_f(\bar{\xi}))}\right)\left[1+\left(\frac{\alpha_1}{\alpha_2 g_3(v_f(\bar{\xi}))}\right)^2\right]^{(n-1)/2} \\ &\cdot\left[1+\alpha_5\left(\frac{v_1}{v_0}+\left(1-\frac{v_1}{v_0}\right)\bar{\xi}\right)\right]\cos(\beta(\bar{\xi}))d\bar{\xi} \\ &+\int_0^{\xi}\left\{4\cdot\int_0^{\bar{\xi}}\alpha_1 g_1(v_f(\xi'))[1+(2\alpha_1 g_2(v_f(\xi')))^2]^{(n-1)/2}d\xi'\right\}\sin(\beta(\bar{\xi}))d\bar{\xi} \\ &+\int_0^{\xi}\alpha_3 g_4(v_f(\bar{\xi}))\sin(\beta(\bar{\xi}))\cos(\beta(\bar{\xi}))d\bar{\xi} \\ &+\int_0^{\xi}\mathscr{F}(\bar{\xi})\mu_f\alpha_3 g_4(v_f(\bar{\xi}))\left[1+\alpha_5\left(\frac{v_1}{v_0}+\left(1-\frac{v_1}{v_0}\right)\bar{\xi}\right)\right]\cos^2(\beta(\bar{\xi}))d\bar{\xi}\end{aligned} \tag{11.27}$$

where $g_1(v_f(\xi))$ and $g_2(v_f(\xi))$ are given by Equations 11.13 and 11.14 and

$$g_3(v_f(\xi)) = 1-\frac{1}{2}\sqrt{\frac{\sqrt{3}\pi v_f(\xi)}{2}} \tag{11.28}$$

$$g_4(v_f(\xi)) = \left(\sqrt{\frac{v_f(\xi)}{v_0}}-1\right)\cdot\left(\sqrt{\frac{v_\infty}{v_f(\xi)}}-1\right)^{-4} \tag{11.29}$$

In Equation (11.27) the first integral represents the viscous resistance, the second the resistance from matrix flow relative to the fibers caused by compaction ("backflow"), the third the resistance from compaction of the fiber bed, and the fourth the resistance from friction.

To obtain the equivalent Newtonian relationship for the pulling resistance (appropriate for thermoset pultrusion), one only has to set the power-law index n equal to unity and $\eta_0 = \mu$ in Equation 11.27; however, for thermoset pultrusion μ_f (the friction coefficient) and $\mathscr{F}$ (the fraction of the composite perimeter that is subject to friction) will vary significantly over the length of the die. $\mathscr{F}$ is small at first but as the matrix at the die wall solidifies it abruptly rises to unity [16,21]. As the matrix at the die wall starts crosslinking (gels), the friction coefficient also changes with degree of crosslinking [16,21]. A fundamental understanding of these dependencies can only be achieved through rigorous heat-transfer and crosslinking-kinetics modeling [16,21]. The pulling resistance model outlined earlier, however, can readily incorporate such dependencies—if known.

In thermoplastic pultrusion, however, μ_f and $\mathscr{F}$ can be assumed approximately constant within a die, although they may change from one die to another. The previous model thus may be initially evaluated without the need for a heat transfer model because it is possible to envision a process where the heat transfer issues in the heated die do not significantly influence the pulling resistance of the die (i.e. where the composite temperature is more or less constant throughout the heated die) [18,19]. For pultrusion of carbon/PEEK, realistic values for the aforementioned five dimensionless groups and the other dimensionless constants required for the predictions that result in Figures 11.17–11.20 are given in Table 11.5 [14]. Thus, one may calculate the *accumulated* resisting contributions of a heated die as a function of the dimensionless length as shown in Figure 11.17, where it has been assumed that half the total length of the die is linearly tapered ($0 \leq \xi < 1$), whereas the second half has a constant cross-section ($1 \leq \xi \leq 2$). From Figure 11.17, where $\alpha_1 = 10$ (the other parameters are given in Table 11.5), it is obvious that the resistance caused by the compaction-induced matrix flow relative to the fibers ("matrix flow") dominates over the other contributions. The dependency of the four resistance contributions on the dimensionless pulling speed, α_1, are illustrated in Figure 11.18, which shows the accumulated pulling resistances (at $\xi = 2$). The figure shows that because the fiber compaction and friction contributions are assumed speed independent, they are negligible for normal processing conditions ($\alpha_1 \geq 10$); the resistance from matrix flow dominates at all realistic pulling speeds. Although this resistance always dominates in thermoplastic pultrusion as long as there is matrix flow relative to the fibers, however, it need not be *as* dominating, as shown in an example in Reference 14. Figure 11.19 shows that for all reasonable fiber-volume-fraction increases, the relationship between the four contributions remains largely unchanged for a final fiber volume fraction of 0.6. Figure 11.20 shows the influence of the shear thinning of the matrix on the total accumulated pulling resistance of the die (at $\xi = 2$). $n = 1$ represents a Newtonian fluid, $n = 0.8$ a moderately shear thinning matrix, such as PEEK, and $n = 0.4$ a strongly shear thinning matrix, such as PP. This figure illustrates that the pulling resistance of the die becomes strongly dependent on the degree of shear thinning of the matrix as pulling speeds increase (recall that realistic pulling speeds are $\alpha_1 \geq 10$).

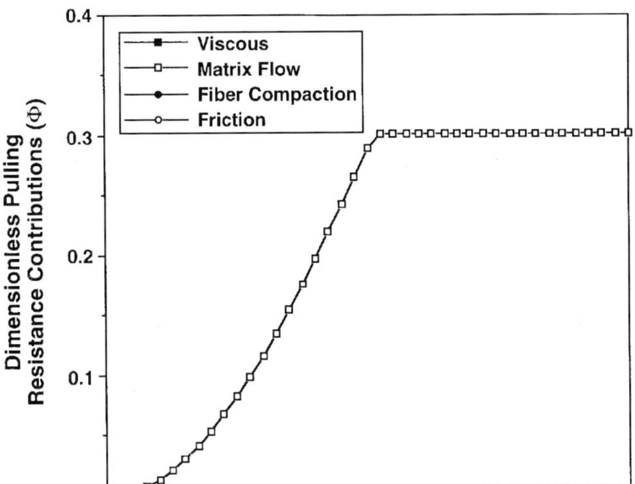

Figure 11.17 Accumulated dimensionless pulling resistance distributions (thermoplastic pultrusion)

Table 11.5 Dimensionless Parameters Used in Predictions Resulting in Figures 11.17–11.20

	Figure			
Parameter	11.17	11.18	11.19	11.20
α_1	10	Varied	10	10
α_2	1	1	1	1
α_3	0.000001	0.000001	0.000001	0.000001
α_4	0.0001	0.0001	0.0001	0.0001
α_5	0.1	0.1	0.1	0.1
β	5°	5°	5°	5°
μ_f	0.5	0.5	0.5	0.5
$\mathscr{F}$	0.1	0.1	0.1	0.1
v_0	0.55	0.55	Varied	0.55
v_1	0.60	0.60	0.60	0.60
v_∞	0.829	0.829	0.829	0.829
n	0.8	0.8	0.8	Varied

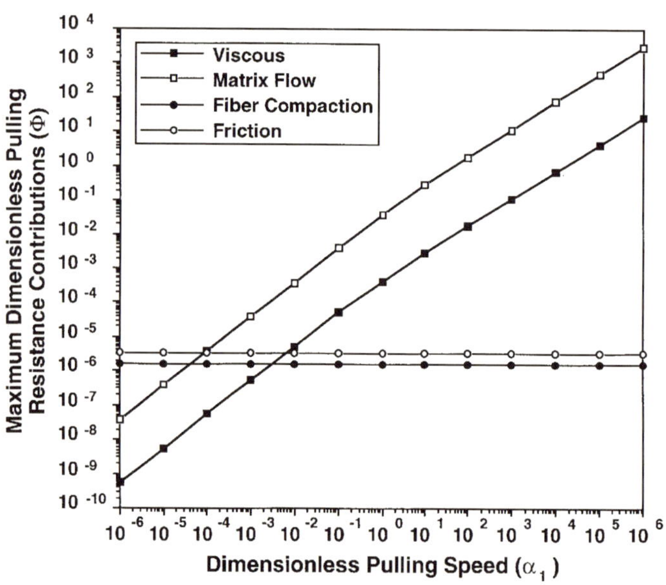

Figure 11.18 Maximum dimensionless pulling resistance contributions as function of dimensionless pulling speed (thermoplastic pultrusion)

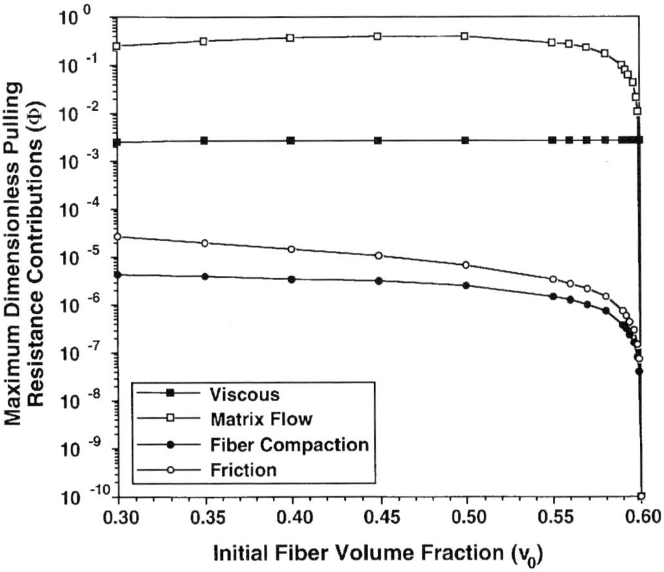

Figure 11.19 Maximum dimensionless pulling resistance contributions as function of initial fiber volume fraction (thermoplastic pultrusion)

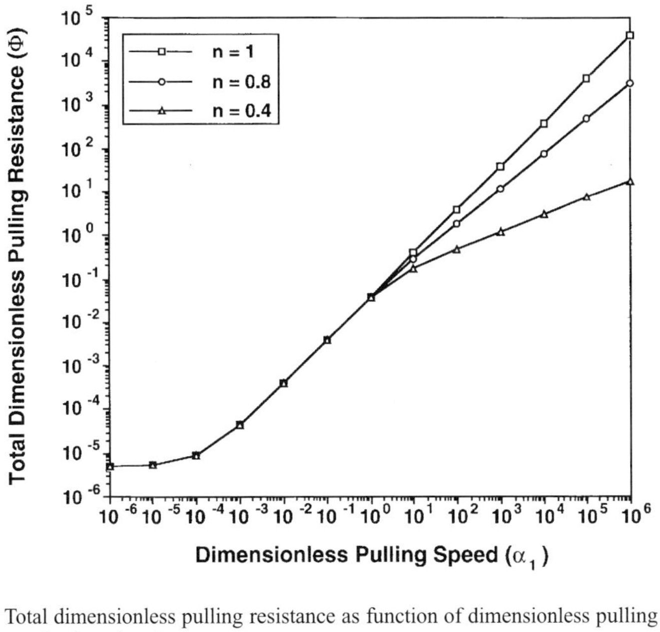

Figure 11.20 Total dimensionless pulling resistance as function of dimensionless pulling speed and power-law index (thermoplastic pultrusion)

11.7 Outlook

From a commercial point of view thermoset pultrusion is a healthy manufacturing technique that provides high-performance composites at low cost. Among developments are ever larger and geometrically more complex components, which if desired may have excellent surface finish (e.g., for use in automotive applications). These process- and product-enhancement developments are likely to continue and may be greatly facilitated by a science-based approach in manufacturing (e.g., through process modeling). Once given the chance, pultruded thermoset composites have proved to be serious competitors to extruded aluminum and steel profiles, as well as concrete structures in civil engineering applications. The main obstacle to a major breakthrough in structural applications lies in the fact that pultruded composites are still more common in applications that are not load-carrying, which is likely to be education and standardization related. The education issue is two-faced: Engineering students tend to learn little or nothing about composites in general and pultrusion in specific, and most already practicing engineers probably learned nothing about composites when they went to school and have likely learned very little on the topic since graduation. Although standards for pultruded composites exist, they are not comparable to the extensive standards and property tables on metals and concrete that engineers have at hand. To improve the future of pultrusion the composites and pultrusion gospels need to be preached at both university and continuing education levels and the work on standardization must continue.

The future of thermoplastic pultrusion is a lot less certain than that of its thermoset counterpart, although there is little doubt that the current technical problems will gradually be eliminated, thus also removing much of the current economical obstacles. However as long as thermoplastic pultrusion is compared with its thermoset relation, which has been shown to be the least expensive of all composites manufacturing processes, thermoplastic pultrusion is nevertheless likely to end up looking unattractive. An important thing to bear in mind is that thermoset pultrusion has limitations in what composites may be manufactured (e.g., in terms of geometry, reinforcement orientations, etc.) and that some of these limitations do not apply to thermoplastic pultrusion. It is unlikely that thermoplastic pultrusion will replace its thermoset counterpart to any appreciable degree, if at all. Within the near future, prepregs are likely to dominate all but the simplest applications, so it is most important to consider which advantages could possibly offset the high raw material cost. Advantages that are of increasing interest are the possibilities to improve toughness, postforming, welding, and recycling. Without offsetting advantages, thermoplastic pultrusion is unlikely to see any widespread interest as long as the raw material is prepregs. Many difficulties could obviously be overcome if the obstacles associated with on-line impregnation could be eliminated. Developments in using low–molecular-weight prepolymers in RIM pultrusion may provide the answer, but this concept has not yet proved commercially viable. Although work on including on-line impregnation stations in thermoplastic pultrusion are underway, this concept must still be considered under development due to the many difficulties encountered.

In the process of further developing and refining the pultrusion processes to take on new challenges, the work to an increasing degree will require a fundamental insight that is obtainable through process modeling only. Whether considering thermoset or thermoplastic pultrusion, the most relevant issue to model is that of heat transfer, which to a certain degree is already employed in process optimization. In thermoset pultrusion the crosslinking kinetics naturally are of utmost relevance, and crystallization kinetics are similarly of importance in thermoplastic pultrusion. The flow issues have so far been neglected to a large degree, but the first-order flow, pressure, and pulling resistance models outlined herein provide a starting point for further developments into the important flow-related mechanisms that to a significant degree govern the pultrusion process.

Nomenclature

A	die cavity surface area	[m^2]
C	spring constant	[Pa, N/m^2]
f	resisting force per unit composite-to-die contact area	[Pa, N/m^2]
$\mathscr{F}$	fraction of composite circumference subject to friction	[dim. less]
F	pulling resistance	[N]
g	a function of the fiber volume fraction	[dim. less]
h	half die height	[m]
K_0	shape factor	[dim. less]
K_1	tortuosity	[dim. less]
$K_0K_1^2$	Kozeny constant	[dim. less]
L	die subdimension	[m]
n	power-law index	[dim. less]
P	matrix pressure, pressure load carried by the fiber bed	[Pa, N/m^2]
q	flow rate per unit area in a die-fixed coordinate system	[m/s]
$\hat{q}$	flow rate per unit area in a fiber-fixed coordinate system	[m/s]
r	radius	[m]
S	permeability	[m^2]
T	temperature	[°C]
U	pulling speed	[m/s]
v	volume fraction	[dim. less]
w	half die width	[m]
x, z	coordinates	[m]
α	dimensionless group in (dimensionless) pulling resistance model	[dim. less]
α_{TE}	coefficient of thermal expansion	[1/°C]
β	taper angle	[dim. less]
$\dot{\gamma}$	shear rate	[1/s]
δ	boundary layer thickness	[m]
η	non-Newtonian matrix shear viscosity	[Pa·s, Ns/m^2]
θ	dimensionless flow rate per unit area in a fiber-fixed coordinate system	[dim. less]
λ	matrix time constant	[s]
μ	Newtonian matrix shear viscosity	[Pa·s, Ns/m^2]
μ_f	coefficient of friction	[dim. less]

ξ	dimensionless length coordinate	[dim. less]
Π	dimensionless matrix pressure	[dim. less]
Φ	dimensionless pulling resistance	[dim. less]
$_{AV}$	average	
$_{c}$	compaction	
$_{f}$	fiber, friction	
$_{L}$	pertaining to distances $x \geq L_1$	
$_{m}$	matrix	
$_{tot}$	total	
$_{TE}$	thermal expansion	
$_{v}$	viscous	
$_{x,z}$	coordinates	
$_{\infty}$	highest obtainable	
$_{0}$	zero shear rate, original, entrance of die	
$_{1}$	pertaining to distances $\xi \geq 1$	
$_{1,\dots,5}$	numbering of constants, dimensionless parameters, etc.	

References

1. Howald, A.M., Meyer, L.S. U.S. Patent 2 571 717 (1951)
2. Barrett, G. "Pultrusion and Pultruded Composites" (1992) Prospectus of the Report of a Multi-Client Research Investigation undertaken by Corporate Development Consultants, Bristol, U.K.
3. Mitrey, J. "Polyester and Vinylester Resins" (1989) In distributed notes at the Pultrusion Process Seminar arranged by Pultrusion Technology, Twinsburg, Ohio
4. Churchman, A.E., Head, P.R., Philips, J.M. *Composites* (French Journal) (1993) 3, p. 103–109
5. Barrett, G.F.C., Johnson, S. EPTA World Pultrusion Conference (1992), 1, Genval (Brussels), Belgium
6. Madenjian, A.R., M.S. thesis (1985) University of Lowell, Mass.; SPI Composites Institute Annual Conference (1985) 40, 11-F
7. Cowen, G., Measuria, U., Turner, R.M. *I. Mech. E.* (1986) C22, p. 105–112
8. O'Connor, J.E. SPI Composites Institute Annual Conference (1987), 42, 1-D; *SAMPE Q.* (1987) 18, p. 32–38
9. Taylor, S.R., Thomas, W.M. International SAMPE Technical Conference (1990) 22, p. 78–87
10. Spoo, K. International SAMPE Symposium (1990) 35, p. 775–781
11. Åström, B.T., Larsson, P.H., Pipes, R.B. *Comp. Manuf.* (1991) 2, p. 114–123
12. Devlin, B.J., Gibson, A.G. Flow Processes in Composite Materials, FPCM '91, Limerick, Ireland (1991); International Conference on Automated Composites, ICAC 91 (1991) The Plastics and Rubber Institute, 3
13. "Pro's Guide to Pultrusion Troubleshooting," (1987) Owens-Corning Fiberglas Corporation, Toledo, Ohio
14. Åström, B.T., Pipes, R.B. *Polym. Comp.* (1993) 14, p. 173–183
15. Batch, G.L., Macosko, C.W. The Manufacturing Science of Composites, (1988) Manufacturing International '88 ASME, Atlanta, p. 57–62
16. Batch, G.L. Ph.D. thesis (1989) University of Minnesota, Minneapolis
17. Kim, Y.R., McCarthy, S.P., Fanucci, J.P. SPE Annual Technical Conference (1991) 49, p. 1966–1969
18. Åström, B.T., Pipes, R.B. SPI Composites Institute Annual Conference (1991) 46, 4-A; *SAMPE Q.* (1991) 22(4), p. 55–64

19. Åström, B.T., Pipes, R.B. *Polym. Comp.* (1993) 14, p. 184–194
20. Bibbo, M.A., Gutowski, T.G. SPE Annual Technical Conference (1986) 44, p. 1430–1432
21. Batch, G.L. Effective Manufacturing Methods of Pultrusion (1990) SME Workshop, Manhattan Beach, Calif.
22. Lee, W.I., Springer, G.S., Smith, F.N. *J. Comp. Mat.* (1991) 25, p. 1632–1652
23. Gasser, K.B. M.S. thesis (1991) Massachusetts Institute of Technology, Cambridge, Mass.
24. Darcy, H. *Les fontaines publiques de la ville de Dijon* (1856) Dalmont, Paris
25. Åström, B.T., Pipes, R.B., Advani, S.G. *J. Comp. Mat.* (1992) 26, p. 1351–1373
26. Bird, R.B., Armstrong, R.C., Hassager, O., *Dynamics of Polymeric Liquids, v. 1* (1987) John Wiley & Sons, New York
27. Kembłowski, Z., Michniewicz, M. *Rheol. Acta* (1981) 20, p. 352–359
28. Gutowski, T.G., Cai, Z. The Manufacturing Science of Composites (1988) Manufacturing International '88, ASME, Atlanta, p. 13–25
29. Gibson, A.G. *Comp. Manuf.* (1992) 3, p. 113–118

12 Principles of Liquid Composite Molding

B. Rikard Gebart and L. Anders Strömbeck

Liquid Composite Molding (LCM) is the name of a class of manufacturing processes that have the common feature that a mold cavity filled with dry reinforcement is injected with liquid resin. Well-known examples of this class of processes are Resin Transfer Molding (RTM) and Structural Reaction Injection Molding (S-RIM). The science and technology of LCM will be reviewed in this chapter. The process steps of preforming, mold filling, and in-mold cure will be discussed in detail, followed by a more cursory discussion of mold design for LCM. In the discussion of preforming the characteristic features of reinforcements relevant to preforming will be described. The most common preforming methods will then be discussed. In mold filling the basic theory and relevant general results will be summar-

ized, followed by a discussion of mold filling problems. Useful rules-of-thumb for different injection strategies will be described and illustrated with an example. For in-mold cure the relevant theory will be summarized followed by a discussion of optimization of cure and possible cure-related problems.

12.1 Introduction

Liquid Composite Molding (LCM) is the common name for several similar processes for the manufacturing of polymeric fiber reinforced composites. Widely used processes that belong to this class are Resin Transfer Molding (RTM), Vacuum-Assisted Resin Injection (VARI), and Structural-Reaction Injection Molding (S-RIM).

Regardless of the variant of LCM the process can be subdivided into a number of steps. The first step is "preforming," which means that dry reinforcement is tailored to the shape of the mold. This can be done in many different ways depending on the complexity of the geometry and the requirements on mechanical performance of the part. The "preform" is then placed in a mold cavity that is subsequently closed. The preform is usually compressed slightly by the mold. The next step is resin injection into the mold cavity until the preform is fully impregnated. The final step is "in-mold cure" (i.e., curing inside the mold until the part is sufficiently stiff to be demolded). A controlled "postcure" is sometimes performed to ensure that optimum properties are obtained.

LCM has successfully been used to manufacture products ranging from cosmetic parts with moderate demands on structural properties up to highly load-bearing parts of military [1] and aerospace quality [2]. One example of the first type of product is body parts for trucks (see Fig. 12.1) that are produced in series typically above 10,000 parts per year. Another example of the latter type of product is a sail yacht rudder shaft (see Fig. 12.2) for which production volumes are typically below 100 parts per year.

The extreme limits of the LCM process exhibit large differences and can almost be said to be a number of different processes. Thus, there is no single recipe for LCM that always leads to successful results. On the other hand, there are several general principles that can be applied in all cases, and that is what we will concentrate on in this chapter.

In this review of LCM we will try to be general when possible, but we are forced to specialize to the RTM process in many cases. In particular the special features of the S-RIM process (high-speed mixing, materials, etc.) are almost completely left out of the discussion. For readers with a special interest in this topic we recommend the book by Macosko [3].

We have chosen to divide the rest of this chapter into three sections on the important topics of preforming, mold filling, and in-mold cure, followed by a section with a brief review of mold design and related topics.

The present state of the art on theory of flow in porous media has been compiled by Kaviany [4], and a detailed derivation of the governing equations through the technique of volume averaging has been performed by Tucker and Dessenberger [5]. The latter text also specializes to the case of processing with LCM. Thus, we see no need to go into the fine

Figure 12.1 Truck part manufactured with RTM by Borealis Industrier AB, Sweden. Figure shows demolding step

Figure 12.2 Sail yacht rudder shaft manufactured with RTM by Nautor Ab, Finland. The shaft is tested for strength and stiffness

details of flow in porous media; however, sufficient detail about the theory will be presented to give the reader a basic understanding of the physics of mold filling.

12.2 Preforming

There are several good reasons to preform the reinforcement before loading the mold. One is to speed up the process and to free the production mold from everything except loading, injection, in-mold cure, and demolding. Another is to improve the quality and reduce part-to-part variations. For fast cycle times the ideal is to make the preform so stiff that it becomes self-locating in the production mold. In other cases when the mechanical properties are of paramount importance one often wants to minimize the amount of preform binder since the mechanical properties can be adversely affected by the binder [6]. In general, a good preform is required to be inexpensive to make and it must be stiff enough to be stacked and handled before injection. The fibers must stay in the direction in which they have been placed during preforming both during handling and injection. To achieve all these goals it is common to apply some form of preforming agent. Both thermoplastic and thermosetting powders are commonly used for this purpose.

An important generic problem to all preforming methods is the tendency for the reinforcement to follow the "inner lane" around corners so that the fiber content becomes low or negligible at the "outer lane" (see Fig. 12.3). This can sometimes lead to mold filling problems because the flow will follow the path of least resistance. One way of reducing the problem is to "compensate" the preform tool and modify the geometry so the preform fills the real mold well at corners. The problem with varying fiber content at the corners is less pronounced for high-fiber volume fractions [7].

The compaction behavior of the preform differs a lot depending on the preforming method and the type of reinforcement that has been used. A typical compaction curve for a carbon-fiber–woven fabric is shown in Figure 12.4. The three curves correspond to

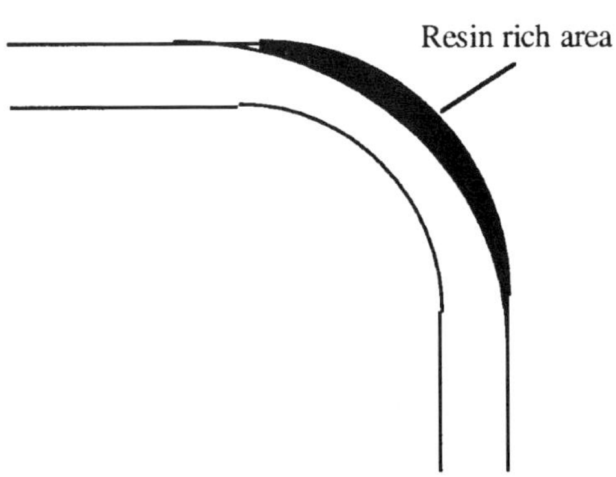

Figure 12.3 Resin-rich areas can easily form at radii if the preform is allowed to follow the "inner curve"

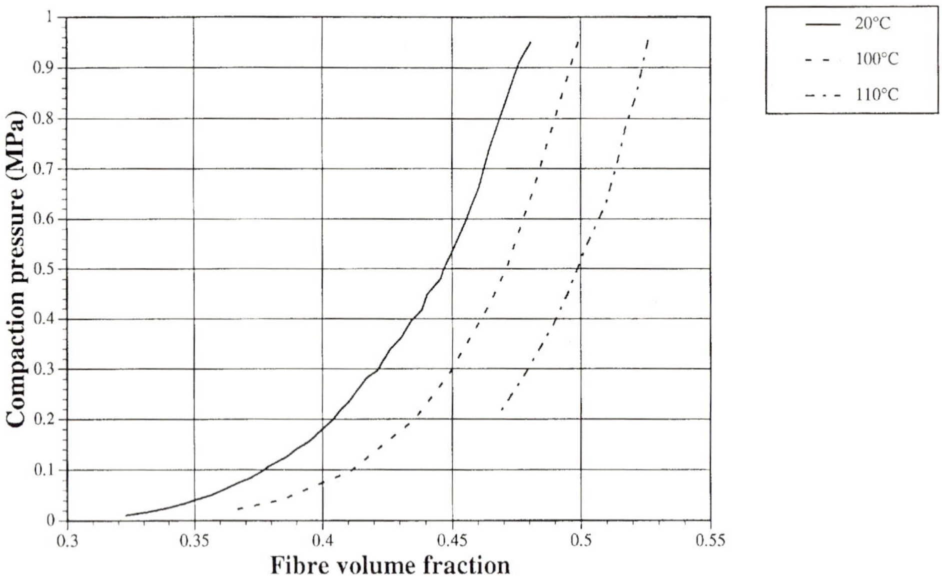

Figure 12.4 Compaction behavior of a woven carbon fiber fabric. The solid curve shows the behavior at room temperature; the other two lines show the behavior at 100°C and 110°C. The fabric was factory treated with an epoxy preform binder

compaction at three different temperatures. At a higher temperature the preforming agent acts as a lubricant, which results in a significantly higher fiber content. On the other hand, the preforming agent will impregnate the tows partly in this case and make it more difficult to achieve a uniform impregnation with resin during mold filling. Other fabrics will exhibit similar behavior to that in Figure 12.4. The levels of fiber volume fraction, however, depend strongly on the fiber architecture. More "crimp" (i.e., more undulation of fibers) and twisted fiber bundles will usually result in a stiffer fabric.

An important observation is that the fiber volume fraction at zero compaction pressure can differ significantly between different fabrics [8]. This represents a lower limit to the range of fiber volume fractions that are acceptable. A lower nominal fiber volume fraction can result in movement of the reinforcement during filling and incomplete impregnation. Another typical feature of the compaction behavior is that all fabrics behave like nonlinear (stiffening) springs and that the possible increase in fiber volume fraction from the value at rest is limited [8] (Fig. 12.4).

In practice, most molds are flexible and will deform when the fiber volume fraction becomes too high (which corresponds to a high compaction pressure). The forces associated with the compaction can be so high that the mold surface becomes dented or even that the entire mold breaks (if the clamping device is powerful enough).

High compaction pressure is a particularly difficult problem when the thickness of the laminate in a part is varying in different positions (stepping of the thickness). In this case even a small error in the placement of the fabric with a corresponding increase in local fiber

Figure 12.5 Glass fiber preforms for truck part made by thermoforming of continuous strand mat. (Borealis Industrier AB, Sweden)

volume fraction can lead to a dramatic increase in compaction pressure. This problem has to be addressed in part, and mold design and sufficient margins for error have to be allowed for.

Both thermosetting and thermoplastic preforming powders are commonly used to stiffen the preform. The preforming agent should ideally not decrease the permeability, the wettability, or the mechanical properties of the finished part, but it should still stiffen the preform so it can be handled. In practice, however, this compromise is difficult to achieve. For example, the mechanical properties can be significantly reduced by the preforming operation. but they can also be close to the value without preform binder with a judicious choice of preforming agent [6]. The raw material suppliers can usually recommend suitable preforming agents for a given matrix system or provide pretreated reinforcement with binder.

The preforming methods can be roughly classified into five basic types:

1. Cut and paste
2. Spray-up of chopped fiber on perforated models
3. Thermoforming (see Fig. 12.5)
4. Weft knitting
5. Braiding

12.2.1 Cut and Paste

Cut and paste is similar to sheet metal forming where the part is subdivided into a number of "simple" shapes that can be easily formed from the reinforcement at hand. The pieces are fit together either with an adhesive or by stitching. If an adhesive is used, it should be remembered that the mechanical properties can be adversely affected by this [6]. The cut-

and-paste method is unfortunately not very efficient, but is presently the only available method for stitch-bonded and many other types of fabrics for high-performance applications.

12.2.2 Spray-Up

Spray-up of chopped fiber on perforated models has been used for many years. One of the difficulties with this method has been creating a uniform and reproducible thickness of the preform. This problem is addressed with the new P4 process [9], where an industrial robot is programmed to hold and move a specially designed spray gun and cutter that sprays the chopped fibers together with a thermoplastic powder on a perforated preform tool. After complete spray-up, hot air is forced through the preform for about 1 min so that the thermoplastic powder melts. After melting, the air stream is switched to cold and the preforming powder solidifies. Advantages with this method are that inexpensive raw material (glass-fiber roving) can be used and it can be automated to a high degree.

12.2.3 Thermoforming

Thermoforming is the most commonly used method for volume production today [10]. In this method a special type of reinforcement that is already impregnated with a preforming powder is heated before it is clamped in a cold preforming tool [11]. The most common type of reinforcement is continuous strand mat which is manufactured by e.g. Vetrotex and Owens Corning.

Woven fabrics are also available, however, with preforming powder (e.g., from Brochier SA). The formability of woven fabrics is limited and only moderately double-curved shapes have been formed commercially. Highly formable woven fabrics suited for thermoforming, however, have recently been introduced on the market (Brochier SA). This may increase the use of woven fabrics in industrial-scale preforming.

Wrinkles and folds formed by draping of woven fabrics can to some extent be predicted with computer simulations [12]. It has also been shown, in the case without wrinkling, that the change in fiber orientation that results from the thermoforming and its influence on permeability and mechanical properties can be computed with good accuracy [13]. Any change in permeability will influence the fill behavior; therefore, it is important to account for the effect of preforming in mold-filling simulations.

Thermoforming is faster and higher fiber contents can be achieved than they can for the spray-up technique. Moreover, if continuous strand mat is used in thermoforming, it can be combined with patches of directional reinforcement. The major drawback with thermoforming is the relatively high level of waste material.

12.2.4 Weft Knitting

Weft knitting (or spiral weaving) is a very old textile technique (Jacquard looms are used). In the composite industry a well known example is woven "socks" for aircraft radomes. The advantage of this technique is the potential for automation, the negligible waste and that directional reinforcement can be achieved. The main disadvantage of the method is the high price of the preforms. Minor disadvantages with the method is that it is difficult to achieve higher fiber volume fractions than app. 45% (with plain weave) and that the preforms have to be "trimmed" before use.

12.2.5 Braiding

Braiding is another old textile technique and is a very good choice for tubular structures [14]. Braids are commercially available in diameters up to about 300 mm and with different types of fibers. The price of braided reinforcements is higher than would be expected based on the potential for automation. The most probable cause for the high price is the present low production volume.

12.3 Mold Filling

The mold-filling step is crucial in RTM because most of the defects are formed in this step. A successful elimination of these problems is only possible with a thorough understanding of the mold-filling step. The theory for mold filling can be derived from first principles, but we will not go into such detail in this section; instead, we aim at providing sufficient background and some useful results to enable process engineers to do their own estimates of fill times, draw conclusions about the origin of defects, optimize process parameters, and so on. For a deeper theoretical treatment of flow in stationary fiber beds we recommend Tucker and Dessenberger [5]. For a description of the state of the art in computer simulations we recommend Advani et al. [15].

12.3.1 Theoretical Considerations

For most of the resins used in RTM the flow through the reinforcement is governed by Darcy's law. Deviations from this law can be expected if the resin is non-Newtonian or if the reinforcement is displaced by the mold filling. The qualitative behavior, however, will generally be as follows.

For flow in one direction Darcy's law predicts that the flow rate per unit area (Q/A) is proportional to the pressure gradient ($\Delta p/L$), and inversely proportional to the viscosity of the resin (μ):

$$\frac{Q}{A} = \frac{K}{\mu}\frac{\Delta p}{L} \tag{12.1}$$

The coefficient of proportionality K is called the *permeability of the reinforcement.* According to theory [5] K is only dependent on the geometry between the fibers in the reinforcement (the "pore space"). Several models for the dependence of K on the fiber volume fraction V_f has been proposed. The most-cited model is the so-called Kozeny-Carman model [16,17], which predicts a quadratic dependence on the fiber radius R in addition to the dependence on V_f

$$K = \frac{R^2}{4k}\frac{(1 - V_f)^3}{V_f^2} \tag{12.2}$$

The constant k is called the Kozeny constant and it attains a value of 0.7 for well-ordered reinforcements with uniformly distributed fibers, (e.g. unidirectional prepreg) [17]. For commonly used fabrics in LCM (e.g., continuous strand mat or weaves) $R^2/4k$ should be seen as an adjustable model parameter with only weak coupling to the fiber or fiber bundle diameter.

Continuous strand mats are approximately isotropic and have almost the same permeability in all directions (in the plane of the fabric). Many other fabrics, however, are strongly anisotropic and have different permeability in different directions. Gebart [18] proposed a model for this class of fabrics derived theoretically from a simplified fiber architecture. The model, which is valid for medium to high fiber volume fractions, was developed for unidirectional fabrics, but it can also be used for other strongly anisotropic fabrics. In this model the permeability in the high permeability direction (which is usually, but not always, in the direction of the majority of fibers) follows the Kozeny-Carman equation (Eq.12.2). In the perpendicular direction, however, it is:

$$K_{\text{transverse}} = CR^2\left(\sqrt{\frac{V_{\max}}{V_f}} - 1\right)^{5/2} \tag{12.3}$$

The model constants are the product CR^2 and $V_{\max}$. $V_{\max}$ is typically in the 0.7–0.9 range, depending on the type of fabric, and it corresponds to a maximum achievable fiber volume fraction that results in complete loss of flow in the slow-flow direction.

The best way to use the Kozeny-Carman model and other permeability models (e.g. the anisotropic model by Gebart) [18], is to use them as interpolation formulas for intermediate volume fractions between known values. Extrapolation should be done with extreme caution because the models are developed for idealized reinforcements. Typical values for the permeability of different types of reinforcement are given in Table 12.1.

The permeability can be determined experimentally in several different ways (e.g., in a radial flow [19] or unidirectional flow experiment [20]). The experiments can also be done with either an advancing flow front (wetting flow) or a fully saturated reinforcement under steady-state conditions. There is some debate in the scientific community whether the

Table 12.1 Typical Permeability Data for Some Reinforcement Materials

Type of material	Fiber volume fraction	Permeability (m^2)
Continuous strand mat	0.25	$1 \cdot 10^{-9}$
Unidirectional glass (along fiber direction, 0°)	0.59	$7.1 \cdot 10^{-11}$
Unidirectional glass (across fiber direction, 90°)	0.59	$1.2 \cdot 10^{-11}$

permeability should differ between the wetting and saturated flow case. The outcome of this discussion still remains to be seen but at least one study reports the same value in both cases [20]. A convenient way to estimate the permeability is to use a rectangular mold where the resin is injected from one of the sides unidirectional toward the opposite side. The other two sides must seal tightly against the reinforcement so that the flow front becomes a straight line.

The permeability is estimated from the fill time $T_f[s]$ of the length $L[m]$:

$$K = \frac{L^2}{T_f} \frac{\mu}{2} \frac{(1 - V_f)}{\Delta p} \tag{12.4}$$

where μ [Pas] denotes the viscosity, V_f the fiber volume fraction, and Δp [Pa] the injection pressure. It is also common to use several samples of the flow front position versus time and a least-square procedure for the permeability [20]. The evaluation of the permeability in the radial flow method is more complicated and involves a nonlinear estimation of parameters. This method is described by Adams et al. [19]. One of the major sources of errors in permeability measurements is mold deflection, and it is a particular nuisance in the radial flow method because the smallest in-plane dimension of the mold (which governs the mold deflection) has to be larger compared with the unidirectional flow method [20]. The major difficulty with the unidirectional flow method is to prevent leakage at the edges [20]. We have also experienced some ambiguity in the results obtained using different experimental fluids (e.g., water, resin, or oil). We therefore recommend the use of real resin in permeability measurements until the ambiguity is eliminated or explained.

The mold-filling time depends both on the permeability of the reinforcement and on the viscosity of the resin. As a rule-of-thumb the catalyst system of the resin and the processing temperature can be chosen so that the gel time is about three times longer than the fill time. The time when the viscosity increases so much that no flow can occur is sometimes referred to as the *noninjection point* or NIP time. The NIP time is related to the gel time of the resin, but it occurs considerably earlier because a moderate increase in viscosity (compared with gellation) will already make further flow impossible. For a further discussion of the rheology of thermosets, see Chapter 2. Most resin systems that have a viscosity below 1 Pas can be resin-transfer molded. Even higher viscosity can be accepted, but the price for this is usually a very long injection and cure time. High-viscosity systems can often be preheated before injection so that the viscosity is reduced sufficiently. A fairly low temperature increase can already be sufficient to reduce the viscosity to the recommended level because the viscosity dependence on temperature is exponential.

For the sake of completeness we will state the equations for a three-dimensional anisotropic medium. In this case Darcy's law can be generalized to [5]:

$$u_i = \frac{-K_{ij}}{\mu} \frac{\partial p}{\partial x_j} \tag{12.5}$$

In this equation u_i should be interpreted as the volumetric flux density (directional flow rate per unit total area). The indexes range from 1 to 3, and repetition of an index indicates summation over that index according to the conventional summation convention for Cartesian tensors. The term "superficial velocity" is often used, but it is in our opinion that it is misleading because u_i is neither equal to the average velocity of the flow front nor to the local velocity in the pores. The permeability K_{ij} is a positive definite tensor quantity and it can be determined both from unidirectional and radial flow experiments [20]. Darcy's law has to be supplemented by a continuity equation to form a complete set of equations. In terms of the flux density this becomes:

$$\frac{\partial u_i}{\partial x_i} = 0 \tag{12.6}$$

Finally, before an attempt at solving the equations can be done, appropriate boundary conditions for the equations have to be defined. These are:

$$p = \text{constant at the moving flow front} \tag{12.7}$$

$$u_i n_i = \frac{-K_{ij}}{\mu} \frac{\partial p}{\partial x_j} n_i = 0 \text{ at the mold in-plane limits} \tag{12.8}$$

where the components of the normal vector to the mold in-plane boundary are n_i.

$$\begin{matrix} p = \text{constant or} \\ u_i = \text{constant} \end{matrix} \quad \text{at the inlet} \tag{12.9}$$

The equations are normally solved with a control volume finite element method. With these methods no boundary conditions are necessary at the outlets (i.e., it is implicitly assumed that there will always be an outlet at the point where the flow front merges and fills the last part of the mold). For a further discussion of boundary conditions and details about the numerical solution of the field equations (Eqs. 12.5 and 12.6) see [15,21,22–24].

12.3.2 Injection Strategies

The mold-filling time and the part quality is affected by the mold-filling strategy (the way in which the resin is introduced into and air is vented out of the mold). The mold-filling strategies can be subdivided into three main types: point injection, edge injection, and peripheral injection. In point injection the resin is introduced through a port in the center of the part, the resin flows essentially radially into the reinforcement, and air is vented at the periphery of the part. Edge injection is accomplished by injection through a film inlet at one edge of the part, the flow is more or less unidirectional over the part, and air is vented at the opposite side. Finally, in peripheral injection resin is introduced in a resin distribution channel around the periphery of the part, the flow is radially inward and air is vented at the center of the part.

The mold-filling time differs considerably between the different strategies, with peripheral injection being much faster than the other two. Depending on other problems that may occur, however, all three alternatives are commonly used. There is also room for innovative methods in special applications (e.g. moving inlet for long hollow parts [25],) but one of the three preceding alternatives is usually used. The three basic strategies can also be combined, as in multipoint injection, to obtain faster filling or better impregnation.

The position of the inlet(s) and outlet(s) is crucial with all three strategies because dry spots or areas with high void content will result if the gates are improperly positioned.

A very important conclusion about the mold-filling can be drawn from an analysis of the governing equations for mold filling [26]. The mold-fill time will theoretically, always be proportional to a particular ratio of the processing parameters, regardless of the shape of the part:

$$T_f = C\frac{\mu L^2 \phi}{\Delta p K} \tag{12.10}$$

where μ is the viscosity of the resin, L is a characteristic length of the mold, ϕ is the porosity ($\phi = 1 - V_f$), Δp is the injection pressure, and $K(\phi)$ is the permeability (or a characteristic magnitude of the permeability tensor). Note that K is a function of ϕ and that normally T_f increases with decreasing ϕ. The dimensionless constant C depends on the shape of the part, the degree of anisotropy of the reinforcement, and the location of inlets and outlets and its value will change if any of these factors change. C can normally be expected to be of order unity or slightly smaller (Table 12.2). Assuming that the permeability dependence on ϕ can be approximated with the Kozeny-Carman equation (Eq. 12.2) one can rewrite Equation 12.10 for small changes of the porosity:

$$T_f = C_2\frac{\mu L^2}{\Delta p}\left|\left(1 - \frac{2}{\phi^3}\left[\frac{1}{\phi} - 1\right]\Delta\phi\right)\right. \tag{12.11}$$

where $\Delta\phi$ is a small change in porosity from the initial value ϕ and C_2 is the product of C_1 and $R^2/4k$ (from the Kozeny-Carman equation). The result in Equation 12.11 can be used for

Table 12.2 Rule-of-Thumb Values for the Constant C in Equation 12.10 for Filling of a Quadratic Plate with Side L

Model constant C	Type of injection
$\frac{1}{2}$	Edge injection
$\frac{1}{16}$	Peripheral injection
$\frac{1}{16}\left(\varepsilon^2 + 2\ln\left(\frac{1}{\varepsilon}\right) - 1\right)$, $\varepsilon \approx \frac{d}{L}$	Point injection (hole diameter d; a value of $\varepsilon = 0.005$ yields $C = 0.6$)

The characteristic length in Eq. 12.10 is equal to the side of the square L.

quick estimates of the influence of changes in the process parameters if the mold-fill time is known for a particular set of parameters. For a change in porosity from 0.5 to 0.45 ($\phi = 0.5$; $\Delta\phi = -0.05$) will result in an 80 percent increase of the fill time. Similar estimates can be made on the influence of the other parameters (i.e., viscosity, part size, and injection pressure).

The formula in Equation 12.11 is a useful rule of thumb, but if more time is available and accurate results are required we recommend computer simulations based on Equations 12.5 and 12.6. Computer programs for this purpose are now available from several vendors.

There are several phenomena that occur in practice that invalidate the assumptions that lead to the fill time formula (Eq. 12.10). Examples of such phenomena are:

- Fiber washing
- "Race tracking" at edges or on top of the reinforcement
- Significant mold deflection
- Significant cure during injection
- Significant pressure drop in resin distribution channels
- Non-Newtonian behavior of the resin
- Binder dissolution in the resin (increases the viscosity)
- Preform variations

Rule-of-thumb estimates of the fill time are available for a few simple cases. One example that is useful in practice is the result for a square plate with side L. The expressions for the constant C for point injection, edge injection, and peripheral injection has been summarized in Table 12.2. These three elementary cases can often be used to get a relatively good estimate of the fill time even in cases with complicated geometry.

Example: Estimation of Fill Time for a Large Composite Part

The following example shows how the results in Table 12.2 together with Equation 12.10 can be used to estimate fill times. A single skin part is chosen with a rectangular shape and a surface area of 2 × 3 m. The permeability of the fabric, the fiber volume fraction, the viscosity of the resin, and the injection pressure are assumed to be known:

$$\begin{aligned} \text{Resin viscosity} &= 0.5 \text{ Pas} \\ \text{Injection pressure} &= 1.5 \text{ bar } (0.15 \text{ MPa}) \\ \text{Permeability} &= 1 \cdot 10^{-9} \text{ m}^2 \\ \text{Fibre volume fraction} &= 0.3 \end{aligned}$$

An order of magnitude estimate (useful in conceptual discussions) can be obtained from Equation 12.10 with a typical length of 3 m (the longest dimension of the part):

$$T_{\text{magn}} = 5.8 \text{ hours} \tag{12.12}$$

Even if the constant C in Equation 12.10 is smaller than unity and the flow distance can be shorter than 3 m, then the fill time will at least be of the order of hours with the given parameters and a simple minded injection strategy. The fill time for a given injection strategy can be reduced by decreasing the viscosity (raise temperature or change resin), increasing the pressure (beware of fiber washing and mold deflection), or changing the reinforcement. The most dramatic change in fill time, however, can be achieved by a reduction of the flow length through additional inlets, resin distribution, or other changes in the injection strategy.

The simplest way to estimate the fill time more carefully is to guess the flow path during filling. The longest flow distance is then estimated from the guess and the fill time can be computed from the formula for unidirectional filling. This method is surprisingly powerful, at least in cases where it is easy to guess the flow path. A useful method to make good guesses is to try to imagine how a heat wave from a sudden temperature rise at the inlet would propagate through the part.

In the preceding example and point injection the flow front will develop as a circular front, starting at the inlet, until it meets the closest side. From then on the front will tend to move unidirectionally in both directions toward the far sides (if there is no leakage at the sides). A reasonable estimate of the fill time is somewhere between the time to fill radially and the time to fill unidirectionally to the far side (flow distance 1.5 m). For the radial flow case L in Equation.12.10 is 3 m (the diameter of a circle touching the wall farthest away). With the assumed alternative with unidirectional flow from the center toward both shorter sides, L in Equation.12.10 is 1.5 m. It is also necessary to estimate an effective inlet radius, in this case an inlet radius of 5 mm is chosen yielding $\varepsilon = 3.3 \cdot 10^{-3}$ that was used in the formula in Table 12.2. The constant C in Equation 12.10 is: 0.5 for the unidirectional flow case and 0.65 for the radial flow case. The upper and lower limit for the fill time in this case are then:

$$T_{\text{radial}} = 1.7 \text{ hours} \tag{12.13}$$

$$T_{\text{unid}} = 44 \text{ minutes} \tag{12.14}$$

The estimate of the fill time shows that a more detailed study of injection strategy and process parameters is necessary because it is difficult to reach an acceptable production economy with a fill time of about 1 h. An obvious action would be to choose a more efficient resin distribution method than a one-point inlet (e.g., peripheral injection or a multipoint injection). In addition, it would be worthwhile to try to lower the viscosity and increase the pressure.

The preceding formulas (Eq.12.10 and Table 12.2) are useful for rough estimates but can sometimes yield significantly shorter fill times than in reality [27,28]. More accurate predictions can be obtained through computer simulations based on Equations 12.5 and 12.6 [15,21–24,29], and this is the recommended method when sufficient time and resources are available. One problem with computer simulations is to obtain realistic values for the material parameters. These depend on the preforming step (fiber reorientation) and to some extent on the loading step (improper location that leads to locally high- or low-fiber volume fraction); however, considerable progress has been made on this problem and it seems likely that the accuracy of simulations will increase further as more progress is made.

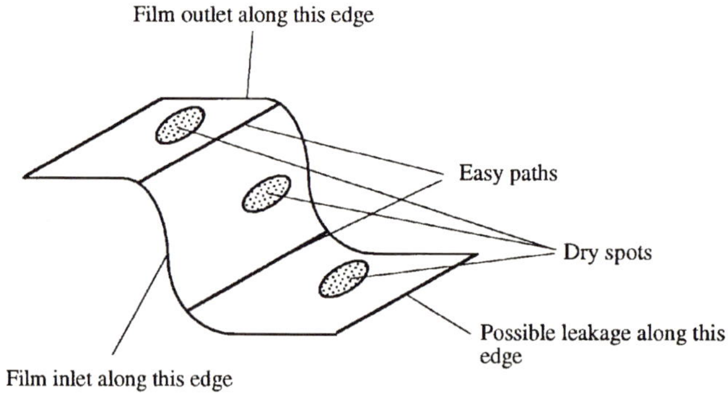

Figure 12.6 Generic curve part with dry spots caused by "race tracking" at radii and at edges

12.3.3 Mold-Filling Problems

There are numerous problems that can occur during mold filling in RTM. Some of them are trivial, but some of them very difficult to eliminate. A few important problems will be discussed in this section.

12.3.3.1 "Race Tracking," Edge Effects

There are many situations in which the permeability in the vicinity of a distinct line is higher than average. The resin will then flow much faster along this line than it does in the rest of the reinforcement. Under unfortunate circumstances the flow along the "race track" will merge with the flow somewhere else and enclose a zone of unimpregnated reinforcement (see Fig. 12.6). The dry region will decrease in size during the rest of the injection due to the increasing pressure around it, but a permanent dry spot will usually result if the enclosed region is large enough. "Race tracks" often occur at regions with large curvature ("corners") and at the edges of the mold.

"Race tracking' is mainly a preforming problem and the best solution is to adjust the preform by increasing the volume fraction at a race track or by compensating the preform geometry. It is also worth considering a change in the injection strategy that makes the formation of dry spots less likely.

12.3.3.2 Flow on Top of the Reinforcement

Poor impregnation will result if the fiber volume fraction is so low that a region without fibers is formed on top of the reinforcement. This can happen, for example, if high performance fabrics with a high uncompacted fiber volume fraction are used at a too-low-fiber volume fraction. This situation is similar to the "race tracking" problem described earlier. Flow on top of the reinforcement can occur, even if the whole cavity thickness is filled with fibers, if

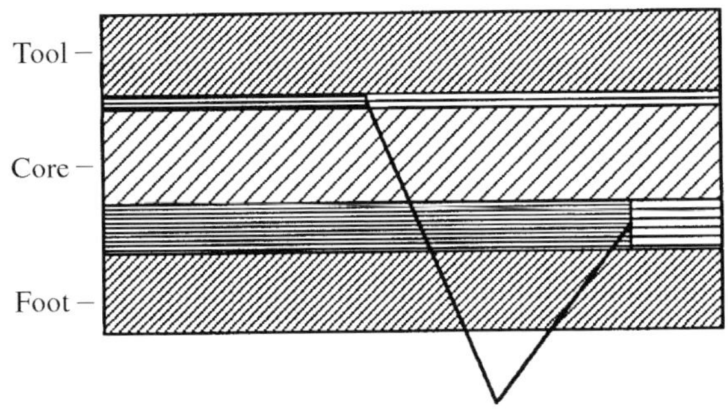

Figure 12.7 Sketch over the flow in a sandwich laminate (the displacement of the core is exaggerated). The flow of resin is from left to right

the compaction pressure on the reinforcement stack is too low and the injection pressure is too high.

Flow on top of the reinforcement is also mainly a preforming problem, and the best solution is to increase the fiber volume fraction or to use a more "springy" reinforcement where the problem is most pronounced. A decrease in injection pressure can also be a way to reduce the problem.

12.3.3.3 Nonuniform Skin Thickness on Sandwich Parts

When the part is of sandwich type there is a potential for a flow instability that will "push" the core toward one of the sides. The resulting part will have a thicker skin on one of the sides.

The flow instability can best be understood by looking at a case with unidirectional flow (see Fig. 12.7). There will always be some nonuniformity between the sides. This will result in the flow front moving faster on one side of the core than it does on the other. The net force on the core from the resin will be higher on the side where the flow front has moved the farthest and as a result the core will be "pushed" away from this side. The displacement of the core will increase the permeability more and the flow front will move even farther ahead on the "fast" side, and so on. The process will reach an equilibrium when the reaction force from the reinforcement becomes large enough to balance the fluid pressure on the other side of the core.

The instability will be more pronounced at low-fiber volume fractions.

There are at least two solutions to this problem. One is to increase the fiber volume fraction; the other is to perforate the core with small holes. The perforated core will allow a redistribution of resin between the sides that will equilibrate the pressure and will result in a better-centered core.

12.3.3.4 Fiber "Washing"

The reinforcement can be displaced significantly by fluid forces if the injection velocity is too high or equivalently if the injection pressure is too high compared with the friction forces between the mold and the reinforcement. The problem will be less pronounced at higher fiber volume fractions when the compaction pressure is higher. As a consequence, the reinforcement is held harder in place. The optimum fiber volume fraction depends on the type of reinforcement used. One way of estimating suitable processing conditions is to measure the "bulk factor" of the fabric (i.e., the fiber volume fraction when the fabric is uncompressed). Fiber volume fractions below this level are completely unacceptable because they will result in fiber washing, flow on top, and so on. On the other hand, too-high fiber volume fraction will make it difficult to close the mold and reduce the permeability so much that it will be difficult to fill the mold in an acceptable time.

12.3.3.5 Dry Spots

Large spots with unimpregnated reinforcement can occur even without race tracking due to improper position of the injection or ventilation gates. The severity of the problem can vary. A longer time for resin flow out of the outlet gate (resin flushing) may be sufficient to solve the problem if the dry spot is close to an outlet. In other cases, it may be necessary to move the outlets in the mold. If the mold is made from composite material (e.g., mass cast or laminated), then this may require the making of a completely new mold. With steel molds new holes can be drilled and old ones can be filled if necessary.

12.3.3.6 High Void Content at Outlet

Voids are usually formed at the flow front during mold filling. The voids move with the resin, but there will always be a region close to the flow front where the void content is higher than it is in the rest of the part [30]. To solve the problem one can either use vacuum assistance during filling or allow a longer time for resin flow (resin flushing) after complete mold filling. Optimization of the process parameters, particularly injection pressure and temperature, can also reduce the problem.

12.3.3.7 High Void Content and Vacuum

Vacuum assistance during mold filling will usually reduce the void content significantly [30]. The void content will be approximately proportional to the absolute pressure in the air inside the mold during mold filling. The mold, the sealings, and all gates, however, must be vacuum tight for this to be true. Even small leaks may be sufficient to give a very high void content. The sensitivity to leaks also depends on the injection strategy. Point injection is generally the most robust method and peripheral injection the most sensitive method (air is "sucked" into the molding).

Another cause of this problem can be the presence of volatile components in the resin. This can be tested by placing a beaker with resin inside a transparent container that is evacuated to the desired vacuum level. It is normal for some dissolved gas to come out of solution, but if gas bubbles continue to form after a long time they are most likely the result of evaporation of a volatile component. For some resins this effect can be so strong that the resin "boils over." The best solution in this case is to change to a resin that can be processed at the desired vacuum level. An alternative solution is to reduce the vacuum level until the problem disappears.

12.3.3.8 General Advice for Low Void Content

Research [30,31] has made it possible to give some hints on how to avoid most of the problems with void formation.

1. Use vacuum assistance in the mold if possible.
2. Degas the resin before injection.
3. Select the vacuum level so low that no volatile components in the resin boils off.
4. Select a good-quality vacuum-tight seal and inspect and clean it before every injection.
5. Select the injection pressure so low that the mold stays completely sealed during the whole injection phase.
6. Bleed off resin in the outlet port until no more gas bubbles can be seen there.
7. Close the outlet port before the injection pressure is released and try to build up a high cure pressure.
8. Do not use excessive temperatures.
9. Select the resin and fabric combination carefully to minimize the void content.
10. Select preforming agent carefully to minimize the void content.
11. Try to increase the bleeding time if the void content is unacceptable. It is an indication of improperly placed gates if this results in a lower void content.
12. Make sure that the mold surface and all gates and feedthroughs are vacuum tight.

The influence from the various process parameters on the void content at the flow front for a particular resin and fabric combination is summarized in Table 12.3.

Table 12.3 Qualitative Dependence of Void Content on Process Parameters for a Combination of BASF Palatal A-430 Vinylester and Brochier Lyvertex 21130 Glass Fiber Fabric

Parameter	Type of change	Average void content	Average length	Total void content
Cure pressure	+	−	No change	−
Vacuum level	+	−	−	−
Injection length	+	Unclear	+	+
Temperature	+	+	No change	+
Injection pressure	+	+	−	Unclear
Flushing time	+	+	−	−
Sizing	Removal	−	−	−
Lay-up	More 90°	+	No change	+

Source: Adapted from Reference [30].

12.4 In-Mold Cure

This section discusses general principles of resin cure in LCM. For a further discussion on the cure and kinetics of thermosets, see Chapter 2 [32].

12.4.1 Fundamentals

The cure of thermoset resins involves the transformation of a liquid resin, first with an increase in viscosity to a gel state (rubber consistency), and finally to a hard solid. In chemical terms, the liquid is a mixture of molecules that reacts and successively forms a solid network polymer. In practice the resin is catalyzed and mixed before it is injected into the mold; thus, the curing process will be initialized at this point. The resin cure must therefore proceed in such a way that the curing reaction is slow or inhibited in a time period that is dictated by the mold fill time plus a safety factor; otherwise, the increase in viscosity will reduce the resin flow rate and prevent a successful mold fill. On completion of the mold filling the rate of cure should ideally accelerate and reach a complete cure in a short time period. There are limitations, however, on how fast the curing can proceed set by the resin itself, and by heat transfer rates to and from the composite part.

An ideal resin processed in optimized conditions should have:

- A suitable and low resin viscosity during mold filling
- A short cure time from completed mold filling to demolding
- No material defects
- No shrinkage

12.4.2 Optimization of Cure

The most common objective for cure optimization is to minimize cycle time. Other important issues are: residual stresses, warping, void formation, and surface quality. All design work for composites need special knowledge, and general design guides are often insufficient to accomplish good design. This is also the case for cure optimization that is a part of composite process design. The reason is the many variations that can be put in to composite design with regard to both materials and processing options.

We can, however, write a list of parameters for cure optimization. To limit the possible combinations, we assume that the composite composition is known (i.e., resin, fibers, geometry, etc., are given, and the mold design and mold material is known). This reduces the number of "buttons" for control of cure to the following:

- Variation of resin cure system
- Temperature control during cure
- Time

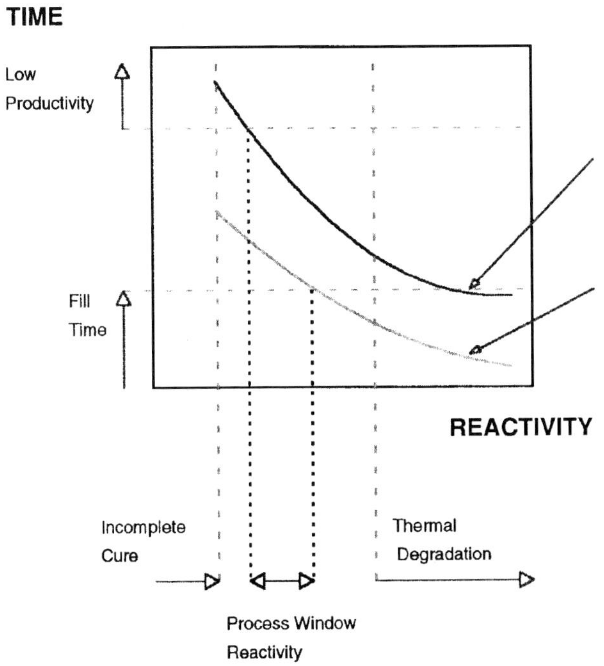

Figure 12.8 Resin reactivity as a function of time for RTM showing process window. The figure is qualitative and the resin reactivity is here a rough measure of the complete reaction rate profile

Although we have limited the parameters to three, there is a larger number of degrees of freedom hidden inside these high-level parameters.

First, it is necessary to identify the primary function in the system we aim to control. From a theoretical point of view the key function is the reaction rate. Thus, we need knowledge of how the reaction rate can be controlled in our system and the limits of the process window. Figure 12.8 is a schematic illustration of the dependence of gel time and complete cure time on resin reactivity. Resin reactivity can be changed by altering the temperature or changing the resin formulation. The process window is formed by the common part of the rectangle formed by the dashed lines and the lines indicating the time to gellation and complete cure. The rectangle is formed by the horizontal lines for fill time and required maximum cycle time and the vertical lines by the requirement for complete cure and no thermal degradation.

Several choices exist for cure optimization. The most straightforward and obvious are:

- Use of existing knowledge from similar material combinations
- Use of guidelines from resin manufacturers
- Small-scale tests to guide in the choice of parameters
- Fundamentally based calculations [32–35]
- Full-scale tests in production mold

The best would be to use all of the preceding actions to find the optimum conditions. Most often, however, the economical constraints and possibly the time put limits to the efforts that can be invested. Moreover the need to find optimum processing conditions may not be critical for many applications. Very often, however, the situation can be the opposite (i.e., the curing process generates material defects or takes too much time, hence actions must be taken to find solutions to these problems).

The use of simulation tools (e.g., computer programs) to calculate curing conditions is an area of great interest and is increasingly used to optimize cure for several processes including LCM [32,35]; however, there are some obstacles to the use of this "advanced" route, such as the need of:

- Skill to use computers
- A computer and a suitable program
- Accurate material properties for all constituents
- Accurate kinetic model for the resin

In addition to the preceding points experience and good knowledge of processing is needed to evaluate the results from the simulation.

The key problem today is often the last point (i.e., to obtain a sufficiently accurate kinetic model). Because these models are different for all resins and also different when the curing system is changed the option is often only one: To obtain the model by one's own measurements. Hence, both skill, knowledge, instrumental capacity, and quite a lot of experimental work is needed just to obtain a model. In addition, the fiber sizing can influence the cure significantly, adding one more factor to take into account [36]. The conclusion is that process optimization by simulation is not a viable route in the common situation in industry. A lot of research work has been done in this area, however, and the state of the art is advancing every year, so it is possible that this may change soon [32,35].

12.4.3 Cure Problems

Cure problems can be divided into problems that give material defects and problems that make the process inefficient. Some examples of "common" problems of the first type will be discussed in this section.

12.4.3.1 Delaminations

Delaminations can occur during cure as a result of high internal stresses. These stresses develop due to resin shrinkage and thermal volume changes. The level of stresses depend on several material properties, such as the Young's modulus, Poisson's ratio, and thermal expansion coefficients of both resin and fibers. In addition, the level of stresses also depends on several conditions, such as fiber orientation, fiber volume fraction, and part geometry.

The strength of the matrix plays a primary role for delaminations to occur, but the solution to delamination problems is usually not to increase matrix strength. In addition, the matrix material is changing during the curing process, and all matrix properties vary with the degree of cure. At a low degree of cure, therefore, both modulus and strength are low regardless of matrix type.

From all of these parameters we can conclude that delaminations can be a very complex problem to control; however, we can identify some important parameters that do affect this problem, and also list some typical cases where delaminations often arise.

High risk for delaminations:

- Thick composite parts
- Geometrically complex parts (varying thickness and closed shapes)
- Resin with high shrinkage
- High temperature variations during cure process

Examples of possible actions to solve delamination problems:

- Reduce temperature and hence prolong cure
- Control cure propagation in thick parts by curing from inside to outside (hollow parts), or cure sequentially along part, which can be done by partial heating of the mold
- Change of resin system to resin with less shrink or use of additives that reduce shrinkage (this will create microvoids)
- Increase pressure during cure

12.4.3.2 Surface Finish

The surface finish can be good in LCM but is usually only good on one side of the laminate. A prerequisite for a good surface finish is to have a high gloss mold surface finish. The mold material and the release agent can also influence the surface finish indirectly.

The primary cause of rough surface is resin shrinkage, provided the mold surface has a high gloss; hence, if a good surface is required some action must be taken to reduce the resin shrinkage.

Examples of possible actions to obtain good surface finish:

- Use a higher temperature on the mold side, where high surface finish is required combined with internal mold pressure during cure
- Use of LP (low profile) resin systems

12.4.3.3 Porous Areas

Porous areas in LCM are often caused by a mold filling problem (see Sec. 12.3.3); however, other causes also exist, (e.g., high temperatures during cure may evaporate resin monomers or dissolved gas that form voids). The actions to be taken to optimize cure for minimum void content are:

- Degassing of resin before injection
- Reducing temperatures during cure

12.4.3.4 Incomplete Cure

The reactions are seldom allowed to go to completion in the mold. The usual practice is to let the part cure until it is sufficiently stiff to demold and then to perform a postcure either in an oven or at room temperature. If the degree of cure is too low at demolding, then a number of problems can occur: It may be impossible to demold without damaging the part; permanent deformation of the part can occur; work place hazards with toxic fumes may occur, and so on.

The solution to the problem is to increase the cure temperature, to let the part cure for a longer time in the mold or to change the resin formulation. It is a good idea to make a serious effort to optimize the cure cycle and resin formulation if incomplete cure should occur.

12.5 Mold Design

The mold design is strongly influenced by the method used for mold making. Mold-making methods can roughly be divided into three classes: direct, indirect, or hybrid. In the direct method the tool cavity is machined directly out of a block of mold material. The machining can be done with modern CAD/CAM methods. In the indirect method a master model is used for casting or for laminating tool faces. In hybrid methods a near net shape mold is cast from materials such as aluminium. The near-net-shape mold is finally adjusted by NC machining to the final tolerances.

12.5.1 General Design Rules

Design for LCM has some specific features such as the need for integration of part design and mold design. This is important because a good composite design does not have to be a good "LCM" molding design. Some of the important features of LCM mold design are therefore discussed here together with design "rules."

12.5.1.1 Master Model

Mass cast tools and composite tools are made from a master model. The surface finish and geometrical tolerance on the mold is then completely determined by the master model. The utmost care must therefore be spent on the master model before the mold is made. This topic is discussed in detail by Morena [33].

12.5.1.2 Radii

Sharp radii in composites are not generally considered a good design because out-of-plane loads tend to be high at these places. It is also important from a process point of view to use radii in bends that are at least three times the thickness of the laminate (it is possible to violate this rule in special cases [7]).

Tight radii can give process problems such as:

- Restrict resin impregnation of fibers
- Difficult to make preform radii to follow mold radii
- Damage of fibers when mold is closed
- Damage of mold edges (plastic molds)

12.5.1.3 Scale of Composite Part

The possibilities with RTM are many and the size of moldings can vary from small to very large. For large components it becomes necessary to take measures to reduce mold closure forces. Large RTM parts, therefore, are usually of low volume fraction because the cost of sufficiently stiff molds and facilities for high closure forces tend to be out of reach for production.

Thicknesses from 1 mm up to at least 50 mm are possible with RTM; however, several difficulties can arise for thick parts, such as the problem of race tracking at edges, air enclosures due to uneven mold filling, and problems connected to cure.

12.5.1.4 Net Shape

One advantage with RTM is the possibility for net-shape manufacturing; however, it should always be evaluated for each specific design if net shape design should be used instead of "trim to size" strategies. Net shape naturally offers the advantage that no trimming is required, but it can add complexity to mold design and choice of mold filling strategy.

12.5.2 Mold Materials

A vast number of different materials are possible alternatives for RTM molds [37]. Each material has its own advantages and disadvantages. Apart from the more obvious properties, such as density, heat capacity, or stiffness of any given material, there are other considerations that have to be taken. The following is a list of some of the other properties that should also be considered in the choice of material:

- Price
- Toughness is a property that is difficult to define quantitatively but which is still very important in practice, particularly the ease with which the surface can be scratched
- Surface finish is another important property in some applications, but some materials can be difficult to polish to a high gloss, or they might corrode from contact with the resin
- Durability is important in large volume production
- The possibility to integrate with other systems may in some cases be important

- The maximum use temperature is of course important in connection with high-temperature resin systems.
- The acquisition time, which can in some cases determine the choice of mold material
- The possibility to repair minor or major damages to the mold surface or the entire mold
- The ease with which a tool made from the material can be modified
- Release properties
- Size limitations
- Chemical compatibility with resins and release agents
- Limits to complexity

The thermal properties (i.e., density, specific heat capacity, and thermal conductivity) have a particularly strong influence on the curing behavior. The exothermal peak temperature is one example: It can differ significantly between a composite mold with low thermal mass and a metal mold [35]. A more thorough discussion of pros and cons of different mold materials can be found in Morena [37].

12.5.3 Stiffness Dimensioning

The tolerances of the molded part are to a large extent determined by the properties of the mold. The most fundamental property for the tolerances is probably the stiffness of the mold. This is particularly important in LCM due to the large forces that can be generated by the injection pressure. As an example, an injection pressure of 4 bar will exert a force of 800 kN or 80 tons on a mold with a surface area of 2 m^2. This force has to be reacted by the clamping device and the mold.

The necessary stiffness of the tool is easy to estimate with standard engineering methods and should therefore always be calculated. The loads on the mold are primarily the internal overpressure from resin and reinforcement. It should be kept in mind that the compaction pressure necessary for closing the mold with dry reinforcement can be as high as or higher than the injection pressure [8]. The mold must be dimensioned for the maximum injection pressure in the whole cavity because this will be the internal over pressure if the outlets are closed while pressure is maintained at the inlet.

The stiffness of the mold must be both high enough to keep a sufficiently uniform cavity thickness and so high that the sealing surfaces are kept sufficiently close to each other, particularly if small diameter O-ring seals are used.

The influence on the total weight of the tool should also be kept under observation during the design process because a high weight can seriously damage the production economy by increasing the cost of handling equipment. If possible, the design of the mold should be done in parallel with the part design so that a less-demanding design can be chosen if it results in excessive tooling weight.

12.5.4 Sealings

The sealings in an RTM mold can take many shapes; however, a very important principle that is recommended is to keep the sealing in one plane because it is very difficult to seal the mold on a curved surface in a reliable way.

The quality of seal that must be used depends on the level of sealing that is necessary, the processing temperature, and the resin type that is used. The highest-quality seal must be used when vacuum is used because any air leakage into the mold will then damage the molded part.

Normal O-rings will usually require very good tolerances on the sealing surface, and there will almost always be resin in the sealing groove after each molding that has to be cleared away. A variation on the O-ring theme is then to use flexible rubber tubes instead. A smaller size groove can then be used, and the large flexible O-ring will then completely fill the groove so that no resin can enter the groove.

Silicon rubber tubes with an inner diameter of 1 mm and an outer diameter of 3 mm have been found to work well in stiff steel tools. The price of this tube is so low that the tube can be replaced every injection if necessary without affecting the part cost significantly.

Hollow O-rings can be pressurized for improved sealing efficiency.

In atmospheric moldings (without vacuum) a larger sealing than the O-rings mentioned above is normally used. One type that has proved to work well is made of silicon rubber and has a rectangular cross section of about $8 \times 18\,\text{mm}$. The sealing is used in a fairly deep groove (app. 13 mm). The sealing in this case must be used with a gap of about 3–4 mm between the sealing surfaces (see Fig. 12.9). This is accomplished with shims of suitable thickness or by allowing some space for the sealing at the matching surface.

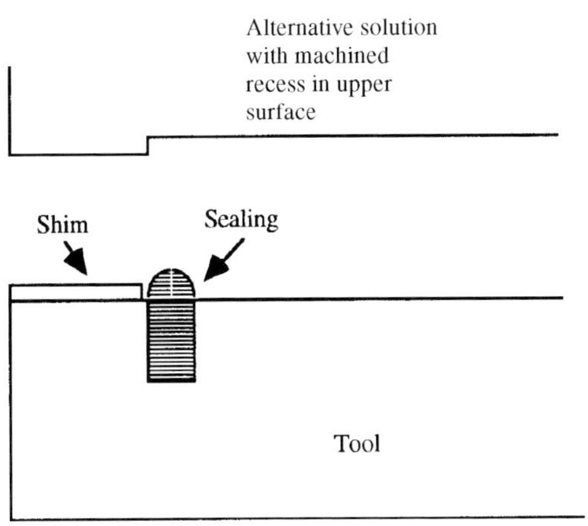

Figure 12.9 Sketch over a technical solution with a rectangular sealing

12.5.5 Clamping

The force needed to close the mold can be substantial. This is a particularly important problem for large area molds with a high-fiber loading because the pressure to compact the fibers can be high. In addition, the clamping system must resist the full injection pressure without leakage at the mold seals.

The choice of clamping system is very much dependent on the rate of production.

Conventional threaded bolts are the simplest choice, but they may also be the most cost efficient for low volume production with cycle times in the range of hours. On the other hand, for high-volume production mold closing as well as mold opening must be achieved much quicker, and the choice may be a hydraulically driven mold carrier.

The choice of mold clamping system is based on:

- Access time for closing the mold
- Weight of mold half to be handled
- The force needed to compress the reinforcement and mold seals
- Maximum pressure in the mold (i.e., injection+compaction pressure)
- Access time for demolding

12.5.6 Heating Systems

The choice of heating system is strongly influenced by the choice of tooling concept. A heavy steel mold that is being used for curing with varying temperature will require a high-capacity convective heater/cooler. On the other hand, a light weight GRP tool used with a fast-reacting polyester system might possibly be used without a heater.

Standard industrial heaters with water or oil have been used extensively with good result. Water-based heaters are available with or without pressurization. With pressurized systems, temperatures up to about 140°C are possible. With pressurized systems, however, special hoses and tubes are necessary, and this will increase the costs for the heating system significantly.

The advantage of water versus oil is that the leakage of heater fluid that always occur during tool change will evaporate or can easily be wiped up with water, but it must be disposed of using absorbents with oil.

Blanket-type electrical heaters are often used in combination with low thermal mass GRP tools (e.g., in the tooling concept advocated by Plastech TT, UK). The drawback with electrical heaters is that locally very high temperatures can occur. This can damage the tool and give an unacceptable variation in the curing process.

When convective heating is used it is important to localize the heating channels on an appropriate distance from the tool surface. The part can be damaged by print-through if the distance between the surface and the channels is too small. The surface temperature may vary and give rise to unacceptable cure variations if the distance is too large. As a rule of thumb the distance to the surface in mass cast tools from aluminum-filled methacrylic resin should be about 15–30 mm, and the spacing between the channels should be about 50 mm. Similar rules can be developed for other tooling materials.

12.6 Conclusions

The most important technical issues for LCM are connected to preforming, mold filling, and in-mold cure. In addition, a robust LCM process needs a proven mold design. Focus in research and development is therefore on these processing areas. There are promising results from modeling of LCM processes, but there are also known difficulties for use in industry. The strong development that is ongoing, however, will change the picture so that simulation is used as a standard tool industrially in a few years.

Nomenclature

Q	Flow rate (in Darcy's law) [m^3/s]
A	Cross-section area (in Darcy's law) [m^2]
K	Permeability [m^2]
μ	Viscosity [Pas]
Δp	Injection pressure [Pa]
L	Injection length [m]
V_f	Fiber volume fraction [dimensionless]
R	Fiber radius [m]
k	Model constant in Kozeny-Carman's model [dimensionless]
C	Model constant in Gebart's permeability model and in Equation 12.10 [dimensionless]
V_{max}	Model constant in Gebart's permeability model [dimensionless]
T_f	Fill time [s]
u_i	Flux density in x_i direction (also called superficial velocity) [m/s]
K_{ij}	ij-component of the permeability tensor [m^2]
n_i	normal vector to the limits of the mold cavity (seals) [m]
ϕ	Porosity ($1 - V_f$) [dimensionless]

Acknowledgments

We want to acknowledge the contribution from Christer Lundemo, Borealis Industrier AB, Sweden, who introduced us to the RTM process and who taught us many of the "secrets" of RTM. We would also like to thank Kevin Potter, SINTEF, Norway, for fruitful discussions and for sharing his experience from RTM mold making. We also want to thank the members of the Brite-Euram project "SimRTM," who took time to read and comment on an early version of this chapter. Finally, we want to acknowledge the important contributions through the years from the staff at the Swedish Institute of Composites.

References

1. Palmqvist, A., *Proc. Forty-Eighth Annual Conf., Composites Inst., SPI* (1993) Cincinnati
2. McCarthy, R. *Coatings Comp. Mat.* (1995) 9, p. 42–47
3. Macosko, C.W. *Fundamentals of Reaction Injection Molding* (1989) Hanser Publishers, Munich
4. Kaviany, M. *Principles of Heat Transfer in Porous Media* (1991) Springer Verlag, New York
5. Tucker, C.L., Dessenberg, R.B. In *Flow and Rheology in Polymer Composites Manufacturing* (1994) Elseiver Science B.V., Amsterdam, p. 257–323
6. Kittelson, J.L., Hacket, S.C. *Proc. Thirty-Ninth Int. SAMPE Symposium* (1994) Anoheim p. 97–107
7. Holmberg, A.J. In *Proc. Durability and Damage Tolerance of Composites* (1995) ASME, San Francisco
8. Lundemo, C.Y., Serrander, T. SICOMP (fax +46-911 744 99) Technical Report 91-020 (1991)
9. Gerard, J.H., Jander, M. *Proc. Forty-Eighth Annual Conf., Composites Inst. SPI* (1993) Cincinnati
10. Gulino, J., Blanc, A., Berthet, G. Technical Report (1988) Vetrotex International
11. Lundström, L., Strömbeck, A., Sandlund, E. SICOMP (fax +46-911 744 99 Technical Report 92-006 (1992)
12. Bergsman, O.K. *Proc. Ninth Int. Conference on Composite Materials* (1993) p. 560–567
13. Long, A.C., Rudd, C.D. *IMechE J. Eng. Manuf.* (1994) 208, p. 269–278
14. Ko, F.K., Pastore, C.M., Head, A.A. *Handbook of Industrial Braiding* (~1988) Atkins & Pearce Covington, Kintuck
15. Advani, S.G., Bruschke, M.V., Parnas, R.S. In *Flow and Rheology in Polymer Composites Manufacturing* (1994) Elsevier Science B.V., Amsterdam, p. 465–515
16. Carman, P.C. *Trans. Int. Chem. Eng.* (1937) 15, p. 150–166
17. Gutowski, T.G., Cai, Z., Bauer, S., Boucher, D., Kingery, J., Wineman, S. *J. Comp. Mat.* (1987) 21, p. 650–669
18. Gebart, B.R. *J. Comp. Mat.* (1992) 26, p. 1100–1133
19. Adams, K.L., Russel, W.B., Rebenfeld, L. *Int. J. Multiphase Flow* (1988) 14, p. 203–215
20. Gebart, B.R., Lidström, P. *Polym. Comp.* (1996) 17, p. 43–51
21. Fracchia, C.A., Castro, J., Tucker, C.L., III. *Proc. American Society for Composites Fourth Technical Conference*, Lancaster, Penn., p. 157–166, 1989
22. Young, W.-B. *J. Comp. Mat.* (1994) 28, p. 1098–1113
23. Lin, R., Lee, L.J., Liou, M. *Int. Polym. Proc.* (1991) 6, p. 356–369
24. Rudd, C.D., Middleton, V., Owen, M.J., Longbottom, A.C., McGeehin, P. *Comp. Manuf.* (1994) 5, p. 177–185
25. Strömbeck, L.A. *Proc. Ninth International Conference on Composite Materials* (1993) Madrid, Span, Vol. 3, p. 497–504
26. Gebart, B.R., Gudmundson, P., Lundemo, C.Y. *Proc. Forty-Seventh Annual Conference* (1992) Composites Inst., SPI, February
27. McGoldrick, C., Wilson, S. Personal Communication (1995) Shorts Brothers Plc, Belfast, Northern Ireland
28. Scott, F.N. Personal Communication (1995) British Aerospace Airbus Ltd., Bristol, England
29. Koorevaar, A. *Proc. of the Techtextil-Symposium, Lecture No. 338* (1995)

30. Lundström, T.S., Gebart, B.R. *Polym. Comp.* (1994) 15, p. 25–33
31. Lundström, T.S. *Composites* (1996) Part A p. 6
32. Calado, V.M.A., Advani, S.G. *Processing of Continuous Fiber Reinforced Composites* Ch. 2, Hanser
33. Lee, D.-S., Han, C.D. *Polym. Eng. Sci.* (1987) 27, p. 955–963
34. Kenny, J.M., Maffezzoli, A.M., Nicolais, L., Mazzola, M. *Comp. Sci. Tech.* (1990) 38, p. 339–358
35. Gebart, B.R., Strömbeck, L.A. *Proceedings Verbundwerk* (1992) Wiesbaden p. D10.1–D10.14
36. Karbhari, V.M., Palmese, G.R., *J. Mat. Sci.*, (1997) 32, p. 5761–5774
37. Morena, J.J. *Advanced Composite Mold Making* (1988) Van Nostrand Reinhold, New York

13 Filament Winding

S.C. Mantell and D. Cohen

Filament winding offers a high-speed, automated, precise technique for manufacture of closed-surface composite structures. Although traditionally applied to cylindrical and spherical parts, the winding technology has also been extended to nonsymmetric parts with complex curvature. In the winding process, a computer-controlled head positions fiber tows on a rotating mandrel. This technique is cost effective, particularly when compared with traditional methods such as autoclave curing. Several aspects of filament winding will be discussed including: winding methodology, winding equipment, cylinder design considerations, process simulation models, and characterization of filament wound parts. An emphasis is placed on the relationship between processing conditions and final part quality for both thermosetting and thermoplastic matrix filament wound cylinders. As such, descriptions of process simulation models and part characterization will be presented in detail.

13.1 Introduction

The filament winding process has emerged as a primary method for manufacturing composite structures at low cost. This automated process offers a high-speed, precise method for laydown of continuous reinforcements to form closed-shape composite structures. During filament winding, continuous fibers that have been impregnated with a thermosetting or thermoplastic polymer matrix are wrapped around a rotating mandrel (Fig. 13.1). The relationship between the cross-head speed V and the mandrel angular velocity ω determines the fiber orientation or winding angle φ_0. The tension of the reinforcing fibers F_o over the mandrel provides a positive radial pressure that serves to compact the previously wound laminate. These fibers are typically wound as tows (gathered strands of fiber), and may be wrapped in groups as adjacent bands (Fig. 13.2). In general, the mandrel is cylindrical, spherical, or cylindrical with two hemispherical ends. More complex shapes can be wound depending on the winding head configuration and control (i.e., robotics head with multiple degrees of freedom) [1].

Filament-wound structures are often used to store fluid under pressure: fuel storage tanks, rocket motor cases, natural gas, and oxygen storage tanks. In some cases, the filament-wound structure is subjected to external pressure: a diving bell or submarine. Filament winding has

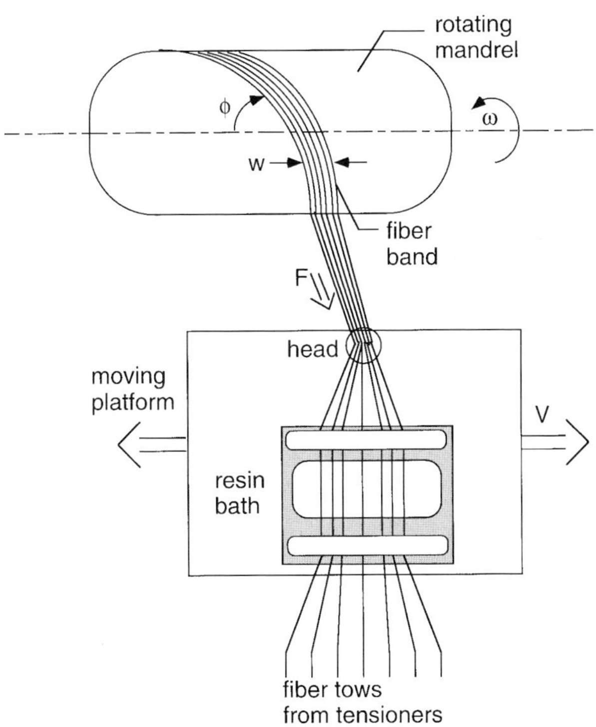

Figure 13.1 Schematic of the filament winding process

Figure 13.2 Full-scale filament-wound case in helical winding phase (courtesy of Alliant Techsystems Inc.)

been selected as a manufacturing process for these structures because the desired shape is cylindrical, and because the process is a cost effective method to produce large-scale, high-strength structures. Moreover, because laminate consolidation can be achieved with no special compaction device (such as an autoclave), the filament-winding process has significant advantages over other composite manufacturing processes:

1. Structure size is not limited to the size of an autoclave.
2. Although oven curing may be required, the cost of a large oven is significantly less than the cost of a similarly sized autoclave.
3. Cylinders may be wound with a wet winding process while maintaining high fiber volume fractions typically seen in parts made from preimpregnated (B staged) materials.

Advantages and disadvantages of the filament-winding process are summarized in Table 13.1. As shown in Table 13.1, the primary advantage of filament-winding is its cost. Filament-winding has a significantly higher rate of material laydown than other composite manufacturing methods (Table 13.2). This high laydown rate, in turn, results in a significantly lower cost (Table 13.3) [2–4]. Several difficulties in filament winding have been the focus of current research. For example, new computer-controlled robotics tape laying equipment has been developed that permits reverse curvature. Pins may be placed on the mandrel to locate the fibers and prevent slipping. New methods for in-line prepregging with feedback control of resin content are being developed to reduce void content on the part. In situ consolidation of thermoplastic matrix cylinders has improved fiber location, allowed for more complex shape

Table 13.1 Advantages and Disadvantages of Filament Winding

Advantages	
1. No autoclave required for processing	a. Labor costs associated with bagging and processing are eliminated b. No limitations on part size
2. Automated process	a. Precise fiber placement b. Continuous fibers with no joints c. Labor cost lower for high volume parts
3. Parts can be fabricated via wet winding	a. Significant cost savings associated with using fiber and resin rather than pepregmaterials b. Minimal material shelf life requirements
Disadvantages	
1. Limitations on part geometry and ply orientation	a. Mandrel removal may be difficult for some part geometries b. Fiber orientation limited within a lamina
2. Obtaining desired fiber volume fraction requires precise control of winding conditions	
3. Cost effective for large batch size (>100 parts)	

Table 13.2 Comparison of Fiber Deployment Methods: Material Laydown Rate

Hand layup	1.5 lb/hr
Preplied broadgoods	3 lb/hr
Filament winding (helical)	100 lb/hr
Tape laying (gantry)	10 lb/hr

Table 13.3 Comparison of Fiber Deployment Methods: Comparative Cost of Filament Winding

Manual production	$353.4
Automated cutting	502.4
Robotic transfer	529.1
Tape layup	471.0
Robotic layup	518.3
Pultrusion	93.5
Filalment winding	226.4

Source: Adapted from References [2–4]

Costs for a 4 ft^2, 24-ply graphite/epoxy part for an annual production of 2,000 units

parts to be wound, and eliminated large oven requirements. Manufacturing process models have emerged as a cost-effective method to study the relationship between processing conditions and final part quality.

There are a number of general references available that contain detailed descriptions of the winding process and equipment itself [5,6]. The purpose of this work, however, is to focus on the relationship between processing conditions and final part quality for both thermosetting and thermoplastic matrix filament wound cylinders. In the subsequent sections, an overview of the process will be presented, followed by detailed descriptions of current process modeling techniques and methods for determining cylinder quality.

13.2 Manufacturing Process

Filament-wound structures are typically cylindrical, spherical, or conical. In the case of cylindrical or conical shapes, there may be domed ends or specially wound flange ends. The fibers and resins can be selected from a wide variety of materials. These material and geometry options make filament winding a versatile manufacturing process.

13.2.1 Winding Techniques

During filament winding, continuous fiber tows are placed on a rotating mandrel by a moving cross-head. Tension is applied to the tows as they are placed. The relationship between the cross-head speed and the mandrel angular velocity determines the fiber orientation or winding angle. The three basic winding patterns are helical ($\varphi_o = \pm\alpha$), hoop ($\varphi_o = 90$), and polar ($\varphi_o = 0$).

Thermosetting matrix materials may be wound by wet winding or prepreg winding techniques. In wet winding, the fiber bands are passed through a resin bath and then immediately wound on a rotating mandrel (Fig. 13.3). In prepreg winding, a preimpregnated tow is placed on the mandrel. Once winding is complete, the cylinder is cured at an elevated temperature and the mandrel is removed. The final cylinder may also undergo a postcure cycle.

For thermoplastic matrix materials, prepreg winding with in situ consolidation is typically employed (Fig. 13.4). As the tow is placed on the mandrel, heat and pressure are locally applied to bond the incoming tow to the substrate laminate (cylinder). Heat sources include hot gas, hot shoes, electrical conduction, induction, microwaves, infrared, and lasers. The winding tension itself may introduce sufficient radial pressure for bonding. Rollers and hot shoes may be alternative pressure sources. Winding conditions and heat and pressure sources ideally are selected such that complete bonding is achieved during the in situ consolidation process. Processing in an oven at elevated temperatures following winding may be desirable to relieve stresses.

(*a*)

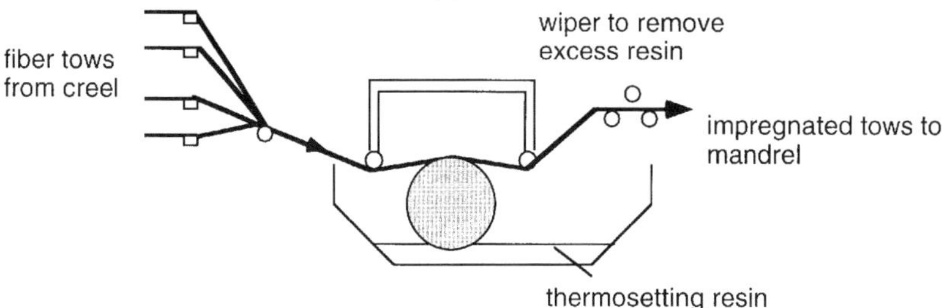

(*b*)

Figure 13.3 Resin impregnation for wet winding: (a) actual equipment (courtesy of Engineering Technology, Inc. "ENTEC") (b) schematic

In general, the process conditions that can be selected and controlled independently during filament winding are:

1. winding speed and hold time between layers
2. initial fiber tension for each layer during winding
3. external temperatures, heating rates, and pressures during the winding, cure, and postcure stages

13.2.2 Fibers and Resins

Glass, organic (aramid, and oriented polyethylene), and graphite are among the fiber materials often used in filament-wound parts. As is the case with any composite manufacturing process, selection of fiber material is based on cost, dimensional stability, impact properties, strength, modulus, durability, and ease of handling. Graphite fibers typically offer the widest range of

Figure 13.4 Hot gas torch for in situ consolidation of thermoplastic matrix cylinders equipment (courtesy of Engineering Technology, Inc., "ENTEC")

strengths and modulii. For filament winding, particularly wet winding of thermosetting matrix composites, the maximum number of filaments per strand is desirable. This will ensure ease of handling and subsequently fast winding times.

The function of the resin matrix material in filament-wound structures is to help distribute the load, maintain proper fiber position, control composite mechanical and chemical properties, and provide interlaminar shear strength. Either a thermosetting or a thermoplastic resin material may be selected. Thermosetting resins may be selected for application in a wet-winding process or as part of a prepreg resin system.

If a thermoset resin is selected for wet winding, desirable characteristics include:

1. the viscosity should be below 2 Pa·s (2,000 cps)
2. the toxicity should be low
3. the pot life should be long (ideally more than 6 h) [5]

It is difficult to control fiber volume fraction during wet winding because the tows are wound under tension and the fiber will move through the wet matrix material. Moreover, the parameters required for resin impregnation in the resin bath (band tension, and band speed through the resin bath) are typically coupled to the material laydown conditions.

The prepreg winding technique offers better control of fiber volume fraction, but at a cost. Material costs are 1.5 to 2 times higher, and there are additional costs associated with storing the preimpregnated (thermosetting matrix) tows. Preimpregnated tows are used almost exclusively for thermoplastic matrix materials, where there are no shelf-life restrictions.

13.3 Equipment

Filament-winding equipment requirements are relatively few in comparison with other composite manufacturing techniques. Tooling consists primarily of a mandrel that provides the internal part geometry. A winding machine and curing oven are the only significant facilities requirements. Additional facilities may be required for storing preimpregnated thermoset tows if prepreg winding is desired. No curing oven is required in the case of in situ consolidation of thermoplastic parts, but the winding machine must have a head with integral heat (and pressure source if required).

In mandrel design and material selection, the following criteria should be considered:

1. cost
2. mandrel reusability (durability)
3. production quantity
4. mandrel material thermal characteristics
5. mandrel strength/ability to resist deflection during winding and cure
6. final part tolerances required
7. dimensional stability

To ease part removal, mandrels may be constructed from water-soluble materials (sand), plaster, or an assemblage of metal shells that is collapsible or segmented [6]. Tube mandrels constructed with a high-quality surface finish and a slight taper are often used for cylindrical parts.

A wide variety of winding machines are commercially available and can be grouped into two general categories: polar winders and helical winders. Winders may be belt or gear driven. Polar winders typically have two axes of rotation (winding arm and mandrel), and they are often used for winding spherical parts (Fig. 13.5). Helical winders are more versatile and can have as many as six axes of motion (Fig. 13.6). In either case, these machines have several key components/features in common: a fiber delivery system, a moving carriage, and head and tail stocks for mounting the mandrel.

A winding machine is typically equipped with computer control, and its programming is done off line. Winding pattern programs are often provided that interface with CAD programs and finite element analysis code. Such tools greatly enhance the process of concurrent engineering. Moreover, computer control allows the precise placement of the fiber on the part such that gap/overlap is minimized and the fiber orientation is controlled precisely. The

Figure 13.5 Tumble winder, a variation of the polar winder, is specifically designed to produce small-scale spheres and tanks at very high speeds (courtesy of Engineering Technology, Inc. "ENTEC")

advent of computer-controlled multiaxis winding machines has allowed manufacturers to locate filaments around curves, openings, and pins precisely.

A complete listing of equipment suppliers and manufacturers may be found in Reference 5.

13.4 Cylinder Design Guidelines

There are several resources available for designing filament-wound cylinders. In general, filament-wound cylinders are classified as cylindrically orthotropic. Adjacent helical plies, $\varphi_o = \pm\alpha$, act as an orthotropic unit. Stresses and strains that result from various loading conditions can be determined by following the principles of laminated plate theory [7]. When applying laminated plate theory, the "plate" consists of the cylinder wall. In this case, the effect of cylinder curvature is neglected, and the θ and z axes are considered the planar axes of the plate. Failure criteria applied in laminated plate theory, such as maximum stress or strain, or the quadratic Tsai-Wu failure criteria [7] may also be applied. Several specialized loading cases have been studied.

Figure 13.6 Multiaxis winder with dual carriages. Axes of motion include: 1. mandrel rotation; 2. carriage traverse; 3. radial motion; 4. vertical motion; 5. eye rotation; and 6. yaw (pitches the wind eye in a 90-degree plane) (courtesy of Engineering Technology, Inc. "ENTEC")

In Tsai [7], an elasticity solution for stresses in a pressurized thick cylindrical vessel is presented. In this analysis, the longitudinal bending deformation due to end closures is neglected, the formulation of the elasticity problem then reduces to a generalized plane strain analysis. The effects of material selection, layup sequence, and winding angles on the burst strength of thick multilayered cylinders are also addressed.

For thin-walled cylinders subject to in-plane (axial and circumferential) loading and axial torsion, Whitney and Halpin [8] have developed an analytic solution for strains. Their analysis is valid in the central region of the cylinder, end support effects are neglected.

Vinson and Sierakowski [9] have studied the effects of domed-end closures on stresses and strains in the cylinder. For example, in pressure vessels edge moments and shear loads will occur at the dome–cylinder interface. They have identified a bending boundary layer in which the discontinuity stresses are significant. These stresses will decay exponentially, with the decay constant being a function of the material properties, layup, and cylinder geometry. For thin cylinders ($r/t > 10$) Pagano and Whitney [10] recommend allowing twice the cylinder radius at each end for the bending boundary layer.

Peters, Humphrey, and Floral [5] describe netting analysis and provide analysis examples for pressure vessels and geodesic dome contours. Several design considerations are outlined in this reference:

1. Orient fibers in direction of loading: For tensile and compressive loads along the axial direction of the cylinder, wind helical plies at the lowest possible angle to the shaft axis; for in-plane shear loads align fibers at ± 45 to the shaft axis

2. for combined loading, use combined angles
3. include hoop windings to minimize diametral dimensional changes in thin cylinders

Additional guidelines that are similar to those for composite plate structures are also provided in Reference [5].

13.5 Filament-Winding Process Models

Process models allow composite case manufacturers to determine the affects of process variable settings on final cylinder quality. Because the cost of a composite cylinder can be as great as $500,000, the ability to simulate filament winding can significantly reduce cost and improve quality. Several computer models of the filament-winding process for both thermoset and thermoplastic matrix materials have been developed. These models are based on engineering principles such as conservation of mass and energy. As such, numerous resin systems and fiber materials can be modeled.

Regardless of matrix or fiber materials, the key process variables for filament winding are temperature, compaction pressure/fiber tension, and laydown rate. Typical measures of final cylinder quality include degree of cure/crystallinity, void volume fraction, fiber volume fraction, and residual stresses and strains.

For filament winding of thermosetting matrix composite cylinders, process models have been developed to study consolidation and fiber motion during winding, changes in fiber tension during winding, composite temperature and subsequent cure, changes in void size during winding, and the final mechanical strength of the cylinder. Each of these models has particular strengths for various applications. Models posed by Cai [11], Agah-Tehrani [12], and Davé [13] focus on the fiber motion process. Springer and coworkers [14–16] have developed extensive models that include thermochemical, fiber motion, stress strain, and void submodels. This research provides a complete model for the entire filament-winding process and has been experimentally validated for prepreg thermosetting matrix cylinders.

For filament winding of thermoplastic matrix composite cylinders, numerous process models for the heating stage during winding have been posed [17–22]. Several researchers [20,21] have modeled the stresses and strains that occur in thermoplastic winding. The consolidation and bonding mechanism between thermoplastic plies has been modeled by Loos [23–26]. Mantell and Springer [21] have extended this bonding model to filament winding and coupled it with thermochemical and stress–strain submodels to provide a comprehensive model of thermoplastic matrix cylinder winding.

In process modeling of filament winding, regardless of matrix material, the process is considered to consist of several simultaneously occurring “subprocesses”: winding, application of heat and/or pressure, consolidation, and void evolution [16]. The process model is consequently broken down into several submodels, each with a distinct function, and each coupled to one another:

1. Thermochemical submodel: The thermochemical submodel provides temperature, viscosity, degree of cure (for thermosets), crystallinity (for thermoplastics), and the time required to complete the cure process.
2. Consolidation/Fiber Motion Submodel: The consolidation and fiber motion submodels evaluate the effects of processing conditions on the interaction between plies. In particular, the consolidation submodel (for thermoplastics) models the bonding between composite plies. The fiber motion submodel (for thermosets) yields the fiber position and fiber volume fraction within the cylinder.
3. Stress/strain submodel: Stresses within the composite that occur during winding, as a result of heating/cooling, or upon mandrel removal are evaluated in the stress/strain submodel.
4. Void Submodel: Changes in void size associated with processing conditions are quantified in this submodel.

Flow charts with relevant inputs and outputs for each submodel are shown in Figures 13.7 and 13.8 for winding of thermosetting and thermoplastic composite cylinders, respectively. The primary differences between process models for thermosetting and thermoplastic cylinders arise in (1) the method of heating, and (2) the mechanics of consolidation/fiber motion.

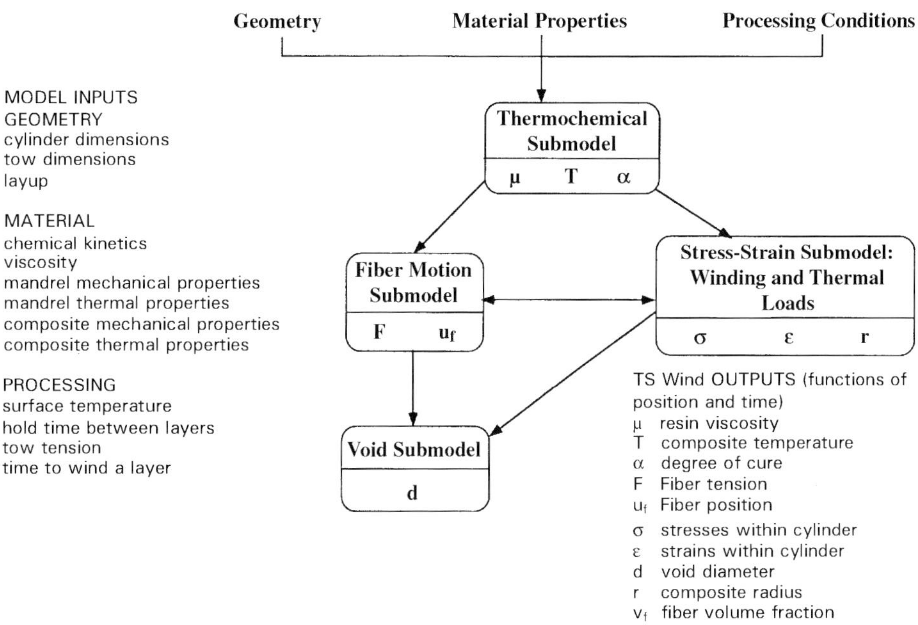

Figure 13.7 Flow chart showing interrelationship of submodels for filament winding with thermosetting matrix materials

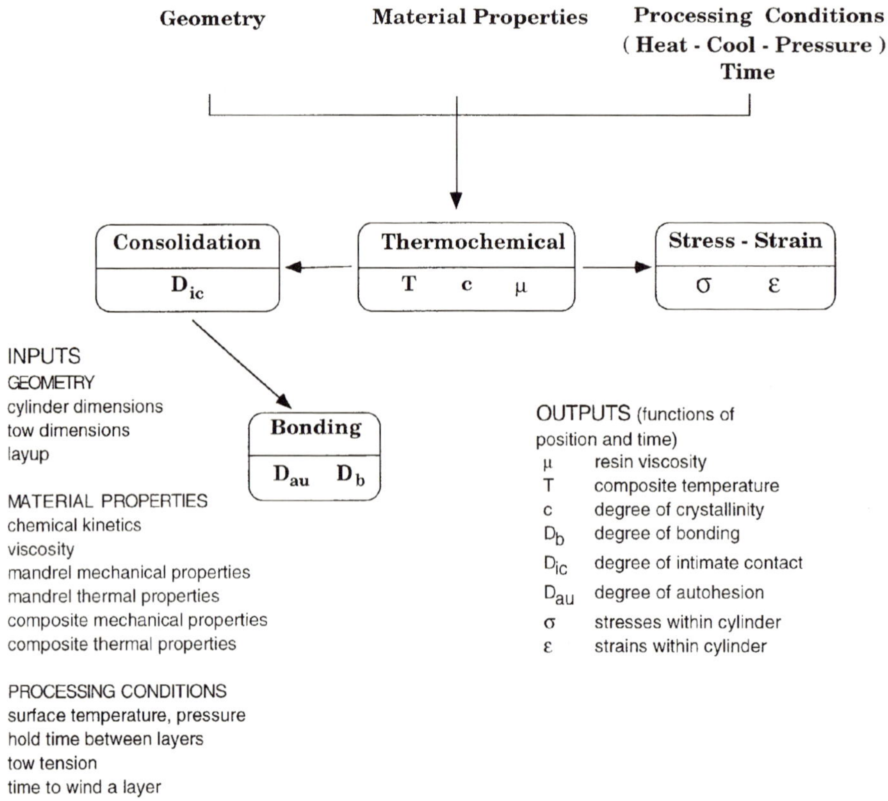

Figure 13.8 Flow chart showing interrelationship of submodels for filament winding with thermoplastic matrix materials

In the following sections, the basic modeling approaches for thermosetting and thermoplastic matrix composite cylinders will be summarized. Differences between the thermosetting and thermoplastic model approaches are highlighted.

13.5.1 Thermochemical Submodel

The energy equation, with temperature varying as a function of radial, circumferential, and axial position and time, is the basis of the thermochemical submodel. The energy absorbed or released during the cure or crystallization of the matrix is included in the energy balance. The appropriate multidimensional energy equation is Equation 13.1,

$$\rho C \frac{\partial T}{\partial t} = \frac{1}{r}\frac{\partial}{\partial r}\left(k_r \frac{\partial T}{\partial r}\right) + \frac{1}{r^2}\frac{\partial}{\partial \theta}\left(k_\theta \frac{\partial T}{\partial \theta}\right) + \frac{\partial}{\partial z}\left(k_z \frac{\partial T}{\partial z}\right) + \rho \dot{Q} \tag{13.1}$$

where t is time, r, θ, and z are the radial, circumferential, and axial coordinates, T is the temperature, ρ is the density, C is the specific heat, and k is the thermal conductivity. $\dot{Q}$ is the rate at which heat is generated or absorbed by chemical reactions. There are no chemical reactions in the fibers, and the last term reduces to Equation 13.2:

$$\rho\dot{Q} = \rho_r v_r \dot{Q}_r \tag{13.2}$$

ρ_r is the matrix density and v_r is the volume fraction of the matrix. The heating rate of the matrix $\dot{Q}_r$ is related to the degree of cure α (thermosets) or crystallinity c (thermoplastics) (Eq. 13.3)

$$\begin{aligned} \dot{Q}_r &= \left(\frac{d\alpha}{dt}\right)H_u \quad \text{thermoset} \\ \dot{Q}_r &= \left(\frac{dc}{dt}\right)H_u \quad \text{thermoplastic} \end{aligned} \tag{13.3}$$

For thermosets, H_u is the total heat of reaction of the matrix. For thermoplastics, H_u is the theoretical ultimate heat of crystallinity.

Once the temperature history is known, the viscosities can be calculated from expressions of the form (Eq. 13.4)

$$\begin{aligned} \mu &= f_1(\alpha, T) \qquad \text{thermoset} \\ \mu &= f_2\left(c, T, \frac{dT}{dt}\right) \quad \text{thermoplastic} \end{aligned} \tag{13.4}$$

Expressions for $d\alpha/dt$, dc/dt, and μ can be found in References 15, 27–29 and 26,30,31 for thermoset and thermoplastic matrix materials, respectively.

In order to complete the problem, the initial and boundary conditions must be given. The temperature and degree of cure or crystallinity must initially (at time zero) be specified at every point inside the composite and the mandrel. For the latter only the temperature is required. As boundary conditions, the temperatures or heat fluxes at the composite outside diameter and mandrel inner diameter must be specified.

Solutions to these equations yield the temperature distribution inside the mandrel and inside the composite as a function of time. Degree of cure or crystallinity and matrix viscosity in the composite as a function of time are also determined. This model is the building block for the other submodels. Viscosity calculations are input to the fiber motion submodel. Temperature and cure calculations are input to the stress submodel. Temperature data are also input to the void submodel.

13.5.2 Fiber Motion Submodel: Thermosetting Matrix Cylinders

The fiber motion submodel yields the fiber position during processing. In filament winding, the fiber position is affected by flow of the resin matrix material, expansion of the mandrel, and expansion of the composite. In the fiber motion submodel, only changes in fiber position caused by flow of the matrix are considered. Changes caused by thermal expansion of the mandrel and composite are included in the stress–strain submodel.

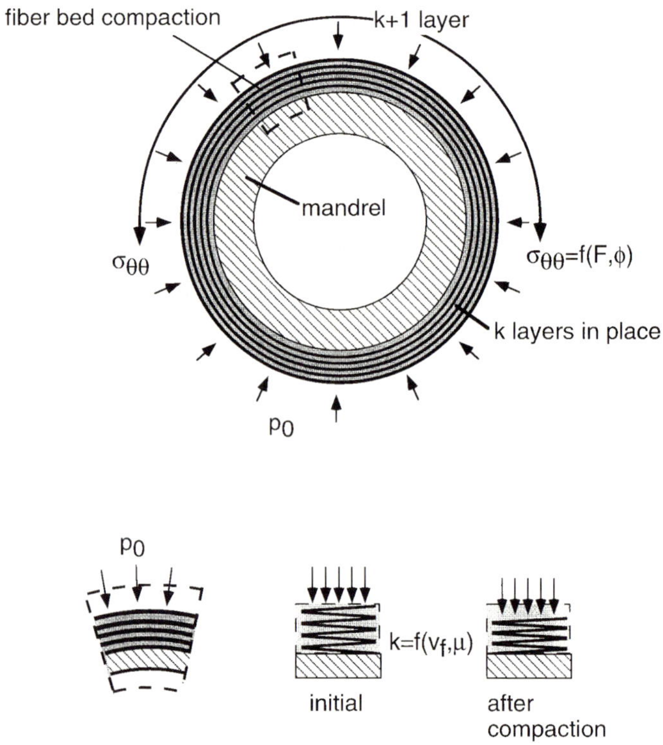

Figure 13.9 Schematic showing compaction of previously wound layers when the $k + 1$ layer is wound. The fiber bed behaves as a nonlinear spring

Two matrix flow submodels have been proposed: the sequential compaction model [15] and the squeezed sponge model [11]. Both flow models are based on Darcy's Law, which describes flow through porous media. Each composite layer is idealized as a fiber sheet surrounded by thermoset resin (Fig. 13.9). By treating the fiber sheet as a porous medium, the matrix velocity $\dot{u}_r$ relative to the fiber sheet is given as (Eq. 13.5):

$$\dot{u}_r = \frac{S}{\mu}\frac{dp}{dr} \tag{13.5}$$

where S is the apparent permeability of the fiber sheet and dp/dr is the pressure drop across the fiber sheet. Note S is a function of the fiber volume fraction [11].

13.5.2.1 Sequential Compaction Model

In the sequential compaction model, once a ply is completely compacted, the adjacent ply may begin compaction. This model assumes that matrix flow normal to the fibers and along the fibers may be decoupled. Another critical assumption is that the matrix supports the entire external pressure, and the fibers provide no support. Combining these considerations along with the winding geometry, the matrix velocity is given as (Eq. 13.6):

$$\dot{u}_r = -\frac{S\,\sigma_f}{\mu\, r_f}\sin^2\phi_o \tag{13.6}$$

The viscosity μ and the fiber tension σ_f may vary with position r and time t.

Once the resin velocity at the layer interfaces is determined from Equation 13.6, the thickness of each ply h is determined from conservation of mass (Eq. 13.7)

$$\frac{\partial}{\partial t}(\rho r_f h) = \rho_r r_i (\dot{u}_r)_{\text{in}} - \rho_r r_o (\dot{u}_r)_{\text{out}} \tag{13.7}$$

r_f is the current fiber sheet position, $(\dot{u}_r)_{\text{in}}$ and $(\dot{u}_r)_{\text{out}}$ refer to the resin velocity at the inner r_i and outer r_o radii of a particular layer. Because the mass of fibers within a fiber sheet remains constant, as the ply thickness changes (with the moving resin) the fiber volume fraction within the ply will change, as will the position of the fiber sheet relative to the mandrel. The fiber tension within each layer σ_f is updated to account for the change in fiber position Δu_f (Eq. 13.8)

$$(\sigma_f)_{t+\Delta t} = (\sigma_f)_t + E_f\left(\frac{\Delta u_f}{r_{fo}}\right) \tag{13.8}$$

where E_f is the fiber modulus and r_{fo} is the initial fiber radial position. Update of the fiber stress is critical both because it is directly related to the pressure gradient, and as it approaches zero fiber buckling may occur within the layer. Solutions are found numerically, by discretizing over time.

13.5.2.2 Squeezed Sponge Model

In the squeezed sponge model [11,13], compaction is not necessarily sequential and the applied pressure is shared by both the matrix and the fiber bed. The fiber bed behaves as a rapidly stiffening nonlinear spring. Pressure in the fiber bed is a function of fiber bed compaction. The rate of change of the fiber volume fraction v_f is related to the pressure drop $dp/d\xi$ [11] (Eq. 13.9)

$$\frac{1}{v_f}\frac{\partial v_f}{\partial t} + \frac{1}{\xi}\frac{\partial}{\partial \xi}\left(\frac{S}{\mu}\xi\frac{\partial p_r}{\partial \xi}\right) = 0 \tag{13.9}$$

This equation incorporates Darcy's Law for fluid flow as well as conservation of mass. The position ξ is related to the original fiber position r by the radial displacement u (Eq. 13.10)

$$\xi = r + u \tag{13.10}$$

The resin pressure p_r is related to the total stress of the composite and the fiber stress σ_f (Eqs. 13.11 and 13.12)

$$\sigma_r = (\sigma_f)_r - p_r \tag{13.11}$$

$$\sigma_\theta = (\sigma_f)_\theta - p_r \tag{13.12}$$

Once the rate of change of the fiber volume fraction is determined, a continuity condition is imposed to find the new deformation variable $\xi(t+\Delta t)$ (Eq. 13.13)

$$\frac{(v_f)_o}{v_f} = \frac{\xi}{r}\frac{\partial \xi}{\partial r} \tag{13.13}$$

This deformation variable determines the strain and subsequent overall stress state. The overall stress state, in turn, is related (through Eqs. 13.11 and 13.12) to the fiber stresses. As in the sequential compaction formulation, solutions for fiber position and fiber volume fraction are found numerically.

Some nonlinearities have been introduced in the squeezed sponge model:

1. The fiber bed stiffness and permeability are functions of the fiber volume fraction
2. The total stress state is shared between the resin and the fiber

As a result, an iterative solution technique is required. Initial values for the fiber volume fraction v_f and resin pressure are assumed and compared with calculated values found by solution of Equations 13.9–13.13. This iterative process is described in detail in Reference [11].

13.5.2.3 Resin Flow Model Comparison

The key assumption in the sequential compaction model (scm) is that consolidation occurs layer by layer. It has been noted in the literature [12,32] that the scm formulation is limited to cases in which the fiber volume fraction is less than 60 percent and to those with "prepreg' winding conditions. Prepreg winding conditions are described as those in which the resin viscosity is high during the winding stage. In Reference [11] a process time constant has been proposed to distinguish between prepreg winding and wet winding. This constant is based on the ratio of the flow time to the wind time for a particular winding operation. The flow time is a measure of the time needed for the resin to flow through one layer. Smith and Poursartip [32] have compared these two well-established resin-flow models and discussed their limitations. Although the squeezed sponge model is the more robust, it also requires more extensive materials characterization and an iterative solution technique. The sequential compaction model is valid if the time scale associated with the winding of each layer is much smaller than the time constant of the resin through the fiber bundle (i.e., prepreg winding) or if the fiber volume fraction is less than 0.60. It should be noted that the sequential compaction model is a limiting case of the squeezed sponge model.

13.5.3 Consolidation Submodel: Thermoplastic Cylinders

Consolidation and development of interlaminar bond strength for thermoplastic matrix composites have been modeled by two mechanisms: intimate contact and autohesion. Intimate contact describes the process by which two irregular ply surfaces become smooth (Fig. 13.10). In areas in which the ply surfaces are in contact, autohesion occurs, and the long thermoplastic polymer chains diffuse across the ply boundaries. Filament winding with thermoplastic matrix materials is considered an on-line consolidation process in that local

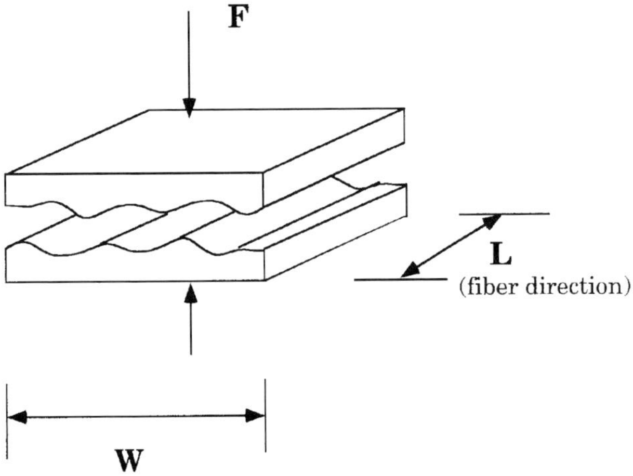

Figure 13.10 The uneven surface of a preimpregnated thermoplastic sheet. The rough surface is oriented relative to the fiber longitudinal axis as shown

heat and pressure are applied to bond the incoming tow to the existing laminate. Models that originally developed for press-consolidation processes [25,26] have been extended to on-line consolidation processes.

A theoretical model for intimate contact for in-situ consolidation has been developed in Reference 21. In this model, the irregular surface of the thermoplastic tow is modeled as a series of rectangular elements, oriented along the fiber axis, which are deformed as local pressure is applied (Fig. 13.11). The amount of "flattening" is quantified as the degree of intimate contact D_{ic} (Eq. 13.14)

$$D_{ic} = \frac{b}{w_o + b_o} \tag{13.14}$$

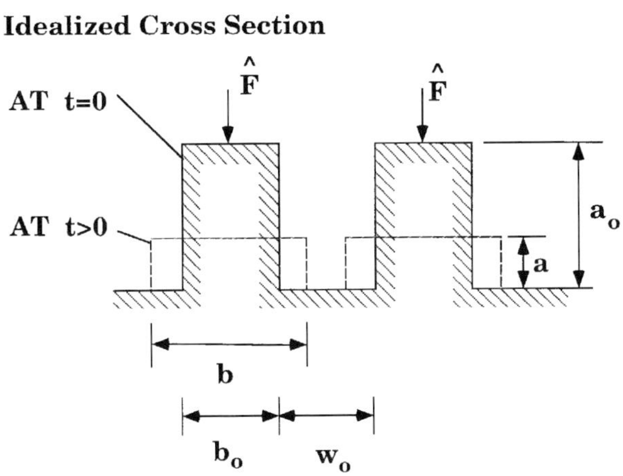

Figure 13.11 Idealization of the uneven thermoplastic prepreg surface as a series of rectangles with relative heights and spacing as shown

where b_o and b are the initial ($t \leq 0$) and instantaneous (at time t) widths of each rectangular element, respectively, and w_o is the initial distance between two adjacent elements. A value of $D_{ic} = 1$ indicates complete contact between the two surfaces.

An expression for the degree of intimate contact can be found by applying conservation of mass and equilibrium of forces (Eq. 13.15)

$$D_{ic} = \frac{1}{1 + \frac{w_o}{b_o}} \left[1 + 5\left(1 + \frac{w_o}{b_o}\left(\left(\frac{a_o}{b_o}\right)^2 \int_0^{t_c} \frac{P_{app}}{\mu_{mf}} dt\right.\right.\right]^{1/5} \tag{13.15}$$

where t_c is the time (contact time) during which pressure p_{app} is applied, μ_{mf} is the viscosity of the fiber–matrix mixture, and a_o is the initial height of a rectangular element. This expression is particularly valuable in that it relates process variables and ply geometry to the intimate contact between plies and it is applicable for on line consolidation and press processing. In Reference [21], Mantell and Springer have simplified Equation 13.15 for consolidation beneath a roller.

The autohesion process starts after intimate contact has been established at any point on the interface. The degree of autohesion D_{au} can be approximated by the expression (Eq. 13.16)

$$D_{au} = \kappa t_a^{n_a} \tag{13.16}$$

where t_a is the time elapsed from the start of the autohesion process, and κ is a constant that depends on temperature. The exponent n_d is a constant. Both κ and n_a must be empirically determined for a particular thermoplastic material. Constants for PEEK thermoplastic are reported in [25,26,33]. The degree of bonding D_b can now be expressed as (Eq. 13.17):

$$D_b = (D_{ic})(D_{au}) \tag{13.17}$$

As is the case for D_{ic}, complete autohesion and complete bonding correspond to $D_{au} = 1$ and $D_b = 1$, respectively. To account for nonisothermal autohesion, which occurs in on-line consolidation processes such as filament winding, the degree of bonding must be calculated at discrete time steps and summed [21].

13.5.4 Stress Submodel

The stress submodel results in the stresses and strains in the composite. Stresses and strains are introduced by fiber tension, thermal expansion or contraction, and chemical changes. The effects of fiber tension, temperature and chemical changes may arise simultaneously; however, the stresses and strains are analyzed separately. The sum of the stresses and strains caused by each of these factors is the actual stress and strain in the composite.

The temperature, fiber tension, stresses, and strains vary only in the radial directions. An elasticity solution is employed to calculate the six components of the stresses and strains. The solution procedure follows the established techniques of elasticity solutions. A displacement field is assumed that satisfies the equilibrium equations and the compatibility conditions. The latter requires that at each interface the displacements and the normal stresses in adjacent layers be the same. For details of the calculations the reader is referred to Reference 15 for thermosetting materials and Reference [20] for thermoplastic materials.

The implementation of mandrel removal is of particular interest in winding model formulation. When the mandrel is removed, there is no radial stress σ_{rr} at the cylinder inner diameter (Eq. 13.18)

$$\sigma_{rr} = 0 \quad \text{at} \quad r = r_m \tag{13.18}$$

This requirement is imposed by adding a radial stress at the cylinder inner diameter that is equal but opposite in magnitude to the contact stress between the mandrel and cylinder. The contact stress corresponds to the radial pressure at the interface at the time the mandrel is removed.

13.5.5 Void Submodel

Voids are introduced in winding during both impregnation, including prepregging operations, and tow placement. In addition to these mechanical means, voids may also be introduced by homogeneous or heterogeneous nucleation (for thermosets). In the void submodel, initiation of voids is not considered; only changes in void size are modeled. The voids are assumed to be spherical and contain a vapor of known composition (i.e., initial concentration of air and water). Void location within the composite cylinder is also assumed to be known. During manufacture, the volume (size) of the void changes because (1) vapor is transported into and out of the void across the void-composite interface, and (2) the pressure changes at the location of the void. The methodology that follows holds for a liquid–gas interface. As the matrix cures or crystallizes, the liquid solidifies and the void size will not change.

For a spherical void of diameter d, the surface tension Γ at the void composite interface is a function of the pressure inside the void P_v and the pressure surrounding the void P (Eq. 13.19)

$$\Gamma = \frac{d(P_v - P)}{4} \tag{13.19}$$

The surface tension is found from an empirical formula and is a function of temperature (determined in the thermochemical submodel). The surrounding pressure P is determined in the resin flow or compaction submodels. The pressure within the void is determined by the partial pressures of the water vapor and air within the void. The mass of water vapor within the void changes during processing and can be described by Fickian diffusion across the void–composite interface [29]. Once the mass of vapor inside the void and the pressure at the location are known, the change in void size can readily be calculated from Equation 13.19. Changes in void size are halted when the resin has solidified.

If the initial void volume fraction and average initial diameter d_o are known, the final void volume fraction v_v can be calculated [34].

$$v_v = \frac{d^3}{(d_o^*)^3 + d^3 - 8} \tag{13.20}$$

where d_o^* is the dimensionless initial diameter of the outer resin shell, scaled with respect to d_o. In Eq. 13.20, the void is assumed to be surrounded by a concentric spherical resin shell of outer diameter d^*, which is related to the initial void volume fraction. Conservation of the mass of resin in the outer shell is used to find the relationship between d^*, d, and d_o^*.

13.6 Filament-Wound Material Characterization

13.6.1 Overview

For typical filament winding applications, the fiber reinforcement provides the stiffness and strength required to maintain structural integrity. Thus, material characterization for filament wound structures focuses on characterizing the fiber dominated stiffness and strength properties of the composite. The stiffness of fiber reinforced plastics (FRPs), in the fiber direction, is dominated by the fiber stiffness characteristics. The strength will be influenced by a number of factors, however, and not all of them are related to the fiber, including:

1. Resin matrix properties
2. Fiber sizing
3. Percentage of fiber volume
4. Manufacturing process variation
5. Fiber alignment
6. Environmental effects
7. Type of loading

In fact, the strength variability of a given fiber–resin combination may be significantly different from the strength variability of a different combination of the same fiber with another resin system [35]. This is particularly the case when environmental and fatigue loading (long-term durability) are considered.

Fiber-reinforced plastics have varying degrees of resistance to adverse environments such as moisture, alkali, acid, and other chemicals. The degree of resistance depends on the fiber–resin system. Moisture absorption and chemical infiltration will be different for different fiber–resin systems. The degradation of composite materials may result from several factors:

1. Loss of reinforcing fiber strength by stress-corrosion
2. Loss of adhesion and interfacial bond strength from degradation of the fiber–matrix interface
3. Chemical degradation of the matrix material
4. Dependence of the matrix modulus and strength on time and temperature
5. Accelerated degradation caused by combined action of temperature and chemical environment.

These environmental factors influence the fiber, matrix material, and interface simultaneously. Thus, the degradation of composites occurs with the degradation of the individual constituents as well as with the loss of interaction between them. Composite properties that are strongly influenced by the matrix and fiber–matrix interface, such as shear and compression, will be most susceptible to moisture exposure and other chemicals that attack the matrix and/or fiber–matrix interface.

Glass fiber can undergo significant strength reduction when exposed to alkaline and/or acidic environments. Aramid fibers are susceptible to attack by certain strong acids and strong bases. Aramid fibers are also susceptible to ultraviolet (UV) degradation. Of the three fiber types (glass, aramid, and carbon), carbon fiber is the most inert to environmental effects. Carbon fiber is not affected by moisture, atmosphere, solvent, bases, or weak acids at room temperature [36]. Carbon fiber, however, reacts with aluminum and titanium through galvanic reaction and must be protected.

Many filament-wound structures are designed to be in service over a long period of time (20–30 years). During this period of time, the filament wound vessels are exposed to sustained static and fatigue loads. The long-term static loads require that the material is resistant to creep-rupture for the applied loads over the lifetime of the structure. Similarly, the material is required to have acceptable resistance to fatigue strength degradation. The creep rupture susceptibility of glass fiber is well known and is a major concern of the American Society of Mechanical Engineering code specifying that the operation pressure of glass–resin vessels be only one sixth that of burst [37]. Although aramid fiber appears to have somewhat longer creep-rupture failure times than glass [38], the aramid fiber long-term strand data exhibit as much scatter as glass, and also shows no indication of an endurance limit. Fiber-reinforced plastics show significant differences in the degree of resiliency to fatigue loads. As with creep–rupture, glass fiber is the most susceptible to fatigue damage, whereas carbon is the most resilient of the three fiber types.

This section discusses test methods applicable to filament wound structures. As discussed earlier, filament wound structures are primarily subjected to internal and/or external pressure that is resisted by the fiber. Greater attention is therefore given to fiber-dominated stiffness/strength-dominated material characterization.

13.6.2 Test Methods

13.6.2.1 Fiber-Dominated Stiffness/Strength Characterization

There are numerous test methods that have been used to characterize the fiber-dominated composite strength and stiffness for filament wound structures. A number of these test methods have been standardized by the ASTM D30 Committee [39]. These standardized tests methods include:

1. ASTM D3379 standard test method for tensile strength and Young's modulus for high-modulus single-filament materials
2. ASTM D4018-81 standard test method for tensile properties of continuous filament carbon and graphite yarns, strands, roving, and tow
3. ASTM D3039 standard test method for tensile properties of fiber–resin composites

4. ASTM D2290 standard test method for apparent tensile strength of ring tubular plastics and reinforced plastics by split disk method
5. ASTM D2291 standard test method for fabrication of ring test specimens for glass-resin composites; and
6. ASTM D2585 standard test method for preparation and tension testing of filament-wound pressure vessels.

Other nonstandardized test methods that have been used to characterize fiber-dominated strength and stiffness properties of filament wound FRP pressure vessels are:

1. Four-in. biaxial composite tube [40–42]
2. Two-in. multiaxial (axial/torsional/internal pressure) composite tube test method [43]
3. Twenty-in. pressurized ring test method [44]

The impregnated fiber strand specimen (ASTM D4018-81), unidirectional lamina tensile specimen (ASTM D3039), NOL (named after the Naval Ordnance Laboratory where this test method was developed) ring specimen (ASTM D2290), and 0.1-m (4-in.) biaxial tube specimen [40–42] are shown in Figure 13.12. The impregnated strand, unidirectional lamina, and NOL ring are small scale specimen geometry which are useful for the initial fiber/resin screening characterization. These test methods are less useful as a characterization technique of the actual filament wound structure.

In the split-disk test method (ASTM D2290) fiber strength is determined from a hoop-wound ring that is loaded to failure in tension by split disks. The failure along the ring circumference is forced to occur at the location that is parallel to the disk split line by a reduced ring section at this location. Under the described loading conditions the stress state at this location is not uniform [45] and fiber failure may not be representative of filament-wound vessel burst strength. The pressurized ring test method (standardized under the ASTM D2291-83) was designed to overcome some of the aforementioned problems associated with the ASTM D2290 split disk test method. Like ASTM D2290, a hoop-wound ring with the identical dimensions to those specified in ASTM D2290 is used (no notch is machined). The ring is internally pressurized to failure using a specially designed test fixture.

The NOL ring test method has some drawbacks, and fiber strength data produced by this test method have limited usefulness for full-scale composite pressure vessel designs. Some of the limitations are: (a) a relatively small specimen width (6.35 mm), which may contribute to a large interaction between edge effects and the ring ultimate strain-to-failure; (b) only a hoop lay-up is used, which is not representative of full-scale composite vessels; (c) a small ring diameter is used, which necessitates a relatively high internal pressure to fail the ring—thus limiting the ring thickness that can be tested; and (d) the delivered fiber strength measured with this test method is only 80 percent of the fiber strength measured in fiber strand tests [46].

The standard ASTM D2585 filament wound pressurized bottle test method utilizes a 0.15-m (5.75-in.)internal diameter filament wound bottle as the test article. This standard test method (with variation in bottle sizes) has been used extensively by the rocket motor industry [47–50] to evaluate glass, aramid, and graphite fiber composite vessel performance. This test method has generally shown good results, but is a relatively expensive test method. Testing of one 0.5-m (20-in.) diameter bottle can cost up to $20K. Other disadvantages are:

(*a*)

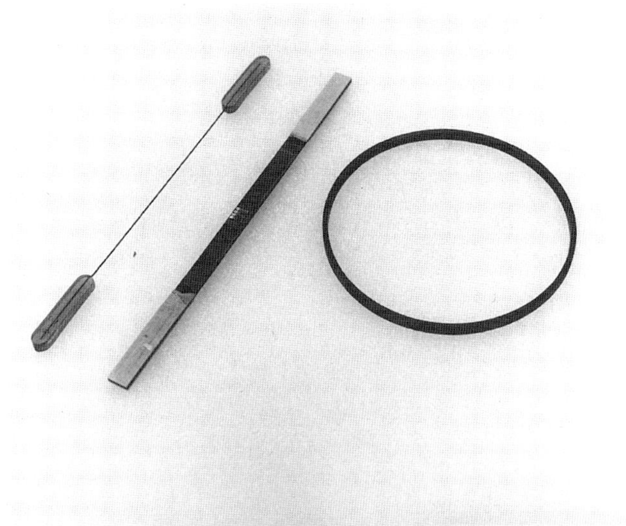

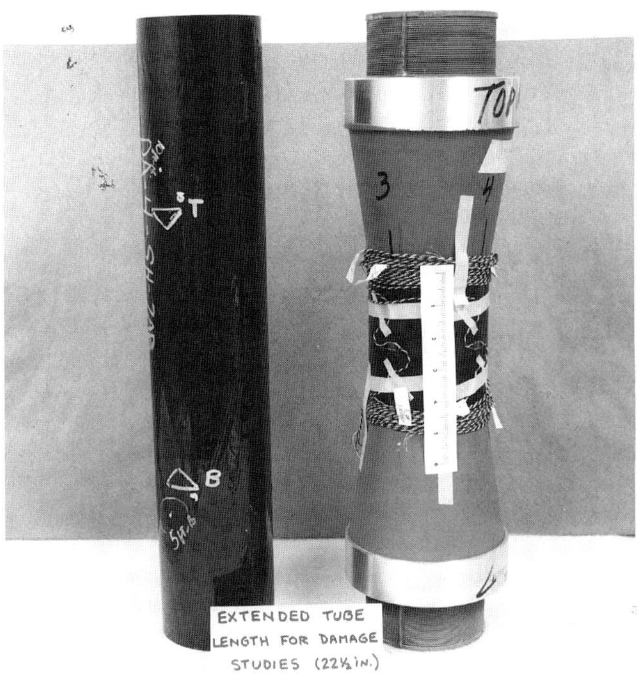

(*b*)

Figure 13.12 Fiber strength test articles. (a) the impregnated fiber strand specimen, unidirectional lamina tensile specimen, and NOL ring; (b) 4-in. biaxial tube test specimen

1. The test article utilizes a relatively thin laminate
2. In many instances failure occurs in the dome or by boss blowout
3. The test method does not allow for the evaluation of the strength variation within the material.

The 0.10-m (4-in.) biaxial cylindrical specimen has been studied extensively by Swanson et al. [40–42]. In this test method a nominal 0.96-m (3.8-in.) inside diameter filament wound tube is cut into 0.33-m to 0.43-m (13-in. to 17-in.) long specimens. An end reinforcement is then added to the specimen, which consists of a combination of fiberglass cloth overwraps, aluminum rings, and low modulus epoxy (Fig. 13.12). The buildup region has a very gradual transition to the gage section in order to reduce the stress concentrations that are associated with the end-griping and pressure-sealing regions. The basic loading of the specimen involves a combination of internal pressure–induced hoop tensile stress and axial loading. The ratio between the axial load–induced stress and the internal pressure–induced hoop stress can be varied. The major advantage of the 0.10-m (4-in.) cylindrical specimen is its capability to characterize the composite under a state of biaxial stress representing the pressure vessel loading conditions. Nevertheless, this test method has some limitations that may be important in the evaluation of full-scale composite pressure vessel material strength and stiffness. Those limitations are:

1. The relatively thin specimen wall thickness 1–2 mm (0.04–0.08-in.)
2. The limited helical fiber angle inflexibility
3. The relatively high cost per specimen (>$1000).

The multiaxial 0.05-m (2-in.) tube test method [43] is a derivative of the 0.10-m (4-in.) cylindrical specimen. In this test method the specimen can also be loaded in torsion, axial tension/compression, and pressure. This type of multiaxial loading conditions allow greater flexibility in investigating generalized three-dimensional failure theory. In pressure vessels, however, where burst strength is dominated by the fiber strength, the maximum strain criterion has been shown to better characterize failure observation. From numerous tests of 0.10-m (4-in.) biaxial tubular specimens, Swanson et al. [40–42] concluded that the maximum strain criterion gave the best agreement with experimentally determined tube burst strength for various stress–ratio and lay-up sequences.

The 0.51-m (20-in.) pressurized composite ring test (Fig. 13.13) was developed to overcome some of the drawbacks discussed. The test fixture is designed to load the specimen by internal pressure using a specially designed rubber bladder. The bladder applies the pressure to the composite ring in a near perfectly uniform distribution. Composite rings with up to 0.025-m (1-in.) wall thickness can be tested. The internally pressurized ring test method is easily converted into an externally pressurized ring test method.

The 0.51-m (20-in.) ring test method was qualified as test analog for full-scale filament-wound composite pressure vessel using three distinct carbon/epoxy 0.51-m (20-in.) inside diameter subscale cylinders. The lay-ups, ply thicknesses, and manufacturing pocedures were representative of full-scale pressure vessels. The full-scale filament-wound vessel diameters ranged between 1.02-m and 3-m (40-in. and 120-in.) and laminate thickness between 6.35 mm and 20.3 mm (0.25-in. and 0.8-in.) A comparison between the fiber strain-to-failure measured from full-scale vessels and subscale ring tests shows good agreement (Table 13.4).

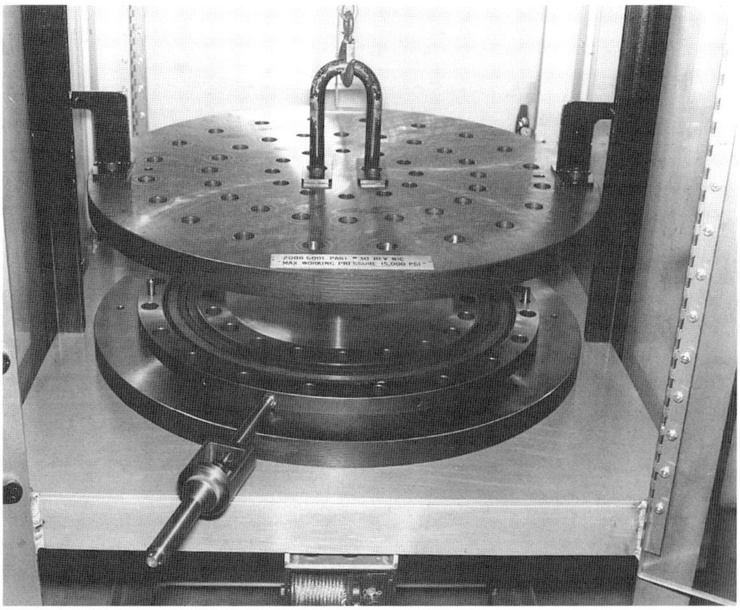

Figure 13.13 Pressurized composite ring test: A 20-in. diameter composite ring is loaded into the test fixture shown and subjected to internal pressure

Table 13.4 Average Fiber Strain-to-Failure as Measured by Full- and Subscale Ring Tests

	Fiber Strain @ Failure (% strain)								
	Full scale				20-in. ring subscale				
Vessel type	R/t	n	Avg.	Cv (%)	R/t	n	Avg.	Cv (%)	Percentage difference
D1	83	30	1.57	3.0	21	7	1.53	3.0	−2.9
D2	80	4	1.46	4.3	41	13	1.52	1.9	+4.1
D3	79	4	1.42	2.7	13	10	1.49	1.4	+4.9

Table 13.4 demonstrates that vessel scale was not a contributing factor in measured fiber strength. The ring test method is unique in that it substantially reduces the cost (<\$200) for determining accurate and statistically significant fiber strength and stiffness for full-scale filament-wound pressure vessels. The method can also be used to characterize strength variability within a cylinder and among several cylinders.

13.6.2.2 Resin Dominated Material Characterization

The composite's resin-dominated material properties are in-plane shear, interlaminar shear, transverse tension/compression, and mode I and II fracture toughness. These properties are less critical to the design of a pressure vessel, although the laminate in-plane and interlaminar shear properties can be critical to the proper design of the domes and attachment regions. Many design applications of filament-wound rocket and booster cases require attachment skirts. Design loads in these regions may include shear, tension, and compression. Superior delamination fracture toughness is also necessary. In many cases the attachment regions also require the design of composite-bolted joints. This section will discuss aspects of resin-dominated material characterization for filament-wound structures.

In-Plane Shear Properties. The basic lamina in-plane shear stiffness and strength is characterized using a unidirectional hoop–wound (90°) 0.1-m nominal internal diameter tube that is loaded in torsion. The test method has been standardized under the ASTM D5448 test method for in-plane shear properties of unidirectional fiber–resin composite cylinders. D5448 provides the specimen and hardware geometry necessary to conduct the test. The lamina in-plane shear curve is typically very nonlinear [51]. The test yields the lamina's in-plane shear strength, τ_{12}, in-plane shear strain at failure, γ_{12}, and in-plane chord shear modulus, G_{12}.

The same test method, although not standardized, can be used to characterize the laminate in-plane shear behavior. This is accomplished by winding a multiorientation (hoop/helical and/or helical only) tube. Other test methods that can be used to measure in-plane shear stiffness/strength of filament wound composites are discussed by Tarnopol'skii and Kincis [45]. These methods include schemes for torsion of intact rings and split rings. Both of these ring test methods are used to evaluate the in-plane shear modulii $G_{\theta r}$ and $G_{\theta z}$ for a filament-wound laminate.

Interlaminar Shear Properties. The interlaminar shear strength test method has been standardized under the ASTM D2344 standard test method. Under this test method a section of a ring specimen is tested in a three-point short-beam shear loading condition. The test specimen can be removed from a unidirectional hoop–wound ring or from the full-scale filament–wound vessel. The specimen geometry and the supporting span length are specified in the standard. Given the specimen geometry and failure load, the apparent shear strength can be calculated. Following the test it must be confirmed visual inspection that failure occurred by interlaminar shear at the midply of the laminate.

13.6.2.3 Transverse Lamina Compressive/Tensile Properties

The lamina's transverse compressive/tensile properties are determined from a hoop wound (90°) cylinder loaded in axial compression and/or tension. The test methods have been standardized under ASTM D5450 and D5449. The tube geometry and manufacturing procedures are similar to those used in the lamina's in-plane shear tests.

The laminate compressive hoop strength/stiffness of a filament-wound vessel can be evaluated using the externally pressurized ring test method [44] or the standardized ASTM D2586 test method. In the ASTM D2586 test method a filament-wound FRP cylinder is tested under hydrostatic pressure to simulate the loading conditions of a pressure vessel under an external pressure load. The main drawback of this test method is the influence of the end constraints on the test results.

13.6.2.4 Interlaminar Fracture Properties

The mode I delamination fracture toughness is measured on flat coupons using the Double Cantilever Beam (DCB) test method. The test method utilizes a unidirectional rectangular composite specimen of uniform thickness with nonadhesive insert (as a delamination initiator) at the midplane. This test method is currently in the process of standardization by ASTM.

13.7 Outlook/Future Applications

In the past, filament-wound parts consisted primarily of axisymmetric cylinders, spheres and domed vessels. Several manufacturing techniques have been developed that allow more complex shapes and curvatures while maintaining the cost effectiveness associated with process automation [52]. These methods have emerged because of advances in programming software. These advances enable precise positioning of the moving head and allow real-time simulation of fiber paths.

In fiber placement, a moving head is mounted to a multidegree of freedom carriage. The fiber placement head is designed to cut and place each tow individually. The head may also include a controlled heat source and compaction roller for in situ consolidation. Tow placement with in situ consolidation has made it possible to place preimpregnated tape along nongeodesic paths, manufacture parts with concave sections, and to locate ply dropoffs precisely.

In tape laying, instead of controlling individual fibers, entire tapes are automatically placed on the tool. A precisely controlled head is again mounted on a multidegree of freedom carriage. This method is typically used for large parts with simple curvatures (up to 30°). Concave parts are difficult to manufacture. By using a wide tape, manufacturing times are shortened and the process is quite cost effective [4]. As in fiber placement, a heat source and compaction roller can be added to the head for in situ consolidation of thermoplastic matrix composite parts.

References

1. Hummler, J., Steiner, K.V. *Experimental Study of Robotic Thermoplastic Filament Winding of Complex Geometries* (1990) Center for Composite Materials, University of Delaware, Newark, Delaware
2. Krolewski, S., Gutowski, T. *SAMPE Q.* (1986) 18, p. 43–51
3. Wang, E., Gutowski, T. *SAMPE J.* (1990) 26, p. 19–26
4. Foley, M. *SAMPE Q.* (1991) 22, p. 61–68
5. Peters, S.T., Humphery, W.D., Foral, R.F. *Filament Winding Composite Structure Fabrication* (1991) Society for the Advancement of Materials and Process Engineering, Corina, California
6. Shibley, A.M. In *Handbook of Composites* Lubin, G. (Ed.) (1982) New York, Van Nostrand Reinhold, p. 449–461

7. Tsai, S. *Composites Design*, Third Ed. (1987) Dayton, Think Composites
8. Whitney, J.M., Halpin, H.C. *J. Comp. Mat.* (1968) 2, p. 360
9. Vinson, J.R., Sierakowski, R.L. *The Behavior of Structures of Composite Materials* (1986) Dordrecht, Martinus Nijhoff
10. Pagano, N.J., Whitney, J.M. *J. Comp. Mat.* (1970) 4, p. 360
11. Cai, Z., Gutowski, T., Allen, S. *J. Comp. Mat.* (1992) 26, p. 1374–1399
12. Agah-Tehrani, A., Teng, H. *Int. J. Solids Structures* (1992) 29, p. 2649–2668
13. Dave, R., Kardos, J.L., Dudukovic, M.P. *Polym. Comp.* (1987) 8, p. 29–38
14. Calius, E.P., Springer, G.S. *Int. J. Solids Structures* (1990) 26, p. 271–297
15. Lee, S.Y., Springer, G.S. *J. Comp. Mat.* (1990) 24, p. 1270–1298
16. Mantell, S.C., Springer, G.S. *Comp. Structures* (1994) 27, p. 141–147
17. Gruber, M.B. *Thermoplastic Tape Laydown and Consolidation* (1986) Society of Manufacturing Engineers, Dearbarn, MI
18. Cirino, M. *Axisymmetric and Cylindrically Orthotropic Analysis of Filament Winding* (1989) Newark, Delaware, University of Delaware
19. Beyeler, E.P., Guceri, S.I. *J. Heat Transfer* (1988) 110, p. 424–430
20. Nejhad, M.N., Cope, R.D., Guceri, S.I. *J. Heat Transfer* (1991) 110, p. 424–430
21. Mantell, S.C., Springer, G.S. *J. Composite Materials* (1992) 26, p. 2348–2377
22. Hwang, S.J., Tucker III, C.L. *J. Thermoplastic Comp. Mat.* (1990) 3, p. 41–51
23. Li, M.C., Loos, A.C. *Autohesion Model for Thermoplastic Composites* (1990) Blacksburg, Virginia, Center for Composite Materials and Structures, Virginia Polytechnic Institute and State University
24. Howes, J.C., Loos, A.C. *Interfacial Strength Development in Thermoplastic Resins and Fiber Reinforced Thermoplastic Composites* (1987) Blacksburg, Virginia, Center for Composite Materials and Structures, Virginia Polytechnic Institute and State University
25. Dara, P.H., Loos, A.C. *Thermoplastic Matrix Composite Processing Model* (1985) Blacksburg, Virginia, Center for Composite Materials and Structures, Virginia Polytechnic Institute and State University
26. Lee, W.I., Springer, G.S. *J. Comp. Mat.* (1987) 21, p. 1017–1055
27. Pearce, E.M., Mijovic, J. *Characterization-Curing-Property Studies of HBRF 55A Resin Formulations* (1985) NASA-Ames Research Center
28. Bhi, S.T., et al. In *Advanced Material Technology '87* (1987) Society for the Advancement of Materials and Process Engineering, Corina, CA
29. Loos, A.C., Springer, G.S. *J. Comp. Mat.* (1983) 17, p. 135–169
30. Maffezzoli, A.M., Kenny, J.M., Nicolais, L. *SAMPE J.* (1989) 25, p. 35–39
31. Velisaris, C.N., Seferis, J.C. *Poly. Eng. Sci.* (1988) 28, p. 583–591
32. Smith, G.D., Poursartip, A. *J. Comp. Mat.* (1993) 27, p. 1695–1711
33. Agarawal, V. *The Role of Molecular Mobility in the Consolidation and Bonding of Thermoplastic Composite Materials* (1991) Center for Composite Materials, University of Delaware, Newark, Delaware
34. Ranganathan, S., Advani, S.G., Lamontia, M.A. *J. Comp. Mat.* (1995) 29, p. 1040–1062
35. Cohen, D., Lowe, K.A. *J. Reinforced Plastics Comp.* (1991) 10, p. 112–131
36. *Composites Engineering Materials Handbook.* Vol. 1. (1987) Metals Park, ASM International
37. *Fiberglass-Reinforced Plastics Pressure Vessels.* ASME Boiler and Pressure Code, Section X, (1971) New York, ASME
38. Chiao, T.T., et al. *Stress Rupture of Strands of an Organic Fiber/Epoxy Matrix, in Composite Materials: Testing and Design (Third Conference), ASTM STP 546* (1974) Philadelphia, ASTM
39. *ASTM Standards and Literature References for Composites Materials*, Second ed. ASTM Committee D-30 on High Modulus Fibers and their Composites (1990) Philadelphia, ASTM
40. Swanson, S.R., Christoforou, A.P. *J. Comp. Mat.* (1988) 20, p. 238–243
41. Swanson, S.R., Trask, B.C. *Comp. Sci. Tech.* (1989) 34, p. 19–34
42. Swanson, S.R., Christoforou, A.P., Colvin, G.E. *Experimental Mechanics* (1988) 20, p. 457–471

43. Groves, S.E., Sanches, R.J., Feng, W.W. *Multiaxial Failure Characterization of Composites* (1991), Lawrence Livermore National Laboratory, Livermore, CA
44. Cohen, D., et al. *J. Comp. Tech. Res.* (1995) 17(4), p. 331–340
45. Tarnopol'skii, Y.M., Kincis, T. *Static Test Methods for Composites* (1981) New York, Van Nostrand Reinhold
46. Cohen, D. *J. Comp. Mat.* (1992) 26, p. 1984–2014
47. Newhouse, N.L., Mumphrey, W.D. In *Seventeenth National SAMPE Technical Conference* (1985) Society for the Advancement of Materials and Process Engineering, Corina, CA
48. Etringer, M.D., Mumford, N.A. In *Composite Case Subcommitee Conference* (1987) JANNAF "Thin-Wall Composite Case Development for SDI Application" JANNAF: Joint Army, Navy, NASA, and Air Force
49. Mumford, N.A. In *Composite Case Subcommittee Conference* (1987) JANNAF "Advanced Materials for Filament Wound Pressure Vessels," JANNAF: Joint Army, Navy, NASA, and Air Force
50. Mumphrey, W.D., Newhouse, N.L. In *Thirty-first International SAMPE Symposium* (1986) Society for the Advancement of Materials and Process Engineering, Corina, CA
51. Cohen, D. *J. Spacecraft and Rockets* (1991) 28, p. 339–346
52. Fisher, K. *High Performance Composites* (1995) 3, p. 23–30

14 Dieless Forming of Thermoplastic–Matrix Composites

Alan K. Miller

Dieless forming is directed at production of large singly curved continuous-fiber composite components, generating *the shape by using an array of small rollers. An initial version focused on tapered components with a variable cross-section along their length. Successes included the use of induction heating for rapid through-thickness temperature change, introduction of bends at a free edge and propagation of them into the laminate, and control of wrinkling on the inside of the bends. Complete success, however, was not attained due to the development of "overcurvature" in the formed components. An energy minimization model showed that the overcurvature was due to an "incompatibility strain" that arises between the formed and not-yet-formed regions, and was inherent to the process. A subsequent version of dieless forming imparted curvature simultaneously across the entire width of the workpiece. This version ("continuous dieless forming") successfully formed laminates and preserved high quality, as measured by ultrasonic examination (6 dB loss) and shear strength tests. Predrying and forming within a vacuum bag were important ingredients, as was manipulation of the temperature gradient across the forming roller. Continuous die-less forming was used to form components with arbitrary and variable curvature, with a good shape accuracy (maximum deviation ± 0.4 mm or 0.016 in.) being attained.*

14.1 Introduction

Advanced continuous-fiber composites have been available for engineering purposes for several decades. Thermosetting systems (e.g., fiberglass/polyester boat hulls, graphite/epoxy missile structures, recreational equipment components) have been the oldest historically and are still the most widely deployed. In some thermoset applications (e.g. filament-wound cylindrical structures) a substantial degree of automation can be employed to minimize the labor costs of fiber or ply placement. In many applications of more complex shape, however, hand labor must be utilized for layup (and often for the vacuum bagging associated with debulk and cure). The costs of fabricated composite structures consequently tend to be high (about \$700/kg or \$1500/lb. for aircraft structures [1]), which limits their utilization. Some properties of thermosetting resins (e.g., moisture absorption and desorption, toughness, fire, and smoke toxicity) also tend to be limiting.

Thermoplastic–matrix composites were developed in an attempt to overcome some of the limitations of thermosetting systems [2]. Their thermoplastic nature is due to their molecular structure, consisting of intertwined long-chain molecules, as opposed to the fully crosslinked molecular structure of thermosets. Their critical properties (with the exception of somewhat lower compression strength because of lower resin modulus) are an improvement over thermoset properties. Also, the inherent capability of thermoplastic matrixes to be softened repeatedly and to be deformed by intermolecular sliding opened up many manufacturing options for composites [3]. Some of these are simple adaptions of thermoset processes (e.g., vacuum bagging with autoclave pressure to fuse the thermoplastic plies into a finished laminate). High processing temperatures complicate these simple adaptations (e.g., raising significant problems with vacuum bag seals). Other processes explored for thermoplastics include matched die forming, superplastic gas-pressure forming, and tape laying [4].

Tooling is a major cost of composite production in all except very high-volume production. The thermoplastic manufacturing processes cited earlier all require fixed tooling of at least the same size as the component. Requirements for thermal expansion compatibility between tooling and workpiece, coupled with the high processing temperatures of thermoplastics, raise the tooling costs further. It would be very desireable to be able to form large components without fixed tooling, somehow generating the shape with an apparatus smaller than the component itself. Pultrusion [5] is one such process that is quite economical, but it requires a low-viscosity matrix and is therefore practiced almost exclusively with thermosetting systems. Roll-forming [6] is available within thermoplastics, but is limited to constant cross-sectional shapes.

Several years ago, the author conceived an approach to the forming of more complex shapes from thermoplastic-matrix, continuous-fiber composites, which would avoid fixed dies and large equipment like presses and autoclaves. As such, it was termed *dieless forming* [7]. The approach actually takes advantage of the continuous fibers that are an impediment in many other processes to flaw-free forming (e.g., forming wrinkles at internal bends). The process has been explored in collaborative work with a number of colleagues. Each contribution will be described in the following sections. The work to date has been partially successful, and the purpose of this chapter is to provide an overall summary of the results, referring the reader to previous publications for many of the details.

14.2 Dieless Forming Concept

The general concept of dieless forming is illustrated in Figure 14.1. The thermoplastic material enters the process as a preconsolidated flat laminate. (Limited attempts to start with a "tacked" layup and accomplish both consolidation and forming, or partial consolidation and preforming to an approximate final shape, were unsuccessful and were not pursued.) The laminate passes through the relatively small active forming region, where it is heated. Within this region it is then bent such that each location receives its correct final radius of curvature.

A number of obvious requirements are associated with this bending process and are listed in Figure 14.1. The curvature of the workpiece can only be controlled where it is contact with

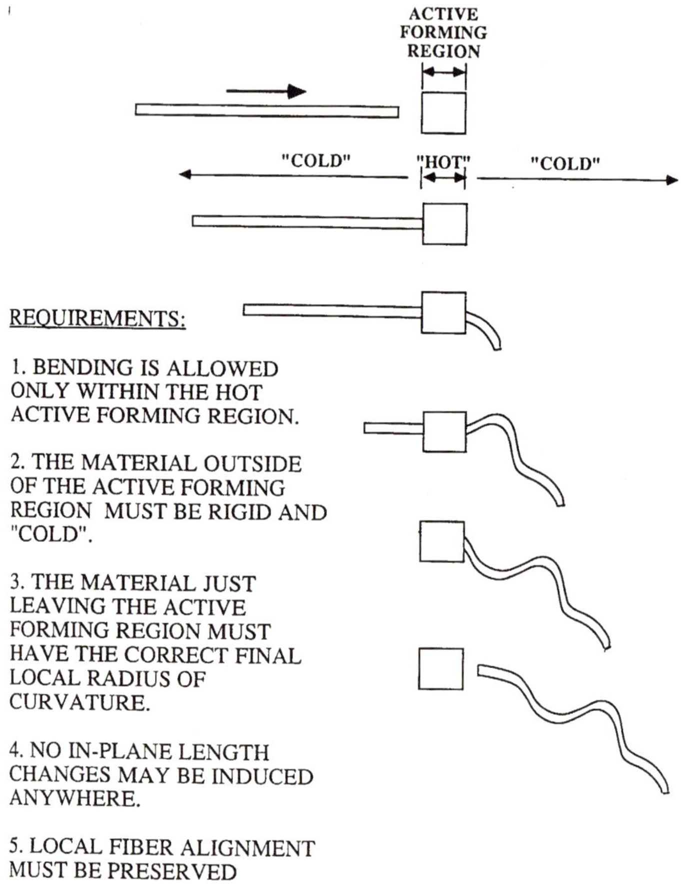

Figure 14.1 General concept of dieless forming. The final shape of the component is *generated* by passing an initially flat workpiece through a compact, heated, active forming zone, bending the workpiece locally only within the active forming zone. The requirements that must be met if such operations are performed on continouous-fiber composites are listed

the apparatus; therefore, it must remain cold and rigid outside the active forming region (Requirements 1 and 2). The only opportunity to give each location the correct local curvature occurs while that location lies within the active forming region (Requirement 3).

The key requirement is the fourth (avoiding any length changes in the plies) which arises from the continuous, inextensible, and incompressible fibers. For singly curved components, this is met by restricting the bending operations to special bending sequences that are termed *kinematically admissible*. This is illustrated in Figure 14.2. While a free edge is hot and inside the active forming region, the total "net bending" contained within a component can be introduced because the plies are free slide over each other at the edge. (This effect is easy to visualize by bending a thick pad of paper near its edge.) Once introduced, this net bending can

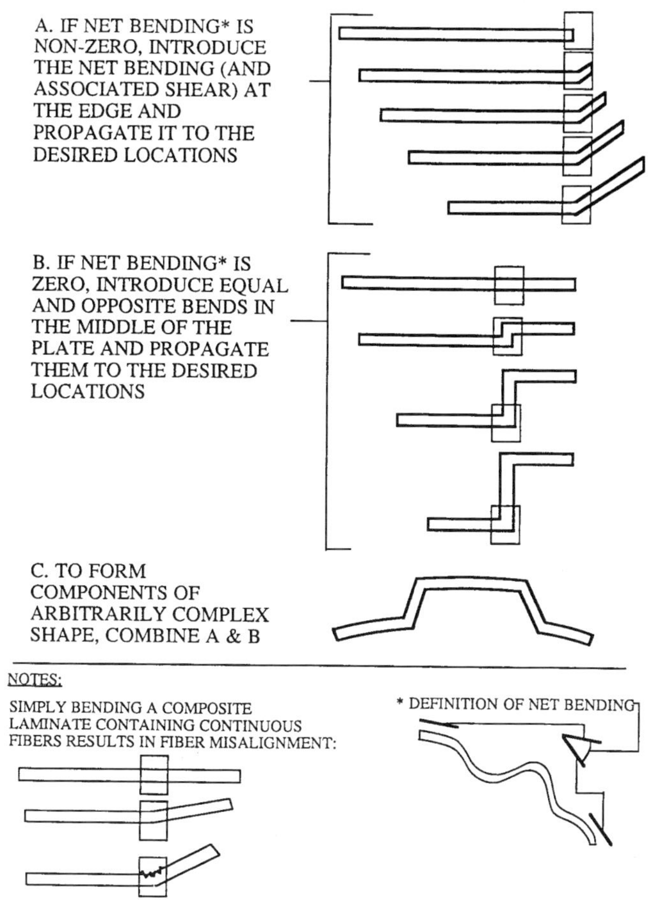

Figure 14.2 Concept of kinematically admissible bending. The "net bending" is introduced while the edge of the plate is within the active forming zone and propagated across the plate while the plate moves through the active forming zone. The local curvature desired at each location is deposited at that location as it leaves the active forming region. This permits formation of arbitrary distributions of curvature by local bending without inducing any length changes in the plies

be propagated within the workpiece without requiring in-plane length changes in the plies. (Try it with a pad of paper.) Bend propagation without affecting the material outside of the active forming region is possible because the "shortness" on the inside of the bend that is *exiting* from the forming region is exactly replaced by the "shortness" on the inside of the bend that is *entering* the forming region. Equal and opposite bends can be initiated in the interior of the laminate without requiring any interply sliding outside of the bend initiation region because the relative slip within one bend cancels that of the other bend. Once introduced, portions of the net bending can be left behind at various locations within the workpiece, thus imparting an arbitrary distribution of curvature.

14.3 Simulations, Shape Categories, and Forming Machine Concepts

As a first step in the work, a computer program (utilizing graphic output) was written to simulate the forming of singly curved components [8]. It allowed the user to input an arbitrary final shape. From that shape the program calculated the specific kinematically admissible bending sequence and simulated the forming operations. These operations were done with three overlapping groups of rollers containing three pairs of rollers each (seven pairs total) (see Fig. 14.3). The first roller group accomodated the curvature of the incoming laminate (generally flat); the third roller group imparted the final curvature to the exiting portion of the laminate. The middle roller group took on a curvature that temporarily stored those portions of the net bending that had not yet been deposited elsewhere. Figure 14.3 illustrates a snapshot within a typical forming "operation," and gives the governing equations. The equations calculate, at each instant in time, what the total net bending contained in the part is (angle A), how much of this bending has already been deposited (angle B), how much bending is contained within the outgoing roller group (angle C), how much net bending must be held within the middle roller group (angle D, equaling A–B–C), and, therefore, what the curvature of this middle roller group must be. Roller positions are controlled so that each roller group takes on the correct curvature and they also share common tangents.

In the absence of long-range fiber-to-fiber sliding, only singly curved components can be formed by the preceding types of operations. Discussions with potential users (especially the aerospace community) led to a clear indication that tapered shapes that were long in their straight direction and of variable cross-section (Fig. 14.4, item 3) were of considerable interest, commonly occurring as stiffeners and skins in tapered structures such as wings and control surfaces. Such shapes are expensive to manufacture because they are not amenable to roll forming or any other "die-less" process.

As a first step in exploring the forming of such tapered shapes, the author wrote a second graphic simulation program [8]. To minimize the size of the forming machine, the component moved back and forth along its long direction; in between longitudinal passes the component was indexed through the apparatus in the transverse direction. Within the machine where the material was heated, two arrays of seven pairs of rollers each imposed the same kinematically

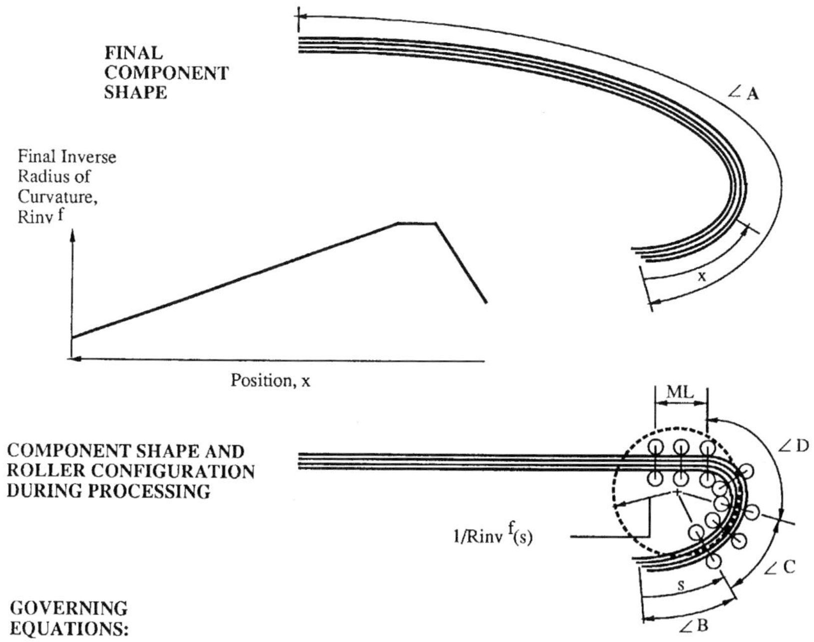

GOVERNING EQUATIONS:

$$\text{Description of Final Shape: } Rinv^{f}(x) = \frac{1}{R^{f}(x)} \quad (1)$$

$$\text{Total Bend in Part} = \int_{x=0}^{L} Rinv^{f}(x)dx = \angle A \quad (2)$$

$$\text{Position of Roller Exit Along Part} = s$$

$$\text{Bend Generated Thus Far} = \angle B = \int_{x=0}^{s} Rinv^{f}(x)dx \quad (3)$$

$$\text{Bend Remaining in Machine} = \angle A - \angle B \quad (4)$$

$$\text{Radius of Curvature of Outgoing Roller Group} = \frac{1}{Rinv^{f}(s)} \quad (5)$$

$$\text{Bend in Outgoing Roller Group} = \angle C = 2^{*}\sin^{-1}\left(\frac{ML^{*}Rinv^{f}(s)}{2}\right) \quad (6)$$

$$\text{Bend in Middle Roller Group} = \angle D = \angle A - \angle B - \angle C \quad (7)$$

$$\text{Radius of Curvature of Middle Roller Group} = \frac{ML/2}{\sin(\angle D/2)} \quad (8)$$

Figure 14.3 Equations governing the kinematically admissible bending algorithm

admissible bending algorithm in the *transverse* direction that was illustrated in Figure 14.3 for bending in the *longitudinal* direction. The component shape evolution associated with this approach is illustrated in Figure 14.5. By varying the positioning of the forming rollers as the workpiece moved longitudinally, variable cross-sections could be formed. Because of the finite transverse indexing motions from pass to pass, the program showed clearly that a temporary shape discontinuity existed between the formed and not-yet-formed regions within

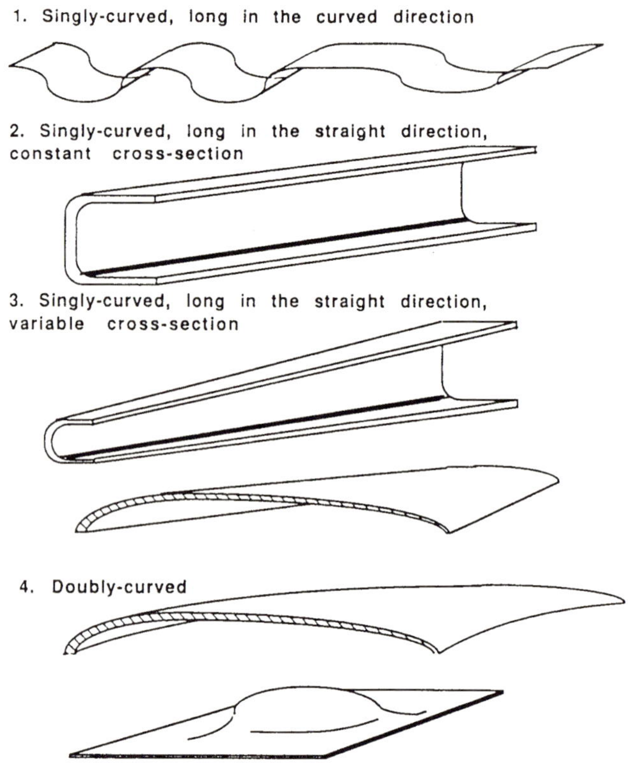

Figure 14.4 Shape categories of components that might be formed from thermoplastic matrix composite materials

any one pass; however, calculations also showed that the magnitude of strain within this region decreased strongly as the transverse feed per pass decreased. This provided a basis for the possibility that by sufficiently small feed per pass, the shape discontinuity could be accomodated by elastic strains, and errors in final formed shape associated with the temporary shape discontinuity could be avoided [8].

Before such an apparatus could be constructed, Ramani discovered that it was not necessary to use seven pairs of rollers in each array. In fact with only four rollers total, arbitrary distributions of single curvature could be imparted to a workpiece. Figure 14.6 shows an example [10]. The operation rests on the assumption that the material does not change length within any one ply; rather the plies can slide over each other in the region of the rollers where the material is heated. The inextensibility and incompressibility of the continuous fibers were essentially converted into an asset instead of a problem. The fact in Figure 14.6 that rollers overlap each other should not be troubling because the rollers were envisioned as part of an apparatus in which the workpiece moved through them longitudinally, thereby enabling the rollers to be offset from one another as shown in Figure 14.7. This figure also illustrates that the rollers were "cluster rollers," which permit both longitudinal and transverse rolling contact with the workpiece.

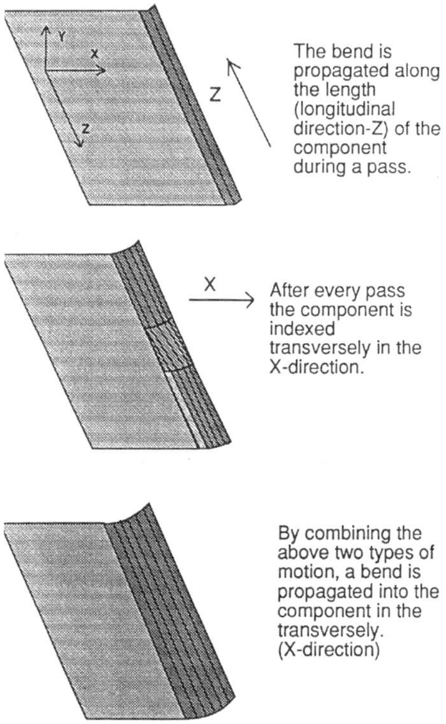

Figure 14.5 Indexing the component in the transverse direction and propagating the bend along the component longitudinally

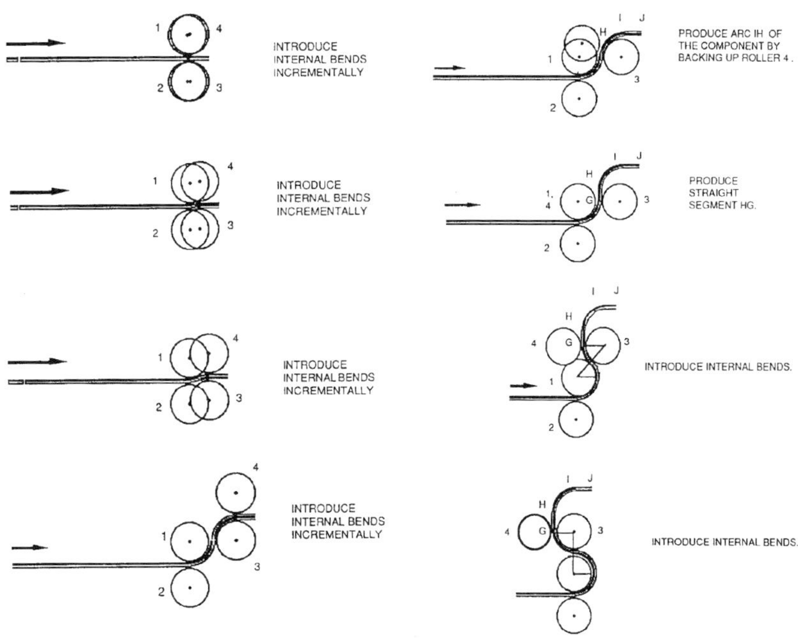

Figure 14.6 Illustration of the use of four rollers to form a component with arbitrary curvature distribution. The process takes advantage of the inextensible nature of the plies, which are only required to slide over each other in the roller region

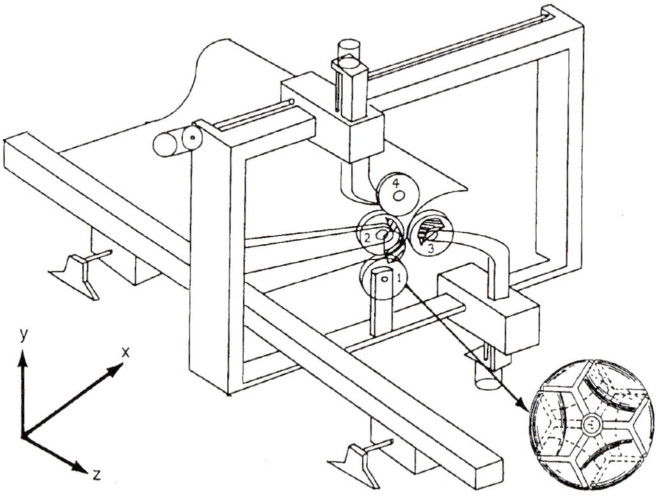

Figure 14.7 Sketch of the design concept for the full dieless forming machine capable of forming arbitrary cross-section tapered components. It utilizes four cluster rollers mounted on X–Y arms, induction heating coils mounted on the arms, and an X–Z translating table to move the workpiece longitudinally during each pass and transversely between passes. Rotation of the cluster rollers is designed to keep the active forming region of the workpiece under tension

14.4 Near-Term Demonstration Machine

The machine illustrated in Figure 14.7 would be fairly complicated. It was judged advisable to explore kinematically admissible bending of thermoplastic laminates first using a simpler machine. To do so, Gur designed the "near-term demonstration machine" [8] illustrated schematically in Figure 14.8a and by photos in Figures 14.8b and c. As Figure 14.8a shows, the initially flat laminate was clamped to an X–Z motion table (computer controlled). The component moved back and forth in the Z direction. In between passes it was indexed to the right in the X direction. The indexing motion forced the right edge of the workpiece to travel between two contoured rollers. The "backing" roller on the bottom had a concave-outward edge, while the roller on the top had a convex-outward edge, and is actually a cluster roller pictured in Figure 14.8b. The general motion to the right combined with the continual back and forth longitudinal motion forced the entire right edge of the laminate to be bent upward by 90 degrees, and propagated this bend backward through the laminate, as in Figure 14.2a.

For thermoplastic laminates the material obviously needs to be heated before entering the roller region. Induction (450 kHz.) was selected as the means for doing this. The basis for the selection was its noncontact nature and its ability to heat through the thickness quickly. Use of induction was also predicated on the fact that most components of interest utilized graphite fibers, which had the required electrical conductivity. A great deal was learned about the

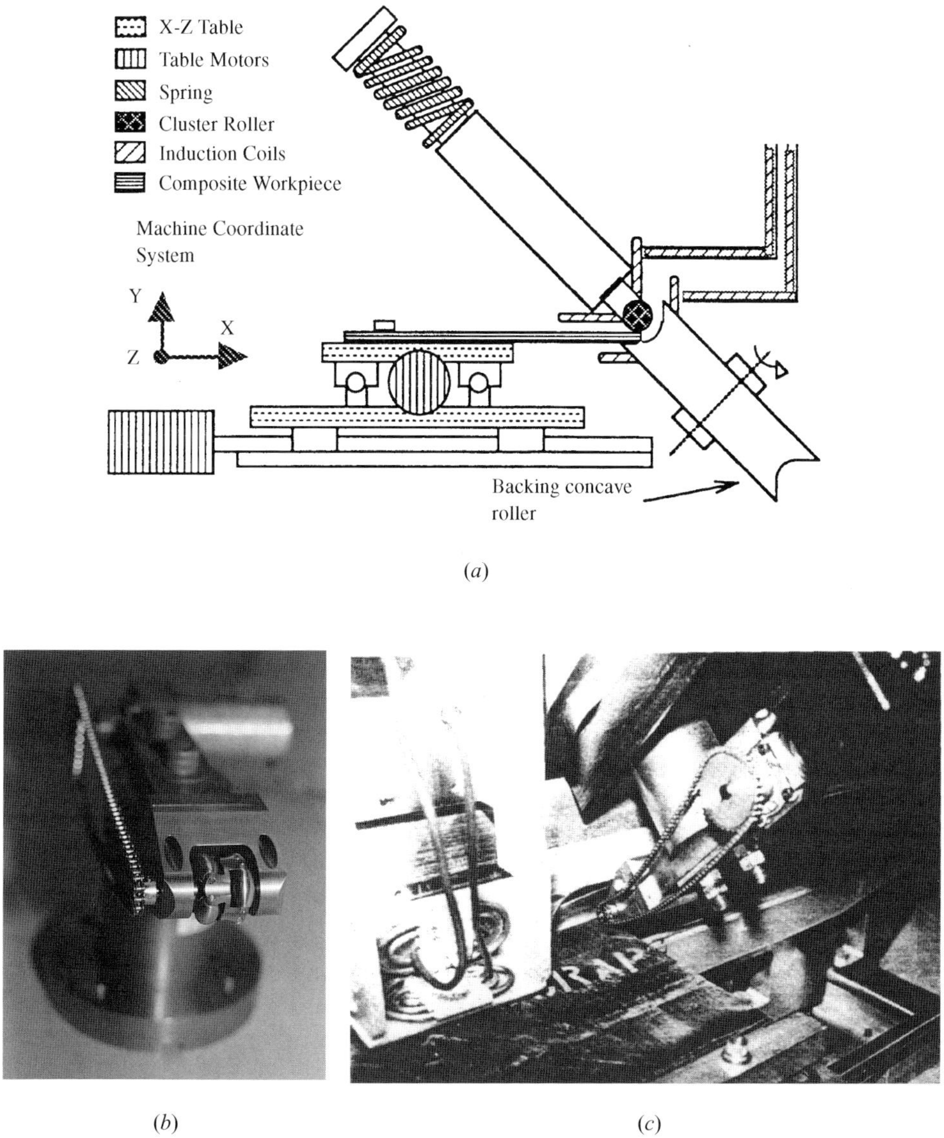

Figure 14.8 (a) Schematic illustration of the near-term demonstration machine. The initially flat composite laminate is moved along the Z axis, and is indexed in between passes in the X-direction, pushing it between the convex cluster roller and the concave backing roller, thus forming a 90-degree bend. (b) Photograph of the cluster roller. (c) Photograph of the forming region of the machine showing the induction coil, the cluster roller, a portion of the backing roller, and a portion of a trial workpiece

Figure 14.9 Photograph of a Gr/PEEK laminate in which a bend was introduced and propagated into the workpiece using the near-term demonstration machine with a variable velocity profile. Fiber alignment, surface quality, and macroscopic shape are good

nature of induction heating in thermoplastic laminates [11], and from this knowledge "Dual-Dee" coils were designed (seen in Figure 14.8c, which also shows more of the machine) that were quite effective for the purpose, producing a narrow stripe of heat in the laminate.

After the customary period of debugging, forming trials with the machine showed initial success [12]. Figure 14.9 shows a Gr/PEEK (polyether ether ketone) laminate16 plies thick, quasi-isotropic layup into which a bend was successfully introduced without damage. Key ingredients in the success included the motor-driven cluster roller that "pulled" the workpiece through the rollers instead of "pushing" it through using the table's X-axis, and a variable longitudinal velocity profile (slower when the ends of the workpiece were within the induction coil) that allowed the ends to heat as much as the central regions.

14.5 Overcurvature—Observations and Model

As such bends were propagated further through the laminate, however, they almost invariably exhibited "overcurvature," as shown in Figure 14.10 [13]. Instead of remaining straight (in accordance with the principle of kinematically admissible bending) the portion of the workpiece that had already emerged from the active forming region accumulated additional bending. This overcurvature destroyed any accuracy in formed shape. It was important to understand whether the overcurvature was solvable by minor improvements in the apparatus, or whether it was inherent to the version of the process being explored.

To start this determination, Ramani and Vinci performed a series of forming experiments using specially made laminates [14]. One laminate had a preponderance of fibers along the longitudinal direction; the other along the transverse direction. The first (Fig. 14.11a) showed

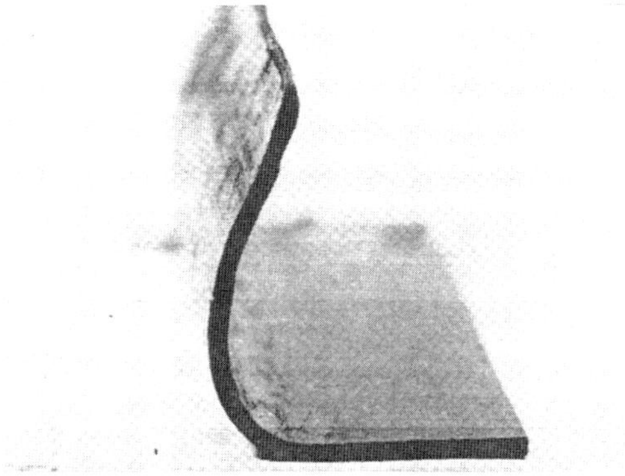

Figure 14.10 Photograph of a quasi-isotropic Gr/PEEK laminate in which a bend was propagated further into the workpiece, showing overcurvature

severe overcurvature; the second showed essentially none. This indicated that there were definite forces at work tending to produce the overcurvature, and that the material's resistance to bending in the transverse direction determined the outcome.

The forces tending to produce the overcurvature turned out to be the temporary shape discontinuity first noted in connection with Figure 14.5. Figure 14.12 illustrates the phenomenon in more detail. Ramani formulated an energy-minimization model that successfully explained the overcurvature [10]. In his model [13], elastic strain energy is imparted to the upper portion of the specimen if it is forced to take on the shape shown in Figure 14.12. Some of the elastic strain energy can be relieved if the upper portion of the workpiece entering the forming zone moves to the left by "plastic hinge" action in the regions

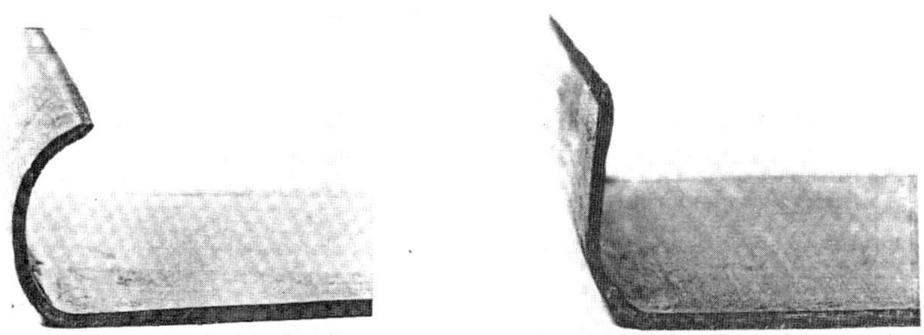

$[(0)_6/(90)_2]_s$ $[(0/90_2)_2(0/90)]_s$

(*a*) (*b*)

Figure 14.11 (a) Photograph of a formed Gr/PEEK laminate containing mostly longitudinal plies. Overcurvature is strong. (b) Photograph of a formed Gr/PEEK laminate containing mostly transverse plies. Overcurvature is absent

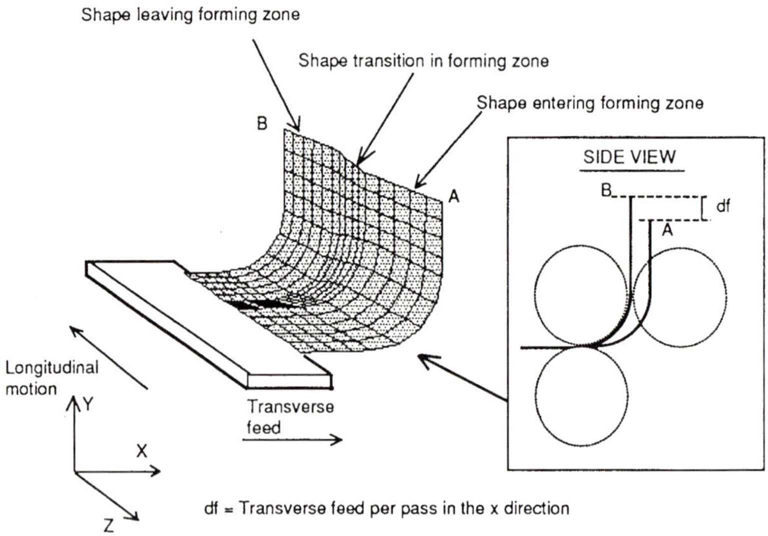

Figure 14.12 Schematic illustrating the discontinuity in formed shape along the length of the component during any one pass due to the finite transverse feed per pass

underneath it (Fig. 14.13). This plastic bending is in excess of that imparted by the forming rollers, and creates the overcurvature. The model successfully simulated the progression of overcurvature in individual workpieces as well as the effects of forming velocity. It was convincing in demonstrating that overcurvature had its basis in the temporary nondevelopable shape that the laminate was forced to take on, and was inherent to the version of dieless forming being explored.

14.6 Continuous Dieless Forming

Given the preceding findings, a version of die-less forming was needed which did not force any temporary nondevelopable shapes (regardless of how small) upon the component. Such a version was explored and demonstrated by Yencho [15]. To emphasize the absence of discontinuities, he termed it *continuous die-less forming* [16].

Yencho essentially returned to the forming sequence first envisioned in Figures 14.1–14.3, in which the component is bent in the longitudinal direction while it travels through the forming apparatus in that same direction. Because the component is bent simultaneously at all locations across its width, no discontinuities are introduced. He designed a new computer-controlled apparatus sketched in Figure 14.14. He simplified the forming rollers even further, reducing them to two (as pictured in Fig. 14.15) and using controlled horizontal and vertical translation of the rightmost roller to create an arbitrary distribution of (single) curvature in the workpiece. Like Ramani, Yencho relied on the inextensible nature of the individual plies

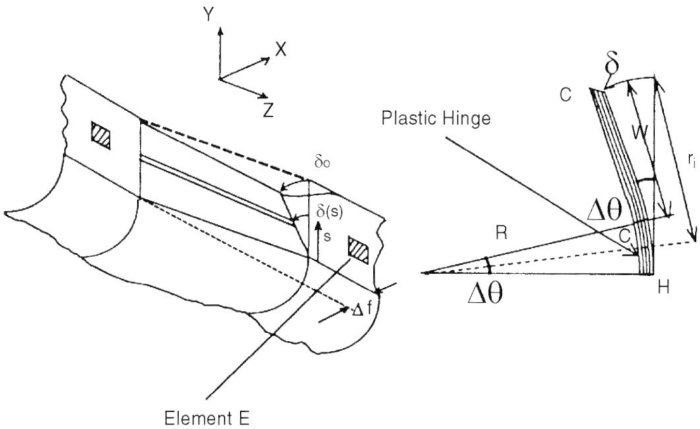

Figure 14.13 Geometry of the plastic hinge that forms due to the shape discontinuity and produces overcurvature

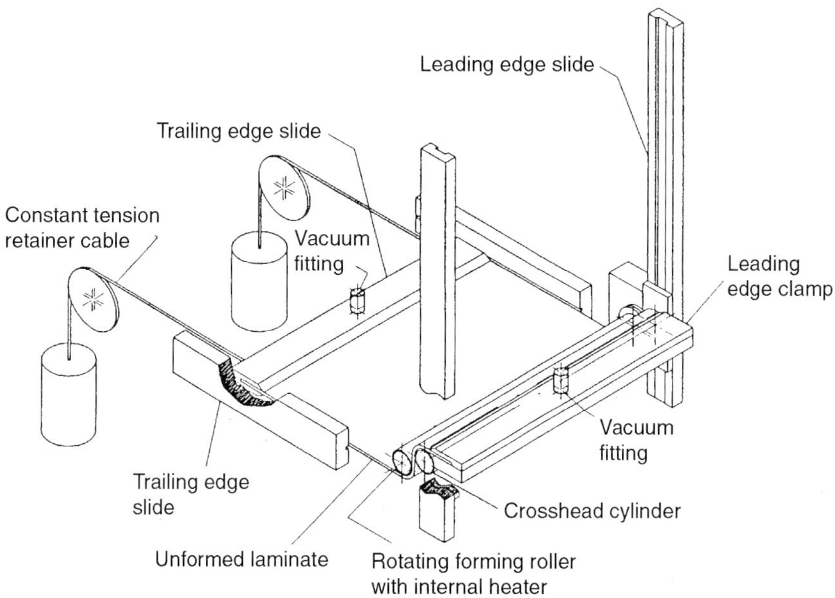

Figure 14.14 Schematic of the "continuous dieless forming" machine showing the vertically moving crosshead cylinder, the fixed forming roller, and the dead weights used for workpiece back tensioning

within the laminate for shape accuracy (i.e., they were only allowed to slide over each other in the portion of the laminate in contact with the forming roller at the left of Fig. 14.15). Yencho predicated the success of the forming operation on having the laminate *not* bend at all after it left the heated forming roller at the left. Two internal, opposing, concentrated 90-degree bends are essentially introduced into the laminate at the two rollers, and one of them is spread

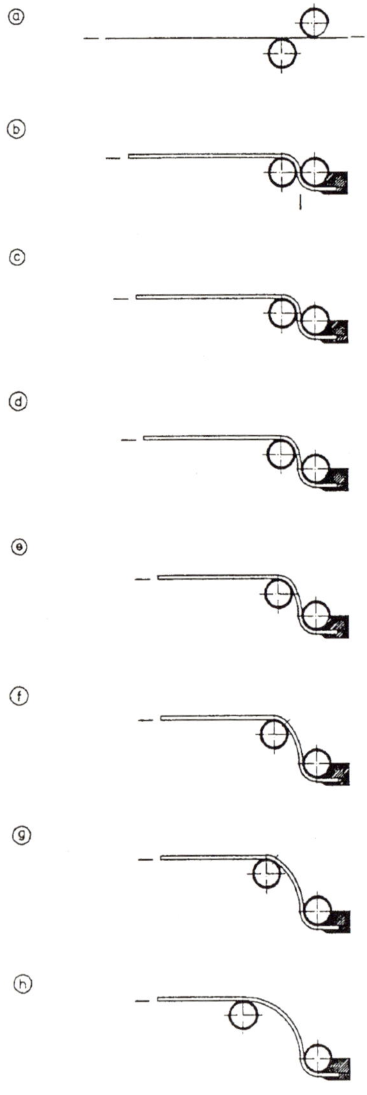

Figure 14.15 General concept of the process of forming a distributed bend using the continuous die-less forming machine in which the cross-head cylinder moves both vertically and horizontally

out into a distributed 90-degree bend by the motion of the machine. Because this is a kinematically admissible forming operation, ply length and ply integrity are maintained.

Because bending is concurrent at all regions across the width, Yencho first had to select a suitable local heating method. Induction was attractive for the reasons given previously. In this version, however, it would be necessary to heat uniformly across the entire width, including the edge regions. Despite substantial effort invested and understanding developed [15], this proved to be impractical, even with the use of an extended-length Dual-Dee coil with flux concentrators (Fig. 14.16). A combination of conduction heating through the forming roller and radiant preheating was employed instead.

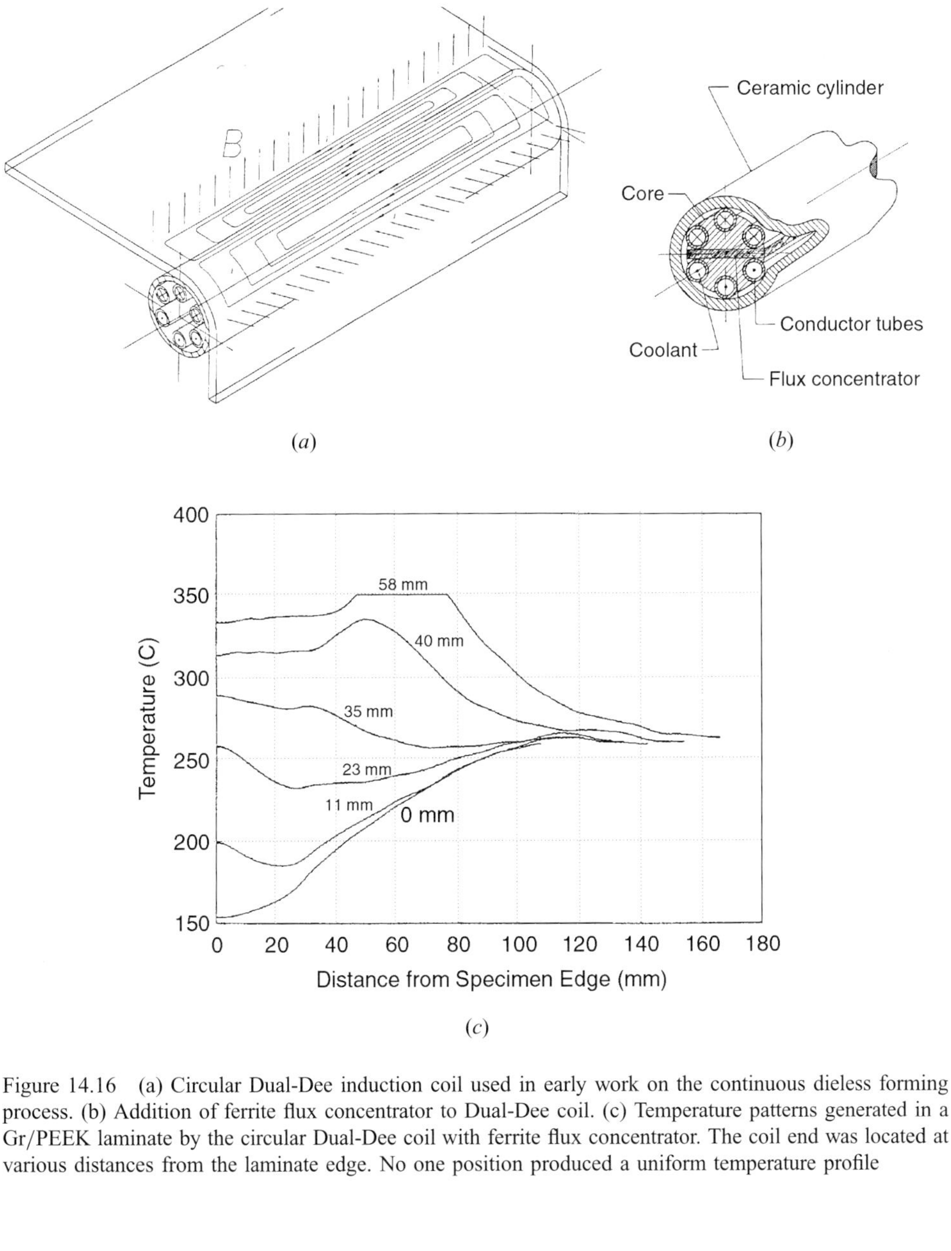

Figure 14.16 (a) Circular Dual-Dee induction coil used in early work on the continuous dieless forming process. (b) Addition of ferrite flux concentrator to Dual-Dee coil. (c) Temperature patterns generated in a Gr/PEEK laminate by the circular Dual-Dee coil with ferrite flux concentrator. The coil end was located at various distances from the laminate edge. No one position produced a uniform temperature profile

Yencho then constructed the initial version of the machine and explored the parameters necessary to successfully form flaw-free parts. A key finding concerned the effects of trapped moisture. Despite the generally hydrophobic nature of PEEK, it was found that small amounts of moisture absorbed by the Gr/PEEK laminates subsequent to their manufacture converted into steam during the forming operation and caused extensive delamination. This was in

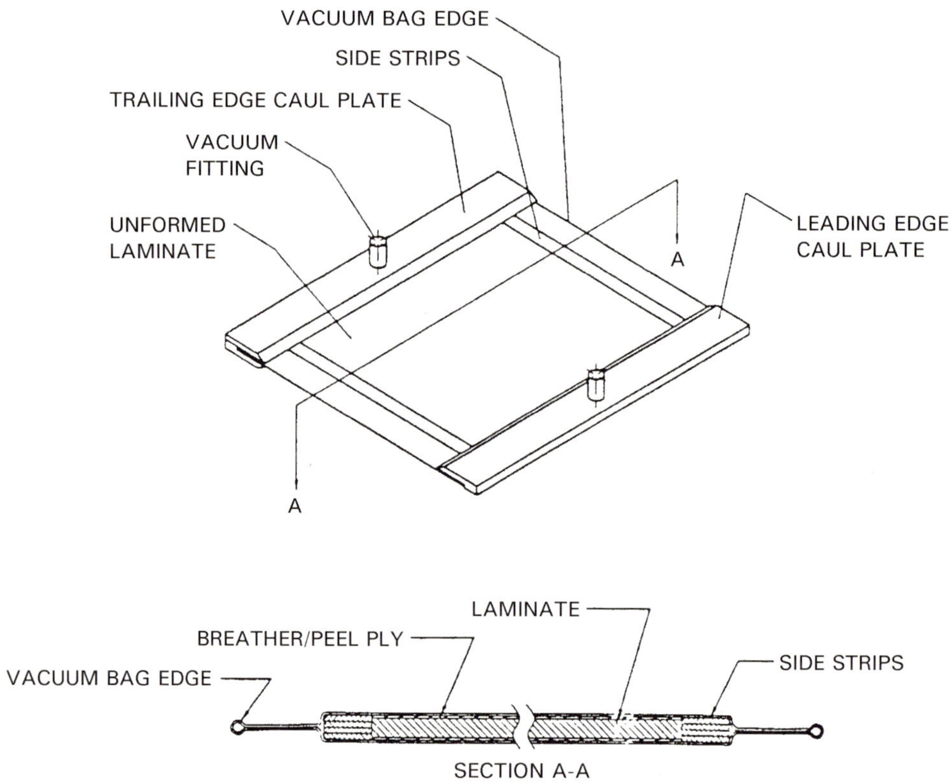

Figure 14.17 Vacuum forming bag for maintaining pressure on the laminate during the forming operation

addition to the natural tendency of thermoplastic composite laminates to delaminate when heated above the melting point because of the stored strain energy in the fibers, which are slightly bent during laminate consolidation [17]. Yencho designed and proved out a formable vacuum bag [15] (Fig. 14.17) that enveloped the part during the entire forming operation, to maintain consolidation. He found that with prior baking of the laminates for about 300 h to drive out the moisture, vacuum pressure was sufficient to maintain consolidation in the laminate and keep the resulting interlaminar shear strength at levels equal to the manufacturer's data sheet [15] (Fig. 14.18).

Yencho studied the process parameters necessary to obtain good quality in detail. In his test specimens, he formed two equal and opposite bends in Gr/PEEK laminates, laying the basis for Gertner's subsequent forming of arbitrary shapes. Yencho used ultrasonic examination as an accurate measure of quality for the detailed optimization of the forming process. He found [16] that by use of a moderate back tension, low forming velocity (0.13 mm/s), and low forming temperature (420°C), together with the prior baking for dry-out, he was able to maintain "aircraft quality" (less than 6-dB loss) in the laminates.

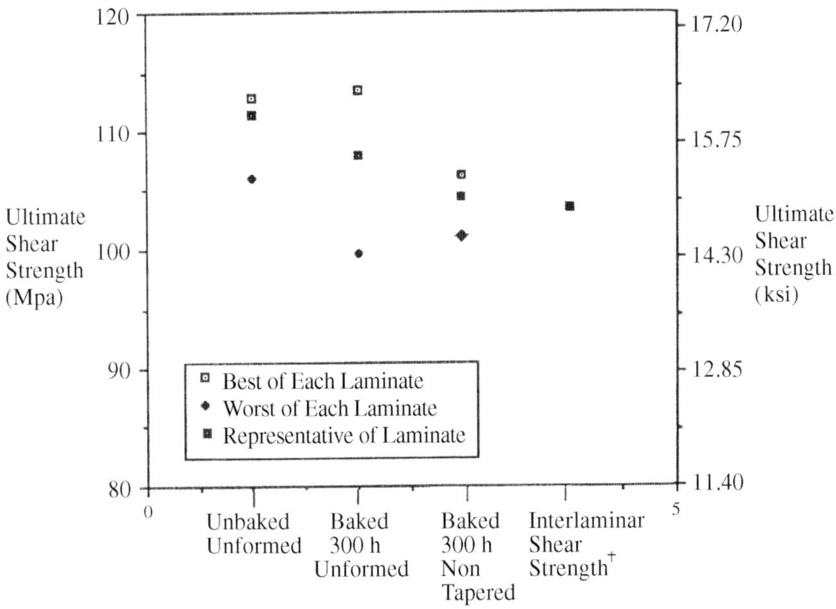

Figure 14.18 Interlaminar shear strength results on predried, formed laminates (third data set from left) compared with unformed laminates (two leftmost sets) and manufacturer's data (right)

14.7 Forming Arbitrary Curved Shapes Without Dies

Gertner extended Yencho's machine to accomplish the original objective of dieless forming, namely the generation of arbitrary curved shapes (different from that of the forming rollers) with good dimensional accuracy [18]. Whereas Yencho's experiments required only the vertical axis of motion of the cross-head (Fig. 14.14), Gertner added the horizontal axis of motion so that the sequences pictured in Figure 14.15 could be carried out. The resulting machine is shown in Figure 14.19. An infrared temperature sensor with rotating scanning mirror was used for laminate temperature measurements (with vacuum bag in place). Gertner [18] worked out the algorithm for converting the desired final shape of the workpiece into the motion of the forming roller crosshead (Fig. 14.20 shows the sequence).

As might be expected, Yencho's assumption that the process could be conducted without having the laminate bend once it had left the forming roller (despite the bending moments introduced) proved most challenging to meet, but ultimately success was attained. The key was maintaining a significant temperature drop from the central portions of the active forming region to the tangent line where the workpiece leaves the forming roller [18] see Figure 14.21. A drop of about 28°C proved sufficient to allow interply sliding in the active forming region while preventing plastic bending at and beyond the tangent line. Figure 14.22 displays this

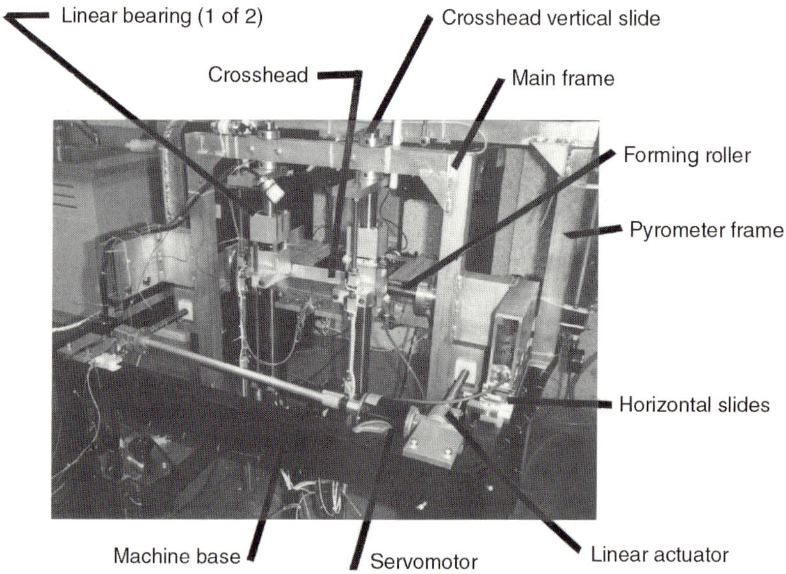

(*a*)

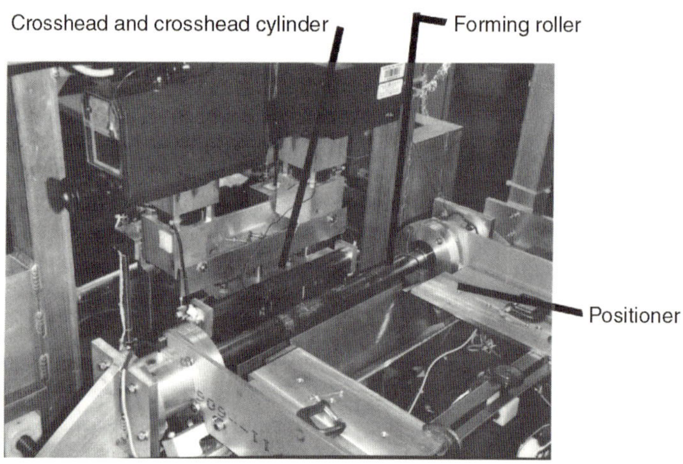

(*b*)

Figure 14.19 (a) Overall view of the continuous dieless forming machine. (b) Close-up view of the continuous dieless forming machine, from the side opposite to that shown in (a)

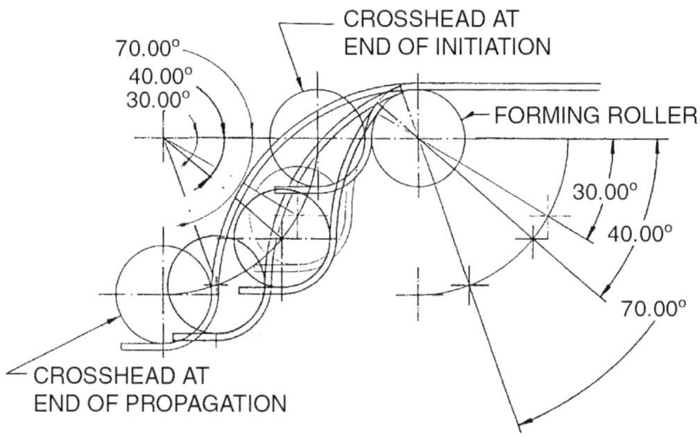

Figure 14.20 Intermediate cross-head locations and corresponding component shapes during forming of a component that has a constant bend radius larger than the forming rollers

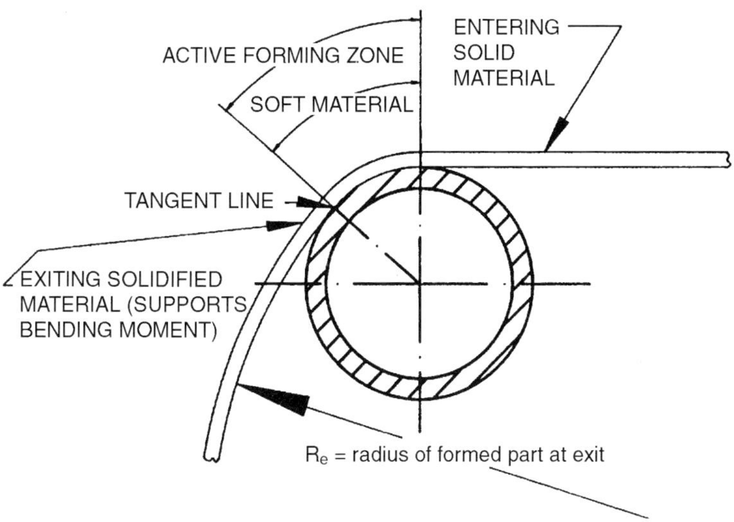

Figure 14.21 Laminate characteristics in the vicinity of the forming roller. The material must be rigid as it leaves the forming roller at the tangent line

requirement alongside the other requirements (e.g., the absolute temperatures required with Gr/PEEK laminates) as a "formability map" for the continuous dieless forming process.

Gertner proved out the process using several shapes of successively increasing difficulty. He first formed constant-radius parts (larger than the forming rollers), then a variable-curvature part. Figure 14.23 shows their shape accuracy, in each case comparing the ideal shape (solid line) against that corresponding actual shape (broken line). The use of low forming velocity and no back tension produced excellent shape accuracy, to within a maximum deviation of ±0.4 mm (0.016 in.). Some tapered parts were also produced.

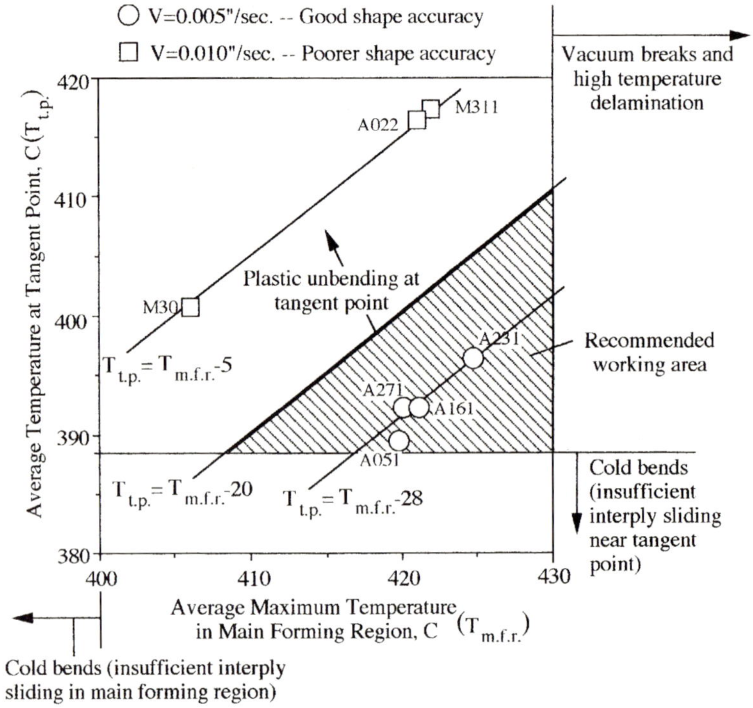

Figure 14.22 "Formability map" that summarizes the essential results of continuous dieless forming. Good shape accuracy and quality are produced in the cross-hatched region where, due to low forming velocity, the tangent point temperature ($T_{t.p.}$) is about 28°C below the main forming region temperature ($T_{m.f.r.}$). This permits adequate interply sliding within the main forming region while avoiding plastic unbending as the material leaves the forming region. Both temperatures are also low enough to avoid vacuum seal leaks and high-temperature delamination

14.8 Summary and Conclusions

Dieless forming is directed at producing large, singly curved continuous-fiber composite components, *generating* the shape using an array of small rollers. An initial version focused on the forming of tapered components, which are straight in their long direction and of variable cross-section along their length. To minimize the size of the forming machine, it imparted curvature in a direction perpendicular to the direction of motion through the machine. The process was simulated graphically on the computer, then explored experimentally with a "near-term demonstration machine." Successful ingredients included:

- Use of induction heating for rapid through-thickness temperature change, with variable velocity profile for temperature control
- Ability to introduce bends at the free edge and propagate them into the laminate

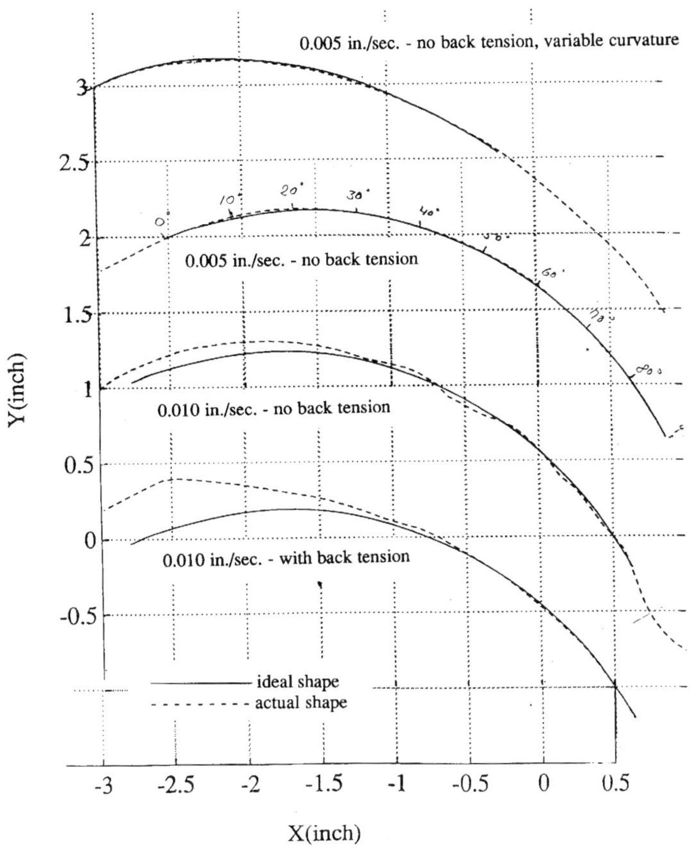

Figure 14.23 Shape accuracy results for continous dieless forming, variable curvature workpiece at the top, and constant-curvature workpieces under three different forming conditions at the bottom. Shape accuracy to about 0.016-in.) is obtained in the top two results

- Ability to control wrinkling on the inside of the bend through the use of a motor-driven "cluster roller"

Complete success, however, was not attained in the initial version. The limiting factor was the development of "overcurvature" in the formed components. This was shown to be due to an "incompatibility strain" which arises between the formed and not-yet-formed regions on any one pass through the forming machine. Using an energy minimization model, the overcurvature was shown to be inherent to the process.

To avoid incompatibility strains, a subsequent version of dieless forming imparted curvature in the same direction as the motion through the forming machine. This version ("continuous dieless forming") successfully formed laminates to high quality, as measured by ultrasonic examination (6-dB loss) and by shear strength tests. Forming within a vacuum bag and oven predrying to remove moisture were important elements.

Continuous dieless forming was then used to form both uniform-curvature and non-uniform curvature components. Good shape accuracy (i.e., maximum deviation of ±0.4 mm or 0.016 in.) was attained. A key ingredient proved to be manipulation of the temperature gradient across the forming roller.

Acknowledgments

This chapter is a summary of work conducted by members of the author's polymer composites research group at Stanford, including Karthik Ramani, Steve Yencho, Micha Gur, Yosef Gertner, Adi Peled, Margaret Talbott, Rick Vinci, Calvin Chang, Eric Menzel, and Alex Payne. Collaboration with Prof. Mark Cutkosky was very helpful. Each individual made many contributions, most of which have been identified in the text, but all are thanked here again.

Initial work on dieless forming was sponsored by the Stanford Institute for Manufacturing and Automation (SIMA). Refinement of the concept by computer modeling and initial roll-bending experiments was sponsored by Lockheed Aeronautical Systems Company under the U.S. Air Force contract, "Manufacturing Science of Complex Shape Thermoplastics," contract #F33615-86-C-5008. Design, construction, and experiments on the two dieless forming machines, including supporting work on induction heating and process modeling, were sponsored by the National Science Foundation Manufacturing Machines and Equipment program under grant #DDM-8608860. The author's continued involvement, after leaving Stanford University, in helping his Stanford colleagues complete the dieless forming work was supported by the Lockheed Missiles and Space Company Independent Research and Development Program.

References

1. "Advanced Airframe Structural Materials: A Primer and Cost Estimating Methodology (1991) RAND Corporation Report R-4016-AF, Santa Monica
2. Cogswell, F.N. *Thermoplastic Aromatic Polymer Composites* (1992) Butterworth–Heinemann
3. *Proceedings of the Fifth Industray/Government Review of Thermoplastic Matrix Composites* (1988) Air Force Wright Aeronautical Laboratories, 8–11 Feb. 1988, San Diego
4. "Manufacturing Science of Complex Shape Thermoplastics (1990) report #LR31816, Lockheed Aeronautical Systems Co., US Air Force Contract #F33615-86-C-5008, Jan. 1990
5. Rolston, A. In *Processing and Fabrication Technology*, Vol. 3, Delaware Composites Design Encyclopedia, Technomic
6. Cattanach, J.B., Harvey, R.C. "Roll Forming of APC PEEK/Carbon Fibre Composite" (1984) Application Note by Imperial Chemical Industries PLC, Welwyn Garden City (England), June 14, 1984
7. Miller, A.K. Process and Compact Apparatus for Forming Fiber Composite Materials, U.S. Patent #4,777,005 (1988)
8. Miller, A.K., Gur, M., Peled, A., Payne, A,, Menzel, E., "Die-Less Forming of Thermosplastic Matrix Continuous-Fiber Composite Materials," *J. Comp. Mat.* (1990) 24, p. 346–381
9. Ramani, K., Miller, A.K., Cutkosky, M.R. *J. Thermoplastic Comp.* (1992) 5, p. 184–201

10. Ramani, K. Ph.D. dissertation (1992) Stanford University, Palo Alto
11. Miller, A.K., Chang, C., Payne, A., Gur, M., Menzel, E., Peled. A. *SAMPE J.* (1990) 26, 37–53
12. Miller, A.K., Gur, M., Peled, A. *Materials and Manufacturing Processes*, (1990) 5, p. 273–300
13. Ramani, K., Miller, A.K., Cutkosky, M.R. *J. Thermoplastic Com.* (1992) 5, p. 202–227
14. Vinci, R.P., Ramani, K., Miller, A.K., Cutkosky, M.R. *Proc. Thirty-Sixth International SAMPE Symposium* (1991) San Diego, April 15–18
15. Yencho, S. *The Continuous Die-Less Forming of Thermoplastic-Matrtix Fiber Composites* (1993) Ph.D. dissertation, Stanford University
16. Yencho, S.A., Miller, A.K. *ASM Conference on Processing, Fabrication, and Applications of Polymeric-Matrix Composites* (1993) Long Beach, Aug. 11
17. Gutowsky, T.G. *SAMPE Q.* (1985) p. 58–64
18. Gertner, J., Miller, A.K. *J Thermoplastic Comp. Mat.* (1995) 9, p. 151–182

15 Intelligent Processing Tools for Composite Processing

F. Abrams

A series of computer aids for the control of polymer processing will be discussed, emphasizing the practical aspects of choosing the proper tool for the application. Where possible, examples of production, or at least developmental experience, will be included. Because there are a variety of processes used in polymer processing, the advantages and disadvantages of each method will be discussed in the context of the types of processes. The chapter is divided into a section on development of process cycles or plans and a section on in-process control. The tools to be discussed include design of experiments, expert systems, models, neural networks, and a variety of combinations of these techniques. The processes to be discussed include injection molding, resin transfer molding, autoclave curing, and prepreg manufacturing. The relative cost and difficulty of developing tools for these applications will be discussed where data is available.

15.1 Introduction

Earlier in this volume (Chap. 9), Joseph and Thomas discuss the theoretical aspects of intelligent control for composites manufacturing. No matter how good the intelligent processing tool, it is still the intelligent selection and use of these aids by human engineers that makes them valuable. None is a panacea, and most processes can benefit from combining methods, first to do an intelligent job of process planning and then to do intelligent process control. The purpose of this chapter is to provide the user with some practical information on computational aids (both methods and tools) that are currently available for the intelligent development and control of polymer matrix composites (PMC) processing. Where possible, examples of production experience are presented; unfortunately, many tools are new or the applications are proprietary, and experience is limited to experimental results. Where production experience with a tool is not available, evaluation is based on past experience with other applications and the types of problems to which the tool appears most suited.

The value of tools discussed in this chapter must be weighed against the costs of implementation. Advantages will depend on the material and labor costs associated with a given part and how much these costs can be reduced by decreased process time and/or decreased rejection rates. Improvements at the process-development stage may result in even greater savings, but they are more difficult to quantify. Implementation costs will include capital equipment, training, development, and recurring costs like disposable sensors and software maintenance. Some methods may not be acceptable in an industrial setting due to complexity of operation or high failure rates. Practical use of intelligent process control requires proper understanding of the benefits and limitations of each method. Where possible, some information on costs and gains will be included in the upcoming discussions. Such costs are variable, however, depending on the relative maturity of the technology, the numbers of systems in use, and the individual demands of the user. It may likewise be shortsighted to base estimates of the benefits on current production requirements. For example, the United States at one time produced the majority of the graphite fibers used in advanced composites; however, foreign fibers have surpassed them in quality, largely due to modernization in process and quality control. For one aerospace application, at least, this variability, along with the general lack of quality, eliminated all domestic fibers from competition [1].

Most of the examples given will be taken from the autoclave curing of thermosets, reflecting the experience of the author; however the strengths and weaknesses of each will be discussed with the intention of guiding the reader in choosing the right tool, or set of tools, for development of any polymer processing.

15.2 The Batch Process Control Problem

The manufacture of polymeric materials, reinforced or unreinforced, is not a single process; rather, it is a sequence of processes. For example, the making of a thermoset composite in an autoclave involves the autoclave cure as well as the various processes required to make the fiber and matrix, the manufacture of prepreg from those ingredients, and the lay-up of the composite and additional support materials (e.g., tooling, bagging and bleeders). Most of the processes used in composite manufacturing are batch processes. Resins are manufactured and mixed in batches. Composites are cured and/or formed in batches. As has been noted by Ghosh [2], batch processes have a number of characteristics that are not encountered in traditional process control. Some of these characteristics are:

1. The process is not in, or intended to be in, equilibrium
2. There are multiple inputs and outputs
3. The state variables are dynamically changing
4. The relationships between inputs and outputs are nonlinear
5. There is an ordered sequence of process steps
6. There are control functions such as safety interlocks, batch management, and scheduling.

Ghosh [2] also lists the requirements for batch process controllers. Batch control methods must include the abilities to:

1. Follow a logical sequence of events
2. Manage multiple dynamic interactions between controls and quality
3. Detect and react to failure conditions
4. Provide event-based data recording and reporting
5. Handle safety interlocks and supervisory control of setpoints
6. Provide predictions of process trends and/or outcomes
7. React to prevent undesirable impacts on quality as well as to control current trends in processing variables

It is desirable in composites manufacturing to control material properties, primary variables, such as strength and chemical state, which define the success of the process and the quality of the product. Many of these primary variables cannot be measured directly during the process. In the manufacture of composites, therefore, as in many batch processes, control decisions have to be made in advance of direct measurement of product quality, creating a control problem where feedback is delayed until after irreversible change has occurred [3]. Some measurements, or combinations of measurements, may be used to infer trends toward the development of desirable, or undesirable, qualities or to recognize landmark inflections in process parameters. The models are incomplete, howevcer, so inferential control [4] has not been widely used in polymer processing.

Even those composite processes that are commonly referred to as "continuous," such as pultrusion, prepregging, and filament winding, are really batch processes in which the location of the material is changing along with material state. Although control may be based on regulating secondary properties (usually temperature) at a given location, the process is not in equilibrium. The state at one station is dependent both on the local conditions, and on the conditions at all previous locations and the rate of transport of a unit volume through those locations. These nonequilibrium "continuous" processes have the same control requirements as any batch process. For the purposes of following discussions, the term *process cycle* will refer to the regulation of secondary variables (e.g., temperature, pressure, vacuum) with time for a given unit volume of material, even though that unit volume may be moving as well.

The typical textbook on chemical engineering process control does not emphasize nonlinear, nonequilibrium control problems [5] although such problems are not unique to the composites industry. There are several methods of "linearizing" nonlinear control problems, but they have not been generally used in composite processing. Control of polymeric processes instead has traditionally involved regulation of secondary variables, such as processing temperatures and pressures, which can be treated as linear over the control range. In the autoclave, for example, a sequence of equilibrium temperature and pressure states called a cure cycle is developed, and the assumption is made that repeating this cure cycle will cause the material to follow the same path from starting materials to final product. "Feedback" is incorporated by executing a series of experimental cure cycles, evaluating the resulting materials, and choosing the cycle that produces the most desirable material. In order for this method to work, variables, such as quality of incoming materials, must be carefully controlled. Postprocess inspection is also necessary, especially in instances where there has been some accidental deviation from the planned process.

An alternative to regulatory control of a prescribed cure cycle is to adjust the cycle, during the cure, in response to the sensed and predicted effects on the primary material variables. These two methods are seldom used completely independently. Regulatory process cycles are generally are run with an operator in attendance who is allowed to modify the cure cycle, at least under emergency conditions. Control systems that develop process cycles in response to the trends of primary variable begin with a process strategy (i.e., essentially, a preplanned default cycle, together with a means of estimating the effects that changes in primary and/or manipulated variables with have on disturbances in that plan).

15.3 Tools for Planning Process Conditions

This section describes some of the tools available for intelligent development of process cycles, such as the time–temperature cycles used in curing composites. Current industrial practice is typically limited to the use of cure cycles. The cycles are based on a series of autoclave temperature and pressure states so that traditional linear, regulatory process control methods can be used. These recipes may not be the ideal method for process control of batch processes because they do not:

1. Manage multiple dynamic interactions between controls and quality (material state)
2. Detect and react to failure conditions
3. Provide event-based data recording and reporting
4. Provide predictions of process trends and/or outcomes
5. React to prevent undesirable impacts on quality as well as to control current trends

Nevertheless, regulatory cure cycles can be used if both materials and process recipes are forgiving of uncontrolled disturbances. Composite manufacturers have learned which polymer systems are forgiving of minor variations in processing [6]. These systems are usually favored over the matrix resins which need more careful process control, even when the latter have better properties. As a result, even when the same brand of resin is used by two different manufacturers, the cure cycles used may be different.

If regulatory control is used, the cure cycles should be developed as efficiently and effectively as possible. The cost of cure cycle development is open-ended because there is no limit on the number of possible variations to cure cycles. There is no guarantee either that the results will be transferable to other processing equipment or materials because the relationship between primary and secondary variables is unpredictable in such complex, path-dependent processes.

15.3.1 Trial and Error

There is a tendency among control and statistics theorists to refer to trial and error as one-variable-at-a-time (OVAT). The results are often treated as if only one variable were controlled at a time. The usual trial, however, involves variation in more than one controlled variable and almost always includes uncontrolled variations. The trial-and-error method is fortunately seldom a random process. The starting cycle is usually based on manufacturers' specifications or experience with a similar process and/or material. Trial variations on the starting cycle are then made, sequentially or in parallel, until an acceptable cycle is found or until funds and/or time run out. The best cycle found, in terms of one or a combination of product qualities, is then selected. Because no process can be repeated exactly in all cases, good cure cycles include some flexibility, called a *process window*, based on equipment limitations and/or experience.

The primary advantage of trial and error is that it has been the common practice for many years. In theory, it does not require prior knowledge of the art to take a suggested process cycle. Vary the conditions, examine the results, and try something new. In reality, an inexperienced engineer could easily flounder in the multitude of potential possible cure cycles while an experienced processor usually finds an acceptable cycle quickly by intelligent choice of trial cycles and evaluation of results. There are many clues to be found upon physical examination of the resulting product that will help the experienced processor select the next trial. For instance, microscopic examination of a polished cross-section of a cured composite can tell the knowledgeable engineer whether the laminate was over- or under-compacted and whether volatiles were a problem.

The disadvantages to trial and error are numerous. It is impractical to try all possible variations on a batch process sequentially because of the time and expense involved, both in making test runs, and in evaluation of results. The most expensive part of process trials is the mechanical and physical testing required to determine product quality. This means that it is quite possible to run a series of tests, choosing conditions at random and never come near the optimum conditions sought. The trial and error method does not provide the sort of understanding useful for in-process control because it usually involves only parameters that can be directly regulated, such as the temperature of the processing equipment. This lack of understanding makes it difficult to extend these cure cycles to new materials, processes, or geometries. Trial and error, like any method for developing preplanned regulatory cure cycles, does not provide the flexibility to deal with perturbations in incoming material properties or process disturbances. The success of this method is very dependent on the selection of trials and the interpretation of results.

The autoclave cure of advanced composites serves as a good illustration of this sort of complex control problem. A typical autoclave cure cycle is shown in Figure 15.1. The temperature is usually increased in stages until some maximum cure temperature is reached. It is then held at maximum cure temperature until complete cure is certain. During this temperature cycle, the pressure is applied to achieve compaction. The 11 items capitalized in Figure 15.1 are all variables. It is clearly not possible to perform an exhaustive study of all variations of all of these variables with finite time and money. In addition, although the preceding template is common, it is not the only possibility. It would also be possible to heat

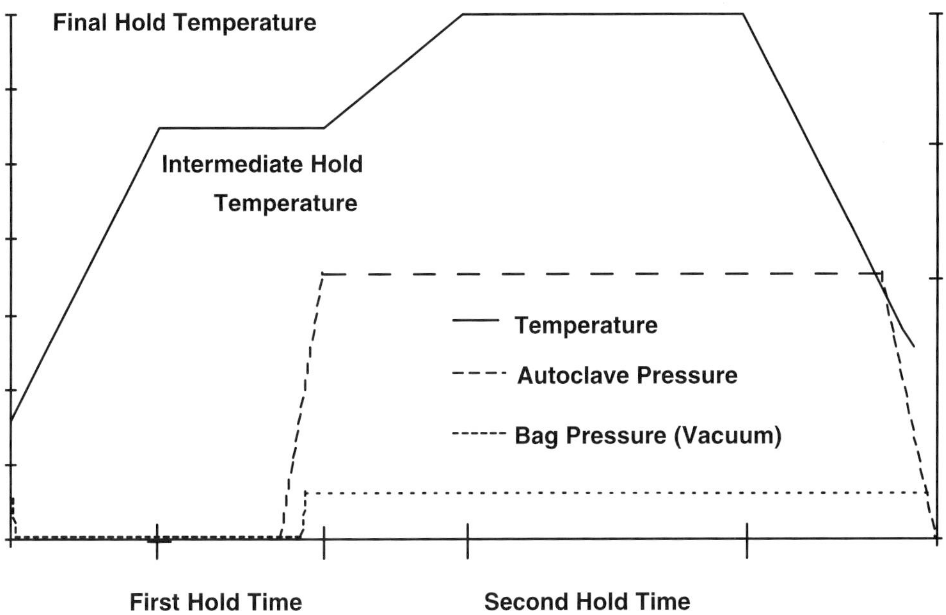

1. Pull vacuum of at least **X inches Hg**
2. Heat @ **Heat-up Rate 1** to **Intermediate Hold Temperature**
3. Hold at **Intermediate Hold Temperature** for at least **Time 1**
4. Raise autoclave pressure to **Y psi**.
   Vent vacuum when pressure reaches 15 psig.
5. Heat @ **Heat-up Rate 2** to **Final Hold Temperature**.
6. Hold at Final Hold Temperature for **Time 2**.
7. Cool slowly at **Cool-down Rate** to **End Cure Temperature**
8. Release pressure.

Figure 15.1 A typical specification for an autoclave cure. The eleven variables in bold type are typical of the parameters the process engineere has to optimize

the composite without holds, to put in more holds, to change the rates more often, or to decrease the autoclave temperature prior to reaching the maximum cure temperature.

To complicate matters further still, the curing process, like many batch processes, is history dependent. For instance, if pressure application in Step 4 were delayed until after gelation of the resin, the laminate would be uncompacted and full of voids, no matter how good the temperature cycle [7]. Less noticeable effects on process-induced stresses and cross-link structure may result from other historical variations and may explain some of the unanticipated variations in mechanical properties and aging that sometimes occur, even when a temperature cycle is rigorously repeated. Other variables that impact the behavior of the material during cure are geometry, the initial cure state of the resin, contaminants in the resin, tooling (materials and geometry), and placement in the loaded autoclave. Most of these variables are not easily controlled.

There are two primary methods to reduce the number of trial process conditions and at the same time gain more information from the experiments done. The first method to be discussed is based on design of experiments (DOE). The second method, process science, is an offshoot of attempts to improve quality control on incoming materials. Each has advantages and often they are combined.

15.3.2 Design of Experiment

DOE is a methodical statistics approach to studying the qualitative effects of process variables. Variables of interest are given a number of values based on the expected relationships [8]. For example, if the relationship is expected to be linear over a range, two variations can be used to approximate the effect of the variable. For effects that are expected to be quadratic, three variations may be needed. These variations are then matrixed to create a set of trials that differentiate and quantify the effect of each variable. If the number of variables is small, then the experiments can be designed as a full factorial. An example of a full two factorial design of an experiment for three variables is shown in Table 15.1.

Table 15.1 A Two-Factorial Design for Experimental Determination of the Effects of Pressure, Heating Rate, and Temperature

Test #	Heating rate (C/min)	First hold temp (°C)	Pressure (N/m^2)
1	2	132	6.89×10^5
2	2	132	5.51×10^5
3	2	121	6.89×10^5
4	2	121	5.51×10^5
5	4	132	6.89×10^5
6	4	132	5.51×10^5
7	4	121	6.89×10^5
8	4	121	5.51×10^5

The full factorial experimental design, however, quickly expands into an unmanageable number of trials. For n variables and p variations on each variable, t, N is calculated by Equation 15.1.

$$N = p^n \tag{15.1}$$

A full three factorial matrix on the 11 variables in the cure cycle shown in Figure 15.1 would mean 177,147 individual trials. A full two factorial design would still mean 2048 trials. Such a design, however, assumes that all interactions, even between all 11 variables, will be important. DOE provides an ordered means of combining variables to reduce the total number of trials. The assumption made is that high-order interactions (i.e., interactions of three or more variables) are rare and/or insignificant. There are several methods for combining variables by DOE. A detailed discussion of these methods is the subject of another book [9].

Courtney [8] used a fractional factorial developed using a Yates design matrix for designing an experiment to evaluate the effects of 10 variables on the injection molding of a cooling fan for an automobile. He was able to reduce the number of experiments to 32 as opposed to the 1024 experiments that would have been generated by a simple two-factorial design. Courtney did note, however, that it was critical to select the four key variables to refine the experimental test matrix. This design was also limited to second-order interactions. In some complex processes, there may be higher order interactions that are significant. Careful selection of methods is important for the correct use of DOE.

There are at least two advantages of this approach over simple trial and error. It is faster, because the trials can be run simultaneously, or at least without waiting for the results of the previous trials. It is also more systematic and proper analysis of the results can indicate if important interactions or variables have been left out. DOE also provides at least some qualitative understanding of the influence of each variable on the process. If an increased temperature results in better mold fill, then it may be that increasing the temperature still further will allow the mold to fill in less time.

DOE shares many of the disadvantages of simple trial and error. It is costly. In fact, if the trials are run simultaneously, then the savings may be eaten up by extra trials that do not contribute to optimization or understanding. This problem can be alleviated by correct selection of the DOE method and the main variables and interactions of concern. The former can be obtained by training that is readily available at most universities and many companies. Selection of variables, however, requires some knowledge of the process. The effect of variables is not necessarily linear, so that the actual "optimum," if there is one, might be between the chosen values for a variable or outside the range studied. Take the example of mold filling time. Up to a point, increasing the temperature will increase the rate of mold fill. There comes a temperature at which reaction rate is so rapid, however, that the polymer gels before the mold is completely full and the mold no longer gets filled, no matter how much the temperature increased. There are statistical means, such as response surface methods, to identify such higher-order dependencies, but data must cover the ranges of interest. It must be remembered, therefore, that any causal effects derived by DOE are valid only within the ranges studied. The relationships are statistical rather than mechanistic.

Like other statistical methods, DOE works best when the results can be easily quantified, the number of parts to be made is large, and the value of these parts is relatively small.

Processes like resin filming for prepregs and adhesives, injection molding of small parts, and shaping of thermoplastics may benefit from DOE for developing cycles or process settings. DOE is less likely to be useful in complex batch processes like autoclave curing where the parts are large and few and quality is multifaceted.

One approach for using DOE on more complex processes is to do the majority of the process development on smaller, representative sections of material, such as test panels, rather than on full-scale parts, and then to scale up with a more limited experimental matrix. There is no guarantee that experience on small-scale test panels will directly translate to large parts because dimensions and thickness of the part are important variables in their own right. Another way to save on costs is to start with a satisfactory process and to continue, via careful monitoring of process variations and results, to extend the range of experience. This method is variously called *statistical process control* or *statistical quality control.*

15.4 Statistical Process Control

SPC or statistical quality control (SQC) is similar to DOE in that it is a statistical, rather than mechanistic, method. Both SPC/SQC and DOE rely on the theory that there is a direct relationship between variations in process controls and resulting changes in product quality. In SPC, however, the experiments are not forced on the process like they are in DOE. The variations in product quality and the random process variations are traced over time instead. The variations in end product are then correlated, if possible, with changes in the process that have occurred during that time. SPC techniques are usually applied to the process after some baseline process has been established by other methods.

The advantage of SPC is its relatively low cost. Many of the parameters that should be tracked for SPC are already part of any good quality assurance and postprocess inspection program. The codes necessary to do analysis are commercially available and do not require additional experts for interpretation. SPC can be used to react to quality drift by adjusting the process. Another advantage is that SPC may pick up unanticipated effectors, such as the time of day or operator performance.

SPC does require statistical quantities of product and automated data tracking. The best processes for SPC are those which are used to make large quantities of inexpensive parts. SPC is also difficult to apply to processes where the number of independent variables is large. Automated data acquisition is a must for SPC, but this is becoming inexpensive and common in the workplace. SPC is also a delayed control method. Many defective parts may be made before SPC corrects the process. The longer it takes to evaluate the results of the process, the more delay in the ability to react to process changes. Another important requirement of SPC is attention to detail on the part of the operator and/or process engineer. No battery of QC tests will detect every variation either in materials and process or in final quality.

An example of successful use of SPC comes from the injection molding industry where part counts are high and part values are generally low. The material in reject parts can sometimes be recycled. The process is very rapid, with few control variables; and the major, often the only, criteria for quality is reproducible shape. For this situation, SPC is an excellent tool for determining acceptable control settings. Wojtaszek, working for an injection molding company, used SPC to reduce the percentage of jobs with greater than 1 percent rejection rates from over 90 percent to less than 15 percent [10]. Hunkar, also working with injection molding, reduced scrap rates by 6.7 percent up to 29.4 percent using SPC, but each test he ran required 100 samples [11]. Neither investigator provides information on the cost of implementation.

An alternate to statistical methods is the analytical study of the material behaviors and mechanisms that link the primary variables to the secondary variables that are more easily controlled. This is often called *processing science*.

15.4.1 Process Science

The process science method for cure cycle development also starts with past experience. A number of characterization methods, often the same methods used to qualify incoming materials, are used to evaluate the cure kinetics and rheology of the materials under simulated cure conditions. The most useful of these is probably dynamic rheological or mechanical analysis (DMA) [12]. This method measures rheological and/or stiffness properties as a function of temperature and time. There are a number of instruments on the market, and the purpose of this chapter is not to discuss the relative merits of characterization equipment. It is important, however, that an instrument being used for cure cycle development be capable of duplicating the range of temperatures and temperature rates (heating and cooling) to be investigated. A number of experiments are usually made at different rates and/or at different constant temperatures. These data can be used to extrapolate, within limits, and to develop a kinetic model.

Examples of three important thermal analysis methods are shown in Figure 15.2. All of the measurements in Figure 15.2 were taken at a constant rate of 2°C/min (3.6°F/min). The top chart is an example of a rheology trace taken from a Rheometrics Dynamic Spectrometer. A specimen of uncured prepreg is loaded in oscillating torsion and the transfer of force through the specimen measured. G' and G'' are the storage and loss components of the shear modulus respectively. One can see both measurements drop initially as the polymer (in this case, a phenolic) is heated. Then, as the reaction cross-links the polymer, the viscosity starts to rise again. The exact point at which flow starts and stops depends on viscosity and a number of other factors. It is safe to say, however, that flow will not happen above some maximum viscosity. The time that a polymer spends below that maximum viscosity is referred to as the "*processing window.*" When the processing window is large, the polymer tends to be easier to process.

Other characterization methods that are of value are dynamic scanning calorimetry (DSC) and thermal gravimetric analysis (TGA). A sample DSC is shown in the middle of Figure 15.2. Most cure reactions are exothermic, and the heat generated by cure can cause excessive heat to build up in the polymer if control is not exercised. DSC measures the generation of heat as a function of time and temperature. This can be used to predict the temperature at which the laminate will begin to cure (the onset of the peak in Fig. 15.2) and the temperature or time at which cure will be complete, further improving the selection of cure cycles to try.

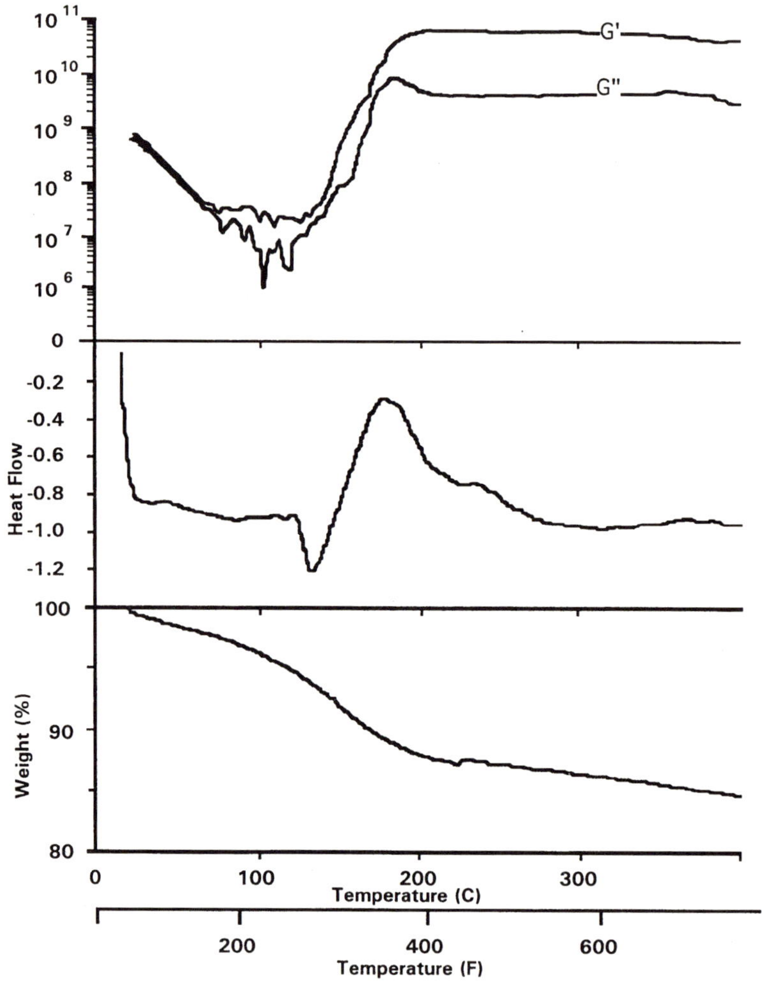

Figure 15.2 Comparison of thermal analysis techniques provides insight into the behavior of the material during the process

This particular DSC has the added feature of an endotherm that is indicated by a valley just prior to the exotherm. The data is for a condensation-curing phenolic, and the endotherm is caused by evaporation of reaction by-products. This evaporation is confirmed by the third trace in Figure 15.2. The TGA shows a rapid weight loss beginning at about the same time as the endotherm in the DSC.

Development of a successful cure cycle uses a combination of the preceding methods. In the autoclave, it is critical to apply pressure within the processing window; but it is better to wait until some of the volatiles have boiled off, especially when dealing with condensation polymers like phenolics and polyimides. Note that these instruments, even though they may duplicate temperature profiles, usually do not have the ability to duplicate pressure and vacuum changes and are done on very small samples. Some experimentation with more realistically sized parts in the actual processing environment is generally needed to validate a process cycle. These and other processing science tools can still be very useful in decreasing the number of trials for cure cycle development.

One advantage of processing science is that it contributes to a mechanistic understanding of the polymer cure. By definition a predetermined cure cycle is not flexible; however, but one that has a good foundation in process science is easier to correct as conditions change. It is also easier to extrapolate knowledge gained by process science to new materials and processes. Further, processing science is the first step toward developing models of the process and toward intelligent in-process control.

The process science approach is still costly. One can expect to spend several thousands of dollars on each new material and probably more on a new process. The equipment for thermal analysis and skilled technicians to do the testing and interpret the results must be obtained or contracted. Basic thermal analysis equipment including a DMA, TGA, and DSC can range from about \$100,000 to 200,000, or more [13]. There are also test facilities that will contract for thermal analysis. For processes with few variables and/or a single quality requirement, the cost may be high compared with DOE or statistical process control. For complex processes with many variables and high-cost parts, such as aerospace components, processing science is a means of achieving the understanding that might be quite costly when approached by random trial or statistical methods. The information gained from testing results in a process model. This model may be a simple correlation in the engineer's head or it may be an analytical model on the lab computer used for simulations and predictions.

15.4.2 Analytical Models

One of the most extensively developed tools for intelligent processing is process modeling. Analytical models use a set of differential equations to represent the various behaviors of the material pertinent to the process as a function of time and temperature or some state variable such as the degree of cure (for thermosetting resins). The sets of differential equations representing the process may be derived from first principles (usually transport phenomena, such as heat and momentum transfer) or from curve fits (kinetics for chemical reactions) [14]. Existing models for the autoclave cure range from simple, one-dimensional representations that predict a few critical material properties [14] to complex, three-dimensional (3D) models that can incorporate the effects of part placement in the autoclave and batching of parts [15]. For processes such as injection molding, there are commercially available models with 3D, color, and animation.

Simulation of the process with analytical models can be used to evaluate the effects of changes in process parameters, providing the limiting assumptions of the model are noted [16]. These parametric studies can then be used to select critical experiments for selecting a cure cycle or to establish rules for process-cycle development [17]. If the simulation is true enough to the actual behavior of the material and processing vessel and provides the necessary predictions of material quality, it can even be used to select a cure cycle [15,18].

Simulations of the process are less expensive and less time consuming than actual runs. If the users understand the assumptions, then they can add to the understanding of the process. Even if the equations are derived from data, the interaction of equations may provide useful insight. Mechanistic models can (theoretically at least) provide even more insight. Further, the development of a process model usually includes some of the same experiments useful for process science, creating an experience base for process development.

Process models are unfortunately often oversold and improperly used. Simulations, by definition, are not the actual process. To model the process, assumptions must be made about the process that may later prove to be incorrect. Further, there may be variables in the material or processing equipment that are not included in the model. This is especially true of complex processes. It is important not to confuse virtual reality with reality. The claim is often made that the model can optimize a cure cycle. The complex sets of differential equations in these models cannot be inverted to "optimize" the multiple properties they predict. It is the intelligent use of models by an experimenter or an optimizing routine that finds a best case among the ones tried. As a consequence, the literature is full of references to the development of process models, but examples of their industrial use in complex batch processes are not common.

Good simulations are not free, either. The development cost of the code may not be high because there are so many examples in the literature. The data to fill in constants for these models, however, may require extensive testing and cost thousands, or hundreds of thousands, of dollars. This data is often highly specific to material and process equipment, nor does it stop with the original development. Software requires maintenance and support if it is to be useful to a variety of people over any period of time. Although many models for the curing of composites are available, few are not commercial products with supporting organizations. Commercial products do exist for processes like injection molding and extrusion that have more extensive application.

Finally, the output of process models may not answer the question of whether a cure cycle is actually "good." Output typically includes the temperature distributions, the degree of reaction, and additional information critical to the application. The prediction of some things, such as local temperature, can be quite accurate, whereas other predictions, like resin flow and void formation, are less accurate. The criterion for composite and plastics "goodness," however, are physical, mechanical, and/or electrical properties, not temperature. Existing models cannot directly predict composite strengths. They are instead they are used to infer that the strength of the material has not been substantially degraded (by voids, overheating, or residual stresses) and that the composite has cured enough to have the desired glass transition temperature (degree of cure).

Probably the first major publication of a process model for the autoclave curing process is one by Springer and Loos [14]. Their model is still the basis, in structure if not in detail, for many autoclave cure models. There is little information about results obtained by the use of this model; only instructions on how to use it for trial and error cure cycle development. Lee [16], however, used a very similar model, modified to run on a personal computer, to do a parametric study on variables affecting the autoclave cure. A cure model developed by Pursley was used by Kays in parametric studies for thick graphite epoxy laminates [18]. Quantitative data on the reduction in cure cycle time obtained by Kays was not available, but he did achieve about a 25 percent reduction in cycle time for thick laminates based on historical experience. A model developed by Davé et al. [17] was used to do parametric studies and develop general rules for the prevention of voids in composites. Although the value of this sort of information is difficult to assess, especially without production trials, there is a potential impact on rejection rates.

An example of the cure cycle optimization is the work of Thomas et al. who used a very sophisticated model, together with a rule-based optimization routine, to pick the shortest cure cycle that met a set of performance criteria [15]. Reductions in cure time using this method ranged up to 36 percent for a single complex part and from 8 percent up to 43 percent for batches of mixed parts. Rejection rates were not increased in any case, and they were actually reduced significantly for one part. This model, although transferred to a number of companies, unfortunately has had limited use because of the lack of support for the code and the cost of qualifying it on new materials.

It is doubtful that any of these models paid for themselves in terms of direct production cost savings. The development costs ranged from hundreds of thousands to millions of dollars, including pertinent data. Some of the models were developed at universities for a substantially lower cost, but they are still largely unsupported. Most organizations prefer to develop models of their own because they then understand the code better and can tailor the inputs and outputs to their needs. Much of the expense goes into the development of material data for the equations and to check the accuracy of the model. In the autoclave cure for instance, for every material used in the lay-up, input properties such as density, thermal conductivity, heat capacity, permeability, thickness, and spring constants are required, depending on the model [15]. These properties may vary as a function of pressure and temperature. Resin may flow, changing the makeup of various material layers. The resin cure acts as another source of heat. By-products of the reaction, or contaminants of the original material, may volatilize, absorbing heat and forming voids. In order to predict the likelihood of voids it is necessary to know the primary volatile constituents and their vapor pressures as a function of temperature and concentration. For models that include the behavior of the autoclave, a thorough characterization of flow patterns and temperature gradients in the autoclave is necessary. Starting conditions and boundary conditions must be included as input into the model. Depending on the model, these may include the pressures and temperatures as a function of time as well as the initial water content, resin content, and thickness of the lay-up, and such things as geometry of the part. Few of these data are available in handbooks. Some data are not even easy to measure. It is possible, at least for engineering considerations, that the expense of obtaining all the data may outweigh the advantages of having the model.

It is much harder to gauge the impact of the major advantage of simulation models that is the reduction in time and effort for process-cycle development. A good simulator may reduce the number of experimental trials necessary to develop a process cycle to a very few, nor does it take an experienced engineer to make runs on the computer. Testing to evaluate properties is not needed because there are no laminates produced by simulation.

Regardless of the quality of the model, process cycles designed with models have all of the same problems of process cycles designed by process science, DOE, or SPC. Preplanned regulation of secondary variables does not allow controlled adaptation to unanticipated disturbances in the cycle.

15.4.3 Knowledge-Based Expert Systems

Expert systems are computer programs that simulate the decision-making process of human experts. The hallmark of expert systems is that decisions are based on heuristics (rules of thumb) when data is incomplete or there is not enough time to consider all possibilities [19]. An expert system can be a set of IF–THEN rules in FORTRAN, or it can be a written in one of the languages designed for expert systems, such as LISP. Expert systems can be used off-line to aid in cure cycle selection, or they can be used as real-time advisers or controllers.

Thomas et al. [15] used an off-line expert system optimizing routine in conjunction with an analytical process model to automate the choice of a cure cycle. The cure model was given load and part information. The optimizer, written in FORTRAN, increases the model setpoint until the model predicts that the part temperature is high enough for flow to occur. It then models various hold lengths until it finds a hold long enough for good compaction. There are other rules about allowable rates of autoclave temperature and pressure changes. The assumption here is that there will be critical steps to each process that need to be accomplished in sequence and that the faster one can accomplish them, without violating any other requirement, the better the process cycle. Using an expert system in this way requires a good model and accurate data for input, so it is at least as expensive a method as analytical modeling. There are many standard optimization routines for nonlinear, multi-variable problems, but the sequential, path-dependent nature of the batch process makes the expert system a good fit; still, this method limits the possibilities. For instance, the referenced system would always produce a constant heating rate to a hold.

Expert systems can also be used off-line as advisory systems. If some leeway is left in the process specifications, operators can even be instructed to make simple variations in the process. Hajicek developed an expert system for troubleshooting injection molding, which was designed to be used by operators to adjust controls as needed whenever part quality suffered [20]. These expert systems do not necessarily plan an entire cycle. As we will see in the section on in-process control, however, they can be used to do so. There are even specifications that include rules for altering the cure cycle in response to the results of quality acceptance tests such as flow and volatiles, a primitive set of heuristics.

The development of expert systems need not be costly. There are several expert system shells commercially available, and an expert in artificial intelligence (AI) is no longer needed to program them. Simple IF–THEN rules can easily be programmed in more commonly used languages like FORTRAN and BASIC. In fact, the McDonnell Aircraft expert system referred to earilier was programmed in FORTRAN. Cost depends on the number of rules and the amount of testing required, but one year is a reasonable estimate for the development of the system itself.

One difficulty with expert systems is that the quality and cost depend a great deal on the quality of the expertise used in developing the rules. Common wisdom may be misleading or in error. For instance, it was generally assumed that the only way to avoid damaging exotherms in thick laminates was to heat them very slowly until Pursley and Kays showed that heating thick laminates rapidly and then cooling the outside allowed the exotherm to provide most of the cure energy [18]. Contrary to expectations, if done properly, this strategy made satisfactory laminates in considerably less time.

15.4.4 Artificial Neural Networks

ANNs are a popular means of mapping one function onto another where theory is too complex or poorly understood to model by first principles. ANNs are ideal for representing nonlinear relationships between input and response data, even for multiple input, multiple output processes like polymer batch processes. A training set of experiments is used to adjust the network so that the response mirrors that of experiment. It is important to choose the correct type of network, number of layers, and mode of propagation [21]. Once a training set of data has been generated, ANNs can theoretically extrapolate from that set and a model of the process has been formed. The advantage of ANNs over most analytical models is that they do not require knowledge of the "first principles" relationships between variables. Further, these models can be optimized on a given variable more easily than models based on differential equations. They can even be used to adapt process cycles. Unlike analytical models, however, ANNs are not based on theory, so they are not good at providing explanations. Some process knowledge is required for choosing inputs to ANNs because inclusion of irrelevant inputs can make training impossible.

Joseph and Wu used a simulation to show that an ANN could be used to optimize the cure cycle for variations in raw material properties. The results were encouraging: an 89 percent decrease in standard deviation for thickness and a 96 percent reduction in void size. The mean thickness was also closer to the target thickness for the neural network–optimized processes. They also used the ANN for in-process adjustment of the cure cycle [22].

15.4.5 Summary of Methods

The preceding tools are best used in combination. Process science techniques combined with DOE can be used to select some preliminary experiments to train an ANN or to gather the data necessary to use an existing process model. This model can then be used as a simulator to do additional experiments and, together with an optimization algorithm and/or an expert system, to select a cure cycle. If the range of incoming material properties is known, the same model and/or expert system can be used to tune the process to incoming properties. All cycles, however, must eventually be tested by making real parts and testing those parts.

The shortcoming of all methods for predetermining cure cycles that regulate secondary variables is that they deal only in expectations and probabilities. No matter how many eventualities are anticipated, there is always one more that is unexpected. Unexpected variations in material properties, process equipment malfunctions, and changes to geometries of tool and part all contribute to the uncertainty of the outcome. As a result, in-process, inferential control is needed for the process environment as well as the boundary conditions and material state. Inferential control is relatively new to the polymer processing industry, especially in complex processes where good models are not yet common.

15.5 Tools for Real-Time Process Control

One of the most desirable aspects of plastics and composites is the ability to make net-shaped parts. The same process that creates the material also creates the structure. The penalty for this advantage is that the process of curing a thermosetting plastic or composite part is irreversible. Any part that is not properly processed represents a loss of part, material and the money and time required to make that part, although larger parts are usually repaired if possible. Proper shape becomes a controlled property in addition to the bulk material properties, such as mechanical (stiffness or strength), physical (density, void content, etc.), chemical (degree of cure or carbonization, chemical resistance), electrical (resistivity, conductivity), or any combination of these.

It is not possible, except in parts with very simple requirements, to control all of the material directly and part properties of concern for the typical composite or plastic part. Many of the these properties cannot be directly measured. Most of them cannot even be linearly related to measurable quantities. In addition, some of the properties may require conflicting actions during the process. For instance, a high cure temperature might make for a material with good mechanical properties at high temperatures, but it may warp or shrink the part dimensions unacceptably. As a result, real-time inferential process control is a balancing act that utilizes knowledge of the relationships between measurable secondary parameters and part requirements and the forcing functions available to drive the secondary parameters.

No control is possible without some sort of preestablished plan of action. This plan can be rigidly followed throughout the cycle or it can be flexible, subject to change when the measured progress of the process indicates that change may be advantageous. The tools used to develop the plan for a process have been discussed earlier. This section presents some of the tools and methods used to follow those plans and provide flexibility where needed.

It is not possible to discuss real-time control without a brief discussion of sensors and the measurements they represent. In traditional process control, the measurements and the properties to be controlled are identical. For instance, one controls the temperature of a fluid using feedback from a thermocouple. There is also generally a fairly predictable relationship between the measurement and the forcing function necessary to change that measurement. Except for unusually simple cases, that is not true of polymer processing. The multiple, complex properties to be controlled cannot be measured and are not always predictably related to the process variables. Sensors for polymer curing that measure product qualities may be available, but they are not always economical. Sensors cost money to buy, maintain, and install. Intrusive sensors can interfere with part quality. Sensors may also measure qualities that cannot be or are not controlled. In general, however, the closer a measurement comes to measuring a part quality, the more useful it is for directing the process.

The main focus of this chapter, however, is on the tools that can use this sensor information for real-time control. These tools are still largely developmental because inferential control is relatively new to the composites industry and change is dependent on both technical merit and on other changes in the culture of the industry.

15.5.1 Supervisory Controllers

The state of the art for many polymer processes is a simple time–temperature recipe for the process equipment. This is often a satisfactory solution for processes in which the process equipment temperature and the part temperature are nearly identical. Temperature control for pultrusion of thin cross-sections, for instance, can be satisfactorily based on the die and barrel temperatures. In processes where there is a major difference between the setpoint and the local conditions, it is now becoming more common to implement supervisory controllers which account for these differences. The autoclave curing of composites, for instance, may have large temperature lags between the autoclave and the part as is shown in Figures 15.3

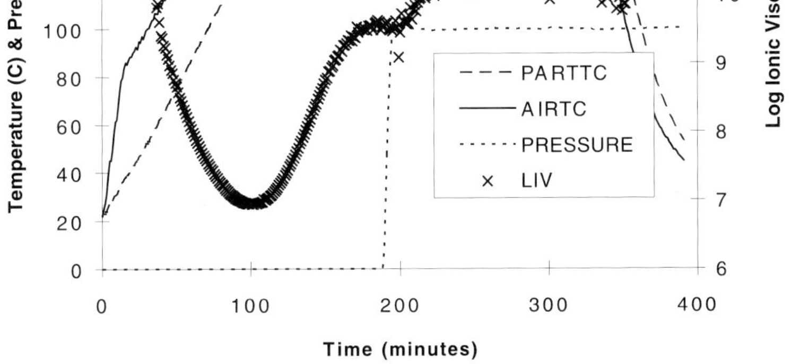

Figure 15.3 The temperature lag between autoclave (AIRTC) and composite part (PARTTC) is large, so a supervisory controller is used to drive the autoclave setpoint up and achieve the desired cure cycle in the composite. Because this cure cycle was developed for the autoclave temperature, however, the resin gels before compaction is complete

and 15.4. Because the temperature of the part has a more direct influence on the cure than the temperature of the autoclave, this process is often now controlled with supervisory controls that incorporate models of the thermal lags between autoclave and parts and adjust the autoclave temperatures accordingly [23–25]. The autoclave temperatures are then manipulated in order to compensate for the thermal lag and give better control over the part temperature cycle. Even good control of the composite temperature, however, does not guarantee good control of composite quality.

In Figure 15.3, a supervisory controller has been used to force the boundary conditions on the laminate. The controller causes the autoclave temperature (AIRTC) to overshoot the hold temperatures intentionally in order to maintain the heating rate on the part (PARTC). Supervisory controllers may also be programmed to recognize alarm conditions, such as over- or underheating, and bring them to the attention of the operator. Most temperature-based cure cycles do not allow for extensive adjustments; however, because of the way specifications are written and interpreted. If a process does not follow the recipe, it is considered suspect and extra inspection and testing may be required.

Most new autoclaves now come with supervisory controllers, and there is no question that these could improve the reliability of temperature-based recipes. The primary advantage is that these controllers allow for control of the condition of the resin rather than the condition of

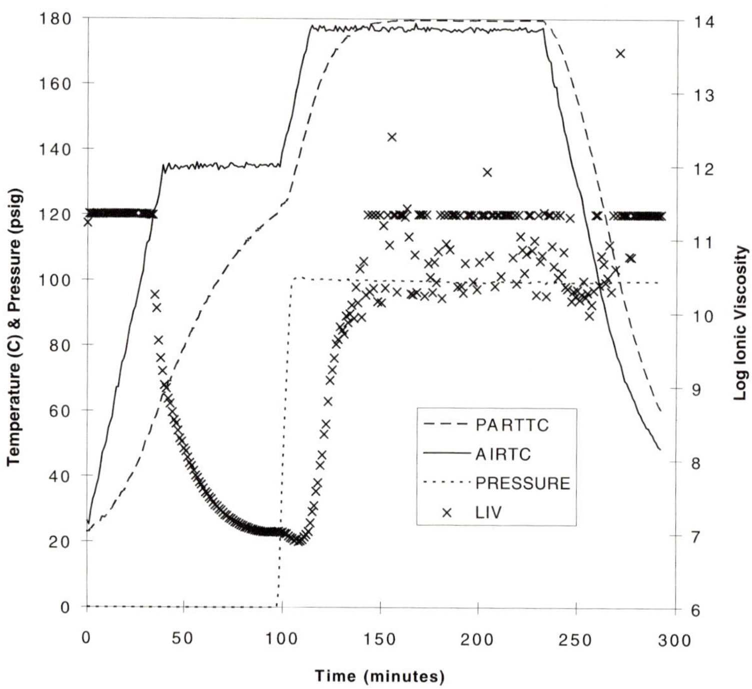

Figure 15.4 The large thermal lag between the part (PARTTC) and autoclave (AIRTC) is not linear or a constant. This cure cycle, however, was written for the autoclave temperature. Pressure is applied at minimum viscosity and compaction is good with few voids

the autoclave, a method that is clearly more relevant to composite quality. These controllers also have the ability to adapt to variations in part geometry, tooling materials, and efficiency of heat transfer. Such controllers can even adjust cure cycles to minimize temperature variations within a load, providing loads are properly instrumented. Most also have at least some ability to recognize and react to failures in equipment or sensors, sometimes preventing the irreversible loss of a part.

The disadvantages of these controllers are relatively few. They are no longer particularly expensive to purchase or install. They can easily be retrofitted to most process equipment. The process specifications can look like specifications for equipment temperatures. It is, however, important to remember that a process specification written for control of the equipment parameters is not likely to be the same as one written for the part parameters. Figures 15.3 and 15.4 illustrate a problem that may occur when using a process specification incorrectly. The specification in question was written for the temperature of the autoclave. In Figure 15.4 the process specification was used to control the autoclave temperature. Note that during the low temperature hold, the part temperature barely has time to reach 132°C (270°F). The log ionic viscosity (LIV) is at a minimum when pressure is applied and the part was well compacted with few voids. When this same cure cycle is interpreted for control of the part temperature as in Figure 15.3, the low temperature hold begins when the part reaches 132°C and (270°F) and the ionic viscosity has risen long before pressure is applied. This part was not compacted. A cure cycle for part temperature was later developed. The pressure application point was changed to the beginning of the hold and the hold was shortened. The resulting cycle was shorter and produced better composites with a more reliable control of fiber volume.

Most supervisory controllers are still used only to regulate part temperatures and heating rates. Even though this is an improvement over regulation of process equipment temperatures, it will not control a process adequately in the presence of unexpected disturbances. If the polymer being cured in Figures 15.3 and 15.4 were aged or had an imbalance in components, the temperature-based cure cycle would not necessarily produce a good part. The true promise of these controllers is the ability to provide inferential control of primary properties. Several methods have been used to utilize the knowledge about relationships between the secondary variables that are easy to measure and control, and the primary variables that are really desired to control.

15.5.2 Knowledge-Based Adaptive Controllers

Most supervisory controllers contain some minimal capability to vary from the process plan, such as the ability to detect faulty thermocouples and react appropriately. Knowledge-based, real-time controllers take this adaptability one step further by using combinations of sensors and/or models to determine the state of the process and predict trends. The controllers then compare the measured state with the desired state, and change the process plan when necessary to adapt to those trends, forcing the outcome toward a desired state. Knowledge-based systems must include some means of converting the sensor data into information and a set of rules to act on that information. Knowledge-based control systems are not always simple tools. The method for converting the data to information may be different from the tool used to determine the proper reaction to that information. Some of the methods for generating and using knowledge about the process in control will be discussed shortly as separate topics. The examples, however, are combinations of these tools in almost every case. In fact, many of the distinctions between methods are hazy.

15.5.3 Expert Systems

Expert systems is a general term used to refer to computer programs that incorporate expert knowledge, usually in the form of rules, to duplicate the behavior of a human expert. Rules can be used to define process states:

IF the past autoclave temperature is less than the current autoclave temperature
THEN the autoclave temperature is rising

This simpler rule can be combined with others to define more complex process states:

IF the middle-laminate temperature is rising
AND the top-laminate temperature is rising
AND middle-laminate temperature is accelerating
THEN accelerated reaction is active [26]

Both of the preceding rules convert data into information. Control rules are then used to convert that information into action. An example control rule uses both of the preceding definitions to decide the action to be taken by the temperature setpoint:

IF an accelerated reaction is active
THEN turn down the temperature setpoint

Note that any of the preceding rules could be written in any computer language. Some languages are designed specifically for expert systems, but the logic of cure control is generally simple enough to use more traditional languages as well.

Expert systems are easy to program and to understand because they usually resemble instructions in English. The time and cost for developing these systems is relatively small. The primary problem usually turns out to be interpreting the sensors. Because first and second derivatives of sensor data are used to find trends and patterns, noise can be a major problem. The rules allow the controller to adapt to the condition of the material and to the geometry of the part. Expert systems make it relatively easy to change to backup plans when sensor or equipment failures occur. In fact, rule-based systems can be quite general and handle a number of materials with little material specific data.

Expert systems, however, have their disadvantages. Perhaps the most significant one is that it is hard to prove rigorously that they will not give an unexpected behavior. This causes production and quality control organizations to be reluctant to adapt to the methods. Writing a specification for expert system control is less exact and, therefore, more difficult to adapt to quality control methods that were developed for time–temperature-based plans. The disadvantage of all knowledge-based systems is that the knowledge has to come from somewhere. Expert knowledge is not always right. For instance, until the work of Kays and Pursley [18], the conventional wisdom was that thicker laminates necessarily required slower heat-up rates and longer cure cycles and all cure cycles consisted of the heat-hold-heat plans typical of the industry. For thick laminates Kays and Pursley showed that a good strategy for cure was to heat the laminate as rapidly as possible until the resin in the center of the laminate began to heat up with the exotherm and then cool the outside of the laminate, allowing the exotherm to finish the cure.

One of the first rule-based control systems for composite curing was written by Baumgartner and Ricker in MBASIC [27]. They used dielectric loss tangent measurements to control the processing of epoxies by manipulating the temperature setpoint to give a constant viscosity hold instead of a constant temperature hold. Because viscosity is more directly related to composite mechanical properties than it is to temperature, this strategy had promise. Baumgartner and Ricker, however, based their strategy on quantitative dielectric values and concluded that variations in dielectric properties with material made this approach impractical for production control. They did not offer any observations on improvements in quality or efficiency.

Hinrichs and Thuen [28] used ultrasonic attenuation to determine the proper time for pressure application during an otherwise traditional pre-established cure cycle. Because dielectric is an electrical property, it is influenced by moisture content and temperature as well as viscosity, so it may vary quantitatively. Ultrasonic measurements are also affected by other parameters (i.e., void content), but they are a mechanical measurement rather than an electric one. The ultrasonic sensors used by Hinrichs unfortunately were less reliable than the dielectric sensors.

All rule-based systems require two elements. First, there must be a plan that can be interpreted in symbolic form. Second, there must be a means of translating the sensor data into symbolic information.

15.5.4 Qualitative Reasoning

Quantitative use of many sensors, like preset recipes, is often dependent on unknown variations in starting materials or equipment; therefore it has limitations as the basis for rule-based control. One way of overcoming this limitation is to use qualitative trends and landmarks from process measurements and controls instead of the quantities themselves. Qualitative process automation (QPA) is based on the qualitative reasoning models of Bobrow, Forbus, deKleer, and others [29]. It simplifies the trends of process variables to three possible states: positive, negative, or zero. The influence of one variable on another is described as positive, negative, or nil. For instance, the autoclave temperature might be said to have a positive effect on the part temperatures, which means that if the autoclave temperature is increasing, then the part temperatures will tend to increase.

The advantage of qualitative models over quantitative models is that it is not necessary to develop the quantitative relationships, only to establish the trends and influences. Qualitative trends can be established with fewer experiments and are less material and process sensitive. Qualitative relationships are also much easier to invert than the differential equations they represent. Much of the information gleaned from process development translates easily into qualitative relationships.

QPA was first incorporated into a controller for autoclave cure by the air force [30]. A commercial version of this controller, Qualitative Process Automation Language (QPAL), was developed by LeClair, Abrams, and Matejka [31]. QPAL starts with a default set of sequenced plans, each of which has an agenda to achieve and a list of states to avoid and conditions for transition to the next plan. The controller then measures progress every 30 s or so and makes a process decision based on the current states and trends. Decision making includes a strategy for resolution of conflicting control recommendations. The most notable feature of the air force system could be its flexibility. The same knowledge base was adequate to cure a wide range of part geometries and materials, including some aged epoxies that could no longer be satisfactorily processed by conventional cycles.

QPAL was first used in production at the air force's McClellan AFB Logistics Center to cure leading edges for the A-10 aircraft. The conventional cure cycle of more than 6 h was reduced to less than 2 h without reducing part quality, which resulted in a return on investment (ROI) of 8.2 based on time savings alone. When reductions in rejection rates and personnel costs were included, the ROI was calculated at greater than 100 over a period of 10 years [32]. Part of this was due to the relatively low cost of development for the expert system; however, this cost estimate did not incorporate some of the prior work on process science and modeling which benefited knowledge-base development. The expert system could have been developed without the benefit of the prior work, but it would have taken more time and cost more. Since then, QPAL has been used at Rohr Industries to cure thick phenolic parts in 60 percent of the cycle time and with an 80 percent decrease in rejection rate [33].

Other researchers have improved on both knowledge base and sensors, often using different programming tools, but with much the same results. Most of these systems have not been transitioned to commercial application because they were developed by researchers who had no interest in the development of user interfaces and the continued support that is necessary for a commercial software package. As a consequence of the success of these experimental systems, however, the developers of commercial supervisory control packages are moving to incorporate support for expert system logic into their software [24,25].

Qualitative reasoning, however, does have limitations. It limits the range of control reactions. Controls can be turned up, down, or held constant, or they can be on or off. Control is not proportional to the deviation from the desired trajectory and does not generally vary with the progress of the process. Qualitative interpretation of sensors and rates can be quite vague. The statement that the autoclave temperature is greater than the part temperature does not provide information on the size of that difference. The control problem can be handled by making frequent decisions, but frequent decisions also introduce more possibility for error. Comparisons can be made more quantitative by increasing the number of "landmarks" for comparison. For instance, instead of just comparing a variable to one rate, it could be compared with two or more rates, being greater than one and less than another. The latter, however, begins to approach another method that has become increasingly popular in process control; fuzzy logic.

15.5.5 Fuzzy Logic

The unfortunately named *fuzzy logic* is similar to qualitative reasoning in that it can handle some of the uncertainty in process measure. Fuzzy relationships are more quantitative than the plus–minus–zero relationships of QR, but they are not as quantitative as analytical models based on differential equations. Fuzzy logic has been enormously successful in the control of other complex processes, such as the movement of Japanese train systems, the operation of household appliances, and the control of some chemical plants [34]. Fuzzy classifications include both the information that a sensor is in some symbolic state or range and the certainty or degree to which it is within that range. Because of this, fuzzy logic is also able to provide a direction for control as well as a degree of control reaction. In effect, fuzzy logic controllers are to QPA what proportional controllers are to on–off controllers.

Fuzzy models may be slightly more difficult to program and develop than qualitative models, but they share all of the advantages. In addition, they provide smoother control. This may be a particular advantage in processes that require more rapid control decisions than autoclave curing. The cost of developing these controllers is relatively low and the data required to develop them is usually available from the development of the original process plan. Further, neural networks can be used to automate the development of control–process relationships.

Wu and Joseph [35] incorporated fuzzy logic into a knowledge-based control system for control of composite curing. They used the fuzzy logic to interpret the sensors and adjust the amount of control reaction on a simulated process. Even though the limitations of the simulator used did not allow full evaluation of the advantages of this system, it did show that the controller could react to material and process variations and improve the process plan.

Whereas fuzzy logic offers an improvement over qualitative reasoning in some aspects, the relationships it uses still must be derived from data. There fortunately, exists a means of automating the development of these relationships, even in the absence of a detailed analytical model.

15.5.6 Artificial Neural Networks

Much attention is being given to the potential of Artificial Neural Networks (ANN)s in process control, and they work very well for continuous processes. Farnsworth and Eitman showed that an ANN, even with an operator in the loop, could decrease variability in resin film thickness by more than 60 percent and greatly reduce setup time for new materials [33]. ANNs are excellent for pattern matching and can easily be trained to relationships between data [36]. Data interpretation in a noisy environment, including detection of sensor failures, is complicated. Simple filtering techniques work satisfactorily for handling noise on most individual sensors, but much more information can be derived from the pattern of a group of sensors. Fusion of sensor information is currently handled by expert rules in systems like QPA but could quite possibly be better done with ANNs.

Applications to batch processes have been less common, but there has been some work done on the use of an ANN to control autoclave curing. Joseph et al. used an ANN successfully to cure a part, reducing cure times and improving qualities, such as thickness control and void content [37]. When more variables were included, however, the computational problem became intractable. This particular approach to using an ANN broke variables down into time, temperature, and pressure recipes, which, as noted in the Section 15.3.2, can lead to exponential growth of necessary training cases.

As computational power increases it is possible that the inclusion of more variables with ANNs may be practical. Further, there are different ANN methods; what works with one method may not work with another. If one attempts to relate the pattern of a batch process, such as the autoclave cure, to the quality of the final product, it is clear that limits must be placed on the number of patterns to be considered in a training set. The number of patterns required for training is proportional to the number of weights within it. The number of variables in the patterns must similarly be limited because those are the inputs to the neural net. One place where ANNs can definitely be used is in the inner loop of process control. Because of the wide range of temperatures, tuning the traditional Proportional–Integral–Derivative (PID) controller is usually a compromise between the high and low ends of the range. ANNs are one way to make a controller self-tuning with regard to the setpoint.

There are two considerations that limit the usefulness of ANNs in batch process control. One is the path-dependent nature of the process, which demands that time be considered a variable. The nature of the batch process itself is a progression of stages. The goal of one stage may not be the goal of another, so there must be a way for neural networks to consider the process as a whole or transition from one stage of behavior to another. The latter requires the interaction of the ANN with a model or an expert system. A second disadvantage of ANNs is that they do not "explain" their decisions with a trace of logic as expert systems and model-based controllers can. This means that ANNs cannot be expected to extrapolate outside of the training set of process conditions reliably. The lack of an explanation is also likely to decrease user confidence and to prevent user learning, although it may not necessarily inhibit machine learning.

15.5.7 Analytical Models

In traditional process control, models are often used to predict the deviation of the controlled variable from the desired state, the process error. This assumes that one knows the desired state. In complex batch processes, the desired state of the process is also dependent on history and changing dynamically. Further, most process models have to predict the outcome of an entire cycle to determine if the product will be good, so predictions are not available in real time, even for a slow process like the autoclave cure; however, partial models have been used as "virtual" sensors to expand on the information available from sensors [38]. Saliba et al. used a kinetic model to predict the degree of cure as a function of time and temperature in a mold and used that predicted degree of cure to time pressure application and determine the completion of cure. Others [39] have used the predictions of models together with the measured progress of the process to predict future trends and even project process outcomes.

Reaction to these trends is then governed by some sort of expert system. The Shrinking Horizon Process Model developed by Joseph and Thomas (Chap. 9) uses similar concepts, but recognizes that control options decrease as the run progresses.

One obvious disadvantage to closed-loop control of polymer processing is the dependence on sensors. This can increase the cost of manufacturing because it must be included in or attached to the part or tool during the process. Further, sensors must be maintained and purchased. There is also the problem of detecting and compensating for the inevitable sensor failure. When the controller is relying on sensor information to make decisions, there must be a backup. There are now some commercial controllers that are capable of handling a variety of sensors and of automatically detecting sensor failure and transferring to backup information sources [24,25]. It is still up to the user to decide what information will be used as a backup. Another sensor may be used if available; or a model may be used to extrapolate from the point of failure.

The major problem for control based on material states, however, is the quality control culture that requires that parts be accepted based on adherence to a preset cycle within specified limits. Because state-based inferential control systems could theoretically come up with a new cure cycle every time, this sort of specification cannot be used with such systems. Specifications instead have to be in terms of the process plan used for the cure. The satisfactory completion of a certain cure history without alarm states would be assumed to produce an acceptable part. Once the culture was able to accept that difference for autoclave curing, production costs at the U.S. air force's McClellan AFB Logistic Center were substantially reduced [32]. This type of specification could also give material review boards a head start on investigations because they would know that a part did not meet specification as well as what sorts of flaws might result from the deviation. The experience at McClellan is that there are fewer parts to review. It is even conceivable that, with improvements to sensors, much of the current postcure nondestructive evaluation used to verify the quality of parts could be incorporated into the process, building quality in rather than inspecting it in after the fact.

15.6 Summary

There are a number of tools available to the process engineer for designing a preset process cycle for regulating secondary variables such as temperature and pressure and for supervisory control of the process cycle based on inferred composite properties. Many of these have been tested in a variety of applications. None of these tools is capable of handling all of the tasks of a batch process controller; but they can be combined, and the resulting systems have potential far beyond that of any one tool by itself.

There is currently an important move underway in the materials industry to go from the preset regulatory cure cycle to supervisory control of more important properties, including the inferred state of the material. The flexibility of the latter approach enables rapid transition to new materials and processing methods without costly development cycles. State-based control also makes it possible to prevent some quality problems rather than inspecting and eliminating unsatisfactory products. This is particularly important for irreversible processes like the casting and cure of thermosetting polymers or composites made from those polymers. Even in the injection molding of low-cost thermoplastic polymers (a reversible process), time and cost savings have been demonstrated. There are significant issues, however, for inferential control of the complex, batch-type processes used to make most composite parts. A control system for polymer manufacturing must have the following capabilities:

1. Follow a logical sequence of events
2. Manage multiple, dynamic interactions between controls and quality
3. Detect and react to failure conditions
4. Provide event-based data recording and reporting
5. Handle safety interlocks and supervisory control of setpoints
6. Provide predictions of process trends and/or outcomes
7. React to prevent undesirable impacts on quality as well as to control current trends

References

1. Burns, P., *et al.* (1996) WL/MLBC, Wright-Patterson AFB, Ohio
2. Ghosh, A. Presentation at AIChE Annual Meeting (1991) Los Angeles, November
3. Pao, Y., Phillips, S.M., Sobajic, D.J. *Int. J. Control* (1992) 56(2), p. 263–289
4. Weber, R., Brosilow, C.B. *AIChE J.* (1972) 18(3), p. 614–623
5. Stephanopoulos, G. *Chemical Process Control: An Introduction to Theory and Practice* (1984) Prentice-Hall, Englewood Cliffs, New Jersey
6. Hancox, N.L. *SAMPE Q.* (1987) 18(2), p. 1–6
7. Browning, C.E., Campbell, F.C., Mallow, A. Paper Presented at Third Annual Conference of Emerging Technologies in Materials (1987) Minneapolis, August
8. Courtney, J.F. *Automotive Challenge, Composites Response: Proceedings of Automotive RETEC '87* (1987) The Society of Plastics Engineers, Brookfield Center, Conn., p. 290–294
9. Box, G., Hunter, W., Hunter, J. *Statistics for Experimenters* (1978) John Wiley & Sons, New York
10. Wojtaszek, H.J. *Automotive Challenge, Composites Response: Proceedings of Automotive RETEC '87* (1987) The Society of Plastics Engineers, Brookfield, Conn., p. 248–251
11. Hunkar, D.B. *Automotive Challenge, Composites Response: Proceedings of Automotive RETEC '87* (1987) The Society of Plastics Engineers, Brookfield Center, Conn., p. 252–256
12. Roberts, R. *Volume 1: Composites—Engineered Materials Handbook* (1987) ASM International, Metals Park, Ohio, p. 745–760
13. Price, W. Personal Conversation, October 1996
14. Loos, A.C., Springer, G.S. *J. Comp. Mat.* (1983) 17(2), p. 135–160
15. Thomas, J.A., et al. WRDC-TR-89-4084 (1989) *WL/MLBC*, Wright-Patterson AFB, Ohio
16. Lee, C.W. Paper Presented at the Air Force Symposium on the Processing Science of Composites (1985) Los Angeles
17. Dave, R., Kardos, J., Dudukovic, M. Proceedings of the American Society for Composites (1986) Technomic Publishing, Lancaster, Penn., p. 137–153
18. Kays, A.O. (1985) AFWAL-TR-85-4090, DTIC# ADB097305L, Wright-Patterson AFB Ohio
19. Townsend, C., Feucht, D. (1986) TAB Books, Inc., Blue Ridge Summit, Penn., p. 11
20. Hajicek, J.D. The Automotive Challenge and Plastics Response, Conference Proceedings (1987) Detroit, November 2–4, 1987, The Society of Plastics Engineers, Brookfield Center CT, p. 120–122
21. Chen, C.T. LeClair, S.R. *J. Reinforced Plastics Comp.* (1991) 10(4)

22. Wu, H.T., Joseph, B. *SAMPE J.* (1990) 26(6), p. 39–54
23. Hinrichs, R.J. *Engineered Materials Handbook: Composites* (1987) ASM Internatioinal, Materials Park, Ohio
24. Product Literature, Aerospace Service Controls, 1994
25. Product Literature, AVPRO, 1993
26. Lee, C.W., Abrams, F.L. Proceedings of the Fifth US–Japan Conference on Composite Materials, June 24–27, 1990, Tokyo, Japan
27. Baumgartner, W.E., Ricker, T. *SAMPE J.* (1983) 19(4), p. 6–16
28. Hindrichs, R.J., Thuen, J., U.S. Patent No. 4,515,545 (1984)
29. Bobrow, D. (Ed.). "Qualitative Reasoning About Physical Systems," MIT Press, Cambridge, MA, 1984
30. LeClair, S.R., Abrams, F.L., Lagnese, T.J., Lee, C.W., Parks, J.B. "Qualitative Process Automation for Autoclave Cure for Composite Parts," (1987) AFWAL-TR-87-4083, AFWAL/MLTC, Wright-Patterson AFB Ohio
31. LeClair, S.R., Abrams, F.L., Matejka, R.F. *Artificial Intelligence for Engineering Design, Analysis, and Manufacturing* (1989) 3(2), p. 125–136
32. Warnock, R., LeClair, S.R. (1992) WL-TR-92-4085, AFWAL/MLIM, Wright-Patterson AFB Ohio
33. Eitman, D., Dirling, R. (1995) WL-TR-96-4007, WL/MLBC, Wright-Patterson AFB Ohio
34. McNeill, D., Freiberger, P. *Fuzzy Logic* (1994) Touchstone Press, New York
35. Wu, H., Joseph, B. *SAMPE J.* (1990) 26(6), p. 39–54
36. Chen, C.T., LeClair, S.R. *J. Reinforced Plastics Comp.* (1991) 10(4)
37. Joseph, B., Hanratty, W.F., Kardos, J. *J. Comp. Mat.* (1995) 29, p. 1000
38. Saliba, T.E., Quinter, S., Abrams, F.L. *Int. J. Comp. Eng.* (1992) 2(2), p. 105–115
39. Pardee, W.J., Shaff, M.A., Hayes-Roth, B. Intelligent Processing of Carbon–Carbon Composites: Final Technical Report (1990) Contract No. N00014-87-C-0724, 135, Office of Naval Research, Arlington Virginia

Index

Raj S. Davé received his Doctor of Science in Chemical Engineering from Washington University, St. Louis, in 1986. After three years of teaching and research at Michigan Molecular Institute, he spent seven years at Monsanto Plastics/Bayer Polymers. In 1996 he received a law degree from the University of Connecticut. He is now a patent attorney at Morrison & Foerster, Washington, specializing in patenting polymers and composite materials, among many other things.

Alfred C. Loos earned his Ph.D. in Mechanical Engineering from the University of Michigan in 1982. Upon graduation, he joined the faculty of Virginia Polytechnic Institute and State University, where he is currently Professor of Engineering Science and Mechanics and Materials Science and Engineering. Professor Loos teaches courses in mechanics of materials, introductory materials science, mechanics of composite materials, and composites manufacturing. His research interests include composite materials processing, environmental effects on organic matrix composites, and mechanics of composite materials. Professor Loos has published more than 100 technical papers and reports. Twenty-eight students have completed graduate degrees under his direction.